Probability and Mathematical Statistics (Continued)

RUBINSTEIN • Simulation and The Monte Carlo Method
SCHEFFE • The Analysis of Variance
SEBER • Linear Regression Analysis
SEN • Sequential Nonparametrics: Invariance Principles and Statistical Inference
SERFLING • Approximation Theorems of Mathematical Statistics
TJUR • Probability Based on Radon Measures
WILLIAMS • Diffusions, Markov Processes, and Martingales, Volume I: Foundations
ZACKS • Theory of Statistical Inference

Applied Probability and Statistics

ANDERSON, AUQUIER, HAUCK, OAKES, VANDAELE, and WEISBERG • Statistical Methods for Comparative Studies
ARTHANARI and DODGE • Mathematical Programming in Statistics
BAILEY • The Elements of Stochastic Processes with Applications to the Natural Sciences
BAILEY • Mathematics, Statistics and Systems for Health
BARNETT • Interpreting Multivariate Data
BARNETT and LEWIS • Outliers in Statistical Data
BARTHOLOMEW • Stochastic Models for Social Processes, *Third Edition*
BARTHOLOMEW and FORBES • Statistical Techniques for Manpower Planning
BECK and ARNOLD • Parameter Estimation in Engineering and Science
BELSLEY, KUH, and WELSCH • Regression Diagnostics: Identifying Influential Data and Sources of Collinearity
BENNETT and FRANKLIN • Statistical Analysis in Chemistry and the Chemical Industry
BHAT • Elements of Applied Stochastic Processes
BLOOMFIELD • Fourier Analysis of Time Series: An Introduction
BOX • R. A. Fisher, The Life of a Scientist
BOX and DRAPER • Evolutionary Operation: A Statistical Method for Process Improvement
BOX, HUNTER, and HUNTER • Statistics for Experimenters: An Introduction to Design, Data Analysis, and Model Building
BROWN and HOLLANDER • Statistics: A Biomedical Introduction
BROWNLEE • Statistical Theory and Methodology in Science and Engineering, *Second Edition*
BURY • Statistical Models in Applied Science
CHAMBERS • Computational Methods for Data Analysis
CHATTERJEE and PRICE • Regression Analysis by Example
CHERNOFF and MOSES • Elementary Decision Theory
CHOW • Analysis and Control of Dynamic Economic Systems
CHOW • Econometric Analysis by Control Methods
CLELLAND, BROWN, and deCANI • Basic Statistics with Business Applications, *Second Edition*
COCHRAN • Sampling Techniques, *Third Edition*
COCHRAN and COX • Experimental Designs, *Second Edition*
CONOVER • Practical Nonparametric Statistics, *Second Edition*
CORNELL • Experiments with Mixtures: Designs, Models and The Analysis of Mixture Data
COX • Planning of Experiments

n back

Introduction to Linear Regression Analysis

About the authors

Douglas C. Montgomery is Professor of Industrial and Systems Engineering at the Georgia Institute of Technology, where he has taught for twelve years. He earned his Ph.D. in Industrial Engineering/Operations Research from Virginia Polytechnic Institute. He is the author of **Design and Analysis of Experiments** (Wiley, 1976); co-author, with W.W. Hines, of **Probability and Statistics in Engineering and Management Science, 2nd Ed.** (Wiley, 1980); and co-author, with L.A. Johnson, of **Forecasting and Time Series Analysis,** and **Operations Research in Production Planning, Scheduling and Inventory Control** (Wiley, 1974).

Elizabeth A. Peck is a Senior Management Science Analyst with The Coca-Cola Company. She received her M.S. in applied statistics from the Georgia Institute of Technology.

INTRODUCTION TO LINEAR REGRESSION ANALYSIS

DOUGLAS C. MONTGOMERY
Georgia Institute of Technology

ELIZABETH A. PECK
The Coca-Cola Company

175 YEARS OF PUBLISHING
1807 1982

JOHN WILEY & SONS,
New York Chichester Brisbane Toronto Singapore

Library of Congress Cataloging in Publication Data:

Montgomery, Douglas C.
Introduction to linear regression analysis.

(Wiley series in probability and mathematical statistics. Applied probability and statistics section, ISSN 0271-6356)
Includes bibliographical references and index.
1. Regression analysis. I. Peck, Elizabeth A., 1953- II. Title. III. Series.

QA278.2.M65 519.5′36 81-11512
ISBN 0-471-05850-5 AACR2

Printed in the United States of America

10 9 8 7 6 5 4 3 2 1

PREFACE

Regression analysis is one of the most widely used statistical techniques for analyzing multifactor data. Its broad appeal results from the conceptually simple process of using an equation to express the relationship between a set of variables. Regression analysis is also interesting theoretically because of the elegant underlying mathematics. Successful use of regression analysis requires an appreciation of both the theory and the practical problems that often arise when the technique is employed with real-world data.

This book is intended as a text for a basic course in linear regression analysis. It contains the standard topics as well as some of the newer and more unconventional ones, and blends both theory and application so that the reader will obtain an understanding of the basic principles necessary to apply regression methods in a variety of practical settings. This book is an outgrowth of lecture notes for a course in regression analysis at the Georgia Institute of Technology. This course is typically taken by seniors and first-year graduate students in various fields of engineering, the physical sciences, applied mathematics, and management. We have also used the material in a number of seminars and short courses on regression for professional audiences. It is assumed that the reader has a basic knowledge of statistics such as that usually obtained from a first course, including familiarity with significance tests, confidence intervals, and the normal, t, χ^2, and F distributions. Some knowledge of matrix algebra is also necessary.

The widespread availability of computers and good software has contributed greatly to the expanding use of regression. This book is oriented towards the analyst who uses computers for problem solution. Throughout the book we have illustrated the output from major regression computer packages, and provided guidelines for relating the information in the computer output displays to the concepts presented in the text.

The book is divided into 10 chapters. Chapters 1 and 2 introduce linear regression models, and provide the standard results for least squares estimation in simple linear regression. Chapter 3 discusses methods for model adequacy checking, including the basic residual plots, testing for lack of fit, detection of outliers, and other diagnostics for investigating departures from the usual regression assumptions. Remedial measures for most of these problems are also discussed, including analytical methods for selecting transformations to stabilize variance and achieve linearity. Chapter 4 is an introduction to least squares fitting in multiple regression. In addition to the standard results, several techniques are presented that we have found extremely helpful in practice, including special types of residual plots, techniques for identifying high-leverage or influential subsets of data, a procedure for obtaining a model-independent estimate of error, identification of interpolation and extrapolation points in prediction, and various methods for scaling residuals. Chapter 5 discusses the special case of polynomials, including a brief introduction to piecewise polynomial fitting using splines. Chapter 6 introduces modeling and analysis considerations when using indicator variables. Variable selection and model building is presented in Chapter 7. Both stepwise-type and all possible regressions selection algorithms are presented along with several summary statistics for evaluating subset regression models.

The first seven chapters form the nucleus of a modern, practically oriented course in linear regression analysis. The last three chapters present a collection of topics that sometimes fall beyond the range of a first course, but that are of increasing importance in the practical use of regression. Chapter 8 focuses on the multicollinearity problem. Included are the sources of multicollinearity, its harmful effects, available diagnostics, and a survey of remedial measures. While we give an extensive discussion of biased estimation, we emphasize that it is not a cureall and its indiscriminant use should be avoided. Chapter 9 introduces several topics, including autocorrelated errors, weighted and generalized least squares, robust regression methods, and simultaneous inference in regression. Problems of multicollinearity, autocorrelation, and nonnormality occur frequently in practice, and we believe that students in a first course should be introduced to them. We postponed their presentation in the text because the concepts can be more easily understood once the reader has a firm grasp of ordinary least squares methods. Chapter 10 introduces model validation, which, as opposed to internal adequacy checking to measure the quality of the fit, is designed to investigate the likely success of the model in its intended operating environment.

Each chapter (except Chapter 1) contains a set of exercises. These exercises include both straightforward computational problems designed to reinforce the reader's understanding of regression methodology and mind expanding problems dealing with more abstract concepts. In most instances the problems utilize real data, or are based on real-world settings that represent typical applications of regression. A computer would be helpful (in some cases necessary) in solving some of these problems and we urge the reader to take full advantage of this resource.

We would like to thank the many individuals who have provided substantial assistance in the preparation of this book. Dr. Ronald G. Askin of the University of Iowa, Dr. Mary Sue Younger of the University of Tennessee, and two anonymous readers offered many helpful suggestions that greatly improved the book. We particularly appreciate the many graduate students at Georgia Tech and the professional practitioners who have suffered through the various drafts that we have used as course material. Their penetrating questions and constructive criticism were significant contributions. Dr. Michael E. Thomas, Director of the School of Industrial and Systems Engineering at Georgia Tech, strongly encouraged and supported the project. We are also grateful to the Georgia Institute of Technology and The Coca-Cola Company for providing computer time and other resources necessary for the development of the manuscript. Ms. Jackie Smith, Ms. Sharon Benton, Ms. Pam Morrison, Ms. Betty Plummer and Ms. Joene Owen efficiently and carefully typed various versions of the manuscript. We are also indebted to the American Statistical Association, John Wiley and Sons, and The Biometrika Trustees for permission to use copyrighted material. The Office of Naval Research has sponsored much of our research in regression analysis. We are grateful for their support.

DOUGLAS C. MONTGOMERY
ELIZABETH A. PECK

Atlanta, Georgia
September 1981

CONTENTS

1. Introduction 1

1.1 Regression and model building, 1
1.2 Uses of regression, 5
1.3 The role of the computer, 6

2. Simple linear regression and correlation 8

2.1 The simple linear regression model, 8
2.2 Least squares estimation of the parameters, 9
2.2.1 Estimation of β_0 and β_1, 9
2.2.2 Properties of the least squares estimators and the fitted regression model, 15
2.2.3 Estimation of σ^2, 17
2.2.4 An alternate form of the model, 19
2.3 Hypothesis testing on the slope and intercept, 20
2.4 Interval estimation in simple linear regression, 26
2.4.1 Confidence intervals on β_0, β_1, and σ^2, 26
2.4.2 Interval estimation of the mean response, 29
2.5 Prediction of new observations, 31
2.6 The coefficient of determination, 33
2.7 Hazards in the use of regression, 34
2.8 Regression through the origin, 38
2.9 Estimation by maximum likelihood, 43
2.10 Correlation, 45
2.11 Sample computer output, 50
Problems, 52

3. Measures of model adequacy 57

3.1 Introduction, 57
3.2 Residual analysis, 58
3.2.1 Definition of residuals, 58

3.2.2 Normal probability plot, 59
3.2.3 Plot of residuals against $\hat{y}_i$, 63
3.2.4 Plot of residuals against x_i, 66
3.2.5 Other residual plots, 67
3.2.6 Statistical tests on residuals, 70
3.3 Detection and treatment of outliers, 70
3.4 A test for lack of fit, 75
3.5 Transformations to a straight line, 79
3.6 Variance stabilizing transformations, 88
3.7 Analytical methods for selecting a transformation, 93
3.7.1 Transformations on y, 94
3.7.2 Transformations on x, 96
3.8 Weighted least squares, 99
Problems, 104

4. Multiple linear regression **109**

4.1 Multiple regression models, 109
4.2 Estimation of the model parameters, 111
4.2.1 Least squares estimation of the regression coefficients, 111
4.2.2 A geometrical interpretation of least squares, 118
4.2.3 Properties of the least squares estimators, 120
4.2.4 Estimation of σ^2, 121
4.2.5 Inadequacy of scatter diagrams in multiple regression, 122
4.3 Confidence intervals in multiple regression, 124
4.3.1 Confidence intervals on the regression coefficients, 124
4.3.2 Confidence interval estimation of the mean response, 127
4.4 Hypothesis testing in multiple linear regression, 128
4.4.1 Test for significance of regression, 128
4.4.2 Tests on individual regression coefficients, 131
4.4.3 Special case of orthogonal columns in $\mathbf{X}$, 137
4.4.4 Testing the general linear hypothesis $\mathbf{T\beta}=\mathbf{0}$, 139
4.5 Prediction of new observations, 141
4.6 Hidden extrapolation, 142
4.7 Measures of model adequacy in multiple regression, 146
4.7.1 The coefficient of multiple determination, 146
4.7.2 Residual analysis, 147
4.7.3 Estimation of pure error from near-neighbors, 154
4.7.4 Detecting influential observations, 160

4.8 Standardized regression coefficients, 167
4.9 Sample computer output, 172
4.10 Computational aspects, 174
Problems, 176

5. Polynomial regression models **181**
5.1 Introduction, 181
5.2 Polynomial models in one variable, 182
5.2.1 Basic principles, 182
5.2.2 Piecewise polynomial fitting (splines), 189
5.3 Polynomial models in two or more variables, 198
5.4 Orthogonal polynomials, 206
Problems, 211

6. Indicator variables **216**
6.1 The general concept of indicator variables, 216
6.2 Comments on the use of indicator variables, 231
6.2.1 Indicator variables versus regression on allocated codes, 231
6.2.2 Indicator variables as a substitute for a quantitative regressor, 232
6.2.3 Models with only indicator variables, 233
6.3 Regression models with an indicator response variable, 233
6.3.1 A linear model, 234
6.3.2 A nonlinear model, 238
Problems, 241

7. Variable selection and model building **244**
7.1 Introduction, 244
7.1.1 The model building problem, 244
7.1.2 Consequences of model misspecification, 245
7.1.3 Criteria for evaluating subset regression models, 249
7.2 Computational techniques for variable selection, 255
7.2.1 All possible regressions, 256
7.2.2 Directed search on t, 266
7.2.3 Stepwise regression methods, 270
7.2.4 Other procedures, 279
7.3 Some final considerations, 282
Problems, 284

8. Multicollinearity **287**

8.1 Introduction, 287
8.2 Sources of multicollinearity, 288
8.3 Effects of multicollinearity, 290
8.4 Multicollinearity diagnostics, 296
8.4.1 Examination of the correlation matrix, 297
8.4.2 Variance inflation factors, 299
8.4.3 Eigensystem analysis of $\mathbf{X}'\mathbf{X}$, 301
8.4.4 Other diagnostics, 305
8.5 Methods for dealing with multicollinearity, 306
8.5.1 Collecting additional data, 306
8.5.2 Model respecification, 307
8.5.3 Ridge regression, 310
8.5.4 Generalized ridge regression, 327
8.5.5 Principal components regression, 334
8.5.6 Latent root regression analysis, 339
8.5.7 Comparison and evaluation of biased estimators, 339
Problems, 342

9. Topics in the use of regression analysis **347**

9.1 Autocorrelation, 347
9.1.1 Sources and effects of autocorrelation, 347
9.1.2 Detecting the presence of autocorrelation, 348
9.1.3 Parameter estimation methods, 353
9.2 Generalized and weighted least squares, 360
9.3 Robust regression, 364
9.3.1 The need for robust estimation, 364
9.3.2 M estimators, 367
9.3.3 R and L estimation, 381
9.3.4 Robust ridge regression, 382
9.4 Why do regression coefficients have the "wrong" sign? 383
9.5 Effect of measurement errors in the X's, 386
9.6 Simultaneous inference in regression, 389
9.6.1 Simultaneous inference on model parameters, 389
9.6.2 Simultaneous estimation of mean response, 396
9.6.3 Prediction of m new observations, 399
9.7 Inverse estimation (calibration or discrimination), 400
9.8 Designed experiments for regression, 405
9.9 The relationship between regression and analysis of variance, 408
Problems, 415

10. Validation of regression models **424**

10.1 Introduction, 424
10.2 Validation techniques, 425
10.2.1 Analysis of model coefficients and predicted values, 426
10.2.2 Collecting fresh data, 427
10.2.3 Data splitting, 430
10.3 Data from planned experiments, 443
Problems, 444

References **447**

Appendix **461**

A. Statistical tables **461**

A.1 Cumulative normal distribution, 462
A.2 Percentage points of the χ^2 distribution, 464
A.3 Percentage points of the t distribution, 465
A.4 Percentage points of the F distribution, 466
A.5 Orthogonal polynomials, 476
A.6 Critical values of the Durbin–Watson statistic, 478
A.7 Percentage points of the Bonferroni t statistic, 479
A.8 Percentage points of the maximum modulus t statistic, 482

B. Data sets for exercises **484**

B.1 National Football League 1976 team performance, 485
B.2 Solar thermal energy test data, 486
B.3 Gasoline mileage performance for 32 automobiles, 487
B.4 Property valuation data, 488
B.5 Belle Ayr liquefaction runs, 489
B.6 Tube-flow reactor data, 490
B.7 Factors affecting CPU time, 491
B.8 Effectiveness of insecticides on cricket eggs, 492

Author Index **497**

Subject Index **501**

Introduction to
Linear Regression Analysis

1

INTRODUCTION

1.1 Regression and model building

Regression analysis is a statistical technique for investigating and modeling the relationship between variables. Applications of regression are numerous and occur in almost every field, including engineering, the physical sciences, economics, management, life and biological sciences, and the social sciences. In fact, regression analysis may be the most widely used statistical technique.

As an example of a problem in which regression analysis may be helpful, suppose that an industrial engineer employed by a soft drink beverage bottler is analyzing the product delivery and service operations for vending machines. He suspects that the time required by a route deliveryman to load and service a machine is related to the number of cases of product delivered. The engineer visits 25 randomly chosen retail outlets having vending machines, and the in–outlet delivery time (in minutes) and the volume of product delivered (in cases) are observed for each. The 25 observations are plotted in Figure 1.1*a*. This graph is called a *scatter diagram*. This display clearly suggests a relationship between delivery time and delivery volume; in fact, the impression is that the data points generally, but not exactly, fall along a straight line. Figure 1.1*b* illustrates this straight line relationship.

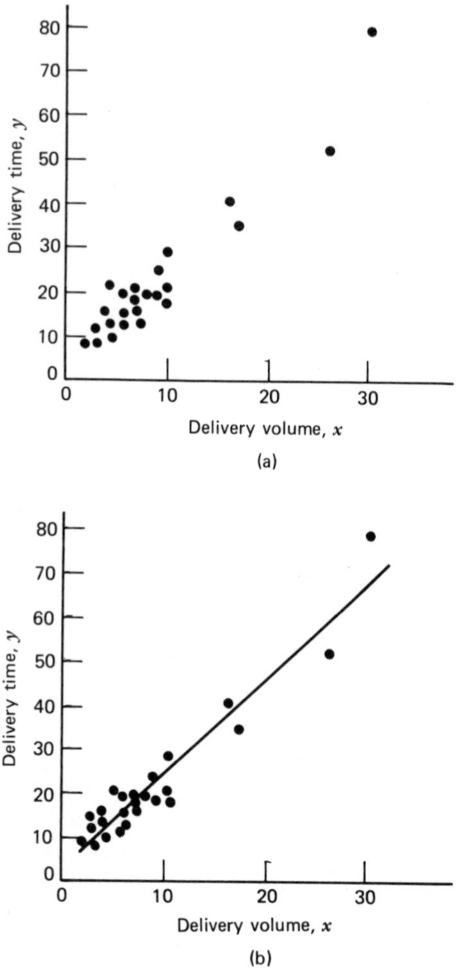

Figure 1.1 (**a**) Scatter diagram for delivery time and delivery volume. (**b**) Straight line relationship between delivery time and delivery volume.

If we let y represent delivery time and x represent delivery volume, then the equation of a straight line relating these two variables is

$$y = \beta_0 + \beta_1 x \tag{1.1}$$

where β_0 is the intercept and β_1 is the slope. Now the data points do not fall exactly on a straight line, so (1.1) should be modified to account for this. Let the difference between the observed value of y and the straight line $(\beta_0 + \beta_1 x)$ be an *error* ε. It is convenient to think of ε as a statistical error;

that is, it is a device that accounts for the failure of the model to fit the data exactly. The error may be made up of the effects of other variables on delivery time, measurement errors, and so forth. Thus a more plausible model for the delivery time data is

$$y=\beta_0+\beta_1 x+\varepsilon \tag{1.2}$$

Equation (1.2) is called a *linear regression model.* Customarily x is called the independent variable and y is called the dependent variable. However, this often causes confusion with the concept of statistical independence, so we refer to x as the *predictor* or *regressor* variable and y as the *response* variable. Because (1.2) involves only one regressor variable it is called a simple linear regression model. In general, the response may be related to k regressors, $x_1, x_2, \ldots, x_k$, so that

$$y=\beta_0+\beta_1 x_1+\beta_2 x_2+\cdots+\beta_k x_k+\varepsilon \tag{1.3}$$

This is called a multiple linear regression model, because more than one regressor is involved. The adjective "linear" is employed to indicate that the model is linear in the parameters $\beta_0, \beta_1, \ldots, \beta_k$, not because y is a linear function of the x's. We shall see subsequently that many models in which y is related to the x's in a nonlinear fashion can still be treated as linear regression models as long as the equation is linear in the β's.

An important objective of regression analysis is to estimate the unknown parameters in the regression model. This process is also called fitting the model to the data. We will study several parameter estimation techniques in this book. One of these techniques is the method of least squares (introduced in Chapter 2). For example, the least squares fit to the delivery time data is

$$\hat{y}=3.3208+2.1762x$$

where $\hat{y}$ is the fitted or estimated value of delivery time corresponding to a delivery volume of x cases. This fitted equation is plotted in Figure 1.1*b*.

The next phase of a regression analysis is called *model adequacy checking*, in which the appropriateness of the model is studied and the quality of the fit ascertained. Through such analyses the usefulness of the regression model may be determined. The outcome of adequacy checking may indicate either that the model is reasonable or that the original fit must be modified. Thus regression analysis is an *iterative* procedure, in which data lead to a model and a fit of the model to the data is produced. The quality of the fit is then investigated, leading either to modification of the model or the fit or to adoption of the model. This process will be illustrated several times in subsequent chapters.

A regression model does not imply a cause and effect relationship between the variables. Even though a strong empirical relationship may exist between two or more variables, this cannot be considered evidence that the regressor variables and the response are related in a cause-effect manner. To establish causality, the relationship between the regressors and the response must have a basis outside the sample data—for example, the relationship may be suggested by theoretical considerations. Regression analysis can aid in confirming a cause-effect relationship, but it cannot be the sole basis of such a claim.

In almost all applications of regression, the regression equation is only an approximation to the true relationship between variables. For example, in Figure 1.2, we have illustrated a situation where a relatively complex relationship between y and x may be well approximated by a linear regression equation. Sometimes a more complex approximating function is necessary, as in Figure 1.3, where a "piecewise-linear" regression function is used to approximate the true relationship between y and x.

Generally regression equations are valid only over the region of the regressor variables contained in the observed data. For example, consider

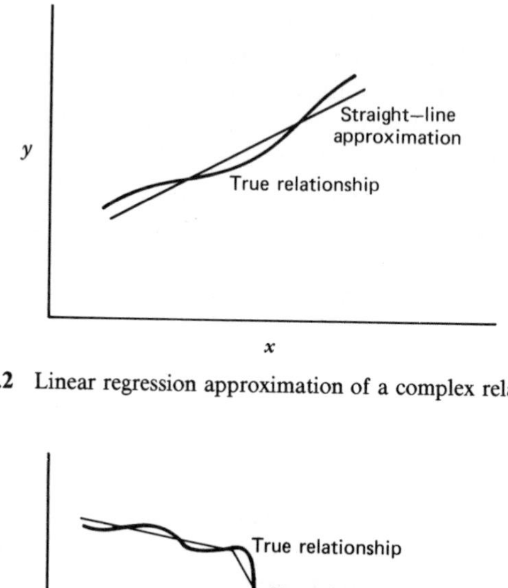

Figure 1.2 Linear regression approximation of a complex relationship.

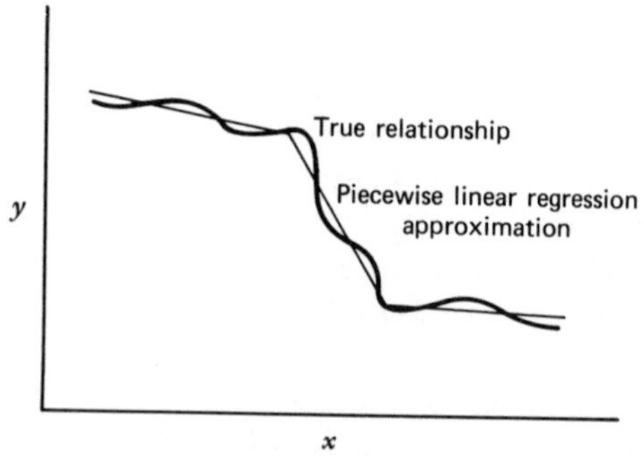

Figure 1.3 Piecewise linear approximation of a complex relationship.

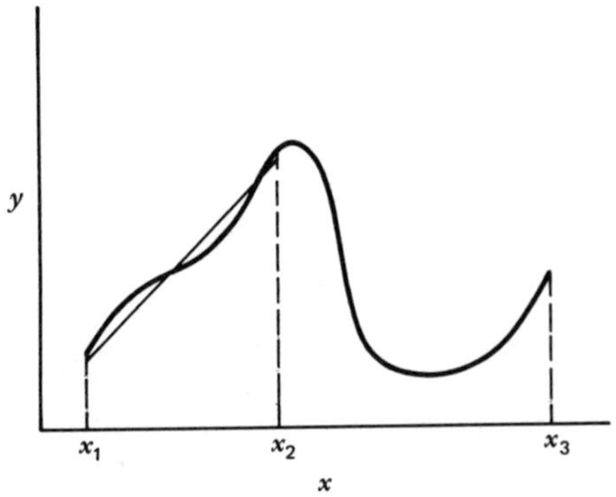

Figure 1.4 The danger of extrapolation in regression.

Figure 1.4. Suppose that data on y and x were collected in the interval $x_1 \leqslant x \leqslant x_2$. Over this interval, the linear regression equation shown in Figure 1.4 is a good approximation of the true relationship. However, suppose this equation were used to predict values of y for values of the regressor variable in the region $x_2 \leqslant x \leqslant x_3$. Clearly, the linear regression model is useless over this range of x because of model error or equation error.

Finally, it is important to remember that regression analysis is part of a broader data-analytic approach to problem solving. That is, the regression equation itself may not be the primary objective of the study. It is usually more important to gain insight and understanding concerning the system generating the data.

An essential aspect of regression analysis is data collection. Because the conclusions from the analysis are conditional on the data, a good data collection effort can have many benefits, including a simplified analysis and a more generally applicable model. The data used in a regression analysis must be representative of the system studied. Without representative data the regression model and conclusions drawn from it are likely to be in error. Care should be devoted to accurate data collection. Many of the techniques used in regression analysis can be seriously distorted by inaccurately recorded data. Preliminary data editing before the regression analysis is conducted often identifies these errors.

1.2 Uses of regression

Regression models are used for several purposes, including

1. Data description.
2. Parameter estimation.
3. Prediction and estimation.
4. Control.

Engineers and scientists frequently use equations to summarize or describe a set of data. Regression analysis is helpful in developing such equations. For example, we may collect a considerable amount of delivery time, delivery volume data, and a regression model would probably be a much more convenient and useful summary of those data than a table or even a graph.

Sometimes parameter estimation problems can be solved by regression methods. For example, suppose that an electrical circuit contains an unknown resistance of R ohms. Several different known currents are passed through the circuit and the corresponding voltages measured. The scatter diagram will indicate that voltage and current are related by a straight line through the origin with slope R (because voltage E and current I are related by Ohm's law $E=IR$). Regression analysis can be used to fit this model to the data, producing an estimate of the unknown resistance.

Many applications of regression involve prediction of the response variable. For example, we may wish to predict delivery time for a specified number of cases of soft drinks to be delivered. These predictions may be helpful in planning delivery activities such as routing and scheduling, or in evaluating the productivity of delivery operations. The dangers of extrapolation when using a regression model for prediction because of model or equation error have been discussed previously (see Figure 1.4). However, even when the model form is correct, poor estimates of the model parameters may still cause poor prediction performance.

Regression models may be used for control purposes. For example, a chemical engineer could use regression analysis to develop a model relating the tensile strength of paper to the hardwood concentration in the pulp. This equation could then be used to control the strength to suitable values by varying the level of hardwood concentration. When a regression equation is used for control purposes, it is important that the variables be related in a causal manner. Note that a cause and effect relationship may not be necessary if the equation is to be used only for prediction. In this case it is only necessary that the relationships that existed in the original data used to build the regression equation are still valid. For example, the daily electricity consumption during August in Atlanta, Georgia may be a good predictor for the maximum daily temperature in August. However, any attempt to reduce the maximum temperature by curtailing electricity consumption is clearly doomed to failure.

1.3 The role of the computer

Building a regression model is an iterative process. Usually several analyses are required as improvements in the model structure and flaws in the data are discovered. A good regression computer program is a necessary tool in the procedure.

The routine application of standard regression computer programs often does not lead to successful results. The computer is *not* a substitute for creative thinking about the problem. Regression analysis requires the *intelligent* and *artful* use of the computer. We must learn how to interpret what the computer is telling us and how to incorporate that information in subsequent models. We will illustrate the use of several different regression computer programs throughout the book. Our objectives will be not to learn the detailed operation of these programs, but rather to understand the information contained in the output. Without this ability, it is virtually impossible to successfully build a regression model.

2

SIMPLE LINEAR REGRESSION AND CORRELATION

2.1 The simple linear regression model

This chapter considers the simple linear regression model; that is, a model with a single regressor x, and the relationship between the response y and x a straight line. This simple linear regression model is

$$y=\beta_0+\beta_1x+\varepsilon \tag{2.1}$$

where the intercept β_0 and the slope β_1 are unknown constants and ε is a random error component. The errors are assumed to have mean zero and unknown variance σ^2. Additionally, we usually assume that the errors are uncorrelated. This means that the value of one error does not depend on the value of any other error.

It is convenient to view the regressor x as controlled by the data analyst and measured with negligible error, while the response y is a random variable. That is, there is a probability distribution for y at each possible value for x. The mean of this distribution is

$$E(y|x)=\beta_0+\beta_1x \tag{2.2a}$$

and the variance is

$$V(y|x)=V(\beta_0+\beta_1 x+\varepsilon)=\sigma^2 \tag{2.2b}$$

Thus the mean of y is a linear function of x although the variance of y does not depend on the value of x. Furthermore, because the errors are uncorrelated, the responses are also uncorrelated.

The parameters β_0 and β_1 are usually called *regression coefficients*. The slope β_1 is the change in the mean of the distribution of y produced by a unit change in x. If the range of data on x includes $x=0$, then the intercept β_0 is the mean of the distribution of the response y when $x=0$. If the range of x does not include zero, then β_0 has no practical interpretation.

2.2 Least squares estimation of the parameters

The parameters β_0 and β_1 are unknown and must be estimated using sample data. Suppose that we have n pairs of data, say $(y_1, x_1), (y_2, x_2), \ldots, (y_n, x_n)$. These data may result either from a controlled experiment designed specifically to collect the data, or from existing historical records.

2.2.1 Estimation of β_0 and β_1

The method of least squares is used to estimate β_0 and β_1. That is, we will estimate β_0 and β_1 so that the sum of the squares of the differences between the observations y_i and the straight line is a minimum. From (2.1) we may write

$$y_i=\beta_0+\beta_1 x_i+\varepsilon_i, \qquad i=1,2,\ldots,n \tag{2.3}$$

Equation (2.1) may be viewed as a *population* regression model while (2.3) is a *sample* regression model, written in terms of the n pairs of data (y_i, x_i), $i=1,2,\ldots,n$. Thus the least squares criterion is

$$S(\beta_0,\beta_1)=\sum_{i=1}^{n}(y_i-\beta_0-\beta_1 x_i)^2 \tag{2.4}$$

The least squares estimators of β_0 and β_1, say $\hat\beta_0$ and $\hat\beta_1$, must satisfy

$$\left.\frac{\partial S}{\partial \beta_0}\right|_{\hat\beta_0,\hat\beta_1}=-2\sum_{i=1}^{n}(y_i-\hat\beta_0-\hat\beta_1 x_i)=0$$

and

$$\left.\frac{\partial S}{\partial \beta_1}\right|_{\hat{\beta}_0, \hat{\beta}_1} = -2\sum_{i=1}^{n}(y_i - \hat{\beta}_0 - \hat{\beta}_1 x_i)x_i = 0$$

Simplifying these two equation yields

$$\begin{aligned} n\hat{\beta}_0 + \hat{\beta}_1 \sum_{i=1}^{n} x_i &= \sum_{i=1}^{n} y_i \\ \hat{\beta}_0 \sum_{i=1}^{n} x_i + \hat{\beta}_1 \sum_{i=1}^{n} x_i^2 &= \sum_{i=1}^{n} y_i x_i \end{aligned} \tag{2.5}$$

Equations (2.5) are called the *least squares normal equations*. The solution to the normal equations is

$$\hat{\beta}_0 = \bar{y} - \hat{\beta}_1 \bar{x} \tag{2.6}$$

and

$$\hat{\beta}_1 = \frac{\sum_{i=1}^{n} y_i x_i - \dfrac{\left(\sum_{i=1}^{n} y_i\right)\left(\sum_{i=1}^{n} x_i\right)}{n}}{\sum_{i=1}^{n} x_i^2 - \dfrac{\left(\sum_{i=1}^{n} x_i\right)^2}{n}} \tag{2.7}$$

where

$$\bar{y} = \frac{1}{n}\sum_{i=1}^{n} y_i \quad \text{and} \quad \bar{x} = \frac{1}{n}\sum_{i=1}^{n} x_i$$

are the averages of y_i and x_i, respectively. Therefore $\hat{\beta}_0$ and $\hat{\beta}_1$ in (2.6) and (2.7) are the least squares estimators of the intercept and slope. The fitted simple linear regression model is then

$$\hat{y} = \hat{\beta}_0 + \hat{\beta}_1 x \tag{2.8}$$

Equation (2.8) gives a point estimate of the mean of y for a particular x.

Since the denominator of (2.7) is the corrected sum of squares of the x_i and the numerator is the corrected sum of cross-products of x_i and y_i, we may write these quantities in a more compact notation as

$$S_{xx}=\sum_{i=1}^{n} x_i^2-\frac{\left(\sum_{i=1}^{n} x_i\right)^2}{n}=\sum_{i=1}^{n}(x_i-\bar{x})^2 \tag{2.9}$$

and

$$S_{xy}=\sum_{i=1}^{n} y_i x_i-\frac{\left(\sum_{i=1}^{n} y_i\right)\left(\sum_{i=1}^{n} x_i\right)}{n}=\sum_{i=1}^{n} y_i(x_i-\bar{x}) \tag{2.10}$$

Thus a convenient way to write (2.7) is

$$\hat{\beta}_1=\frac{S_{xy}}{S_{xx}} \tag{2.11}$$

The difference between the observed value y_i and the corresponding fitted valued $\hat{y}_i$ is a *residual*. Mathematically, the ith residual is

$$e_i=y_i-\hat{y}_i=y_i-(\hat{\beta}_0+\hat{\beta}_1 x_i), \qquad i=1,2,\ldots,n \tag{2.12}$$

Residuals play an important role in investigating the adequacy of the fitted regression model and in detecting departures from the underlying assumptions. This topic will be discussed in Chapter 3.

- **Example 2.1** A rocket motor is manufactured by bonding an igniter propellant and a sustainer propellant together inside a metal housing. The shear strength of the bond between the two types of propellant is an important quality characteristic. It is suspected that shear strength is related to the age in weeks of the batch of sustainer propellant. Twenty observations on shear strength and the age of the corresponding batch of propellant have been collected, and are shown in Table 2.1. The scatter diagram, shown in Figure 2.1, suggests that there is a strong statistical relationship between shear strength and propellant age, and the tentative assumption of the straight-line model $y=\beta_0+\beta_1 x+\varepsilon$ appears to be reasonable.

Table 2.1 Data for Example 2.1

Observation i	Shear Strength (psi), y_i	Age of Propellant (weeks), x_i
1	2158.70	15.50
2	1678.15	23.75
3	2316.00	8.00
4	2061.30	17.00
5	2207.50	5.50
6	1708.30	19.00
7	1784.70	24.00
8	2575.00	2.50
9	2357.90	7.50
10	2256.70	11.00
11	2165.20	13.00
12	2399.55	3.75
13	1779.80	25.00
14	2336.75	9.75
15	1765.30	22.00
16	2053.50	18.00
17	2414.40	6.00
18	2200.50	12.50
19	2654.20	2.00
20	1753.70	21.50

To estimate the model parameters, first calculate

$$S_{xx} = \sum_{i=1}^{n} x_i^2 - \frac{\left(\sum_{i=1}^{n} x_i\right)^2}{n}$$

$$= 4{,}677.69 - \frac{71{,}422.56}{20}$$

$$= 1{,}106.56$$

and

$$S_{xy} = \sum_{i=1}^{n} x_i y_i - \frac{\sum_{i=1}^{n} x_i \sum_{i=1}^{n} y_i}{n}$$

$$= 528{,}492.64 - \frac{(267.25)(42{,}627.15)}{20}$$

$$= -41{,}112.65$$

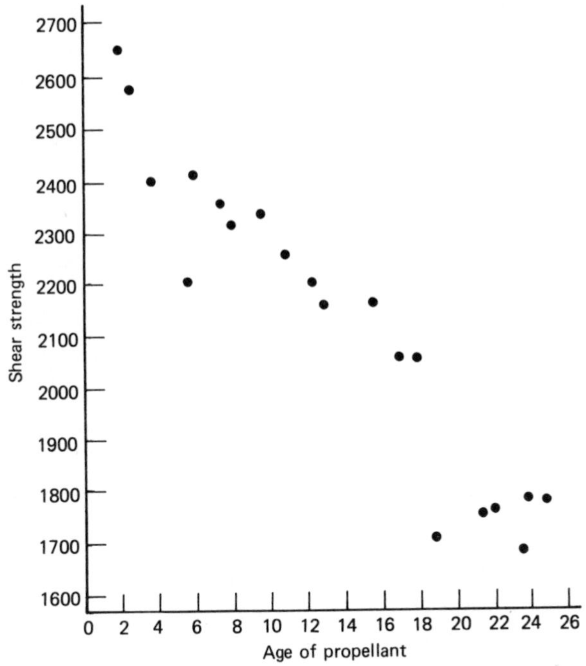

Figure 2.1 Scatter diagram of shear strength versus propellant age, Example 2.1.

Therefore, from (2.11) and (2.6), we find that

$$\hat{\beta}_1 = \frac{S_{xy}}{S_{xx}} = \frac{-41,112.65}{1,106.56} = -37.15$$

and

$$\hat{\beta}_0 = \bar{y} - \hat{\beta}_1 \bar{x} = 2131.3575 - (-37.15)13.3625 = 2,627.82$$

The least squares fit is

$$\hat{y} = 2,627.82 - 37.15x$$

We may interpret the slope -37.15 as the average weekly decrease in propellant shear strength due to the age of the propellant. Since the lower limit of the x's is near the origin, the intercept 2627.82 represents the shear strength in a batch of propellant immediately following manufacture. Table 2.2 displays the observed values y_i, the fitted values $\hat{y}_i$, and the residuals.

Table 2.2 Data, fitted values, and residuals for Example 2.1

Observed Value, y_i	Fitted Value, $\hat{y}_i$	Residual, e_i
2158.70	2051.94	106.76
1678.15	1745.42	−67.27
2316.00	2330.59	−14.59
2061.30	1996.21	65.09
2207.50	2423.48	−215.98
1708.30	1921.90	−213.60
1784.70	1736.14	48.56
2575.00	2534.94	40.06
2357.90	2349.17	8.73
2256.70	2219.13	37.57
2165.20	2144.83	20.37
2399.55	2488.50	−88.95
1779.80	1698.98	80.82
2336.75	2265.58	71.17
1765.30	1810.44	−45.14
2053.50	1959.06	94.44
2414.40	2404.90	9.50
2200.50	2163.40	37.10
2654.20	2553.52	100.68
1753.70	1829.02	−75.32
$\Sigma y_i = 42627.15$	$\Sigma \hat{y}_i = 42627.15$	$\Sigma e_i = 0.00$

After obtaining the least squares fit, a number of interesting questions come to mind, including:

1. How well does this equation fit the data?
2. Is the model likely to be useful as a predictor?
3. Are any of the basic assumptions (such as constant variance and uncorrelated errors) violated, and if so, how serious is this?

All these issues must be investigated before the model is finally adopted for use. As noted previously, the residuals play a key role in evaluating model adequacy. Residuals can be viewed as realizations of the model errors ε_i. Thus to check the constant variance and uncorrelated errors assumption, we must ask ourselves if the residuals look like a random sample from a distribution with these properties. We will return to these questions in

Chapter 3, where the use of residuals in model adequacy checking is explored.

2.2.2 Properties of the least squares estimators and the fitted regression model

The least squares estimators $\hat{\beta}_0$ and $\hat{\beta}_1$ have several important statistical properties. First, note from Equations (2.6) and (2.7) that $\hat{\beta}_0$ and $\hat{\beta}_1$ are linear combinations of the observations y_i. For example,

$$\hat{\beta}_1 = \frac{S_{xy}}{S_{xx}} = \sum_{i=1}^{n} c_i y_i$$

where $c_i = (x_i - \bar{x})/S_{xx}$ for $i = 1, 2, \ldots, n$.

Consider the bias property for $\hat{\beta}_0$ and $\hat{\beta}_1$. We have, for $\hat{\beta}_1$

$$\begin{aligned} E(\hat{\beta}_1) &= E\left(\sum_{i=1}^{n} c_i y_i\right) \\ &= \sum_{i=1}^{n} c_i E(y_i) \\ &= \sum_{i=1}^{n} c_i (\beta_0 + \beta_1 x_i) \\ &= \beta_0 \sum_{i=1}^{n} c_i + \beta_1 \sum_{i=1}^{n} c_i x_i \end{aligned}$$

since $E(\varepsilon_i) = 0$ by assumption. Furthermore we can show directly that

$$\sum_{i=1}^{n} c_i = 0 \quad \text{and} \quad \sum_{i=1}^{n} c_i x_i = 1$$

Therefore

$$E(\hat{\beta}_1) = \beta_1$$

That is, assuming that $E(y_i) = \beta_0 + \beta_1 x_i$, $\hat{\beta}_1$ is an *unbiased estimator* of β_1. Similarly, we may show that $\hat{\beta}_0$ is an unbiased estimator of β_0, or

$$E(\hat{\beta}_0) = \beta_0$$

The variance of $\hat{\beta}_1$ is found as

$$V(\hat{\beta}_1) = V\left(\sum_{i=1}^{n} c_i y_i\right)$$

$$= \sum_{i=1}^{n} c_i^2 V(y_i) \tag{2.13}$$

because the observations y_i are uncorrelated, and so the variance of the sum is just the sum of the variances. The variance of each term in the sum is $c_i^2 V(y_i)$, and we have assumed that $V(y_i) = \sigma^2$; consequently,

$$V(\hat{\beta}_1) = \sigma^2 \sum_{i=1}^{n} c_i^2 = \frac{\sigma^2 \sum_{i=1}^{n} (x_i - \bar{x})^2}{S_{xx}^2}$$

$$= \frac{\sigma^2}{S_{xx}} \tag{2.14}$$

The variance of $\hat{\beta}_0$ is

$$V(\hat{\beta}_0) = V(\bar{y} - \hat{\beta}_1 \bar{x})$$

$$= V(\bar{y}) + \bar{x}^2 V(\hat{\beta}_1) - 2\bar{x}\,\mathrm{Cov}(\bar{y}, \hat{\beta}_1)$$

Now the variance of $\bar{y}$ is just $V(\bar{y}) = \sigma^2/n$, and the covariance between $\bar{y}$ and $\hat{\beta}_1$ can be shown to be zero. (See Problem 2.14). Thus

$$V(\hat{\beta}_0) = V(\bar{y}) + \bar{x}^2 V(\hat{\beta}_1)$$

$$= \sigma^2 \left[\frac{1}{n} + \frac{\bar{x}^2}{S_{xx}}\right] \tag{2.15}$$

An important result concerning the quality of the least squares estimators $\hat{\beta}_0$ and $\hat{\beta}_1$ is the *Gauss–Markov* theorem, which states that for the regression model (2.1) with the assumptions $E(\varepsilon) = 0$, $V(\varepsilon) = \sigma^2$, and uncorrelated errors, the least squares estimators are unbiased and have minimum variance when compared with all other unbiased estimators that are linear combinations of the y_i. We often say that the least squares estimators are *best linear unbiased estimators*, where "best" implies minimum variance.

There are several other useful properties of the least squares fit. These include:

1. The sum of the residuals in any regression model that contains an intercept β_0 is always zero; that is,

$$\sum_{i=1}^{n} (y_i - \hat{y}_i) = \sum_{i=1}^{n} e_i = 0$$

This property follows directly from the first normal equation in (2.5), and is demonstrated in Table 2.2 for the residuals from Example 2.1. Rounding errors may affect the sum.

2. The sum of the observed values y_i equals the sum of the fitted values $\hat{y}_i$, or

$$\sum_{i=1}^{n} y_i = \sum_{i=1}^{n} \hat{y}_i$$

Table 2.2 demonstrates this result for Example 2.1.

3. The least squares regression line always passes through the *centroid* [the point $(\bar{y}, \bar{x})$] of the data.

4. The sum of the residuals weighted by the corresponding value of the regressor variable always equals zero; that is,

$$\sum_{i=1}^{n} x_i e_i = 0$$

5. The sum of the residuals weighted by the corresponding fitted value always equals zero; that is,

$$\sum_{i=1}^{n} \hat{y}_i e_i = 0$$

2.2.3 Estimation of σ^2

In addition to estimating β_0 and β_1, an estimate of σ^2 is required to test hypotheses and construct interval estimates pertinent to the regression model. Ideally, we would like this estimate not to depend on the adequacy of the fitted model. This is only possible when there are several observations on y for at least one value of x (see Section 3.4), or when prior information concerning σ^2 is available. When this approach cannot be used, the estimate of σ^2 is obtained from the residual or error sum of squares

$$SS_E = \sum_{i=1}^{n} e_i^2 = \sum_{i=1}^{n} (y_i - \hat{y}_i)^2 \tag{2.16}$$

A convenient computing formula for SS_E may be found by substituting $\hat{y}_i = \hat{\beta}_0 + \hat{\beta}_1 x_i$ into (2.16) and simplifying, yielding

$$SS_E = \sum_{i=1}^{n} y_i^2 - n\bar{y}^2 - \hat{\beta}_1 S_{xy} \tag{2.17}$$

But

$$\sum_{i=1}^{n} y_i^2 - n\bar{y}^2 = \sum_{i=1}^{n} (y_i - \bar{y})^2 \equiv S_{yy}$$

is just the corrected sum of squares of the observations, so

$$SS_E = S_{yy} - \hat{\beta}_1 S_{xy} \tag{2.18}$$

The residual sum of squares has $n-2$ degrees of freedom, because two degrees of freedom are associated with the estimates $\hat{\beta}_0$ and $\hat{\beta}_1$ involved in obtaining $\hat{y}_i$. Now the expected value of SS_E is $E(SS_E) = (n-2)\sigma^2$, so an unbiased estimator of σ^2 is

$$\hat{\sigma}^2 = \frac{SS_E}{n-2} = MS_E \tag{2.19}$$

The quantity MS_E is called the error mean square or the *residual mean square*. The square root of $\hat{\sigma}^2$ is sometimes called the *standard error of regression*, and it has the same units as the response variable y. Because $\hat{\sigma}^2$ depends on the residual sum of squares, any violation of the assumptions on the model errors or any misspecification of the model form may seriously damage the usefulness of $\hat{\sigma}^2$ as an estimate of σ^2.

• **Example 2.2** To estimate σ^2 for the data in Example 2.1, first find

$$S_{yy} = \sum_{i=1}^{n} y_i^2 - n\bar{y}^2 = \sum_{i=1}^{n} y_i^2 - \frac{\left(\sum_{i=1}^{n} y_i\right)^2}{n}$$

$$= 92{,}547{,}433.45 - \frac{(42{,}627.15)^2}{20}$$

$$= 1{,}693{,}737.60$$

From (2.19), the error sum of squares is

$$SS_E = S_{yy} - \hat{\beta}_1 S_{xy}$$

$$= 1{,}693{,}737.60 - (-37.15)(-41{,}112.65)$$

$$= 166{,}402.65$$

Therefore, the estimate of σ^2 is computed from (2.19) as

$$\hat{\sigma}^2 = \frac{SS_E}{n-2} = \frac{166{,}402.65}{18} = 9244.59$$

• Remember that this estimate of σ^2 is *model-dependent*.

2.2.4 An alternate form of the model

There is an alternate form of the simple linear regression model that occasionally proves useful. Suppose that we redefine the regressor variable x_i as the deviation from its own average, say $x_i - \bar{x}$. The regression model then becomes

$$y_i = \beta_0 + \beta_1(x_i - \bar{x}) + \beta_1\bar{x} + \varepsilon_i$$

$$= (\beta_0 + \beta_1\bar{x}) + \beta_1(x_i - \bar{x}) + \varepsilon_i$$

$$= \beta_0' + \beta_1(x_i - \bar{x}) + \varepsilon_i \qquad (2.20)$$

Note that redefining the regressor variable in (2.20) has shifted the origin of the x's from zero to $\bar{x}$. In order to keep the fitted values the same in both the original and transformed models, it is necessary to modify the original intercept. The relationship between the original and transformed intercept is

$$\beta_0' = \beta_0 + \beta_1\bar{x} \qquad (2.21)$$

The least squares normal equations for this form of the model are

$$n\hat{\beta}_0' = \sum_{i=1}^{n} y_i$$

$$\hat{\beta}_1 \sum_{i=1}^{n} (x_i - \bar{x})^2 = \sum_{i=1}^{n} y_i(x_i - \bar{x})$$

and the resulting least squares estimators are

$$\hat{\beta}_0' = \bar{y} \tag{2.22a}$$

$$\hat{\beta}_1 = \frac{\sum_{i=1}^{n} y_i(x_i - \bar{x})}{\sum_{i=1}^{n} (x_i - \bar{x})^2} = \frac{S_{xy}}{S_{xx}} \tag{2.22b}$$

Thus in this form of the model the intercept is estimated by $\bar{y}$ and the slope is unaffected by the transformation.

Several advantages are associated with this alternate form of the linear regression model. First, the normal equations are easier to solve than were Equations (2.5), because the cross-product terms have vanished. Second, the least squares estimators $\hat{\beta}_0' = \bar{y}$ and $\hat{\beta}_1 = S_{xy}/S_{xx}$ are *uncorrelated*; that is, $\text{Cov}(\hat{\beta}_0', \hat{\beta}_1) = 0$. This will make some applications of the model easier, such as finding confidence intervals on the mean of y (see Section 2.4.2). Finally, the fitted model is

$$\hat{y} = \bar{y} + \hat{\beta}_1(x - \bar{x}) \tag{2.23}$$

Although (2.23) and (2.8) are equivalent (they both produce the same value of $\hat{y}$ for the same value of x), (2.23) directly reminds the analyst that the regression model is only valid over the range of x in the *original data*. This region is centered at $\bar{x}$.

2.3 Hypothesis testing on the slope and intercept

We are often interested in testing hypotheses and constructing confidence intervals about the model parameters. Hypothesis testing is discussed in this section, and Section 2.4 deals with confidence intervals. These procedures require that we make the additional assumption that the model errors ε_i are normally distributed. Thus the complete assumptions are that the errors are normally and independently distributed with mean 0 and variance σ^2, abbreviated NID$(0, \sigma^2)$. In Chapter 3 we discuss how these assumptions can be checked through *residual analysis*.

Suppose that we wish to test the hypothesis that the slope equals a constant, say β_{10}. The appropriate hypotheses are

$$H_0\colon \beta_1 = \beta_{10}$$

$$H_1\colon \beta_1 \neq \beta_{10} \tag{2.24}$$

where we have specified a two-sided alternative. Since the errors ε_i are NID$(0, \sigma^2)$, the observations y_i are NID$(\beta_0 + \beta_1 x_i, \sigma^2)$. Now $\hat{\beta}_1$ is a linear combination of the observations, so $\hat{\beta}_1$ is normally distributed with mean β_1 and variance σ^2/S_{xx}, using the mean and variance of $\hat{\beta}_1$ found in Section 2.2.2. Therefore the statistic

$$Z_0 = \frac{\hat{\beta}_1 - \beta_{10}}{\sqrt{\sigma^2/S_{xx}}}$$

is distributed $N(0,1)$ if the null hypothesis $H_0\colon \beta_1 = \beta_{10}$ is true. If σ^2 were known, we could use Z_0 to test the hypotheses (2.24). However, the residual mean square MS_E is an unbiased estimator of σ^2, and the distribution of $(n-2)MS_E/\sigma^2$ is χ^2_{n-2}. Furthermore, MS_E and $\hat{\beta}_1$ are independent random variables. These conditions imply that if we replace σ^2 in Z_0 by $\hat{\sigma}^2 = MS_E$, the statistic

$$t_0 = \frac{\hat{\beta}_1 - \beta_{10}}{\sqrt{MS_E/S_{xx}}} \tag{2.25}$$

is distributed as t with $n-2$ degrees of freedom if the null hypothesis H_0: $\beta_1 = \beta_{10}$ is true. The degrees of freedom on t_0 are the number of degrees of freedom associated with MS_E. The statistic t_0 is used to test H_0: $\beta_1 = \beta_{10}$ by comparing the observed value of t_0 from (2.25) with the upper $\alpha/2$ percentage point of the t_{n-2} distribution $(t_{\alpha/2, n-2})$ and rejecting the null hypothesis if

$$|t_0| > t_{\alpha/2, n-2} \tag{2.26}$$

A similar procedure can be used to test hypotheses about the intercept. To test

$$H_0\colon \beta_0 = \beta_{00}$$

$$H_1\colon \beta_0 \neq \beta_{00} \tag{2.27}$$

we would use the statistic

$$t_0 = \frac{\hat{\beta}_0 - \beta_{00}}{\sqrt{MS_E\left[\frac{1}{n} + \frac{\bar{x}^2}{S_{xx}}\right]}} \tag{2.28}$$

and reject the null hypothesis if $|t_0| > t_{\alpha/2, n-2}$.

A very important special case of (2.24) is

$$H_0: \beta_1 = 0$$

$$H_1: \beta_1 \neq 0 \tag{2.29}$$

This hypothesis relates to the *significance of regression*. Failing to reject H_0: $\beta_1 = 0$ implies that there is no linear relationship between x and y. This situation is illustrated in Figure 2.2. Note that this may imply either that x is of little value in explaining the variation in y and that the best estimator of y for any x is $\hat{y} = \bar{y}$ (Figure 2.2a), or that the true relationship between x and y is not linear (Figure 2.2b).

Alternatively, if H_0: $\beta_1 = 0$ is rejected, this implies that x is of value in explaining the variability in y. This is illustrated in Figure 2.3. However, rejecting H_0: $\beta_1 = 0$ could mean either that the straight-line model is adequate (Figure 2.3a), or that even though there is a linear effect of x, better results could be obtained with the addition of higher-order polynomial terms in x (Figure 2.3b).

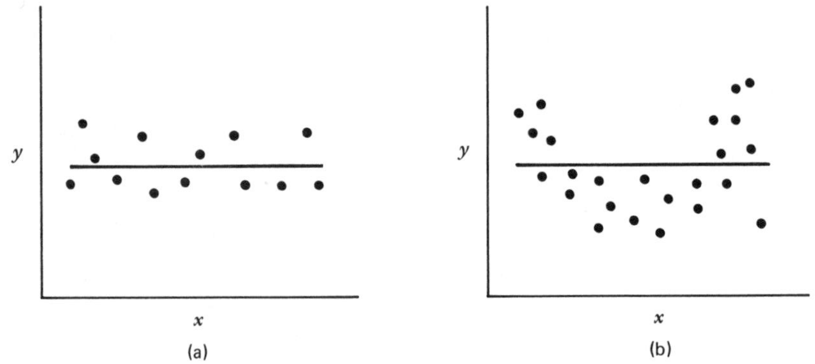

Figure 2.2 Situations where the hypothesis $H_0: \beta_1 = 0$ is not rejected.

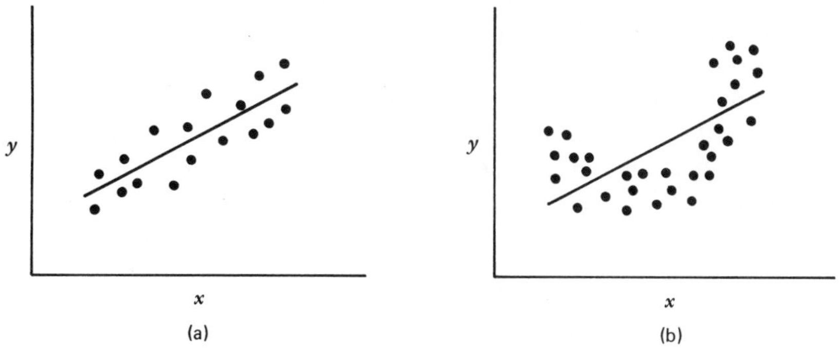

Figure 2.3 Situations where the hypothesis $H_0: \beta_1 = 0$ is rejected.

The test procedure for H_0: $\beta_1 = 0$ may be developed from two approaches. The first approach starts with the identity

$$y_i - \bar{y} = (\hat{y}_i - \bar{y}) + (y_i - \hat{y}_i) \tag{2.30}$$

Squaring both sides of (2.30) and summing over all n observations produces

$$\sum_{i=1}^{n} (y_i - \bar{y})^2 = \sum_{i=1}^{n} (\hat{y}_i - \bar{y})^2 + \sum_{i=1}^{n} (y_i - \hat{y}_i) + 2\sum_{i=1}^{n} (\hat{y}_i - \bar{y})(y_i - \hat{y}_i) \tag{2.31}$$

Note that the third term on the right-hand side of (2.31) can be rewritten as

$$\begin{aligned} 2\sum_{i=1}^{n} (\hat{y}_i - \bar{y})(y_i - \hat{y}_i) &= 2\sum_{i=1}^{n} \hat{y}_i(y_i - \hat{y}_i) - 2\bar{y}\sum_{i=1}^{n} (y_i - \hat{y}_i) \\ &= 2\sum_{i=1}^{n} \hat{y}_i e_i - 2\bar{y}\sum_{i=1}^{n} e_i \\ &= 0 \end{aligned}$$

since the sum of the residuals is always zero (Property 1, Section 2.2.2), and the sum of the residuals weighted by the corresponding fitted value $\hat{y}_i$ is also zero (Property 5, Section 2.2.2). Therefore (2.31) becomes

$$\sum_{i=1}^{n} (y_i - \bar{y})^2 = \sum_{i=1}^{n} (\hat{y}_i - \bar{y})^2 + \sum_{i=1}^{n} (y_i - \hat{y}_i)^2 \tag{2.32}$$

The left-hand side of (2.32) is the corrected sum of squares of the observations, S_{yy}, which measures the total variability in the observations. The two components of S_{yy} measure, respectively, the amount of variability in the observations y_i accounted for by the regression line, and the residual variation left unexplained by the regression line. We recognize $SS_E = \Sigma_{i=1}^n (y_i - \hat{y}_i)^2$ as the residual sum of squares from (2.16). It is customary to call $SS_R = \Sigma_{i=1}^n (\hat{y}_i - \bar{y})^2$ the *regression* sum of squares. Thus (2.32) may be written as

$$S_{yy} = SS_R + SS_E \tag{2.33}$$

Comparing (2.33) with (2.18) we see that the regression sum of squares may be computed as

$$SS_R = \hat{\beta}_1 S_{xy} \tag{2.34}$$

The degree of freedom breakdown is determined as follows. S_{yy} has $n-1$ degrees of freedom, because one degree of freedom has been "lost" due to the constraint $\Sigma_{i=1}^n (y_i - \bar{y}) = 0$ on the deviations $y_i - \bar{y}$. SS_R has 1 degree of freedom, because SS_R is completely determined by one parameter, namely $\hat{\beta}_1$ [see (2.34)]. Finally, we have noted previously that SS_E has $n-2$ degrees of freedom because two constraints are imposed on the deviations $y_i - \hat{y}_i$ as a result of estimating $\hat{\beta}_0$ and $\hat{\beta}_1$. Note that the degrees of freedom have an additive property:

$$n-1 = 1 + (n-2) \tag{2.35}$$

To test the hypothesis H_0: $\beta_1 = 0$, an *analysis of variance* procedure is used. The test statistic is

$$F_0 = \frac{\dfrac{SS_R}{1}}{\dfrac{SS_E}{(n-2)}} = \frac{MS_R}{MS_E} \tag{2.36}$$

where MS_R is the regression mean square and MS_E is the residual mean square. The expected values of these mean squares are

$$E(MS_E) = \sigma^2$$

$$E(MS_R) = \sigma^2 + \beta_1^2 S_{xx}$$

Table 2.3 Analysis of variance for testing significance of regression

Source of Variation	Sum of Squares	Degrees of Freedom	Mean Square	F_0
Regression	$SS_R = \hat{\beta}_1 S_{xy}$	1	MS_R	MS_R / MS_E
Residual	$SS_E = S_{yy} - \hat{\beta}_1 S_{xy}$	$n-2$	MS_E	
Total	S_{yy}	$n-1$		

Furthermore MS_R and MS_E are independent random variables. Then if the null hypothesis H_0: $\beta_1 = 0$ is true, the test statistic F_0 in (2.36) follows the $F_{1,n-2}$ distribution. The expected mean squares indicate that if the observed value of F_0 is large, then it is likely that the slope $\beta_1 \neq 0$. Therefore, to test the hypothesis H_0: $\beta_1 = 0$, compute the test statistic F_0 and reject H_0 if

$$F_0 > F_{\alpha,1,n-2}$$

The test procedure is summarized in Table 2.3.

- **Example 2.3** We will test for significance of regression in the model developed in Example 2.1 for the rocket propellant data. The fitted model is $\hat{y} = 2627.82 - 37.15x$, $S_{yy} = 1{,}693{,}737.60$, and $S_{xy} = -41{,}112.65$. The regression sum of squares is computed from (2.34) as

$$SS_R = \hat{\beta}_1 S_{xy} = (-37.15)(-41{,}112.65) = 1{,}527{,}334.95.$$

The analysis of variance is summarized in Table 2.4. The observed $F_0 = 165.21$, and from Appendix Table A.4, $F_{.01,1,18} = 8.29$, so we reject H_0: $\beta_1 = 0$. •

Table 2.4 Analysis of variance table for Example 2.1

Source of Variation	Sum of Squares	Degrees of Freedom	Mean Square	F_0
Regression	1,527,334.95	1	1,527,334.95	165.21
Residual	166,402.65	18	9,244.59	
Total	1,693,737.60	19		

The test for significance of regression may also be performed using the t-test Equation (2.24) with $\beta_{10}=0$, say

$$t_0 = \frac{\hat{\beta}_1}{\sqrt{MS_E/S_{xx}}} \tag{2.37}$$

However, note that upon squaring both sides of (2.37), we obtain

$$t_0^2 = \frac{\hat{\beta}_1^2 S_{xx}}{MS_E} = \frac{\hat{\beta}_1 S_{xy}}{MS_E} = \frac{MS_R}{MS_E} \tag{2.38}$$

Thus t_0^2 in (2.38) is identical to F_0 in (2.36). In general the square of a t random variable with f degrees of freedom is an F random variable with one and f degrees of freedom in the numerator and denominator, respectively. Although the t-test for H_0: $\beta_1=0$ is equivalent to the F-test, the t-test is somewhat more adaptable, as it could be used for one–sided alternative hypotheses (either H_1: $\beta_1<0$ or H_1: $\beta_1>0$), while the F-test considers only the two-sided alternative. Regression computer programs routinely produce both the analysis of variance in Table 2.3 and the t- statistic in (2.37).

Finally, remember that deciding that $\beta_1=0$ is a very important conclusion that is only *aided* by the t or F-test. The inability to show that the slope is not statistically different from zero may not necessarily mean that y and x are unrelated. It may mean that our ability to detect this relationship has been obscured by the variance of the measurement process, or that the range of values of x is inappropriate. A great deal of nonstatistical evidence and knowledge of the subject matter in the field is required to conclude that $\beta_1=0$.

2.4 Interval estimation in simple linear regression

In this section, we consider confidence interval estimation of the regression model parameters. We also discuss interval estimation of the mean response $E(y)$ for given values of x. The normality assumptions introduced in Section 2.3 continue to apply.

2.4.1 Confidence intervals on β_0, β_1, and σ^2

In addition to point estimates of β_0, β_1, and σ^2, we may also obtain confidence interval estimates of these parameters. The width of these confidence intervals is a measure of the overall quality of the regression line.

If the errors are normally and independently distributed, then the sampling distribution of both

$$\frac{\hat{\beta}_1 - \beta_1}{\sqrt{\dfrac{MS_E}{S_{xx}}}} \quad \text{and} \quad \frac{\hat{\beta}_0 - \beta_0}{\sqrt{MS_E\left[\dfrac{1}{n} + \dfrac{\bar{x}^2}{S_{xx}}\right]}}$$

is t with $n-2$ degrees of freedom. Therefore a 100 $(1-\alpha)$ percent confidence interval on the slope β_1 is given by

$$\hat{\beta}_1 - t_{\alpha/2, n-2}\sqrt{\frac{MS_E}{S_{xx}}} \leqslant \beta_1 \leqslant \hat{\beta}_1 + t_{\alpha/2, n-2}\sqrt{\frac{MS_E}{S_{xx}}} \tag{2.39}$$

and a 100 $(1-\alpha)$ percent confidence interval on the intercept β_0 is

$$\hat{\beta}_0 - t_{\alpha/2, n-2}\sqrt{MS_E\left[\frac{1}{n} + \frac{\bar{x}^2}{S_{xx}}\right]} \leqslant \beta_0 \leqslant \hat{\beta}_0 + t_{\alpha/2, n-2}\sqrt{MS_E\left[\frac{1}{n} + \frac{\bar{x}^2}{S_{xx}}\right]} \tag{2.40}$$

These confidence intervals have the usual frequency interpretation. That is, if we were to take repeated samples of the same size at the same x levels, and construct, for example, 95 percent confidence intervals on the slope for each sample, then 95 percent of those intervals will contain the true value of β_1.

The quantity

$$\text{se}(\hat{\beta}_1) = \sqrt{\frac{MS_E}{S_{xx}}}$$

in (2.39) is called the *standard error* of the slope $\hat{\beta}_1$. It is a measure of how precisely the slope has been estimated. Similarly,

$$\text{se}(\hat{\beta}_0) = \sqrt{MS_E\left[\frac{1}{n} + \frac{\bar{x}^2}{S_{xx}}\right]}$$

in (2.40) is the standard error of the intercept $\hat{\beta}_0$. Regression computer programs usually report the standard errors of the regression coefficients.

If the errors are normally and independently distributed, the sampling distribution of

$$\frac{(n-2)MS_E}{\sigma^2}$$

is chi-square with $n-2$ degrees of freedom. Thus

$$P\left\{\chi^2_{1-\alpha/2,\,n-2} \leqslant \frac{(n-2)MS_E}{\sigma^2} \leqslant \chi^2_{\alpha/2,\,n-2}\right\} = 1-\alpha$$

and consequently a 100 $(1-\alpha)$ percent confidence interval on σ^2 is

$$\frac{(n-2)MS_E}{\chi^2_{\alpha/2,\,n-2}} \leqslant \sigma^2 \leqslant \frac{(n-2)MS_E}{\chi^2_{1-\alpha/2,\,n-2}} \tag{2.41}$$

- **Example 2.4** We will construct 95 percent confidence intervals on β_1 and σ^2 using the rocket propellant data from Example 2.1. The standard error of $\hat{\beta}_1$ is

$$\text{se}(\hat{\beta}_1) = \sqrt{\frac{MS_E}{S_{xx}}} = \sqrt{\frac{9244.59}{1106.56}} = 2.89$$

and from Appendix Table A.3, $t_{.025,18} = 2.101$. Therefore, using (2.39), we find

$$\hat{\beta}_1 - t_{.025,18}\sqrt{\frac{MS_E}{S_{xx}}} \leqslant \beta_1 \leqslant \hat{\beta}_1 + t_{.025,18}\sqrt{\frac{MS_E}{S_{xx}}}$$

$$-37.15 - (2.101)(2.89) \leqslant \beta_1 \leqslant -37.15 + (2.101)(2.89)$$

or

$$-43.22 \leqslant \beta_1 \leqslant -31.08$$

In other words, 95 percent of such intervals will include the true value of the slope.

If we had chosen a different value for α, the width of the resulting confidence interval would have been different. For example, the 90 percent confidence interval on β_1 is $-42.16 \leqslant \beta_1 \leqslant -32.14$, which is narrower than the 95 percent confidence interval. The 99 percent confidence interval is $-45.49 \leqslant \beta_1 \leqslant -28.81$, which is wider than the 95 percent confidence interval. In general, the larger the confidence coefficient $(1-\alpha)$, the wider the confidence interval.

The 95 percent confidence interval on σ^2 is found from (2.41) as follows:

$$\frac{(n-2)MS_E}{\chi^2_{.025,\,n-2}} \leqslant \sigma^2 \leqslant \frac{(n-2)MS_E}{\chi^2_{.975,\,n-2}}$$

$$\frac{18(9244.59)}{\chi^2_{.025,18}} \leqslant \sigma^2 \leqslant \frac{18(9244.59)}{\chi^2_{.975,18}}$$

From Appendix Table A.2, $\chi^2_{.025,18}=31.5$ and $\chi^2_{.975,18}=8.23$. Therefore the desired confidence interval becomes

$$\frac{18(9244.59)}{31.5}\leqslant\sigma^2\leqslant\frac{18(9244.59)}{8.23}$$

or

$$5{,}282.62\leqslant\sigma^2\leqslant 20{,}219.03$$

•

2.4.2 Interval estimation of the mean response

A major use of a regression model is to estimate the mean response $E(y)$ for a particular value of the regressor variable x. For example, we might wish to estimate the mean shear strength of the propellant bond in a rocket motor made from a batch of sustainer propellant that is 10 weeks old. Let x_0 be the level of the regressor variable for which we wish to estimate the mean response, say $E(y|x_0)$. We assume that x_0 is any value of the regressor variable within the range of the original data on x used to fit the model. An unbiased point estimator of $E(y|x_0)$ is found from the fitted model as

$$\widehat{E(y|x_0)}\equiv\hat{y}_0=\hat{\beta}_0+\hat{\beta}_1x_0 \tag{2.42}$$

To obtain a $100(1-\alpha)$ percent confidence interval on $E(y|x_0)$, first note that $\hat{y}_0$ is a normally distributed random variable because it is a linear combination of the observations y_i. The variance of $\hat{y}_0$ is

$$\begin{aligned}V(\hat{y}_0)&=V(\hat{\beta}_0+\hat{\beta}_1x_0)\\&=V[\bar{y}+\hat{\beta}_1(x_0-\bar{x})]\\&=\frac{\sigma^2}{n}+\frac{\sigma^2(x_0-\bar{x})^2}{S_{xx}}\\&=\sigma^2\left[\frac{1}{n}+\frac{(x_0-\bar{x})^2}{S_{xx}}\right]\end{aligned}$$

since (as noted in Section 2.2.4) $\mathrm{Cov}(\bar{y},\hat{\beta}_1)=0$. Thus the sampling distribution of

$$\frac{\hat{y}_0-E(y|x_0)}{\sqrt{MS_E\left(\dfrac{1}{n}+\dfrac{(x_0-\bar{x})^2}{S_{xx}}\right)}}$$

is t with $n-2$ degrees of freedom. Consequently, a $100(1-\alpha)$ percent confidence interval on the mean response at the point $x=x_0$ is

$$\hat{y}_0 - t_{\alpha/2,n-2}\sqrt{MS_E\left(\frac{1}{n}+\frac{(x_0-\bar{x})^2}{S_{xx}}\right)}$$

$$\leqslant E(y|x_0) \leqslant \hat{y}_0 + t_{\alpha/2,n-2}\sqrt{MS_E\left(\frac{1}{n}+\frac{(x_0-\bar{x})^2}{S_{xx}}\right)} \tag{2.43}$$

Note that the width of the confidence interval for $E(y|x_0)$ is a function of x_0. The interval width is a minimum for $x_0=\bar{x}$ and widens as $|x_0-\bar{x}|$ increases. Intuitively this is reasonable, as we would expect our best estimates of y to be made at x-values near the center of the data, and for the precision of estimation to deteriorate as we move to the boundary of the x-space.

- **Example 2.5** Consider finding a 95 percent confidence interval on $E(y|x_0)$ for the rocket propellant data in Example 2.1. This confidence interval is found from (2.43) as

$$\hat{y}_0 - t_{\alpha/2,n-2}\sqrt{MS_E\left(\frac{1}{n}+\frac{(x_0-\bar{x})^2}{S_{xx}}\right)} \leqslant E(y|x_0) \leqslant$$

$$\hat{y}_0 + t_{\alpha/2,n-2}\sqrt{MS_E\left(\frac{1}{n}+\frac{(x_0-\bar{x})^2}{S_{xx}}\right)}$$

$$\hat{y}_0 - (2.101)\sqrt{9244.59\left(\frac{1}{20}+\frac{(x_0-13.3625)^2}{1106.56}\right)} \leqslant$$

$$E(y|x_0) \leqslant \hat{y}_0 + (2.101)\sqrt{9244.59\left(\frac{1}{20}+\frac{(x_0-13.3625)^2}{1106.56}\right)}$$

If we substitute values of x_0 and the fitted value $\hat{y}_0$ at that value of x_0 into this last equation, we will obtain the 95 percent confidence interval on the mean response at $x=x_0$. For example, if $x_0=\bar{x}=13.3625$, then $\hat{y}_0=2131.40$, and the confidence interval becomes

$$2086.230 \leqslant E(y|13.3625) \leqslant 2176.571.$$

Table 2.5 Confidence limits on $E(y|x_0)$ for several values of x_0

Lower Confidence Limit	x_0	Upper Confidence Limit
2438.919	3	2593.821
2341.360	6	2468.481
2241.104	9	2345.836
2136.098	12	2227.942
2086.230	$\bar{x}=13.3625$	2176.571
2024.318	15	2116.822
1905.890	18	2012.351
1782.928	21	1912.412
1657.395	24	1815.045

Table 2.5 contains the 95 percent confidence limits on $E(y|x_0)$ for several other values of x_0. Note that the width of the confidence interval increases as $|x_0-\bar{x}|$ increases.

The probability statement associated with the confidence interval (2.43) holds only when a single confidence interval on the mean response is to be constructed. Procedures for constructing several confidence intervals that, considered jointly, have a specified confidence level is a *simultaneous* statistical inference problem. These problems will be discussed in Chapter 9.

2.5 Prediction of new observations

An important application of the regression model is prediction of new observations y corresponding to a specified level of the regressor variable x. If x_0 is the value of the regressor variable of interest, then

$$\hat{y}_0=\hat{\beta}_0+\hat{\beta}_1 x_0 \tag{2.44}$$

is the point estimate of the new value of the response y_0.

Now consider obtaining an interval estimate of this future observation y_0. The confidence interval on the mean response at $x=x_0$ [Equation (2.42)] is inappropriate for this problem because it is an interval estimate on the *mean* of y (a parameter) not a probability statement about future observations from that distribution. We will develop a *prediction interval* for the future observation y_0. Note that the random variable

$$\psi=y_0-\hat{y}_0$$

is normally distributed with mean zero and variance

$$V(\psi) = V(y_0 - \hat{y}_0) = \sigma^2\left[1 + \frac{1}{n} + \frac{(x_0 - \bar{x})^2}{S_{xx}}\right]$$

because the future observation y_0 is independent of $\hat{y}_0$. If we use $\hat{y}_0$ to predict y_0, then the standard error of $\psi = y_0 - \hat{y}_0$ is the appropriate statistic upon which to base a prediction interval. Thus the $100(1-\alpha)$ percent prediction interval on a future observation at x_0 is

$$\hat{y}_0 - t_{\alpha/2, n-2}\sqrt{MS_E\left(1 + \frac{1}{n} + \frac{(x_0 - \bar{x})^2}{S_{xx}}\right)} \leqslant y_0 \leqslant \hat{y}_0 + t_{\alpha/2, n-2}\sqrt{MS_E\left(1 + \frac{1}{n} + \frac{(x_0 - \bar{x})^2}{S_{xx}}\right)} \tag{2.45}$$

The prediction interval (2.45) is of minimum width at $x_0 = \bar{x}$, and widens as $|x_0 - \bar{x}|$ increases. By comparing (2.45) with (2.43), we observe that the prediction interval at x_0 is always wider than the confidence interval at x_0, because the prediction interval depends on both the error from the fitted model, and the error associated with future observations.

- **Example 2.6** We will find a 95 percent prediction interval on a future value of propellant shear strength in a motor made from a batch of sustainer propellant that is 10 weeks old. Using (2.45), we find that the prediction interval is

$$\hat{y}_0 - t_{\alpha/2, n-2}\sqrt{MS_E\left(1 + \frac{1}{n} + \frac{(x_0 - \bar{x})^2}{S_{xx}}\right)} \leqslant y_0 \leqslant \hat{y}_0 + t_{\alpha/2, n-2}\sqrt{MS_E\left(1 + \frac{1}{n} + \frac{(x_0 - \bar{x})^2}{S_{xx}}\right)}$$

$$2256.32 - (2.101)\sqrt{9244.59\left(1 + \frac{1}{20} + \frac{(10 - 13.3625)^2}{1106.56}\right)} \leqslant y_0 \leqslant 2256.32 + (2.101)\sqrt{9244.59\left(1 + \frac{1}{20} + \frac{(10 - 13.3625)^2}{1106.56}\right)}$$

which simplifies to

$$2048.32 \leqslant y_0 \leqslant 2464.32$$

Therefore a new motor made from a batch of 10–week old sustainer propellant could reasonably be expected to have a propellant shear strength between 2048.32 psi and 2464.32 psi.

We may generalize (2.45) somewhat to find a $100(1-\alpha)$ percent prediction interval on the *mean* of m future observations on the response at $x = x_0$. Let $\bar{y}_0$ be the mean of m future observations at $x = x_0$. A point estimator of $\bar{y}_0$ is $\hat{y}_0 = \hat{\beta}_0 + \hat{\beta}_1 x_0$. The $100(1-\alpha)$ percent prediction interval on $\bar{y}_0$ is

$$\hat{y}_0 - t_{\alpha/2, n-2}\sqrt{MS_E\left(\frac{1}{m} + \frac{1}{n} + \frac{(x_0 - \bar{x})^2}{S_{xx}}\right)} \leqslant \bar{y}_0 \leqslant \hat{y}_0 + t_{\alpha/2, n-2}\sqrt{MS_E\left(\frac{1}{m} + \frac{1}{n} + \frac{(x_0 - \bar{x})^2}{S_{xx}}\right)} \tag{2.46}$$

2.6 The coefficient of determination

The quantity

$$R^2 = \frac{SS_R}{S_{yy}} = 1 - \frac{SS_E}{S_{yy}} \tag{2.47}$$

is called the *coefficient of determination*. Since S_{yy} is a measure of the variability in y without considering the effect of the regressor variable x, and SS_E is a measure of the variability in y remaining after x has been considered, R^2 is often called the proportion of variation explained by the regressor x. Because $0 \leqslant SS_E \leqslant S_{yy}$, it follows that $0 \leqslant R^2 \leqslant 1$. Values of R^2 that are close to 1 imply that most of the variability in y is explained by the regression model. For the data in Example 2.1, we have

$$R^2 = \frac{SS_R}{S_{yy}} = \frac{1{,}527{,}334.95}{1{,}693{,}737.60} = 0.9018$$

that is, 90.18 percent of the variability in the data is accounted for by the regression model.

The statistic R^2 should be used with caution, since it is always possible to make R^2 large by adding enough terms to the model. For example, if there are no repeat points (more than one y value at the same x value) a polynomial of degree $n-1$ will give a "perfect" fit ($R^2=1$) to n data points. When there are repeat points, R^2 can never be exactly equal to 1, because the model cannot explain the variability due to "pure" error.

Although R^2 increases if we add a regressor variable to the model, this does not necessarily mean the new model is superior to the old one. Unless the error sum of squares in the new model is reduced by an amount equal to the original error mean square, the new model will have a larger error mean square than the old one because of the loss of one degree of freedom for error. Thus the new model will actually be worse than the old one.

The magnitude of R^2 also depends on the range of variability in the regressor variable. Generally, R^2 will increase as the spread of the x's increases and decrease as the spread of the x's decreases, provided the assumed model form is correct. Hahn [1973] observes that the expected value of R^2 from a straight-line regression is approximately

$$E(R^2) \simeq \frac{\hat{\beta}_1^2 S_{xx}}{\hat{\beta}_1^2 S_{xx} + \sigma^2}$$

Clearly, the expected value of R^2 will increase (decrease) as S_{xx} (a measure of the spread of the x's) increases (decreases). Thus a large value of R^2 may result simply because x has been varied over an unrealistically large range. On the other hand R^2 may be small because the range of x was too small to allow its relationship with y to be detected.

There are several other misconceptions about R^2. In general R^2 does not measure the magnitude of the slope of the regression line. A large value of R^2 does not imply a steep slope. Furthermore R^2 does not measure the appropriateness of the linear model, for R^2 will often be large even though y and x are nonlinearly related. For example, R^2 for the regression equation in Figure 2.3*b* will be relatively large, even though the linear approximation is poor. Remember that although R^2 is large, this does not necessarily imply that the regression model will be an accurate predictor.

2.7 Hazards in the use of regression

Regression analysis is widely used and, unfortunately, it is frequently *misused*. There are several common abuses of regression that should be mentioned. They include:

1. Regression models are intended as interpolation equations over the range of the regressor variable(s) used to fit the model. They may not be valid for extrapolation outside of this range. Refer to Figure 1.4.

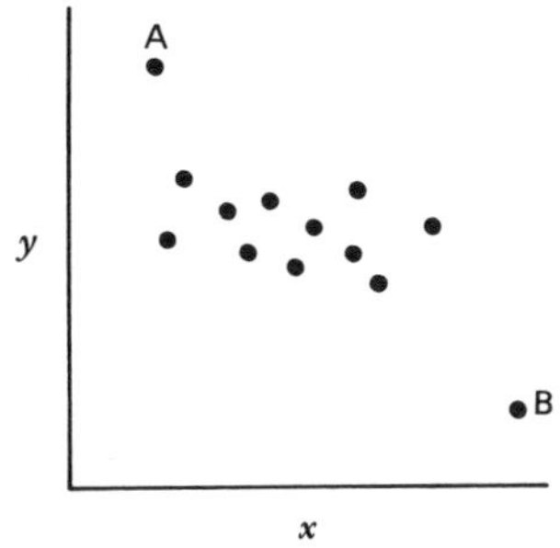

Figure 2.4 Two influential observations.

2. The disposition of the x-values plays an important role in the least squares fit. While all points have equal weight in determining the height of the line, the slope is more strongly influenced by the remote values of x. For example, consider the data in Figure 2.4. The slope in the least squares fit depends heavily on either or both of the points A and B. Furthermore the remaining data would give a very different estimate of the slope if A and B were deleted. Situations such as this require corrective action, such as further analysis and possible deletion of the unusual points, estimation of the model parameters with some technique that is less seriously influenced by these points than least squares, or restructuring the model, possibly by introducing further regressors.

A somewhat different situation is illustrated in Figure 2.5, where one of the 18 observations is very remote in x-space. In this example, the slope is largely determined by the extreme point. If this point is deleted the slope estimate is probably zero. Because of the gap between the two clusters of points, we really have only two distinct information units with which to fit the model. Thus there are effectively far fewer than the apparent 16 degrees of freedom for error.

Situations such as these seem to occur fairly often in practice. In general we should be aware that in some data sets one point (or a small cluster of points) may control key model properties.

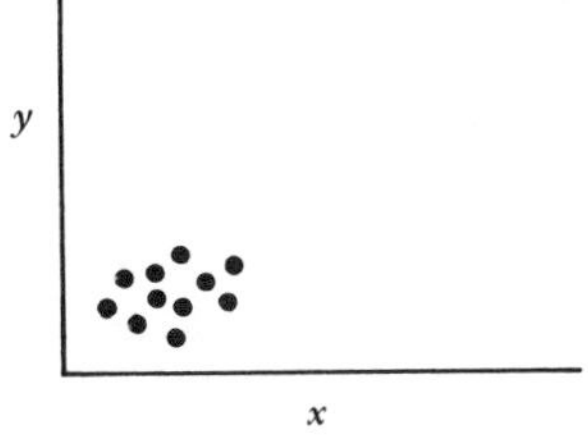

Figure 2.5 A point remote in x-space.

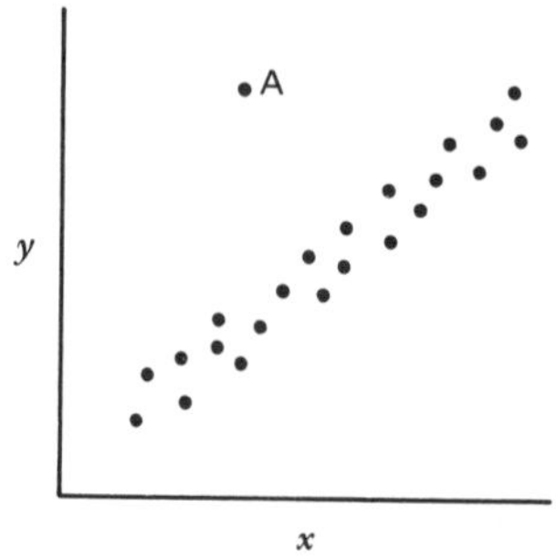

Figure 2.6 An outlier.

3. Outliers or bad values can seriously disturb the least squares fit. For example, consider the data in Figure 2.6. Observation A seems to be an "outlier" or "bad value" because it falls far from the line implied by the rest of the data. If this point is really an outlier, then the estimate of the intercept may be incorrect and the residual mean square may be an inflated estimate of σ^2. On the other hand the data point may not be a bad value, and may be a highly useful piece of evidence concerning the process under investigation. Methods for detecting and dealing with outliers are discussed more completely in Chapter 3.

4. As mentioned in Chapter 1, just because a regression analysis has indicated a strong relationship between two variables, this does not imply that the variables are related in any causal sense. Our expectations of discovering cause and effect relationships from regression should be modest.

As an example of a "nonsense" relationship between two variables, consider the data in Table 2.6. This table presents the number of certified mental defectives in the United Kingdom per 10,000 of estimated population (y), the number of radio receiver licenses issued (x_1), and the first name of the President of the United States (x_2), for the years 1924–1937. We can show that the regression equation relating y to x_1 is

$$\hat{y} = 4.582 + 2.204x_1$$

The t-statistic for testing H_0: $\beta_1 = 0$ for this model is $t_0 = 27.312$ (significant at $\alpha = .001$), and the coefficient of determination is $R^2 = 0.9842$. That is, 98.42 percent of the variability in the data is explained by the number of radio receiver licenses issued. Clearly, this is a "nonsense" relationship, as it is highly unlikely that the number of mental defectives in the population is functionally related to the number of radio receiver licenses issued. The reason for this strong statistical relationship is that y and x_1 are monotonically related [two sequences of numbers are monotonically related if as one

Table 2.6 Data illustrating "nonsense" relationships between variables

Year	Number of Certified Mental Defectives per 10,000 of Estimated Population in the U.K. (y)	Number of Radio Receiver Licenses Issued (millions) in the U.K. (x_1)	First name of President of the U.S. (x_2)
1924	8	1.350	Calvin
1925	8	1.960	Calvin
1926	9	2.270	Calvin
1927	10	2.483	Calvin
1928	11	2.730	Calvin
1929	11	3.091	Calvin
1930	12	3.647	Herbert
1931	16	4.620	Herbert
1932	18	5.497	Herbert
1933	19	6.260	Herbert
1934	20	7.012	Franklin
1935	21	7.618	Franklin
1936	22	8.131	Franklin
1937	23	8.593	Franklin

Source: Kendall and Yule [1950], and Tufte [1974].

sequence increases (for example), the other always either increases or decreases]. In this example y is increasing because diagnostic procedures for mental disorders are becoming more refined over the years represented in the study and x_1 is increasing because of the emergence and low-cost availability of radio technology over the years.

Any two sequences of numbers that are monotonically related will exhibit similar properties. To illustrate this further, suppose we regress y on the number of letters in the first name of the U.S. President in the corresponding year. The model is

$$\hat{y} = -26.442 + 5.900x_2$$

with $t_0 = 8.996$ (significant at $\alpha = .001$) and $R^2 = 0.8709$. Clearly this is a "nonsense" relationship as well.

5. In some applications of regression the value of the regressor variable x required to predict y is unknown. For example, consider predicting maximum daily load on an electric power generation system from a regression model relating the load to the maximum daily temperature. To predict tomorrow's maximum load, we must first predict tomorrow's maximum temperature. Consequently, the prediction of maximum load is *conditional* on the temperature forecast. The accuracy of the maximum load forecast depends on the accuracy of the temperature forecast. This must be considered when evaluating model performance.

Other abuses of regression will be discussed in subsequent chapters. For further reading on this subject, see Box [1966] and Box, Hunter, and Hunter [1978].

2.8 Regression through the origin

Some regression situations seem to imply that a straight line passing through the origin should be fit to the data. A "no-intercept" model often seems appropriate in analyzing data from chemical and other manufacturing processes. For example, the yield of a chemical process is zero when the process operating temperature is zero.

The no-intercept model is

$$y=\beta_1 x+\varepsilon \tag{2.48}$$

Given n observations (y_i, x_i), $i=1,2,\ldots,n$, the least squares function is

$$S(\beta_1)=\sum_{i=1}^{n}(y_i-\beta_1 x_i)^2$$

The only normal equation is

$$\hat{\beta}_1\sum_{i=1}^{n}x_i^2=\sum_{i=1}^{n}y_i x_i \tag{2.49}$$

and the least squares estimator of the slope is

$$\hat{\beta}_1=\frac{\sum_{i=1}^{n}y_i x_i}{\sum_{i=1}^{n}x_i^2} \tag{2.50}$$

The estimator $\hat{\beta}_1$ is unbiased for β_1, and the fitted regression model is

$$\hat{y}=\hat{\beta}_1 x \tag{2.51}$$

The estimator of σ^2 is

$$\hat{\sigma}^2 \equiv MS_E = \frac{\sum_{i=1}^{n} (y_i - \hat{y}_i)^2}{n-1} = \frac{\sum_{i=1}^{n} y_i^2 - \hat{\beta}_1 \sum_{i=1}^{n} y_i x_i}{n-1} \tag{2.52}$$

with $n-1$ degrees of freedom.

Making the normality assumption on the errors, we may test hypotheses and construct confidence and prediction intervals for the no-intercept model. The $100(1-\alpha)$ percent confidence interval on β_1 is

$$\hat{\beta}_1 - t_{\alpha/2, n-1} \sqrt{\frac{MS_E}{\sum_{i=1}^{n} x_i^2}} \leqslant \beta_1 \leqslant \hat{\beta}_1 + t_{\alpha/2, n-1} \sqrt{\frac{MS_E}{\sum_{i=1}^{n} x_i^2}} \tag{2.53}$$

A $100(1-\alpha)$ percent confidence interval on $E(y|x_0)$, the mean response at $x = x_0$ is

$$\hat{y}_0 - t_{\alpha/2, n-1} \sqrt{\frac{x_0^2 MS_E}{\sum_{i=1}^{n} x_i^2}} \leqslant E(y|x_0) \leqslant \hat{y}_0 + t_{\alpha/2, n-1} \sqrt{\frac{x_0^2 MS_E}{\sum_{i=1}^{n} x_i^2}} \tag{2.54}$$

The $100(1-\alpha)$ percent prediction interval on a future observation at $x = x_0$, for example y_0, is

$$\hat{y}_0 - t_{\alpha/2, n-1} \sqrt{MS_E \left(1 + \frac{x_0^2}{\sum_{i=1}^{n} x_i^2}\right)} \leqslant y_0 \leqslant \hat{y}_0$$

$$+ t_{\alpha/2, n-1} \sqrt{MS_E \left(1 + \frac{x_0^2}{\sum_{i=1}^{n} x_i^2}\right)} \tag{2.55}$$

Both the confidence interval (2.54) and the prediction interval (2.55) widen as x_0 increases. Furthermore the length of the confidence interval (2.54) at $x = 0$ is zero, because the model assumes that the mean of y at $x = 0$ is known with certainty to be zero. This behavior is considerably different from that observed in the intercept model. The prediction interval (2.55) has nonzero length at $x_0=0$ because the random error in the future observation must be taken into account.

It is relatively easy to misuse the no-intercept model, particularly in situations where the data lie in a region of x-space remote from the origin. For example, consider the no-intercept fit in the scatter diagram of chemical process yield (y) and operating temperature (x) in Figure 2.7*a*. Although over the range of the regressor variable $100°F \leqslant x \leqslant 200°F$, yield and temperature seem to be linearly related, forcing the model to go through the origin provides a visibly poor fit. A model containing an intercept, such as illustrated in Figure 2.7*b*, provides a much better fit in the region of x-space where the data were collected.

Frequently the relationship between y and x is quite different near the origin than it is in the region of x-space containing the data. This is

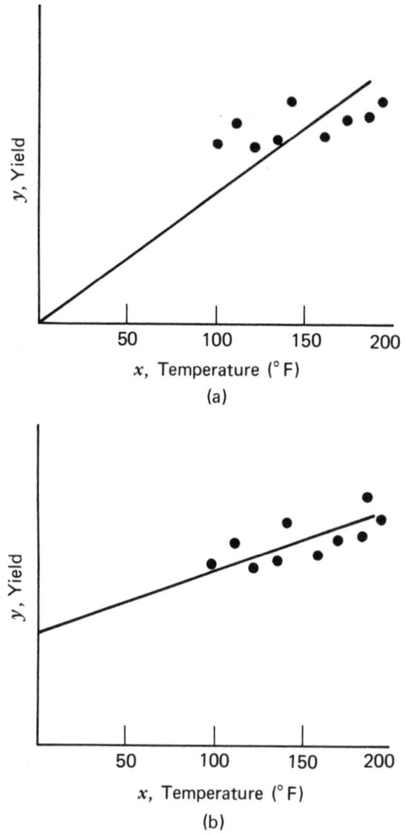

Figure 2.7 Scatter diagrams and regression lines for chemical process yield and operating temperature. **(a)** No intercept model. **(b)** Intercept model.

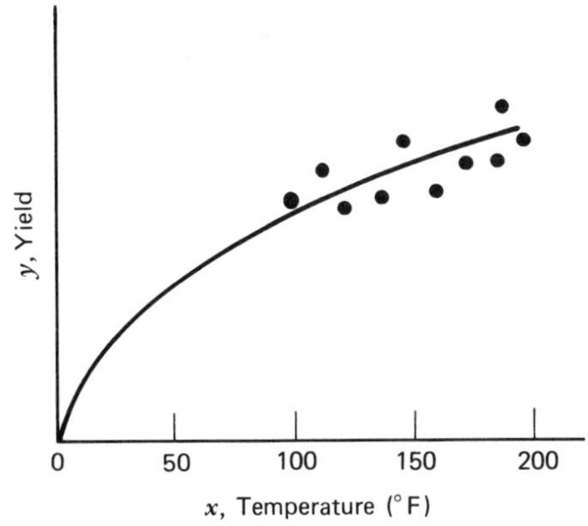

Figure 2.8 True relationship between yield and temperature.

illustrated in Figure 2.8 for the chemical process data. Here it would seem that either a quadratic or a more complex nonlinear regression model would be required to adequately express the relationship between y and x over the entire range of x. Such a model should only be entertained if the range of x in the data is sufficiently close to the origin.

The scatter diagram sometimes provides guidance in deciding whether or not to fit the no-intercept model. Alternatively, we may fit both models and choose between them based on the quality of the fit. If the hypothesis $\beta_0 = 0$ cannot be rejected in the intercept model, this is an indication that the fit may be improved by using the no-intercept model. The residual mean square is a useful way to compare the quality of fit. The model having the smaller residual mean square is the best fit, in the sense that it minimizes the estimate of the variance of y about the regression line.

Generally, R^2 is not a good comparative statistic for the two models. For the intercept model R^2 is the proportion of variability, measured by the sum of squares about the mean $\bar{y}$, accounted for by regression, while for the no-intercept model, R^2 measures the proportion of variability in the y's about the *origin* explained by regression.* We could encounter an apparently contradictory situation in which the residual mean square for the intercept model is less than the residual mean square for the no-intercept model, yet R^2 for the no-intercept model exceeds R^2 for the intercept model. Similarly, the F-tests for the two models are also not directly comparable.

- **Example 2.7** The time required for a merchandiser to stock a grocery store shelf with a soft drink product as well as the number of cases of product stocked is shown in Table 2.7. The scatter diagram shown in Figure 2.9

*If R^2 for the no-intercept model is computed based on S_{yy}, a negative R^2 could result since the variation about the regression line (SS_E) could exceed S_{yy}.

Table 2.7 Shelf stocking data for Example 2.7

Time, y (minutes)	Cases Stocked, x
10.15	25
2.96	6
3.00	8
6.88	17
0.28	2
5.06	13
9.14	23
11.86	30
11.69	28
6.04	14
7.57	19
1.74	4
9.38	24
0.16	1
1.84	5

suggests that a straight line passing through the origin could be used to express the relationship between time and the number of cases stocked. Furthermore since if the number of cases $x=0$, then shelf stocking time $y=0$, this model seems intuitively reasonable. Note also that the range of x is close to the origin.

The slope in the no-intercept model is computed from (2.50) as

$$\hat{\beta}_1 = \frac{\sum_{i=1}^{n} y_i x_i}{\sum_{i=1}^{n} x_i^2} = \frac{1841.98}{4575.00} = 0.4026$$

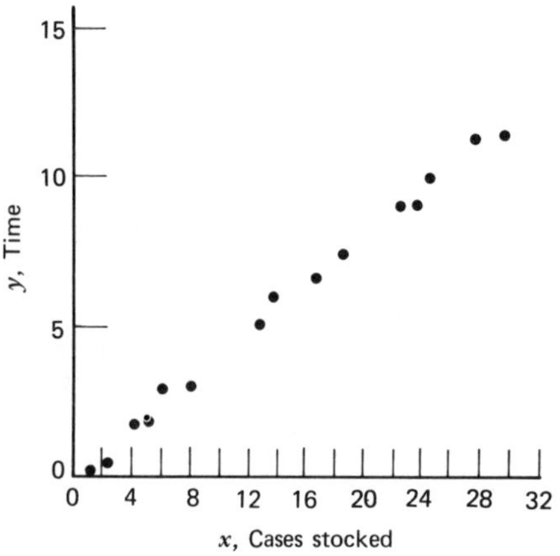

Figure 2.9 Scatter diagram of shelf stocking data.

Therefore the fitted equation is

$$\hat{y}=0.4026x$$

The residual mean square for this model is $MS_E=0.0893$ and $R^2=0.9883$. Furthermore, the t-statistic for testing H_0: $\beta_1=0$ is $t_0=91.13$, which is significant at $\alpha=.01$. These summary statistics do not reveal any startling inadequacy in the no-intercept model.

We may also fit the intercept model to the data for comparative purposes. This results in

$$\hat{y}=-0.0938+0.4071x$$

The t–statistic for testing H_0: $\beta_0=0$ is $t_0=-0.65$, which is not significant, implying that the no-intercept model may provide a superior fit. The residual mean square for the intercept model is $MS_E=0.0931$ and $R^2=0.9997$. Since MS_E for the no-intercept model is smaller than MS_E for the intercept model, we conclude that the no-intercept model is superior. As noted previously, the R^2 statistics are not directly comparable.

2.9 Estimation by maximum likelihood

The method of least squares can be used to estimate the parameters in a linear regression model regardless of the form of the distribution of the errors ε. Least squares produces best linear unbiased estimators of β_0 and β_1. Other statistical procedures, such as hypothesis testing and confidence interval construction, assume that the errors are normally distributed. If the form of the distribution of the errors is known, an alternative method of parameter estimation, the method of maximum likelihood, can be used.

Consider the data (y_i, x_i), $i=1,2,\ldots,n$. If we assume that the errors in the regression model are NID$(0,\sigma^2)$, then the observations y_i in this sample are normally and independently distributed random variables with mean $\beta_0+\beta_1x_i$ and variance σ^2. The likelihood function is found from the joint distribution of the observations. If we consider this joint distribution with the observations given and the parameters β_0, β_1, and σ^2 unknown constants, we have the likelihood function. For the simple linear regression model with normal errors, the likelihood function is

$$\begin{aligned} L(y_i, x_i, \beta_0, \beta_1, \sigma^2) &= \prod_{i=1}^{n} (2\pi\sigma^2)^{-1/2} \exp\left[-\frac{1}{2\sigma^2}(y_i-\beta_0-\beta_1x_i)^2\right] \\ &= (2\pi\sigma^2)^{-n/2} \exp\left[-\frac{1}{2\sigma^2}\sum_{i=1}^{n}(y_i-\beta_0-\beta_1x_i)^2\right] \end{aligned}$$

The maximum likelihood estimators are the parameter values, say $\tilde{\beta}_0$, $\tilde{\beta}_1$, and $\tilde{\sigma}^2$, that maximize L, or equivalently, $\ln L$. Thus

$$\ln L(y_i, x_i, \beta_0, \beta_1, \sigma^2) = -\left(\frac{n}{2}\right)\ln 2\pi - \left(\frac{n}{2}\right)\ln \sigma^2 - \left(\frac{1}{2\sigma^2}\right)\sum_{i=1}^{n}(y_i - \beta_0 - \beta_1 x_i)^2$$

and the maximum likelihood estimators $\tilde{\beta}_0$, $\tilde{\beta}_1$, and $\tilde{\sigma}^2$ must satisfy

$$\left.\frac{\partial \ln L}{\partial \beta_0}\right|_{\tilde{\beta}_0, \tilde{\beta}_1, \tilde{\sigma}^2} = \frac{1}{\tilde{\sigma}^2}\sum_{i=1}^{n}(y_i - \tilde{\beta}_0 - \tilde{\beta}_1 x_i) = 0 \tag{2.56a}$$

$$\left.\frac{\partial \ln L}{\partial \beta_1}\right|_{\tilde{\beta}_0, \tilde{\beta}_1, \tilde{\sigma}^2} = \frac{1}{\tilde{\sigma}^2}\sum_{i=1}^{n}(y_i - \tilde{\beta}_0 - \tilde{\beta}_1 x_i)x_i = 0 \tag{2.56b}$$

and

$$\left.\frac{\partial \ln L}{\partial \sigma^2}\right|_{\tilde{\beta}_0, \tilde{\beta}_1, \tilde{\sigma}^2} = -\frac{n}{2\tilde{\sigma}^2} + \frac{1}{2\tilde{\sigma}^4}\sum_{i=1}^{n}(y_i - \tilde{\beta}_0 - \tilde{\beta}_1 x_i)^2 = 0 \tag{2.56c}$$

The solution to (2.56) is

$$\tilde{\beta}_0 = \bar{y} - \tilde{\beta}_1 \bar{x} \tag{2.57a}$$

$$\tilde{\beta}_1 = \frac{\sum_{i=1}^{n} y_i(x_i - \bar{x})}{\sum_{i=1}^{n}(x_i - \bar{x})^2} \tag{2.57b}$$

$$\tilde{\sigma}^2 = \frac{\sum_{i=1}^{n}(y_i - \tilde{\beta}_0 - \tilde{\beta}_1 x_i)^2}{n} \tag{2.57c}$$

Notice that the maximum likelihood estimators of the intercept and slope, $\tilde{\beta}_0$ and $\tilde{\beta}_1$, are identical to the least squares estimators of these parameters. Also, $\tilde{\sigma}^2$ is a biased estimator of σ^2. This biased estimator is related to the unbiased estimator $\hat{\sigma}^2$ [Equation (2.19)] by $\tilde{\sigma}^2 = [(n-1)/n]\hat{\sigma}^2$. The bias is small if n is moderately large. Generally the unbiased estimator $\hat{\sigma}^2$ is used.

In general, maximum likelihood estimators have better statistical properties than least squares estimators. The maximum likelihood estimators are unbiased (including $\tilde{\sigma}^2$, which is *asymptotically* unbiased, or unbiased as n becomes large) and have minimum variance when compared to *all* other unbiased estimators. They are also *consistent* estimators (consistency is a large sample property indicating that the estimators differ from the true parameter value by a very small amount as n becomes large), and they are a set of *sufficient* statistics (this implies that the estimators contain all the "information" in the original sample of size n). For more information on maximum likelihood estimation in regression models, see Graybill [1961, 1976], Searle [1971], and Seber [1977].

2.10 Correlation

Our discussion of regression analysis thus far has assumed that x is a controllable variable, measured with negligible error, and that y is a random variable. Many applications of regression analysis involve situations where *both* x and y are random variables and the levels of x cannot be controlled. In these situations, we usually assume that the observations (y_i, x_i), $i = 1, 2, \ldots, n$ are jointly distributed random variables. For example, suppose we wish to develop a regression model relating soft drink sales to the maximum daily temperature. Clearly, we cannot control the maximum daily temperature. We would randomly select n days and observe a maximum temperature (x_i) and a sales level (y_i) for each. Thus (y_i, x_i) are jointly distributed random variables. In such models the inferences that result are conditional on the observed x's.

We usually assume that the joint distribution of y and x is the bivariate normal distribution. That is,

$$f(y,x)=\frac{1}{2\pi\sigma_1\sigma_2\sqrt{1-\rho^2}}$$

$$\times\exp\left\{-\frac{1}{2(1-\rho^2)}\left[\left(\frac{y-\mu_1}{\sigma_1}\right)^2+\left(\frac{x-\mu_2}{\sigma_2}\right)^2-2\rho\left(\frac{y-\mu_1}{\sigma_1}\right)\left(\frac{x-\mu_2}{\sigma_2}\right)\right]\right\} \tag{2.58}$$

where μ_1 and σ_1^2 are the mean and variance of y, μ_2 and σ_2^2 are the mean and variance of x, and

$$\rho=\frac{E(y-\mu_1)(x-\mu_2)}{\sigma_1\sigma_2}=\frac{\sigma_{12}}{\sigma_1\sigma_2}$$

is the correlation coefficient between y and x. σ_{12} is the covariance of y and x.

The conditional distribution of y for a given value of x is

$$f(y|x)=\frac{1}{\sqrt{2\pi}\,\sigma_{1.2}}\exp\left[-\frac{1}{2}\left(\frac{y-\beta_0-\beta_1 x}{\sigma_{1.2}}\right)^2\right] \tag{2.59}$$

where

$$\beta_0=\mu_1-\mu_2\rho\frac{\sigma_1}{\sigma_2} \tag{2.60a}$$

$$\beta_1=\frac{\sigma_1}{\sigma_2}\rho \tag{2.60b}$$

and

$$\sigma_{1.2}^2=\sigma_1^2(1-\rho^2) \tag{2.60c}$$

That is, the conditional distribution of y given x is normal with mean

$$E(y|x)=\beta_0+\beta_1 x \tag{2.61}$$

and variance $\sigma_{1.2}^2$. Note that the mean of the conditional distribution of y given x is a straight-line regression model. Furthermore there is a relationship between the correlation coefficient ρ and the slope β_1. From (2.60b) we see that if $\rho=0$ then $\beta_1=0$, which implies that there is no linear regression of y on x. That is, knowledge of x does not assist us in predicting y.

The method of maximum likelihood may be used to estimate the parameters β_0 and β_1. It may be shown that the maximum likelihood estimators of these parameters are

$$\hat{\beta}_0=\bar{y}-\hat{\beta}_1\bar{x} \tag{2.62a}$$

and

$$\hat{\beta}_1=\frac{\sum_{i=1}^{n} y_i(x_i-\bar{x})}{\sum_{i=1}^{n}(x_i-\bar{x})^2}=\frac{S_{xy}}{S_{xx}} \tag{2.62b}$$

The estimators of the intercept and slope in (2.62) are identical to those given by the method of least squares in the case where x was assumed to be a controllable variable. In general the regression model with y and x jointly normally distributed may be analyzed by the methods presented previously for the model with x a controllable variable. This follows because the random variable y given x is independently and normally distributed with mean $\beta_0 + \beta_1 x$ and constant variance $\sigma_{1.2}^2$. These results will also hold for *any* joint distribution of y and x such that the conditional distribution of y given x is normal.

It is possible to draw inferences about the correlation coefficient ρ in this model. The estimator of ρ is the *sample correlation coefficient*

$$r = \frac{\sum_{i=1}^{n} y_i (x_i - \bar{x})}{\left[\sum_{i=1}^{n} (x_i - \bar{x})^2 \sum_{i=1}^{n} (y_i - \bar{y})^2 \right]^{1/2}}$$

$$= \frac{S_{xy}}{\left[S_{xx} S_{yy} \right]^{1/2}} \tag{2.63}$$

Note that

$$\hat{\beta}_1 = \left(\frac{S_{yy}}{S_{xx}} \right)^{1/2} r \tag{2.64}$$

so that the slope $\hat{\beta}_1$ is just the sample correlation coefficient r multiplied by a scale factor which is the square root of the "spread" of the y's divided by the "spread" of the x's. Thus $\hat{\beta}_1$ and r are closely related, although they provide somewhat different information. The sample correlation coefficient r is a measure of the association between y and x, while $\hat{\beta}_1$ measures the predicted change in y for a unit change in x. In the case of a controllable variable x, r has no meaning because the magnitude of r depends on the choice of spacing for x. We may also write, from (2.64)

$$r^2 = \hat{\beta}_1^2 \frac{S_{xx}}{S_{yy}}$$

$$= \frac{\hat{\beta}_1 S_{xy}}{S_{yy}}$$

$$= \frac{SS_R}{S_{yy}}$$

$$= R^2$$

which we recognize from (2.47) as the coefficient of determination. That is, the coefficient of determination R^2 is just the square of the correlation coefficient between y and x.

While regression and correlation are closely related, regression is a more powerful tool in many situations. Correlation is only a measure of association, and is of little use in prediction. However, regression methods are useful in developing quantitative relationships between variables, which can be used in prediction.

It is often useful to test the hypothesis that the correlation coefficient equals zero; that is,

$$H_0\colon \rho=0$$
$$H_1\colon \rho\neq 0 \tag{2.65}$$

The appropriate test statistic for this hypothesis is

$$t_0=\frac{r\sqrt{n-2}}{\sqrt{1-r^2}} \tag{2.66}$$

which follows the t distribution with $n-2$ degrees of freedom if $H_0\colon \rho=0$ is true. Therefore we would reject the null hypothesis if $|t_0|>t_{\alpha/2,\,n-2}$. This test is equivalent to the t-test for $H_0\colon \beta_1=0$ given in Section 2.3. This equivalence follows directly from (2.64).

The test procedure for the hypothesis

$$H_0\colon \rho=\rho_0$$
$$H_1\colon \rho\neq\rho_0 \tag{2.67}$$

where $\rho_0\neq 0$ is somewhat more complicated. For moderately large samples (e.g., $n\geqslant 25$) the statistic

$$Z=\operatorname{arctanh} r=\frac{1}{2}\ln\frac{1+r}{1-r} \tag{2.68}$$

is approximately normally distributed with mean

$$\mu_Z=\operatorname{arctanh}\rho=\frac{1}{2}\ln\frac{1+\rho}{1-\rho}$$

and variance

$$\sigma_Z^2=(n-3)^{-1}$$

Therefore to test the hypotheses H_0: $\rho=\rho_0$ we may compute the statistic

$$Z_0=(\operatorname{arctanh} r-\operatorname{arctanh}\rho_0)(n-3)^{1/2} \tag{2.69}$$

and reject H_0: $\rho=\rho_0$ if $|Z_0|>Z_{\alpha/2}$.

It is also possible to construct a $100(1-\alpha)$ percent confidence interval for ρ, using the transformation (2.68). The $100(1-\alpha)$ percent confidence interval is

$$\tanh\left(\operatorname{arctanh} r-\frac{Z_{\alpha/2}}{\sqrt{n-3}}\right)\leqslant\rho\leqslant\tanh\left(\operatorname{arctanh} r+\frac{Z_{\alpha/2}}{\sqrt{n-3}}\right), \tag{2.70}$$

where $\tanh u=(e^u-e^{-u})/(e^u+e^{-u})$.

- **Example 2.8.** Consider the soft drink delivery time data introduced in Chapter 1. The 25 observations on delivery time y and delivery volume x are listed in Table 2.8. The scatter diagram shown in Figure 1.1 indicates a strong linear relationship between delivery time and delivery volume. The sample correlation coefficient between delivery time y and delivery volume x is

$$r=\frac{S_{xy}}{[S_{xx}S_{yy}]^{1/2}}=\frac{2473.3440}{[(1136.5600)(5784.5426)]^{1/2}}=0.9646$$

Table 2.8 Data for Example 2.8

Observation	Delivery Time, y	Number of Cases, x	Observation	Delivery Time, y	Number of Cases, x
1	16.68	7	14	19.75	6
2	11.50	3	15	24.00	9
3	12.03	3	16	29.00	10
4	14.88	4	17	15.35	6
5	13.75	6	18	19.00	7
6	18.11	7	19	9.50	3
7	8.00	2	20	35.10	17
8	17.83	7	21	17.90	10
9	79.24	30	22	52.32	26
10	21.50	5	23	18.75	9
11	40.33	16	24	19.83	8
12	21.00	10	25	10.75	4
13	13.50	4			

If we assume that delivery time and delivery volume are jointly normally distributed, we may test the hypothesis

$$H_0: \rho=0$$
$$H_1: \rho\neq 0$$

using the test statistic

$$t_0=\frac{r\sqrt{n-2}}{\sqrt{1-r^2}}=\frac{0.9646\sqrt{23}}{\sqrt{1-0.9305}}=17.55$$

Since $t_{.025,23}=2.069$, we reject H_0 and conclude that the correlation coefficient $\rho\neq 0$. Finally, we may construct an approximate 95 percent confidence interval on ρ from (2.70). Since $\operatorname{arctanh} r=\operatorname{arctanh} 0.9646=2.0082$, (2.70) becomes

$$\tanh\left(2.0082-\frac{1.96}{\sqrt{22}}\right)\leqslant\rho\leqslant\tanh\left(2.0082+\frac{1.96}{\sqrt{22}}\right)$$

which reduces to

$$0.9202\leqslant\rho\leqslant 0.9845$$

Although we know that delivery time and delivery volume are highly correlated, this information is of little use in predicting delivery time for example, as a function of the number of cases of product delivered. This would require a regression model. The straight-line fit (shown graphically in Figure 1.1*b*, page 2) relating delivery time to delivery volume is

$$\hat{y}=3.3208+2.1762x$$

Further analysis would be required to determine if this equation is an adequate fit to the data, and if it is likely to be a successful predictor.

2.11 Sample computer output

Computer programs are an indispensable part of modern regression analysis and many excellent programs are available. The information obtained from one such program [the SAS (Statistical Analysis System) General Linear Models Procedure] for the data in Example 2.8 is shown in Figure 2.10. Note that we have named the response variable **TIME** and the regressor variable **CASES**. A great many decimal places are shown in the output, even though not all of these decimal places are meaningful. Most other codes produce similar results.

GENERAL LINEAR MODELS PROCEDURE

DEPENDENT VARIABLE: TIME

SOURCE	DF	SUM OF SQUARES	MEAN SQUARE	F VALUE
MODEL	1	5382.40879702	5382.40879702	307.85
ERROR	23	402.13380298	17.48407839	PR > F
CORRECTED TOTAL	24	5784.54260000		0.0001

R-SQUARE	C.V.	STD DEV	TIME MEAN
0.930461	18.6803	4.18139670	22.38400000

PARAMETER	ESTIMATE	T FOR H0: PARAMETER=0	PR > |T|	STD ERROR OF ESTIMATE
INTERCEPT	3.32077990	2.42	0.0237	1.37107367
CASES	2.17616668	17.55	0.0001	0.12402956

OBSERVATION	OBSERVED VALUE	PREDICTED VALUE	RESIDUAL
1	16.68000000	18.55394665	-1.87394665
2	11.50000000	9.84927993	1.65072007
3	12.03000000	9.84927993	2.18072007
4	14.88000000	12.02544661	2.85455339
5	13.75000000	16.37777997	-2.62777997
6	18.11000000	18.55394665	-0.44394665
7	8.00000000	7.67311325	0.32688675
8	17.83000000	18.55394665	-0.72394665
9	79.24000000	68.60578025	10.63421975
10	21.50000000	14.20161329	7.29838671
11	40.33000000	38.13944675	2.19055325
12	21.00000000	25.08244668	-4.08244668
13	13.50000000	12.02544661	1.47455339
14	19.75000000	16.37777997	3.37222003
15	24.00000000	22.90628000	1.09372000
16	29.00000000	25.08244668	3.91755332
17	15.35000000	16.37777997	-1.02777997
18	19.00000000	18.55394665	0.44605335
19	9.50000000	9.84927993	-0.34927993
20	35.10000000	40.31561343	-5.21561343
21	17.90000000	25.08244668	-7.18244668
22	52.32000000	59.90111354	-7.58111354
23	18.75000000	22.90628000	-4.15628000
24	19.83000000	20.73011332	-0.90011332
25	10.75000000	12.02544661	-1.27544661

SUM OF RESIDUALS	0.00000000
SUM OF SQUARED RESIDUALS	402.13380298
SUM OF SQUARED RESIDUALS - ERROR SS	-0.00000000
FIRST ORDER AUTOCORRELATION	0.52069930
DURBIN-WATSON D	0.94582346

Figure 2.10 Sample computer output (SAS), Example 2.8.

The computer program produces the analysis of variance for testing H_0: $\beta_1=0$ (including the α value at which this hypothesis is rejected), along with R^2, the standard error of regression $\sqrt{MS_E}$ (called **STD DEV** on the computer printout), the mean of the response variable, and the coefficient of variation (called **C.V.** on the printout). The coefficient of variation is computed as $(\sqrt{MS_E}/\bar{y})100$, and it expresses the unexplained variability remaining in the data relative to the mean response. For each model parameter, we are shown the least squares estimate, the t-statistic for the hypothesis that the parameter equals zero along with the α value at which the hypothesis is rejected, and the standard error of the regression coefficient. The t-statistic is computed as $t=\hat{\beta}/\text{se}(\hat{\beta})$.

The computer program also prints the observed values of y, along with the corresponding predicted values and residuals. The sum of the residuals and the sum of the squared residuals (which should equal SS_E) are also displayed. Regression computer programs will also produce plots of the data and residuals useful for model diagnostic checking. These plots will be illustrated in Chapter 3.

Problems

2.1 Appendix Table B.1 gives data concerning the performance of the 26 National Football League teams in 1976. It is suspected that the number of yards gained rushing by opponent (x_8) has an effect on the number of games won by a team (y).

a. Fit a simple linear regression model relating games won y to yards gained rushing by opponents x_8.

b. Construct the analysis of variance table and test for significance of regression.

c. Find a 95 percent confidence interval on the slope.

d. What percent of the total variability in y is explained by this model?

e. Find a 95 percent confidence interval on the mean number of games won if opponents' yards rushing is limited to 2000 yards.

2.2 Suppose we would like to use the model developed in Problem 2.1 to predict the number of games a team will win if it can limit opponents' yards rushing to 1800 yards. Find a point estimate of the number of games won when $x_8=1800$. Find a 90 percent prediction interval on the number of games won.

2.3 Appendix Table B.2 presents data collected during a solar energy project at Georgia Tech.

a. Fit a simple linear regression model relating total heat flux y(kw) to the radial deflection of the deflected rays x (milliradians).

b. Construct the analysis of variance table and test for significance of regression.

c. Find a 99 percent confidence interval on the slope.

d. Calculate R^2.

e. Find a 95 percent confidence interval on the mean heat flux when the radial deflection is 16.5 milliradians.

2.4 Appendix Table B.3 presents data on the gasoline mileage performance of 32 different automobiles.

a. Fit a simple linear regression model relating gasoline mileage y (MPG) to engine displacement x_1 (in.^{3}).

b. Construct the analysis of variance table and test for significance of regression.

c. What percent of the total variability in gasoline mileage is accounted for by the linear relationship with engine displacement?

d. Find a 95 percent confidence interval on the mean gasoline mileage if the engine displacement is 275 in.^{3}.

e. Suppose that we wish to predict the gasoline mileage obtained from a car with a 275 in.^{3} engine. Give a point estimate of mileage. Find a 95 percent prediction interval on the mileage.

f. Compare the two intervals obtained in (d) and (e). Explain the difference between them. Which one is wider, and why?

2.5 Consider the gasoline mileage data in Appendix Table B.3. Repeat Problem 2.4 (Parts a, b, and c) using vehicle weight x_{10} as the regressor variable. Based on comparing the two models, can you conclude that x_1 is a better choice of regressor than x_{10}?

2.6 Appendix Table B.4 presents data for 27 houses sold in Erie, Pennsylvania.

a. Fit a simple linear regression model relating selling price of the house to the current taxes (x_1).

b. Test for significance of regression.

c. What percent of the total variability in selling price is explained by this model?

d. Find a 95 percent confidence interval on β_1.

e. Find a 95 percent confidence interval on the mean selling price of a house for which the current taxes are \$750.00.

2.7 The purity of oxygen produced by fractionation is thought to be related to the percentage of hydrocarbons in the main condenser of the processing unit. Twenty samples are shown below.

Purity (%)	Hydrocarbon (%)	Purity (%)	Hydrocarbon (%)
86.91	1.02	96.73	1.46
89.85	1.11	99.42	1.55
90.28	1.43	98.66	1.55
86.34	1.11	96.07	1.55
92.58	1.01	93.65	1.40
87.33	0.95	87.31	1.15
86.29	1.11	95.00	1.01
91.86	0.87	96.85	0.99
95.61	1.43	85.20	0.95
89.86	1.02	90.56	0.98

a. Fit a simple linear regression model to the data.
b. Test the hypothesis H_0: $\beta_1 = 0$.
c. Calculate R^2.
d. Find a 95 percent confidence interval on the slope.
e. Find a 95 percent confidence interval on the mean purity when the hydrocarbon percentage is 1.00.

2.8 Consider the oxygen plant data in Problem 2.7, and assume that purity and hydrocarbon percentage are jointly normally distributed random variables.

a. What is the correlation between oxygen purity and hydrocarbon percentage?
b. Test the hypothesis that $\rho = 0$.
c. Construct a 95 percent confidence interval for ρ.

2.9 The number of pounds of steam used per month at a plant is thought to be related to the average monthly ambient temperature. The past year's usages and temperatures are shown below.

Month	Temp.	Usage/1000	Month	Temp.	Usage/1000
Jan.	21	185.79	Jul.	68	621.55
Feb.	24	214.47	Aug.	74	675.06
Mar.	32	288.03	Sep.	62	562.03
Apr.	47	424.84	Oct.	50	452.93
May	50	454.68	Nov.	41	369.95
Jun.	59	539.03	Dec.	30	273.98

a. Fit a simple linear regression model to the data.

b. Test for significance of regression.

c. Plant management believes that an increase in average ambient temperature of one degree will increase average monthly steam consumption by 10,000 pounds. Does the data support this statement?

d. Construct a 99 percent prediction interval on steam usage in a month with average ambient temperature of 58°.

2.10 Consider the data in Problem 2.9, and assume that steam usage and average temperature are jointly normally distributed.

a. Find the correlation between steam usage and monthly average ambient temperature.

b. Test the hypothesis that $\rho=0$.

c. Find a 99 percent confidence interval for ρ.

2.11 Consider the soft drink delivery time data in Table 2.9. After examining the original regression model (Example 2.8), one analyst claimed that the model was invalid because the intercept was not zero. He argued that if zero cases were delivered, the time to stock and service the machine would be zero, and the straight-line model should go through the origin. What would you say in response to his comments? Fit a no-intercept model to these data, and determine which model is superior.

2.12 The weight and systolic blood pressure of 26 randomly selected males in the age group 25–30 are shown below. Assume that weight and blood pressure are jointly normally distributed.

Subject	Weight	Systolic BP	Subject	Weight	Systolic BP
1	165	130	14	172	153
2	167	133	15	159	128
3	180	150	16	168	132
4	155	128	17	174	149
5	212	151	18	183	158
6	175	146	19	215	150
7	190	150	20	195	163
8	210	140	21	180	156
9	200	148	22	143	124
10	149	125	23	240	170
11	158	133	24	235	165
12	169	135	25	192	160
13	170	150	26	187	159

a. Find a regression line relating systolic blood pressure to weight.
b. Estimate the correlation coefficient.
c. Test the hypothesis that $\rho=0$.
d. Test the hypothesis that $\rho=0.6$.
e. Find a 95 percent confidence interval for ρ.

2.13 Consider the weight and blood pressure data in Problem 2.12. Fit a no-intercept model to the data, and compare it to the model obtained in Problem 2.12. Which model would you conclude is superior?

2.14 Consider the simple linear regression model $y=\beta_0+\beta_1 x+\varepsilon$, with $E(\varepsilon)=0$, $V(\varepsilon)=\sigma^2$, and the ε uncorrelated.
a. Show that $\text{Cov}(\hat{\beta}_0, \hat{\beta}_1)=-\bar{x}\sigma^2/S_{xx}$.
b. Show that $\text{Cov}(\bar{y}, \hat{\beta}_1)=0$.

2.15 Consider the simple linear regression model $y=\beta_0+\beta_1 x+\varepsilon$, with $E(\varepsilon)=0$, $V(\varepsilon)=\sigma^2$, and the ε uncorrelated.
a. Show that $E(MS_E)=\sigma^2$.
b. Show that $E(MS_R)=\sigma^2+\beta_1^2 S_{xx}$.

2.16 Suppose that we have fitted the straight-line regression model $\hat{y}=\hat{\beta}_0+\hat{\beta}_1 x_1$, but the response is affected by a second variable x_2 such that the true regression function is

$$E(y)=\beta_0+\beta_1 x_1+\beta_2 x_2$$

Is the estimator of the slope in the original simple linear regression model unbiased?

2.17 Suppose that we are fitting a straight line and we wish to make the standard error of the slope as small as possible. Where should the observations $x_1, x_2, \ldots, x_n$ be taken? Discuss the practical aspects of this data collection plan.

2.18 Prove that the maximum value of R^2 is less than 1 if the data contain repeat observations on y at the same value of x.

3

MEASURES OF MODEL ADEQUACY

3.1 Introduction

The major assumptions that we have made thus far in our study of regression analysis are:

1. The relationship between y and x is linear, or at least it is well-approximated by a straight line.

2. The error term ε has zero mean.

3. The error term ε has constant variance σ^2.

4. The errors are uncorrelated.

5. The errors are normally distributed.

Assumptions 4 and 5 imply that the errors are independent random variables. Assumption 5 is required for hypothesis testing and interval estimation.

We should always consider the validity of these assumptions to be doubtful, and conduct analyses to examine the adequacy of the model we have tentatively entertained. The types of model inadequacies discussed here have potentially serious consequences. Gross violations of the assumptions may yield an unstable model, in the sense that a different sample could lead to a totally different model with opposite conclusions. We usually cannot detect departures from the underlying assumptions by examination of the standard summary statistics, such as the t or F-statistics, or R^2. These are "global" model properties, and as such they do not insure model adequacy.

In this chapter we present several methods useful for diagnosing and treating violations of the basic regression assumptions. While our discussion is limited to the simple linear regression model, the same general approach is used for models with several regressors.

3.2 Residual analysis

3.2.1 Definition of residuals

We have defined the residuals as

$$e_i = y_i - \hat{y}_i, \qquad i = 1, 2, \ldots, n \tag{3.1}$$

where y_i is an observation and $\hat{y}_i$ is the corresponding fitted value. Since a residual may be viewed as the deviation between the *data* and the *fit*, it is a measure of the variability not explained by the regression model. It is also convenient to think of the residuals as the realized or observed values of the errors. Thus any departures from the underlying assumptions on the errors should show up in the residuals. Analysis of the residuals is an effective method for discovering several types of model deficiencies.

The residuals have several important properties. They have zero mean and their approximate average variance is

$$\frac{\sum_{i=1}^{n} (e_i - \bar{e})^2}{n-2} = \frac{\sum_{i=1}^{n} e_i^2}{n-2} = \frac{SS_E}{n-2} = MS_E$$

The residuals are not independent, however, as the n residuals have only $n-2$ degrees of freedom associated with them. This nonindependence of the residuals has little effect on their use for investigating model adequacy as long as n is not small.

Sometimes it is useful to work with the *standardized residuals*

$$d_i = \frac{e_i}{\sqrt{MS_E}}, \qquad i = 1, 2, \ldots, n \tag{3.2}$$

The standardized residuals have zero mean and approximately unit variance. Equation (3.2) is not the only way to scale or transform the residuals. We will discuss some of the other methods in Chapter 4.

We will now present several residual plots that are useful for detecting model inadequacies. These methods are both simple and effective, and, consequently, we recommend that they be incorporated into every regression analysis problem. Most standard regression computer programs will produce these plots on request, so little additional effort is usually required to obtain these diagnostics.

3.2.2 Normal probability plot

Although small departures from normality do not affect the model greatly, gross nonnormality is potentially more serious as the t or F-statistics and confidence and prediction intervals depend on the normality assumption. Furthermore if the errors come from a distribution with thicker or heavier tails than the normal, the least squares fit may be sensitive to a small subset of the data. Heavy-tailed error distributions often generate outliers that "pull" the least squares fit too much in their direction. In these cases other estimation techniques (discussed in Chapter 9) should be considered.

A very simple method of checking the normality assumption is to plot the residuals on normal probability paper. This is graph paper designed so that the cumulative normal distribution will plot as a straight line. Let $e_{[1]} < e_{[2]} < \cdots < e_{[n]}$ be the residuals ranked in increasing order. If we plot $e_{[i]}$ against the cumulative probability $P_i = (i - \frac{1}{2})/n$, $i = 1, 2, \ldots, n$ on normal probability paper, the resulting points should lie approximately on a straight line. The straight line is usually determined visually, with emphasis on the central values (e.g., the .33 and .67 cumulative probability points) rather than the extremes. Substantial departures from a straight line indicate that the distribution is not normal. When normal probability plots are constructed automatically by computer, the ranked residual $e_{[i]}$ is usually plotted against the "expected normal value" $\Phi^{-1}[(i - \frac{1}{2})/n]$, where Φ denotes the standard normal cumulative distribution. This follows from the fact that $E(e_{[i]}) \simeq \Phi^{-1}[(i - \frac{1}{2})/n]$ (see David [1970]). Sometimes the expected normal values are called *rankits*.

Figure 3.1a displays an "idealized" normal probability plot. Notice that the points lie approximately along a straight line. Figures 3.1b, c, d, and e present other typical problems. Figure 3.1b shows a sharp upward and downward curve at both extremes, indicating that the tails of this distribution are too heavy for it to be considered normal. Conversely, Figure 3.1c shows flattening at the extremes, which is a pattern typical of samples from

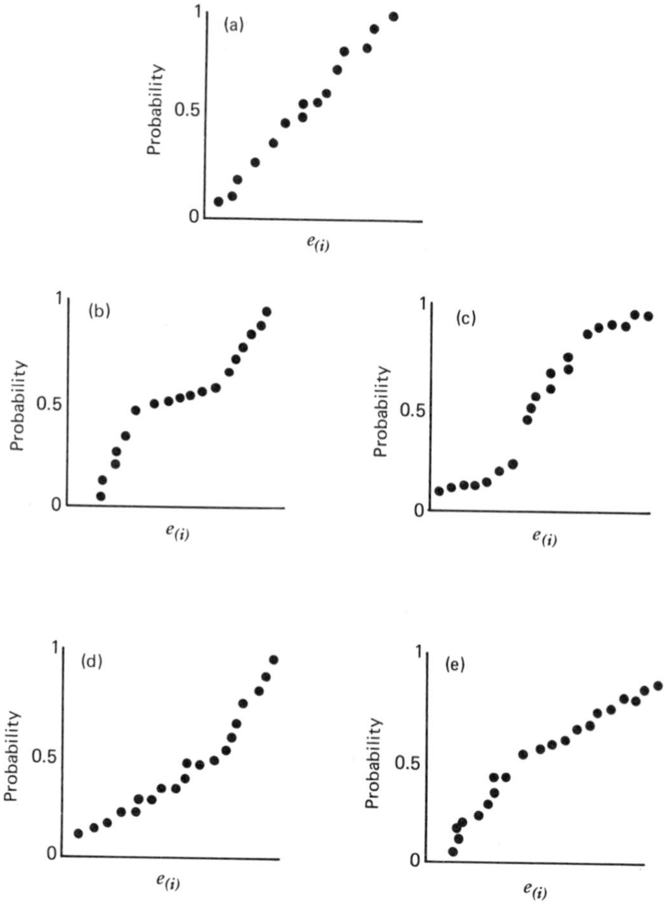

Figure 3.1 Normal probability plots. **(a)** Ideal. **(b)** Heavy-tailed distribution. **(c)** Light-tailed distribution. **(d)** Positive skew. **(e)** Negative skew.

a distribution with thinner tails than the normal. Figures 3.1d and e exhibit patterns associated with positive and negative skew, respectively.*

Because samples taken from a normal distribution will not plot exactly as a straight line, some experience is required to interpret normal probability plots. Daniel and Wood [1980] present normal probability plots for sample sizes 8–384. Study of these plots is helpful in acquiring a feel for how much

*These interpretations assume that the ranked residuals are plotted on the horizontal axis. If the residuals are plotted on the vertical axis, as some computer systems do, the interpretation is reversed.

deviation from the straight line is acceptable. Small sample sizes ($n \leq 16$) often produce normal probability plots that deviate substantially from linearity. For larger sample sizes ($n \geq 32$) the plots are much better behaved. Usually about 20 points are required to produce normal probability plots that are stable enough to be easily interpreted.

Andrews [1979] and Gnanadesikan [1977] note that normal probability plots often exhibit no unusual behavior even if the errors ε_i are not normally distributed. This problem occurs because the residuals are not a simple random sample; they are the remnants of a parameter estimation process. Fitting the parameters tends to destroy the evidence of nonnormality in the residuals, and consequently we cannot always rely on the normal probability plot to detect departures from normality.

A common defect that shows up on the normal probability plot is the occurrence of one or two large residuals. Sometimes this is an indication that the corresponding observations are outliers. Methods for detecting and dealing with outliers are discussed in Section 3.3.

• **Example 3.1** A normal probability plot of the residuals from the rocket propellant data in Example 2.1 is shown in Figure 3.2. This graph was constructed using a BMD–P Library program, and it plots the expected normal value as the vertical scale. If we visualize a straight line passing through these points emphasizing the .33 and .67 cumulative probability points (roughly the -0.5 and $+0.5$ expected normal value), we note that there are two large negative residuals that lie quite far from the rest (Observations 5 and 6 in Table 2.1). These points are potential outliers, and their effect on the model will be considered in Section 3.3. These two points tend to give the residual plot the appearance of one for skewed data. •

The normality assumption may also be checked by constructing a histogram of the residuals. Often however the number of residuals is too small to allow easy visual identification of the shape of the normal distribution.

The standardized residuals are also useful in detecting departures from normality. If the errors are normally distributed, then approximately 68 percent of the standardized residuals should fall between -1 and $+1$, and approximately 95 percent of them should fall between -2 and $+2$. Substantial deviation from these limits indicates potential violation of the normality assumption. If n is small, we may replace the limits ± 1 and ± 2 with the corresponding values from the t_{n-2} distribution. Examination of the standardized residuals in this manner is also helpful in identifying outliers. Finally, we note that some analysts prefer to construct normal probability plots using the standardized residuals d_i rather than the e_i.

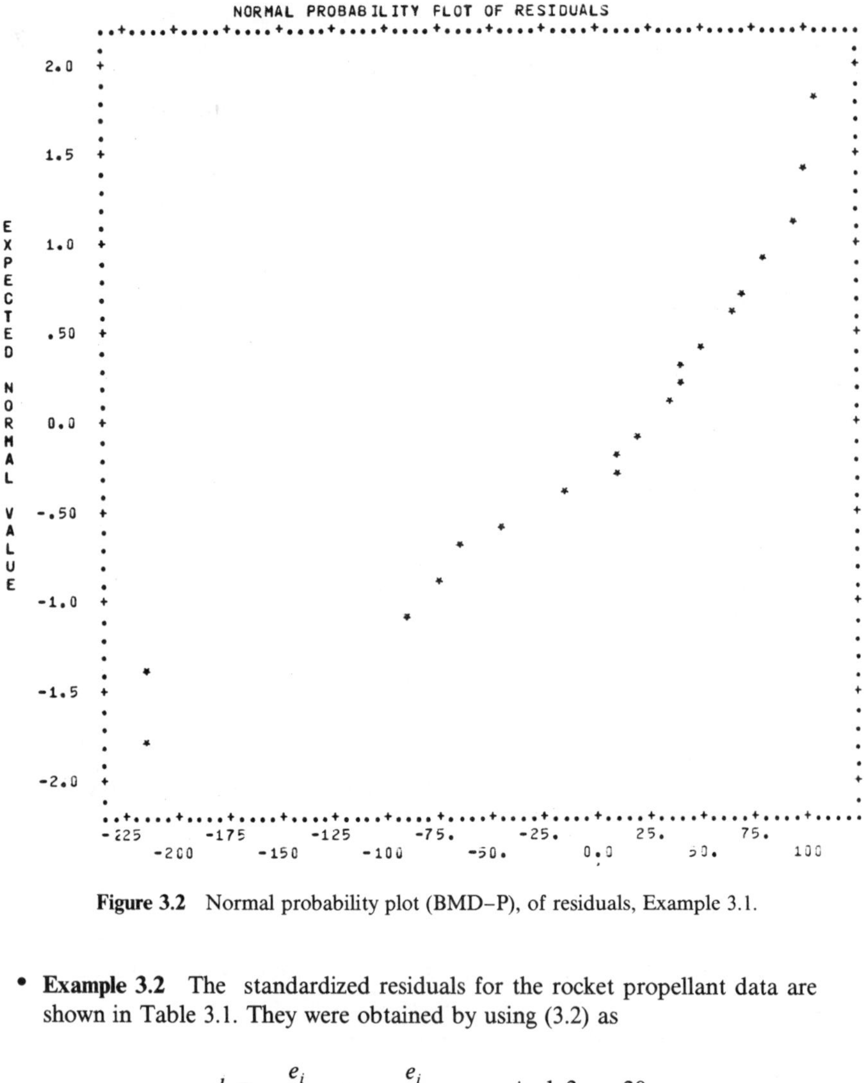

Figure 3.2 Normal probability plot (BMD–P), of residuals, Example 3.1.

- **Example 3.2** The standardized residuals for the rocket propellant data are shown in Table 3.1. They were obtained by using (3.2) as

$$d_i = \frac{e_i}{\sqrt{MS_E}} = \frac{e_i}{\sqrt{9244.59}} \qquad i=1,2,\ldots,20$$

The standardized residuals for Observations 5 and 6 exceed two standard deviations, and the standardized residuals for Observations 1 and 19 exceed one standard deviation. Observations 5 and 6 are the two points producing suspicious residuals on the normal probability plot in Figure 3.1. Examination of the standardized residuals reinforces the earlier indication that Observations 5 and 6 may be outliers. •

Table 3.1 Standardized residuals for the rocket propellant data in Example 3.1

Observation Number, i	e_i	d_i
1	106.76	1.11
2	−67.27	−0.70
3	−14.59	−0.15
4	65.09	0.68
5	−215.98	−2.25
6	−213.60	−2.22
7	48.56	0.51
8	40.06	0.42
9	8.73	0.09
10	37.57	0.39
11	20.37	0.21
12	−88.95	−0.93
13	80.82	0.84
14	71.17	0.74
15	−45.15	−0.47
16	94.44	0.98
17	9.50	0.10
18	37.10	0.39
19	100.68	1.05
20	−75.32	−0.78

3.2.3 Plot of residuals against $\hat{y}_i$

A plot of the residuals e_i versus the corresponding fitted values $\hat{y}_i$ is useful for detecting several common types of model inadequacies*. If this plot resembles Figure 3.3a, which indicates that the residuals can be contained in a horizontal band, then there are no obvious model defects. Plots of e_i versus $\hat{y}_i$ that resemble any of the patterns in Figures 3.3b, c and d are symptomatic of model deficiencies.

The patterns in Figures 3.3b and c indicate that the variance of the errors is not constant. The outward-opening funnel pattern in Figure 3.3b implies that the variance is an increasing function of y (an inward-opening funnel is also possible, indicating that $V(\varepsilon)$ increases as y decreases). The double bow

*The residuals should be plotted versus the fitted values $\hat{y}_i$ and not the observed values y_i because the e_i and the $\hat{y}_i$ are uncorrelated while the e_i and the y_i are usually correlated. See Problem 3.19.

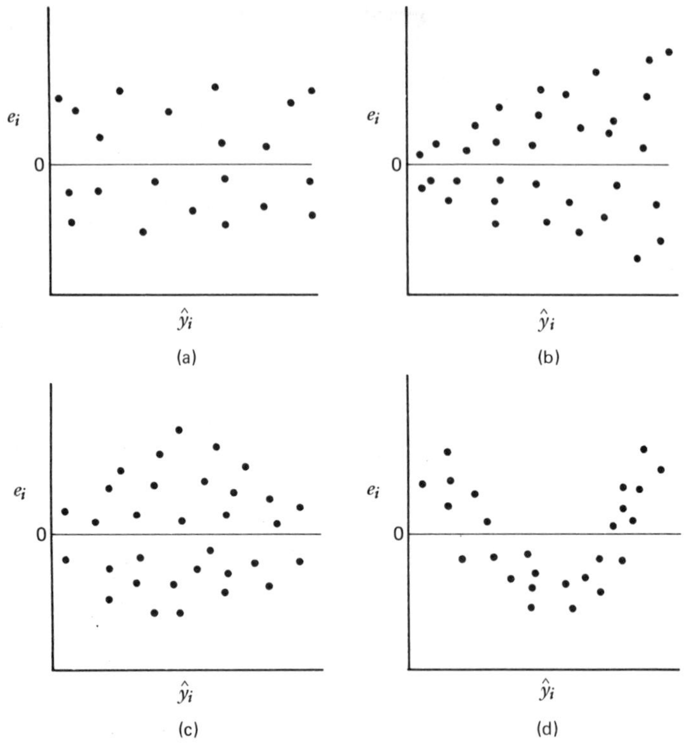

Figure 3.3 Patterns for residual plots. (**a**) Satisfactory. (**b**) Funnel. (**c**) Double bow. (**d**) Nonlinear.

pattern in Figure 3.3c often occurs when y is a proportion between zero and one. The variance of a binomial proportion near 0.5 is greater than one near 0 or 1. The usual approach to dealing with inequality of variance is to apply a suitable transformation to either the regressor or the response variable (see Sections 3.6 and 3.7), or to use the method of weighted least squares (Section 3.8). In practice, transformations on the response are generally employed to stabilize variance.

A curved plot such as in Figure 3.3d indicates nonlinearity. This could mean that other regressor variables are needed in the model. For example, a squared term may be necessary. Transformations on the regressor and/or the response variable may also be required.

A plot of the residuals against $\hat{y}_i$ may also reveal one or more unusually large residuals. These points are, of course, potential outliers. Large residuals that occur at the extreme $\hat{y}_i$ values could also indicate that either the variance is not constant or the true relationship between y and x is not linear. These possibilities should be investigated before the points are considered outliers.

• **Example 3.3** A plot of the residuals e_i versus the fitted values $\hat{y}_i$ for the rocket propellant data is shown in Figure 3.4. This plot reveals the two large residuals for Observations 5 and 6 noted previously in Figure 3.2. Apart from these two points, the plot gives no other obvious indication of curvature or inequality of variance. •

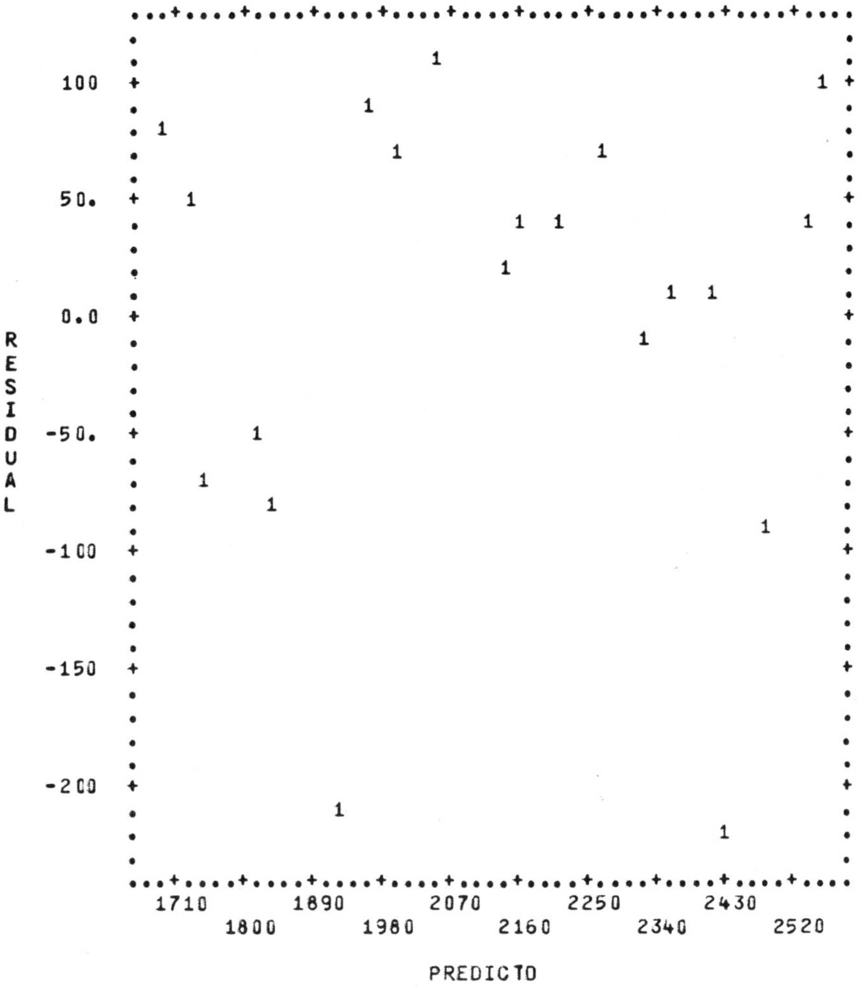

Figure 3.4 Plot (BMD–P) of residuals e_i versus fitted values $\hat{y}_i$ for the rocket propellant data.

3.2.4 Plot of residuals against x_i

Plotting the residuals against the corresponding values of the regressor variable is also helpful. These plots often exhibit patterns such as those in Figure 3.3, except that the horizontal scale is x_i rather than $\hat{y}_i$. Once again an impression of a horizontal band containing the residuals is desirable. The funnel and double bow patterns in Figures 3.3b and c indicate nonconstant variance. The curved band in Figure 3.3d implies that possibly other regressors should be included or that a transformation is necessary.

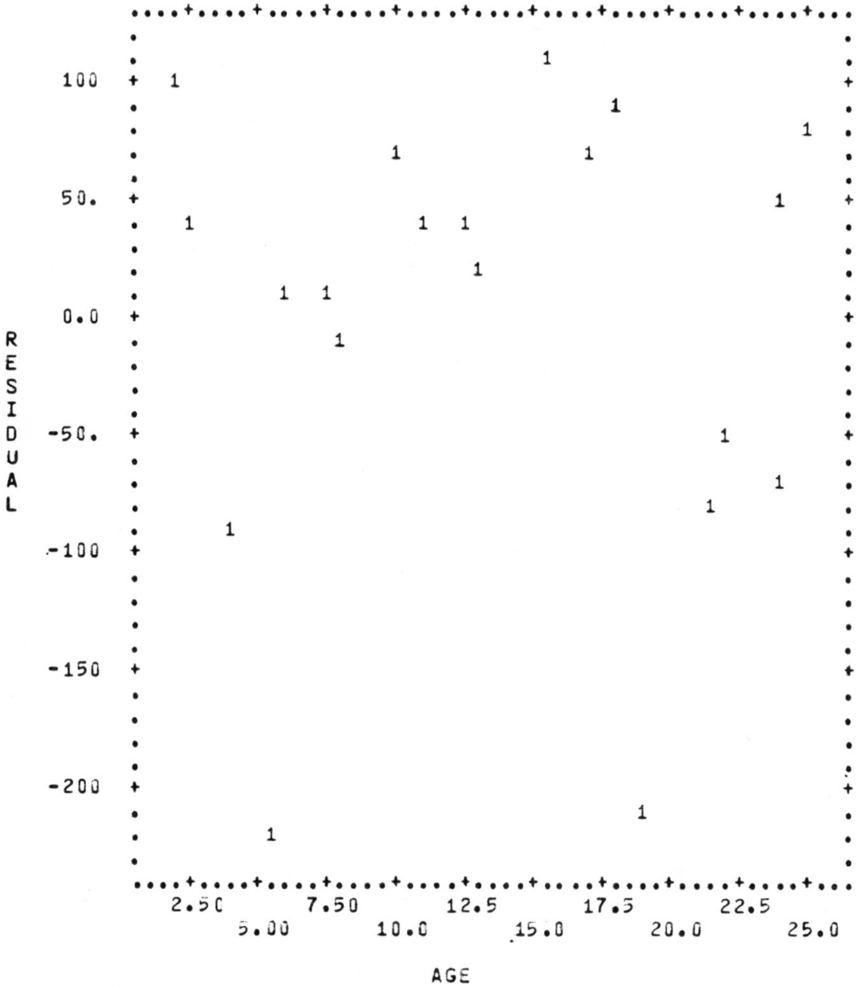

Figure 3.5 Plot (BMD–P) of residuals e_i versus age x_i for the rocket propellant data.

- **Example 3.4** Figure 3.5 presents a plot of the residuals e_i versus age x_i for the rocket propellant data. Here too the large negative residuals for Observations 5 and 6 are apparent. Apart from these two points, there is no strong indication of violation of the assumptions in this plot. •

3.2.5 Other residual plots

In addition to the basic residual plots discussed in Sections 3.2.2, 3.2.3, and 3.2.4, there are several others that are occasionally useful. For example, if the time sequence in which the data were collected is known, it may be instructive to plot the residuals against time order. If such a plot resembles the patterns in Figure 3.3b, c, or d, this may indicate that the variance is changing with time, or that linear or quadratic terms in time should be added to the model. The time sequence plot of residuals may indicate that the errors at one time period are correlated with those at other time periods. The correlation between model errors at different time periods is called autocorrelation. A display such as Figure 3.6a indicates positive autocorrelation, while Figure 3.6b is typical of negative autocorrelation. The presence of autocorrelation in the errors is a serious violation of the basic regression

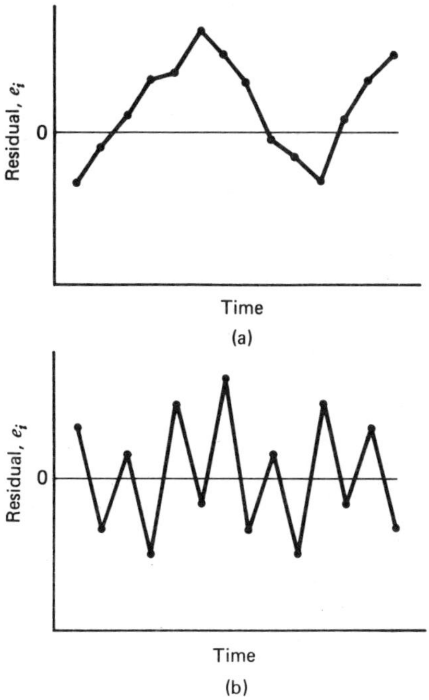

Figure 3.6 Prototype residual plots against time displaying autocorrelation in the errors. (**a**) Positive autocorrelation. (**b**) Negative autocorrelation.

assumptions. Statistical tests for detecting autocorrelation and methods for dealing with it are discussed in Chapter 9.

Plotting the residuals against any omitted regressor can also reveal model inadequacies. Of course, such a plot can be constructed only if the levels of the omitted regressor are known. Any systematic pattern exhibited by this plot indicates that the model may be improved by addition of the new regressor.

The problem situation often suggests other types of residual plots. For example, consider the delivery time data in Example 2.8. The 25 observations were collected on truck routes in four different cities. Observations 1–7 were collected in San Diego, Observations 8–17 in Boston, Observations 18–23 in Austin, and Observations 24 and 25 in Minneapolis. We might suspect that there is a difference in delivery operations from city to city, due to such factors as different types of equipment, different levels of

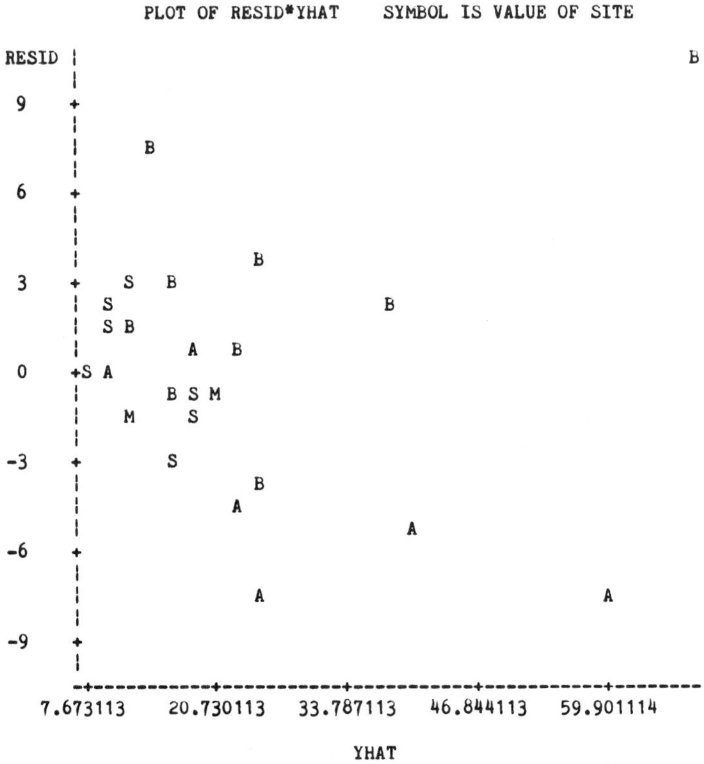

Figure 3.7 Plot (SAS) of residuals e_i versus fitted values $\hat{y}_i$ for the delivery time data (1 observation hidden).

crew training and experience, or motivational factors influenced by management policies. These factors could result in a "site" effect that is not incorporated in the present equation. To investigate this we plot the residuals against $\hat{y}_i$ in Figure 3.7, and identify each residual by city. We see from this plot that 5 of the 6 Austin residuals are negative while 7 of the 10 Boston residuals are positive. This indicates that the model has a tendency to overpredict delivery times in Austin and underpredict delivery times in Boston. This could happen because of the site-dependent factors mentioned above, or because one or more important regressors have been omitted from the model.

A plot of residuals versus sites is shown in Figure 3.8. This display also shows the imbalance of positive and negative residuals in Boston and Austin

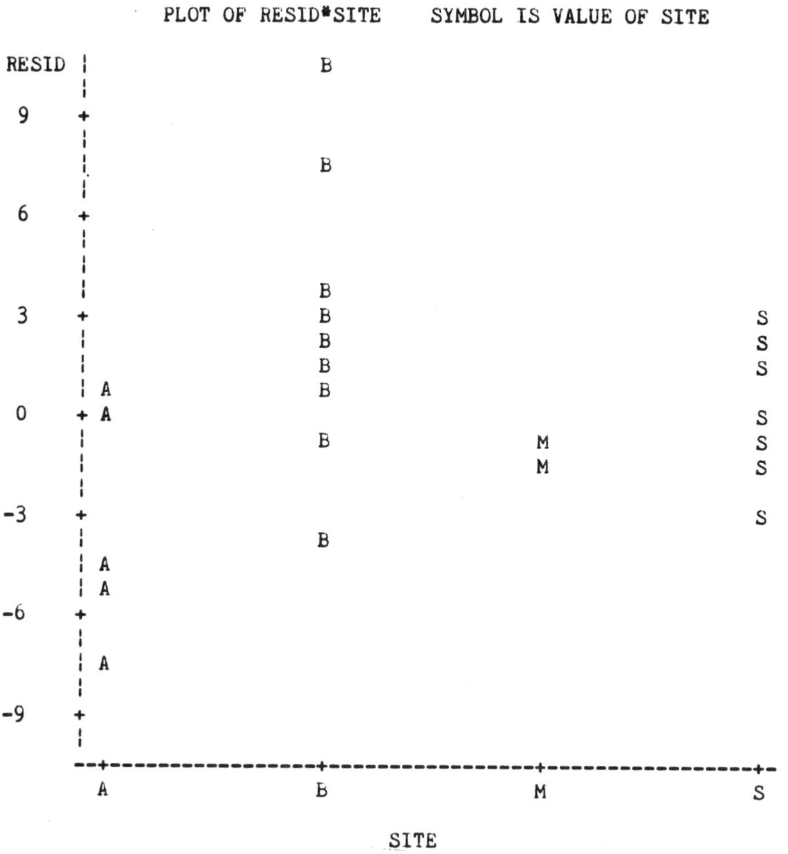

Figure 3.8 Plot (SAS) of residuals e_i versus sites for the delivery time data (2 observations hidden).

noted above. There is also some relatively mild indication that the variance may not be the same at all four sites, as the spread of the residuals from Boston is somewhat larger than the spread of the residuals from the other three sites. However, there is not enough data at each site to state conclusively that this is a problem. Note that the plot of residuals versus $\hat{y}_i$ in Figure 3.7 also indicates some potential inequality of variance.

3.2.6 Statistical tests on residuals

We may apply statistical tests to the residuals to obtain quantitative measures of some of the model inadequacies discussed above. For example, see Anscombe [1961, 1967], Anscombe and Tukey [1963], and Andrews [1971]. Several of these statistics are briefly summarized in Draper and Smith [1981].

These statistical tests are not widely used. In most practical situations, the residual plots are more informative than the corresponding tests. However, since residual plots do require skill and experience to interpret, the statistical tests may occasionally prove useful. For a good example of the use of statistical tests in conjunction with plots see Feder [1974].

3.3 Detection and treatment of outliers

An outlier is an extreme observation. Residuals that are considerably larger in absolute value than the others, say three or four standard deviations from the mean, are potential outliers. Outliers are data points that are not typical of the rest of the data. Depending on their location in the x-space, outliers can have moderate to severe effects on the regression model (for example, see Figures 2.4, 2.5, and 2.6). Residual plots against $\hat{y}_i$ and the normal probability plot are helpful in identifying outliers, as is inspection of the standardized residuals. An excellent general treatment of the outlier problem is in Barnett and Lewis [1978].

Outliers should be carefully investigated to see if a reason for their unusual behavior can be found. Sometimes outliers are "bad" values, occurring as a result of unusual but explainable events. Examples include faulty measurement or analysis, incorrect recording of data, or failure of a measuring instrument. If this is the case, then the outlier should be corrected (if possible) or deleted from the data set. Clearly, discarding "bad" values is desirable because least squares pulls the fitted equation towards the outlier as it minimizes the residual sum of squares. However, we emphasize that there should be strong nonstatistical evidence that the outlier is a "bad" value before it is discarded.

Sometimes we find that the outlier is an unusual but perfectly plausible observation. Deleting these points to "improve the fit of the equation" can be dangerous, as it can give the user a false sense of precision in estimation or prediction. Occasionally we find that the outlier is more important than the rest of the data, because it may control many key model properties. Outliers may also point out inadequacies in the model, such as failure to fit the data well in a certain region of x-space. If the outlier is a point of particularly desirable response (low cost, high yield, etc.), knowledge of the regressor values when that response was observed may be extremely valuable. Identification and follow-up analysis of outliers often result in process improvement, or new knowledge concerning factors whose effect on the response was previously unknown.

Various statistical tests have been proposed for detecting and rejecting outliers. For example, see Anscombe [1960], Anscombe and Tukey [1963], Ellenberg [1976], and Rosner [1975]. Stefansky [1971, 1972] has proposed an approximate test for identifying outliers based on the maximum normed residual $|e_i|/\sqrt{\Sigma_{i=1}^n e_i^2}$ that is particularly easy to apply. Examples of this test and other related references are in Cook and Prescott [1981], Daniel [1976], and Williams [1973]. While these tests may be useful for identifying outliers, they should not be interpreted to imply that the points so discovered should be automatically rejected. As we have noted, these points may be important clues containing valuable information.

The effect of outliers on the regression model may be easily checked by dropping these points and refitting the regression equation. We may find that the values of the regression coefficients or the summary statistics such as the t or F-statistics, R^2, and the residual mean square may be very sensitive to the outliers. Situations in which a relatively small percentage of the data has a significant impact on the model may not be acceptable to the user of the regression equation. Generally, we are happier about assuming that a regression equation is valid if it is not overly sensitive to a few observations. We would like the regression relationship to be embedded in all the observations and not merely an artifice of a few points.

- **Example 3.5** We have noted previously (Examples 3.1, 3.2, 3.3, and 3.4) that Observations 5 and 6 in the rocket propellant data from Example 2.1 have large negative residuals, indicating that the model overpredicts shear strength at these points. Note that Observation 5 occurs at a relatively low value of age (5.5 weeks) and Observation 6 occurs at a relatively high value of age (19 weeks). Thus these two points are widely separated in the x-space and occur near the extreme values of x, and they may be influential in determining model properties. Although neither residual is excessively large, the overall impression from the normal probability plot (Figure 3.2) and plots of the residuals versus $\hat{y}_i$ and x_i (Figures 3.4 and 3.5) is that these two observations are distinctly different from the others.

To investigate the influence of these two points on the model, a new regression equation is obtained with Observations 5 and 6 deleted. A comparison of the summary statistics from the two models is given below.

	Observation 5 and 6 in	Observations 5 and 6 out
$\hat{\beta}_0$	2627.82	2658.97
$\hat{\beta}_1$	−37.15	−37.69
R^2	0.9018	0.9578
MS_E	9244.59	3964.63
se($\hat{\beta}_1$)	2.89	1.98

Deleting Points 5 and 6 has almost no effect on the estimates of the regression coefficients. There has however, been a dramatic reduction in the residual

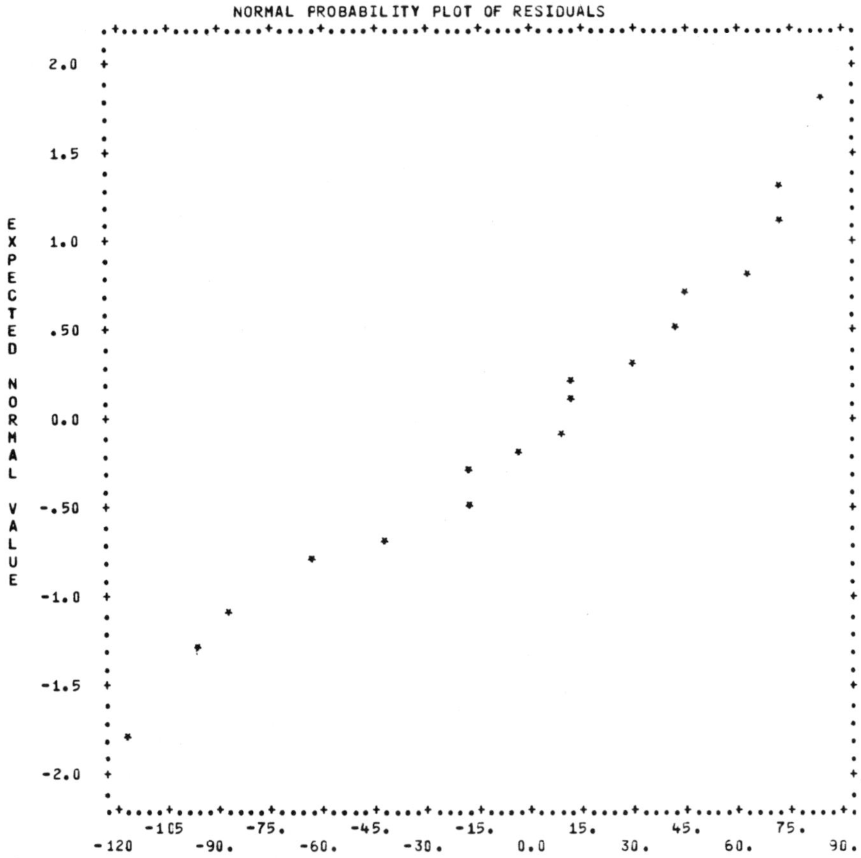

Figure 3.9 Normal probability plot (BMD–P) of the residuals, Example 3.5.

mean square, a moderate increase in R^2, and approximately a one–third reduction in the standard error of $\hat{\beta}_1$.

Since the estimates of the parameters have not changed dramatically, we conclude that Points 5 and 6 are not overly influential. They lie somewhat off the line passing through the other 18 points, but they do not control the slope and intercept. However, these two residuals make up approximately 56 percent of the residual sum of squares. Thus if these points are truly "bad" values and should be deleted, the precision of the parameter estimates would be improved and the widths of confidence and prediction intervals could be substantially decreased. Figures 3.9, 3.10, and 3.11 show the normal probabil-

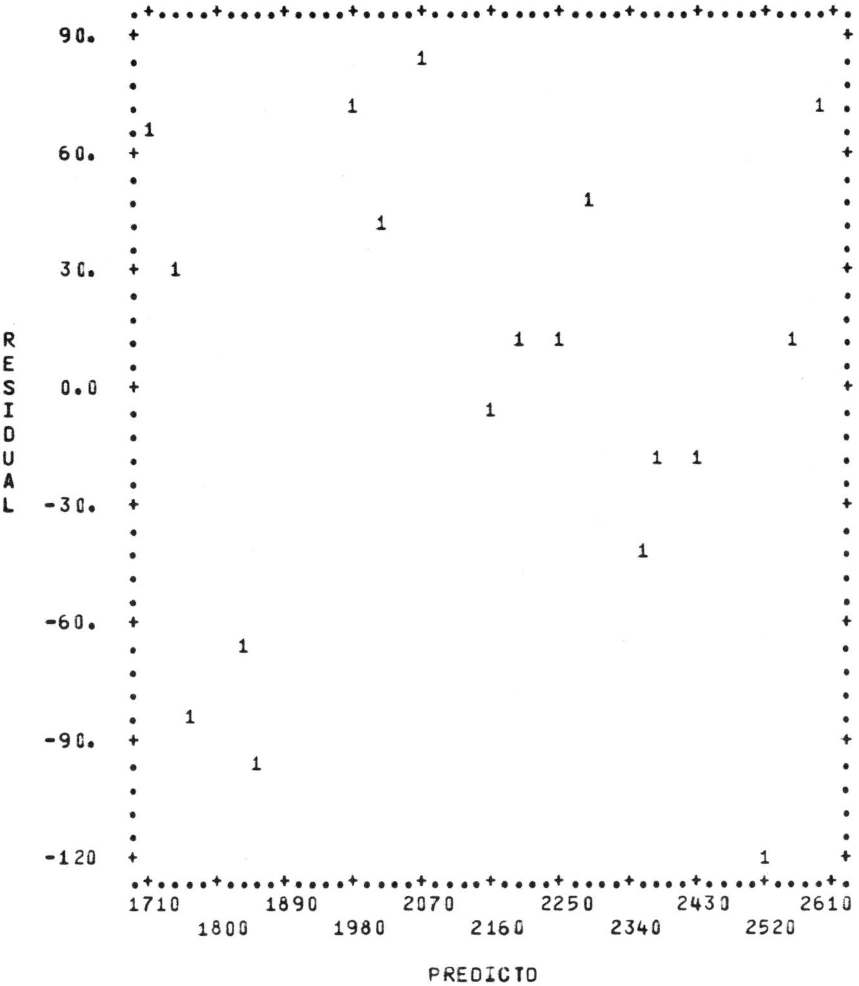

Figure 3.10 Plot (BMD–P) of residuals e_i versus fitted values $\hat{y}_i$, Example 3.5.

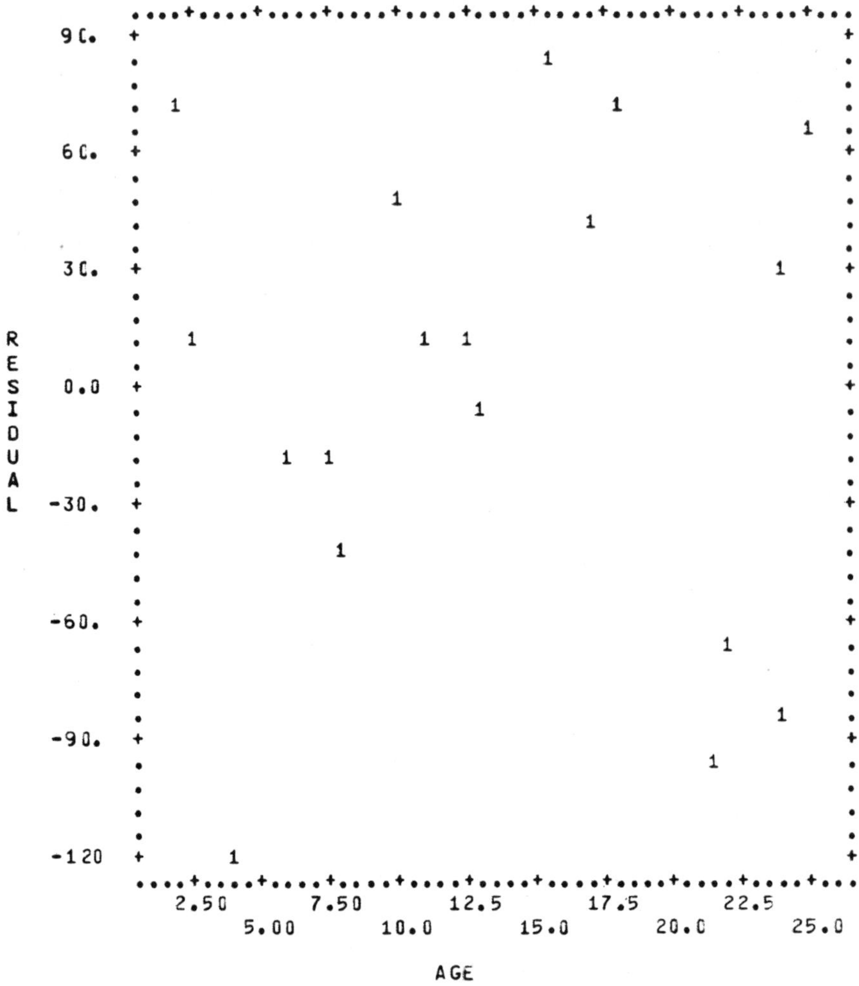

Figure 3.11 Plot (BMD–P) of residuals e_i versus age x_i, Example 3.5.

ity plot of the residuals, the plot of residuals versus $\hat{y}_i$, and the plot of residuals versus x_i, respectively, for the model with Points 5 and 6 deleted. These plots do not indicate any serious departures from assumptions.

Further examination of Points 5 and 6 fails to reveal any reason for the unusually low propellant shear strengths obtained. Therefore we should not discard these two points. However, we feel relatively confident that including them will not seriously limit the use of the model. •

3.4 A test for lack of fit

We will now present a formal statistical test for the lack-of-fit of a regression model. The procedure assumes that the normality, independence, and constant variance requirements are met, and that only the first-order or straight-line character of the relationship is in doubt. For example, consider the data in Figure 3.12. There is some indication that the straight-line fit is not very satisfactory, and it would be helpful to have a test procedure to determine if there is systematic curvature present.

The lack-of-fit test requires that we have repeat observations on the response y for at least one level of x. Draper and Smith [1981] emphasize that these repeat observations should be actual replications, not just duplicate readings or measurements of y. For example, suppose that y is product viscosity and x is temperature. True replication consists of running n_i separate experiments at $x=x_i$ and observing viscosity, not just running a single experiment at x_i and measuring viscosity n_i times. The readings obtained from the latter procedure provide information only on the variability of the method of measuring viscosity. The error variance σ^2 includes this measurement error and the variability associated with reaching and maintaining the same temperature level in different experiments. These repeat points are used to obtain a model-independent estimate of σ^2.

Suppose that we have n_i observations on the response at the ith level of the regressor x_i, $i=1,2,\ldots,m$. Let y_{ij} denote the jth observation on the response at x_i, $i=1,2,\ldots,m$ and $j=1,2,\ldots,n_i$. There are $n=\sum_{i=1}^{m} n_i$ total

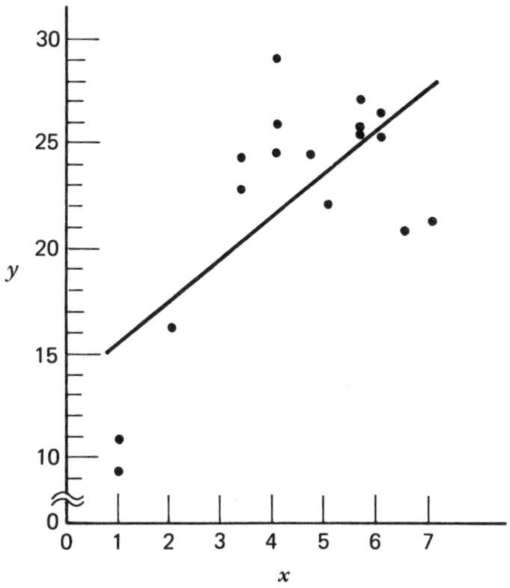

Figure 3.12 Lack of fit of the straight-line model.

observations. The test procedure involves partitioning the residual sum of squares into two components, say

$$SS_E = SS_{PE} + SS_{LOF} \tag{3.3}$$

where SS_{PE} is the sum of squares due to pure error and SS_{LOF} is the sum of squares due to lack of fit.

To develop this partitioning of SS_E, note that the ijth residual is

$$y_{ij} - \hat{y}_i = (y_{ij} - \bar{y}_i) + (\bar{y}_i - \hat{y}_i) \tag{3.4}$$

where $\bar{y}_i$ is the average of the n_i observations at x_i. Squaring both sides of (3.4) and summing over i and j yields

$$\sum_{i=1}^{m} \sum_{j=1}^{n_i} (y_{ij} - \hat{y}_i)^2 = \sum_{i=1}^{m} \sum_{j=1}^{n_i} (y_{ij} - \bar{y}_i)^2 + \sum_{i=1}^{m} n_i (\bar{y}_i - \hat{y}_i)^2 \tag{3.5}$$

since the cross-product term equals zero.

The left-hand side of (3.5) is the usual residual sum of squares. The two components on the right-hand side measure pure error and lack of fit. We see that the pure error sum of squares

$$SS_{PE} = \sum_{i=1}^{m} \sum_{j=1}^{n_i} (y_{ij} - \bar{y}_i)^2 \tag{3.6}$$

is obtained by computing the corrected sum of squares of the repeat observations at each level of x, and then pooling over the m levels of x. If the assumption of constant variance is satisfied, this is a *model-independent* measure of pure error since only the variability of the y's at each x level is used to compute SS_{PE}. Since there are $n_i - 1$ degrees of freedom for pure error at each level x_i, the total number of degrees of freedom associated with the pure error sum of squares is

$$\sum_{i=1}^{m} (n_i - 1) = n - m \tag{3.7}$$

The sum of squares for lack of fit

$$SS_{LOF} = \sum_{i=1}^{m} n_i (\bar{y}_i - \hat{y}_i)^2 \tag{3.8}$$

is a weighted sum of squared deviations between the mean response $\bar{y}_i$ at each x level and the corresponding fitted value. If the fitted values $\hat{y}_i$ are close to the corresponding average responses $\bar{y}_i$, then there is a strong indication that the regression function is linear. If the $\hat{y}_i$ deviate greatly from the $\bar{y}_i$, then it is likely that the regression function is not linear. There are $m-2$ degrees of freedom associated with SS_{LOF}, since there are m levels of x and 2 degrees of freedom are lost because 2 parameters must be estimated to obtain the $\hat{y}_i$. Computationally, we usually obtain SS_{LOF} by subtracting SS_{PE} from SS_E.

The test statistic for lack of fit is

$$F_0 = \frac{SS_{LOF}/(m-2)}{SS_{PE}/(n-m)} = \frac{MS_{LOF}}{MS_{PE}} \tag{3.9}$$

The expected value of MS_{PE} is σ^2, and the expected value of MS_{LOF} is

$$E(MS_{LOF}) = \sigma^2 + \frac{\sum_{i=1}^{m} n_i[E(y_i) - \beta_0 - \beta_1 x_i]^2}{m-2} \tag{3.10}$$

If the true regression function is linear, then $E(y_i) = \beta_0 + \beta_1 x_i$, and the second term of (3.10) is zero, resulting in $E(MS_{LOF}) = \sigma^2$. However, if the true regression function is not linear, then $E(y_i) \neq \beta_0 + \beta_1 x_i$, and $E(MS_{LOF}) > \sigma^2$. Furthermore if the true regression function is linear, then the statistic F_0 follows the $F_{m-2, n-m}$ distribution. Therefore to test for lack of fit, we would compute the test statistic F_0 and conclude that the regression function is not linear if $F_0 > F_{\alpha, m-2, n-m}$.

This test procedure may be easily introduced into the analysis of variance conducted for significance of regression. If we conclude that the regression function is not linear, then the tentative model must be abandoned and attempts made to find a more appropriate equation. Alternatively, if F_0 does not exceed $F_{\alpha, m-2, n-m}$, there is no strong evidence of lack of fit, and MS_{PE} and MS_{LOF} are often combined to estimate σ^2.

Ideally, we find that the F-ratio for lack of fit is not significant, and the hypothesis of significance of regression (H_0: $\beta_1 = 0$) is rejected. Unfortunately, this does not guarantee that the model will be satisfactory as a prediction equation. Unless the variation of the predicted values is large relative to the random error, the model is not estimated with sufficient precision to yield satisfactory predictions. That is, the model may have been fitted to the errors only. Some analytical work has been done on developing criteria for judging the adequacy of the regression model from a prediction point of view. See Box and Wetz [1973], Ellerton [1978], Gunst and Mason [1979], Hill, Judge, and Fomby [1978], and Suich and Derringer [1977]. The Box and Wetz work suggests that the observed F–ratio must be at least four or five times the critical value from the F–table if the regression model is to be useful as a predictor; that is, if the spread of predicted values is to be large relative to the noise.

A relatively simple measure of potential prediction performance is found by comparing the range of the fitted values $\hat{y}_i$ (that is, $\hat{y}_{max} - \hat{y}_{min}$) to their average standard error. It can be shown that, regardless of the form of the model, the average variance of the fitted values is

$$\overline{V(\hat{y})} = \frac{1}{n} \sum_{i=1}^{n} V(\hat{y}_i) = \frac{p\sigma^2}{n} \tag{3.11}$$

where p is the number of parameters in the model. In general, the model is not likely to be a satisfactory predictor unless the range of the fitted values $\hat{y}_i$ is large relative to their average estimated standard error $\sqrt{(p\hat{\sigma}^2)/n}$, where $\hat{\sigma}^2$ is a model-independent estimate of the error variance.

• **Example 3.6** The data in Figure 3.12 is shown below:

x	1.0	1.0	2.0	3.3	3.3	4.0	4.0	4.0	4.7	5.0
y	10.84	9.30	16.35	22.88	24.35	24.56	25.86	29.16	24.59	22.25
x	5.6	5.6	5.6	6.0	6.0	6.5	6.9			
y	25.90	27.20	25.61	25.45	26.56	21.03	21.46			

The straight-line fit is $\hat{y} = 13.301 + 2.108x$, with $S_{yy} = 487.6126$, $SS_R = 234.7087$, and $SS_E = 252.9039$. Note that there are 10 distinct levels of x, with repeat points at $x = 1.0$, $x = 3.3$, $x = 4.0$, $x = 5.6$, and $x = 6.0$. The pure error sum of squares is computed using the repeat points as follows:

Level of x	$\Sigma_j(y_{ij} - \bar{y}_i)^2$	Degrees of Freedom
1.0	1.1858	1
3.3	1.0805	1
4.0	11.2467	2
5.6	1.4341	2
6.0	0.6161	1
Totals:	15.5632	7

Table 3.2 Analysis of variance for Example 3.6

Source of Variation	Sum of Squares	Degrees of Freedom	Mean Square	F_0
Regression	234.7087	1	234.7087	
Residual	252.9039	15	16.8603	
(Lack of fit)	237.3407	8	29.6676	13.34
(Pure error)	15.5632	7	2.2233	
Total	487.6126	16		

The lack of fit sum of squares is found by subtraction as

$$SS_{LOF} = SS_E - SS_{PE}$$

$$= 252.9039 - 15.5632 = 237.3407$$

with $m-2=10-2=8$ degrees of freedom. The analysis of variance incorporating the lack-of-fit test is shown in Table 3.2. The lack-of-fit test statistic is $F_0 = 13.34$, and since $F_{.25,8,7} = 1.70$, we reject the hypothesis that the tentative model adequately describes the data. •

3.5 Transformations to a straight line

The assumption of a straight-line relationship between y and x is the usual starting point in regression analysis. Occasionally we find that the straight-line fit is inappropriate. Nonlinearity may be detected via the lack-of-fit test described in Section 3.4 or from scatter diagrams and residual plots. Sometimes prior experience or theoretical considerations may indicate that the relationship between y and x is not linear. In some cases a nonlinear function can be expressed as a straight line by using a suitable transformation. Such nonlinear models are called *intrinsically linear*.

Several linearizable functions are shown in Figure 3.13. The corresponding nonlinear functions, transformations, and the resulting linear forms are shown in Table 3.3. When the scatter diagram of y against x indicates curvature, we may be able to match the observed behavior of the plot to one of the curves in Figure 3.13 and use the linearized form of the function to represent the data.

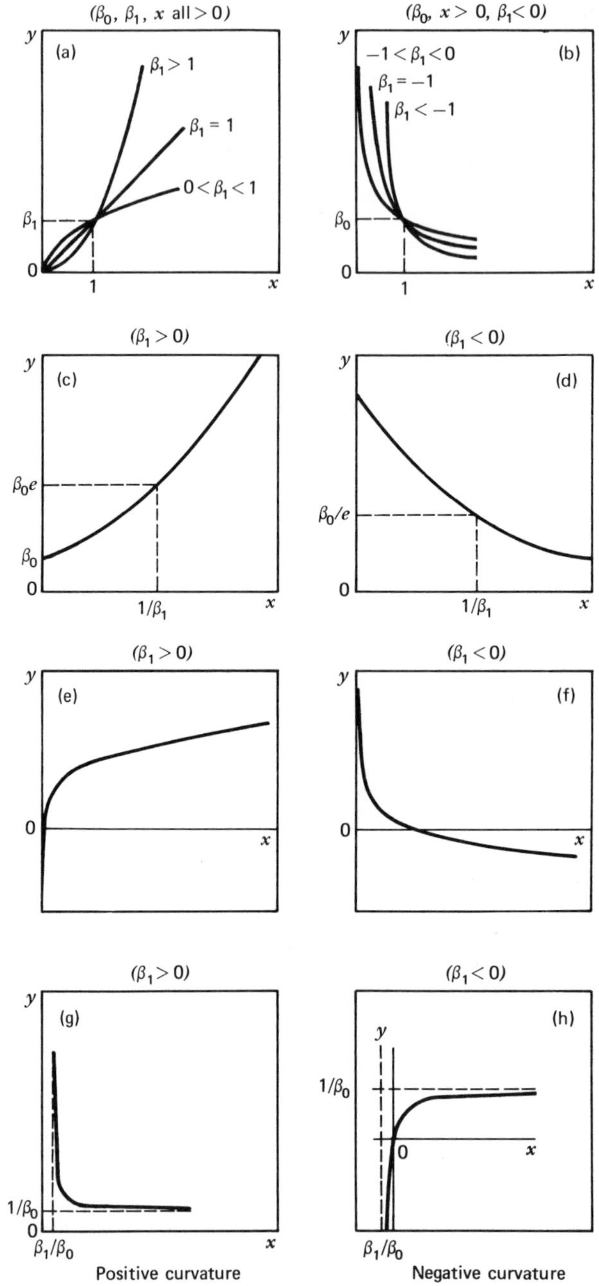

Figure 3.13 Linearizable functions. (From Daniel and Wood [1980], used with permission of the publisher.)

Table 3.3 Linearizable functions and corresponding linear form

Figure	Linearizable Function	Transformation	Linear Form
3.13a,b	$y=\beta_0 x^{\beta_1}$	$y'=\log y,\ x'=\log x$	$y'=\log\beta_0+\beta_1 x'$
3.13c,d	$y=\beta_0 e^{\beta_1 x}$	$y'=\ln y$	$y'=\ln\beta_0+\beta_1 x$
3.13e,f	$y=\beta_0+\beta_1\log x$	$x'=\log x$	$y'=\beta_0+\beta_1 x'$
3.13g,h	$y=\dfrac{x}{\beta_0 x-\beta_1}$	$y'=\dfrac{1}{y},\ x'=\dfrac{1}{x}$	$y'=\beta_0-\beta_1 x'$

To illustrate a nonlinear model that is intrinsically linear, consider the exponential function

$$y=\beta_0 e^{\beta_1 x}\varepsilon$$

This function is intrinsically linear, since it can be transformed to a straight line by a logarithmic transformation

$$\ln y=\ln\beta_0+\beta_1 x+\ln\varepsilon$$

or

$$y'=\beta_0'+\beta_1 x+\varepsilon'$$

as shown in Table 3.3. This transformation requires that the transformed error terms $\varepsilon'=\ln\varepsilon$ are normally and independently distributed with mean 0 and variance σ^2. This implies that the multiplicative error ε in the original model is log-normally distributed. We should look at the residuals from the transformed model to see if the assumptions are valid. Generally if x and/or y are in the proper metric, the usual least squares assumptions are more likely to be satisfied.

Various types of reciprocal transformations are also useful. For example the model

$$y=\beta_0+\beta_1\left(\frac{1}{x}\right)+\varepsilon$$

can be linearized by using the reciprocal transformation $x'=1/x$. The resulting linearized model is

$$y=\beta_0+\beta_1 x'+\varepsilon$$

Other models that can be linearized by reciprocal transformations are

$$\frac{1}{y}=\beta_0+\beta_1 x+\varepsilon$$

and

$$y=\frac{x}{\beta_0 x-\beta_1+\varepsilon}$$

This last model is illustrated in Figure 3.13g and h.

When transformations such as those described above are employed, the least squares estimators $\hat{\beta}_0$ and $\hat{\beta}_1$ have least squares properties with respect to the transformed data, not the original data. For additional reading on transformations, see Box, Hunter and Hunter [1978], Dolby [1963], Mosteller and Tukey [1977, Ch. 4,5,6], Smith [1972], and Tukey [1957].

• **Example 3.7** A research engineer is investigating the use of a windmill to generate electricity. He has collected data on the DC output from his windmill and the corresponding wind velocity. The data are plotted in Figure 3.14 and listed in Table 3.4.

Inspection of the scatter diagram indicates that the relationship between DC output (y) and wind velocity (x) may be nonlinear. However, we initially fit a straight-line model to the data. The regression model is

$$\hat{y}=0.1309+0.2411x$$

The summary statistics for this model are $R^2=0.8745$, $MS_E=0.0557$, and $F_0=160.26$ (significant at one percent). Panel A of Table 3.5 shows the fitted

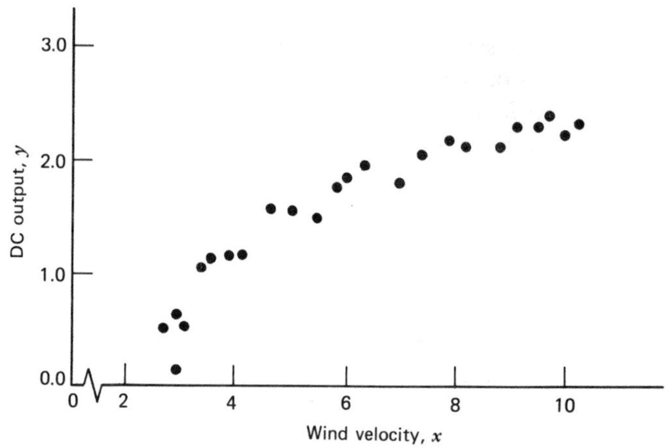

Figure 3.14 Plot of DC output y versus wind velocity x for the windmill data.

Table 3.4 Observed values y_i and regressor variable x_i for Example 3.7

Observation Number, i	Wind Velocity (MPH), x_i	DC Output, y_i
1	5.00	1.582
2	6.00	1.822
3	3.40	1.057
4	2.70	0.500
5	10.00	2.236
6	9.70	2.386
7	9.55	2.294
8	3.05	0.558
9	8.15	2.166
10	6.20	1.866
11	2.90	0.653
12	6.35	1.930
13	4.60	1.562
14	5.80	1.737
15	7.40	2.088
16	3.60	1.137
17	7.85	2.179
18	8.80	2.112
19	7.00	1.800
20	5.45	1.501
21	9.10	2.303
22	10.20	2.310
23	4.10	1.194
24	3.95	1.144
25	2.45	0.123

values and residuals obtained from this model. In Table 3.5, the observations are arranged in order of increasing windspeed. The residuals show a distinct pattern; that is, they move systematically from negative to positive and back to negative again as windspeed increases.

Plots of the residuals versus $\hat{y}_i$ and x_i are shown in Figures 3.15 and 3.16, respectively. Both plots indicate model inadequacy, and imply that the linear relationship has not captured all the information in the windspeed variable. Note that the curvature that was apparent in the scatter diagram 3.14 is greatly amplified in the residual plots. Clearly some other model form must be considered.

We might initially consider using a quadratic model such as

$$y=\beta_0+\beta_1 x+\beta_2 x^2+\varepsilon$$

Table 3.5 Observations y_i ordered by increasing wind velocity, fitted values $\hat{y}_i$, and residuals e_i for both models for Example 3.7

Wind Velocity, x_i	DC Output, y_i	A. Straight-Line Model $\hat{y}=\hat{\beta}_0+\hat{\beta}_1 x$		B. Transformed Model $\hat{y}=\hat{\beta}_0+\hat{\beta}_1(1/x)$	
		$\hat{y}_i$	e_i	$\hat{y}_i$	e_i
2.45	0.123	0.7217	−0.5987	0.1484	−0.0254
2.70	0.500	0.7820	−0.2820	0.4105	0.0895
2.90	0.653	0.8302	−0.1772	0.5876	0.0654
3.05	0.558	0.8664	−0.3084	0.7052	−0.1472
3.40	1.057	0.9508	0.1062	0.9393	0.1177
3.60	1.137	0.9990	0.1380	1.0526	0.0844
3.95	1.144	1.0834	0.0606	1.2233	−0.0793
4.10	1.194	1.1196	0.0744	1.2875	−0.0935
4.60	1.562	1.2402	0.3218	1.4713	0.0907
5.00	1.582	1.3366	0.2454	1.5920	−0.0100
5.45	1.501	1.4451	0.0559	1.7065	−0.2055
5.80	1.737	1.5295	0.2075	1.7832	−0.0462
6.00	1.822	1.5778	0.2442	1.8231	−0.0011
6.20	1.866	1.6260	0.2400	1.8604	0.0056
6.35	1.930	1.6622	0.2678	1.8868	0.0432
7.00	1.800	1.8189	−0.0189	1.9882	−0.1882
7.40	2.088	1.9154	0.1726	2.0418	0.0462
7.85	2.179	2.0239	0.1551	2.0955	0.0835
8.15	2.166	2.0962	0.0698	2.1280	0.0380
8.80	2.112	2.2530	−0.1410	2.1908	−0.0788
9.10	2.303	2.3252	−0.0223	2.2168	0.0862
9.55	2.294	2.4338	−0.1398	2.2527	−0.1472
9.70	2.386	2.4700	−0.0840	2.2640	0.1220
10.00	2.236	2.5424	−0.3064	2.2854	−0.0494
10.20	2.310	2.5906	−0.2906	2.2990	0.0110

to account for the apparent curvature. However, Figure 3.14 suggests that as windspeed increases, DC output approaches an upper limit of approximately 2.5 amps. This is also consistent with the theory of windmill operation. Since the quadratic model will eventually bend downward as windspeed increases, it would not be appropriate for this data. A more reasonable model for the windmill data that incorporates an upper asymptote would be

$$y=\beta_0+\beta_1\left(\frac{1}{x}\right)+\varepsilon$$

Figure 3.17 is a scatter diagram with the transformed variable $x'=1/x$. This

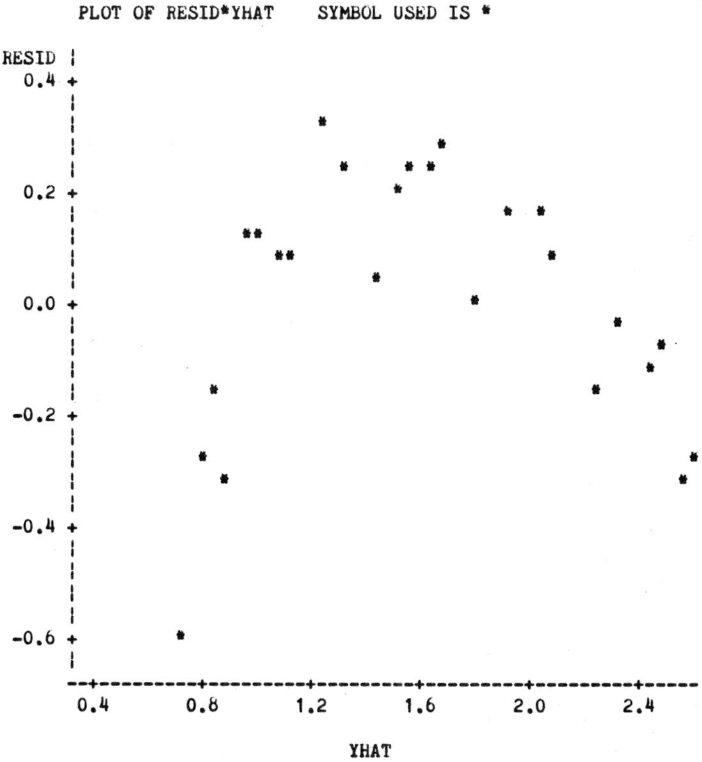

Figure 3.15 Plot (SAS) of residuals e_i versus fitted values $\hat{y}_i$ for the windmill data.

plot appears linear, indicating that the reciprocal transformation is appropriate. The fitted regression model is

$$\hat{y}=2.9789-6.9345x'$$

The summary statistics for this model are $R^2=0.9800$, $MS_E=0.0089$, and $F_0=1128.43$ (significant at one percent).

The fitted values and corresponding residuals from the transformed model are shown in Panel B of Table 3.5. Plots of these residuals versus $\hat{y}$ and $1/x$ are shown in Figures 3.18 and 3.19, respectively. These plots do not reveal any serious model inadequacy. The normal probability plot, shown in Figure 3.20, gives a mild indication that the errors come from a distribution with heavier tails than the normal (notice the slight upward and downward curve at the extremes). Since there is no strong signal of model inadequacy, we conclude that the transformed model is satisfactory.

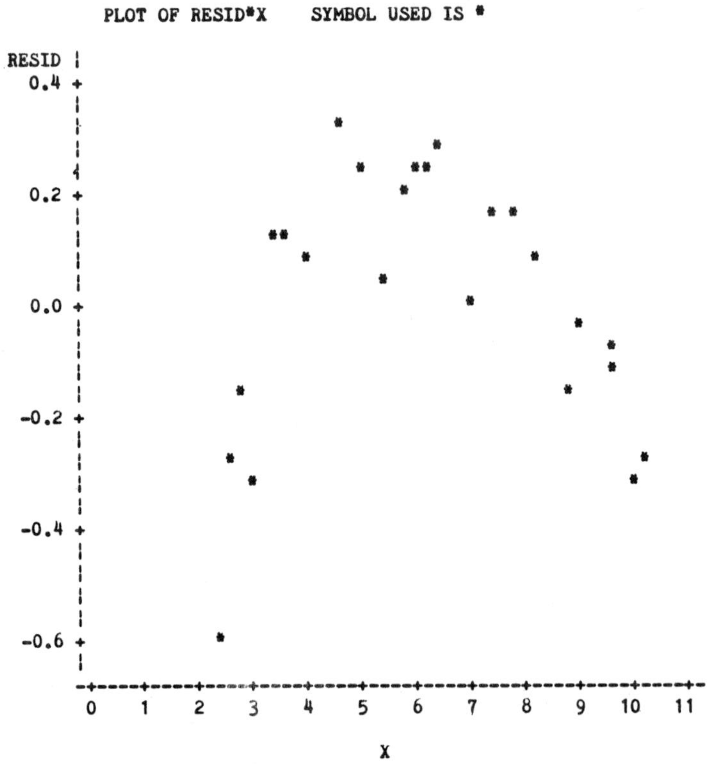

Figure 3.16 Plot (SAS) of residuals e_i versus wind velocity x_i for the windmill data.

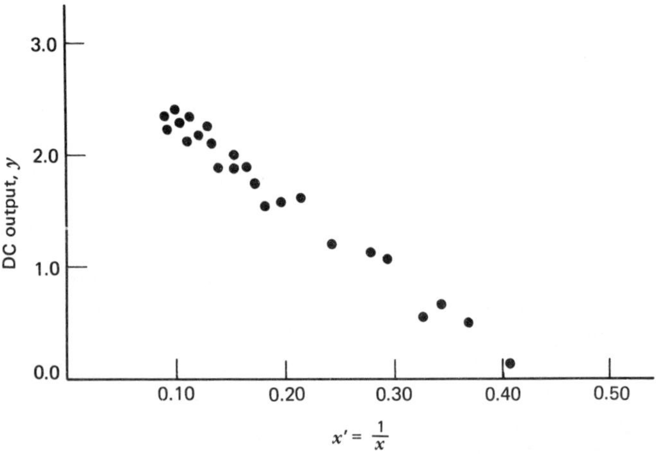

Figure 3.17 Plot of DC output versus $x' = 1/x$ for the windmill data.

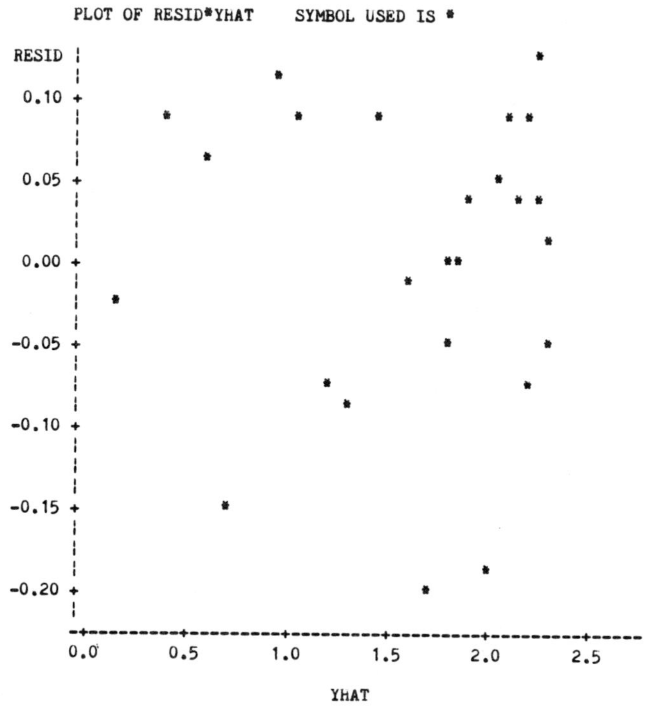

Figure 3.18 Plot (SAS) of residuals e_i versus fitted values $\hat{y}_i$ for the transformed model for the windmill data.

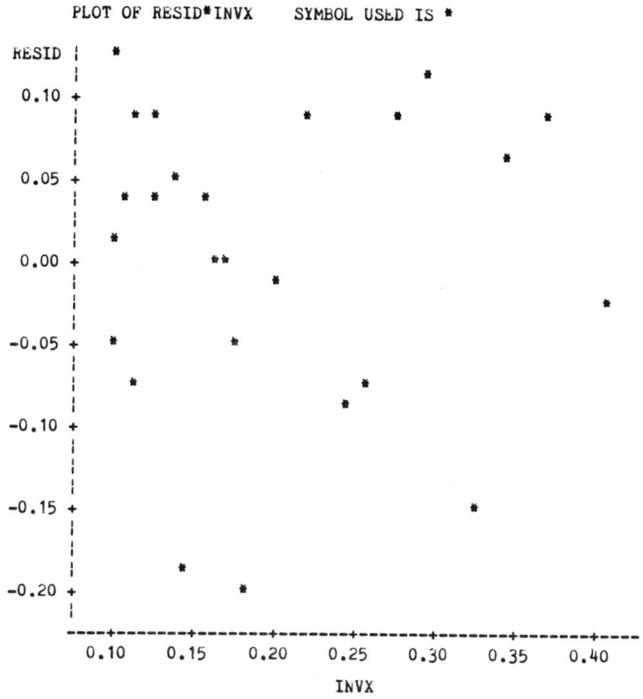

Figure 3.19 Plot (SAS) of residuals e_i versus reciprocal of windspeed for the windmill data.

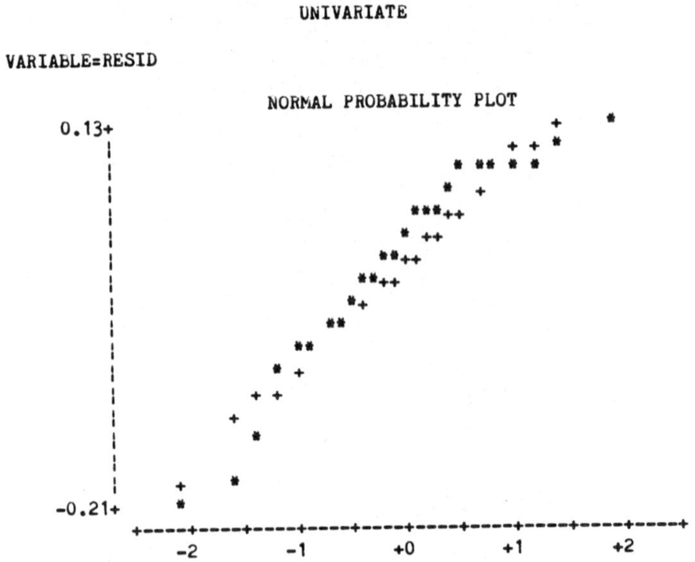

Figure 3.20 Normal probability plot (SAS) of the residuals for the transformed model for the windmill data. The plotting symbol + represents the normal cumulative distribution and * represents the residual cumulative distribution.

3.6 Variance stabilizing transformations

The assumption of constant variance is a basic requirement of regression analysis. A common reason for the violation of this assumption is for the response variable y to follow a probability distribution in which the variance is functionally related to the mean. For example, if y is a Poisson random variable, then the variance of y is equal to the mean. Since the mean of y is related to the regressor variable x, the variance of y will be proportional to x. Variance stabilizing transformations are often useful in these cases. Thus if the distribution of y is Poisson, we could regress $y' = \sqrt{y}$ against x, since the variance of the square root of a Poisson random variable is independent of the mean. As another example, if the response variable is a proportion $(0 \leqslant y_i \leqslant 1)$ and the plot of the residuals versus $\hat{y}_i$ has the double bow pattern of Figure 3.2c, then the arcsin transformation $y' = \sin^{-1}(\sqrt{y})$ is appropriate.

Several commonly used variance stabilizing transformations are summarized below:

Relationship of σ^2 to $E(y)$	Transformation
$\sigma^2 \propto \text{constant}$	$y'=y$ (no transformation)
$\sigma^2 \propto E(y)$	$y'=\sqrt{y}$ (square root; Poisson data)
$\sigma^2 \propto E(y)[1-E(y)]$	$y'=\sin^{-1}(\sqrt{y})$ (arcsin; binomial proportions $0 \leq y_i \leq 1$)
$\sigma^2 \propto [E(y)]^2$	$y'=\ln(y)$ (log)
$\sigma^2 \propto [E(y)]^3$	$y'=y^{-1/2}$ (reciprocal square root)
$\sigma^2 \propto [E(y)]^4$	$y'=y^{-1}$ (reciprocal)

The strength of a transformation depends on the amount of curvature that it induces. The transformations given above range from the relatively mild square root to the relatively strong reciprocal. Generally speaking, a mild transformation applied over a relatively narrow range of values (y_{max}/y_{min} less than two or three, for example) has little effect. On the other hand a strong transformation over a wide range of values will have a dramatic effect on the analysis.

Sometimes we can use prior experience or theoretical considerations to guide us in selecting an appropriate transformation. However, in many cases we have no a priori reason to suspect that the error variance is not constant. Our first indication of the problem is from inspection of the scatter diagram or residual analysis. In these cases the appropriate transformation must be selected empirically.

It is important to detect and correct a nonconstant error variance. If this problem is not eliminated the least squares estimators will still be unbiased, but they will no longer have the minimum variance property. This means that the regression coefficients will have larger standard errors than necessary. The effect of the transformation is usually to give more precise estimates of the model parameters and increased sensitivity for the statistical tests.

When the response variable has been reexpressed, the predicted values are in the transformed scale. It is often necessary to convert the predicted values back to the original units. Unfortunately, applying the inverse transformation directly to the predicted values gives an estimate of the median of the distribution of the response instead of the mean. It is usually possible to devise a method for obtaining unbiased predictions in the original units. Procedures for producing unbiased point estimates for several standard transformations are given by Neyman and Scott [1960]. Confidence or prediction intervals may be directly converted from one metric to another, as these interval estimates are percentiles of a distribution and percentiles are unaffected by transformation. However, there is no assurance that the resulting intervals in the original units are the shortest possible intervals. For further discussion, see Land [1974].

• **Example 3.8** An electric utility is interested in developing a model relating peak hour demand (y) to total energy usage during the month (x). This is an important planning problem, because while most customers pay directly for energy usage (in KWH), the generation system must be large enough to meet the maximum demand imposed. Data for 53 residential customers for the month of August, 1979 are shown in Table 3.6 and a scatter diagram is given in Figure 3.21. As a starting point, a simple linear regression model is

Table 3.6 Demand (y) and energy usage (x) data for 53 residential customers, August 1979

Customer	x(KWH)	y(KW)	Customer	x(KWH)	y(KW)
1	679	.79	27	837	4.20
2	292	.44	28	1748	4.88
3	1012	.56	29	1381	3.48
4	493	.79	30	1428	7.58
5	582	2.70	31	1255	2.63
6	1156	3.64	32	1777	4.99
7	997	4.73	33	370	.59
8	2189	9.50	34	2316	8.19
9	1097	5.34	35	1130	4.79
10	2078	6.85	36	463	.51
11	1818	5.84	37	770	1.74
12	1700	5.21	38	724	4.10
13	747	3.25	39	808	3.94
14	2030	4.43	40	790	.96
15	1643	3.16	41	783	3.29
16	414	.50	42	406	.44
17	354	.17	43	1242	3.24
18	1276	1.88	44	658	2.14
19	745	.77	45	1746	5.71
20	435	1.39	46	468	.64
21	540	.56	47	1114	1.90
22	874	1.56	48	413	.51
23	1543	5.28	49	1787	8.33
24	1029	.64	50	3560	14.94
25	710	4.00	51	1495	5.11
26	1434	.31	52	2221	3.85
			53	1526	3.93

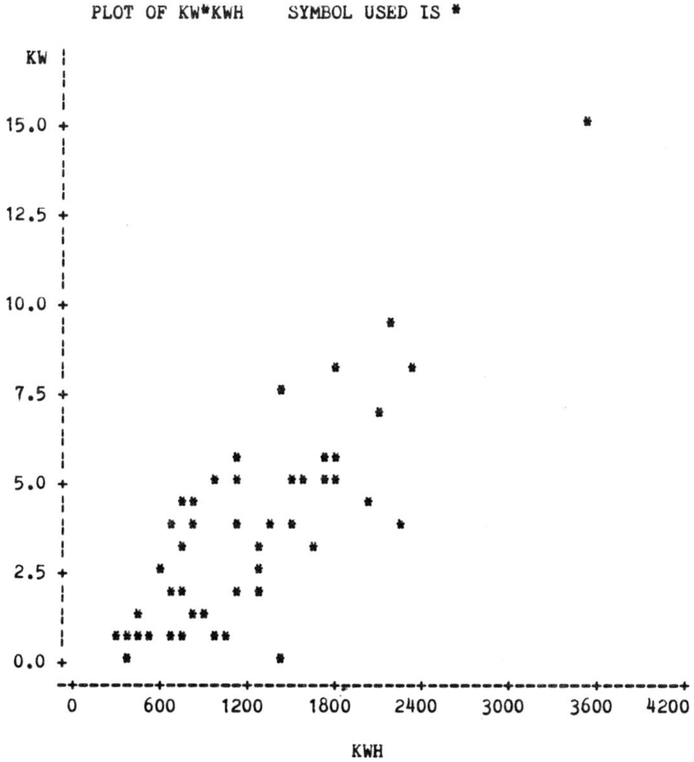

Figure 3.21 Scatter diagram (SAS) of the energy demand (KW) versus energy usage (KWH), Example 3.8 (7 observations hidden).

assumed and the least squares fit is

$$\hat{y} = -0.8313 + 0.00368x$$

The analysis of variance is shown in Table 3.7. For this model $R^2 = 0.7046$; that is, about 70 percent of the variability in demand is accounted for by the straight-line fit to energy usage. The summary statistics do not reveal any obvious problems with this model.

A plot of the residuals versus the fitted values $\hat{y}_i$ is shown in Figure 3.22. The residuals form an outward-opening funnel, indicating that the error variance is increasing as energy consumption increases. A transformation may be helpful in correcting this model inadequacy. To select the form of the transformation, note that the response variable y may be viewed as a "count" of the number of kilowatts used by a customer during a particular hour. The simplest probabilistic model for count data is the Poisson distribution. This suggests regressing $y' = \sqrt{y}$ on x as a variance stabilizing transformation. The

Table 3.7 Analysis of variance for regression of y on x for Example 3.8

Source of Variation	Sum of Squares	Degrees of Freedom	Mean Square	F_0
Regression	303.6331	1	302.6331	121.66[a]
Residual	126.8660	51	2.4876	
Total	429.4991	52		

[a]Significant at 1 percent.

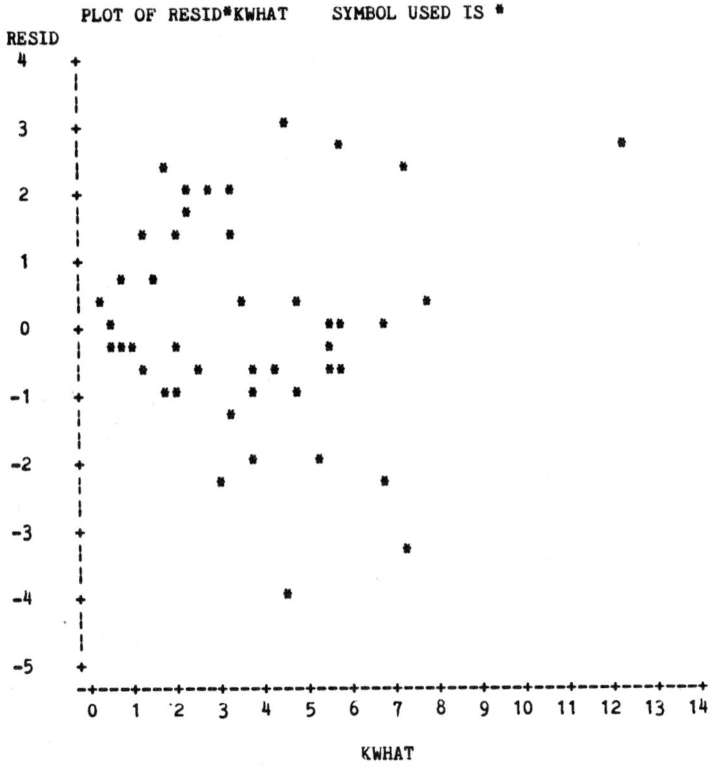

Figure 3.22 Plot (SAS) of residuals e_i versus fitted values $\hat{y}_i$, Example 3.8 (9 observations hidden).

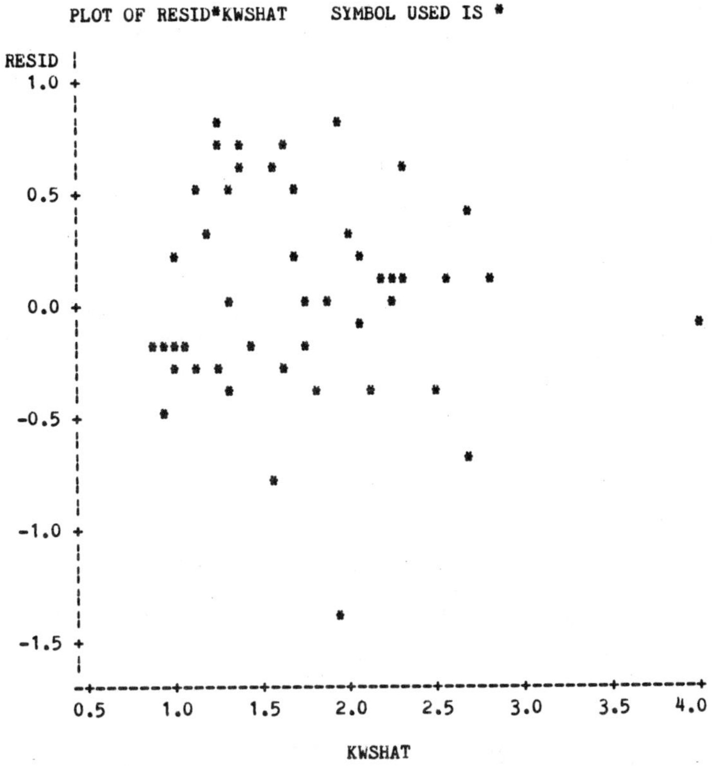

Figure 3.23 Plot (SAS) of residuals e_i versus fitted values $\hat{y}_i'$ for the transformed data, Example 3.8 (7 observations hidden).

resulting least squares fit is

$$\hat{y}' = 0.5822 + 0.0009529x$$

The residuals from this least squares fit are plotted against $\hat{y}_i'$ in Figure 3.23. The impression from examining this plot is that the variance is stable; consequently, we conclude that the transformed model is adequate. Note that there is one suspiciously large residual (Customer 26) and one customer whose energy usage is somewhat large (Customer 50). The effect of these two points on the fit should be studied further before the model is released for use. •

3.7 Analytical methods for selecting a transformation

While in many instances transformations are selected empirically, more formal, objective techniques can be applied to help specify an appropriate transformation. This section will discuss and illustrate analytical procedures for selecting transformations on y and x.

3.7.1 Transformations on y

Suppose that we wish to transform y to correct nonnormality and/or nonconstant variance. A useful class of transformations is the power transformation y^λ, where λ is a parameter to be determined (for example $\lambda=\frac{1}{2}$ means use $\sqrt{y}$ as the response, and $\lambda=0$ implies the log transformation ln y). Box and Cox [1964] show how the parameters of the regression model and λ can be estimated simultaneously using the method of maximum likelihood.

The procedure consists of performing a standard least squares fit using

$$y^{(\lambda)}=\begin{cases}\dfrac{y^\lambda-1}{\lambda\dot{y}^{\lambda-1}}, & \lambda\neq 0\\ \dot{y}\ln y, & \lambda=0\end{cases} \tag{3.12}$$

where $\dot{y}=\ln^{-1}[(1/n)\Sigma_{i=1}^{n}\ln y_i]$ is the geometric mean of the observations. The maximum likelihood estimate of λ corresponds to the value of λ for which the residual sum of squares from the fitted model $SS_E(\lambda)$ is a minimum. This value of λ is usually determined by fitting a model to $y^{(\lambda)}$ for various values of λ, plotting $SS_E(\lambda)$ versus λ, and then reading the value of λ that minimizes $SS_E(\lambda)$ from the graph. Usually 10–20 values of λ are sufficient for estimation of the optimum value. A second iteration can be performed using a finer mesh of values if desired. Notice that we *cannot* select λ by *directly* comparing residual sums of squares from the regressions of y^λ on x, because for each λ the residual sum of squares is measured on a different scale. Equation (3.12) scales the responses so that the residual sums of squares are directly comparable. We recommend that the analyst use simple choices for λ, as the practical difference in the fits for $\lambda=0.5$ and $\lambda=0.596$ is likely to be small, but the former is much easier to interpret.

An approximate $100(1-\alpha)$ percent confidence interval on λ can be found by computing

$$SS^*=SS_E(\lambda)\left(1+\frac{t^2_{\alpha/2,\nu}}{\nu}\right) \tag{3.13}$$

where ν is the number of residual degrees of freedom ($\nu=n-2$ for simple linear regression), and reading the corresponding confidence limits on λ from the graph. If this confidence interval includes the value $\lambda=1$, this suggests that no transformation is necessary.

• **Example 3.9** Recall the electric utility data introduced in Example 3.8. We will use the Box and Cox procedure to select a variance stabilizing transformation. The values of $SS_E(\lambda)$ for various values of λ are shown below:

λ	$SS_E(\lambda)$
−2	34,101.0381
−1	986.0423
−0.5	291.5834
0	134.0940
0.125	118.1982
0.25	107.2057
0.375	100.2561
0.5	96.9495
0.625	97.2889
0.75	101.6869
1	126.8660
2	1,275.5555

This display indicates that $\lambda=0.5$ (the square root transformation) is very close to the optimum value. Notice that we have used a finer "grid" on λ in the vicinity of the optimum. This is helpful in locating the optimum λ more precisely, and in plotting the residual sum of squares function.

A graph of the residual sum of squares versus λ is shown in Figure 3.24. If we take $\lambda=0.5$ as the optimum value, then an approximate 95 percent

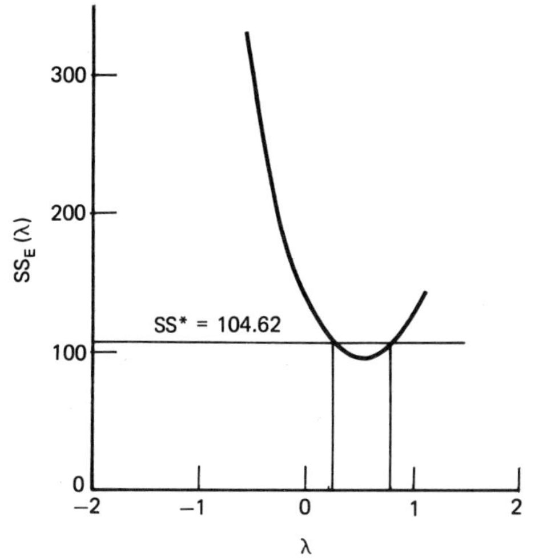

Figure 3.24 Plot of residual sum of squares $SS_E(\lambda)$ versus λ.

confidence interval for λ may be found by calculating the critical sum of squares

$$SS^* = SS_E(\lambda)\left(1 + \frac{t^2_{.025,\nu}}{\nu}\right)$$

$$= 96.9495\left(1 + \frac{(2.095)^2}{51}\right)$$

$$= 104.62$$

The corresponding values of $\lambda^- = 0.26$ and $\lambda^+ = 0.80$ read from the curve give the lower and upper confidence limits for λ, respectively. Since these limits do not include the value 1 (implying no transformation), we conclude that a transformation is helpful. Furthermore, the square root transformation that was used in Example 3.8 has an analytic justification.

3.7.2 Transformations on x

Suppose that the relationship between y and x is nonlinear, but that the usual assumptions of normally and independently distributed responses with constant variance are at least approximately satisfied. We would like to select an appropriate transformation on the regressor variable so that the relationship between y and the transformed regressor is as simple as possible. Box and Tidwell [1962] describe an analytical procedure for determining the form of the transformation on x. While their procedure may be used in the general regression situation, we will present its application to the simple linear regression model.

Assume that the response variable y is related to a power of the regressor, say $\xi = x^\alpha$, as

$$E(y) = f(\xi, \beta_0, \beta_1) = \beta_0 + \beta_1 \xi$$

where

$$\xi = \begin{cases} x^\alpha, & \alpha \neq 0 \\ \ln x, & \alpha = 0 \end{cases}$$

and β_0, β_1, and α are unknown parameters. Suppose that α_0 is an initial guess of the constant α. Usually this first guess is $\alpha_0 = 1$, so that $\xi_0 = x^{\alpha_0} = x$, or that no transformation at all is applied in the first iteration. Expanding about the initial guess in a Taylor series and ignoring terms of higher than first order gives

$$E(y)=f(\xi_0,\beta_0,\beta_1)+(\alpha-\alpha_0)\left\{\frac{df(\xi,\beta_0,\beta_1)}{d\alpha}\right\}_{\substack{\xi=\xi_0\\ \alpha=\alpha_0}}$$

$$=\beta_0+\beta_1 x+(\alpha-1)\left\{\frac{df(\xi,\beta_0,\beta_1)}{d\alpha}\right\}_{\substack{\xi=\xi_0\\ \alpha=\alpha_0}} \tag{3.14}$$

Now if the term in braces in (3.14) were known, it could be treated as an additional regressor variable, and it would be possible to estimate the parameters β_0, β_1, and α in (3.14) by least squares. The estimate of α could be taken as an improved estimate of the transformation parameter. The term in braces in (3.14) can be written as

$$\left\{\frac{df(\xi,\beta_0,\beta_1)}{d\alpha}\right\}_{\substack{\xi=\xi_0\\ \alpha=\alpha_0}}=\left\{\frac{df(\xi,\beta_0,\beta_1)}{d\xi}\right\}_{\xi=\xi_0}\left\{\frac{d\xi}{d\alpha}\right\}_{\alpha=\alpha_0}$$

and since the form of the transformation is known, that is, $\xi=x^{\alpha}$, we have $d\xi/d\alpha=x\ln x$. Furthermore,

$$\left\{\frac{df(\xi,\beta_0,\beta_1)}{d\xi}\right\}_{\xi=\xi_0}=\frac{d(\beta_0+\beta_1 x)}{dx}=\beta_1$$

This parameter may be conveniently estimated by fitting the model

$$\hat{y}=\hat{\beta}_0+\hat{\beta}_1 x \tag{3.15}$$

by least squares. Then an "adjustment" to the initial guess $\alpha_0=1$ may be computed by defining a second regressor variable as $w=x\ln x$, estimating the parameters in

$$E(y)=\beta_0^*+\beta_1^* x+(\alpha-1)\beta_1 w$$

$$=\beta_0^*+\beta_1^* x+\gamma w \tag{3.16}$$

by least squares*, giving

$$\hat{y}=\hat{\beta}_0^*+\hat{\beta}_1^*+\hat{\gamma}w \tag{3.17}$$

and taking

$$\alpha_1=\frac{\hat{\gamma}}{\hat{\beta}_1}+1 \tag{3.18}$$

*Equation (3.16) is a multiple regression model. The details of fitting these models are explained in Chapter 4.

as the revised estimate of α. Note that $\hat{\beta}_1$ is obtained from (3.15) and $\hat{\gamma}$ from (3.17); generally $\hat{\beta}_1$ and $\hat{\beta}_1^*$ will differ. This procedure may now be repeated using a new regressor $x'=x^{\alpha_1}$ in the calculations.

Box and Tidwell [1962] note that this procedure usually converges quite rapidly, and often the first-stage result α_1 is a satisfactory estimate of α. They also caution that round-off error is potentially a problem and successive values of α may oscillate wildly unless enough decimal places are carried. Convergence problems may be encountered in cases where the error standard deviation σ is large, or when the range of the regressor is very small compared to its mean. This situation implies that the data do not support the need for any transformation.

- **Example 3.10** We will illustrate this procedure using the windmill data in Example 3.7. The scatter diagram in Figure 3.14 suggests that the relationship between DC output (y) and windspeed (x) is not a straight line, and that some transformation on x may be appropriate.

We begin with the initial guess $\alpha_0=1$, and fit a straight-line model, giving $\hat{y}=0.1309+0.2411x$. Then defining $w=x\ln x$ we fit (3.16), and obtain

$$\begin{aligned}\hat{y}&=\hat{\beta}_0^*+\hat{\beta}_1^*x+\hat{\gamma}w\\&=-2.4168+1.5344x-0.4626w\end{aligned}$$

From (3.18) we calculate

$$\alpha_1=\frac{\hat{\gamma}}{\hat{\beta}_1}+1=\frac{-0.4626}{0.2411}+1=-0.92$$

as the improved estimate of α. Note that this estimate of α is very close to -1, so that the reciprocal transformation on x actually used in Example 3.7 is supported by the Box–Tidwell procedure.

To perform a second iteration, we would define a new regressor variable $x'=x^{-0.92}$ and fit the model

$$\begin{aligned}\hat{y}&=\hat{\beta}_0+\hat{\beta}_1x'\\&=3.1039-6.6784x'\end{aligned}$$

Then a second regressor $w'=x'\ln x'$ is formed and we fit

$$\begin{aligned}\hat{y}&=\hat{\beta}_0^*+\hat{\beta}_1^*x'+\hat{\gamma}w'\\&=3.2409-6.445x'+0.5994w'\end{aligned}$$

The second–step estimate of α is thus

$$\alpha_2 = \frac{\hat{\gamma}}{\hat{\beta}_1} + \alpha_1 = \frac{0.3994}{(-6.6784)} + (-0.92) = -1.01$$

- which again supports the use of the reciprocal transformation on x.

3.8 Weighted least squares

Linear regression models with nonconstant error variance can also be fitted by the method of *weighted least squares*. In this method of estimation, the deviation between the observed and expected values of y_i is multiplied by a weight w_i chosen inversely proportional to the variance of y_i. The weighted least squares function is

$$S(\beta_0, \beta_1) = \sum_{i=1}^{n} w_i (y_i - \beta_0 - \beta_1 x_i)^2 \tag{3.19}$$

The resulting least squares normal equations are

$$\begin{aligned} \hat{\beta}_0 \sum_{i=1}^{n} w_i + \hat{\beta}_1 \sum_{i=1}^{n} w_i x_i &= \sum_{i=1}^{n} w_i y_i \\ \hat{\beta}_0 \sum_{i=1}^{n} w_i x_i + \hat{\beta}_1 \sum_{i=1}^{n} w_i x_i^2 &= \sum_{i=1}^{n} w_i x_i y_i \end{aligned} \tag{3.20}$$

Solving (3.20) will produce weighted least squares estimates of β_0 and β_1.

To use weighted least squares, the weights w_i must be known. In some problems, the weights may be easily determined. For example, if the observation y_i is actually an average of n_i observations at x_i, and if all *original* observations have constant variance σ^2, then the variance of y_i is $V(y_i) = V(\varepsilon_i) = \sigma^2/n_i$, and we would choose the weights as $w_i = n_i$. Sometimes the variance of y_i may be a function of the regressor, for example, $V(y_i) = V(\varepsilon_i) = \sigma^2 x_i$. In that case, we would use $w_i = 1/x_i$ as the weights. When the primary source of error is measurement error, and different observations are measured by different instruments of unequal but known accuracy, weighted least squares may be employed with the weights chosen being inversely proportional to the variances of measurement error. In many problems, we will not know the weights initially and will have to estimate them based on the results of an ordinary (unweighted) least squares fit.

A more general treatment of weighted least squares will be given in Chapter 9 (Section 9.2). We now give an example of weighted least squares illustrating one approach to estimating the weights.

- **Example 3.11** The average monthly income from food sales and the corresponding annual advertising expenses for 30 restaurants is shown in columns (a) and (b) of Table 3.8. Management is interested in the relationship between these variables, and so a linear regression model relating food sales y to

Table 3.8 Restaurant food sales data

Obs. i	(a) Income, y_i	(b) Advertising Expense, x_i	(c) $\bar{x}$	(d) s_y^2	(e) Weights, w_i
1	81,464	3,000			6.21771E-08
2	72,661	3,150	3,078.3	26,794,620	5.79507E-08
3	72,344	3,085			5.97094E-08
4	90,743	5,225	5,287.5	30,722,010	2.98667E-08
5	98,588	5,350			2.90195E-08
6	96,507	6,090			2.48471E-08
7	126,574	8,925			1.60217E-08
8	114,133	9,015			1.58431E-08
9	115,814	8,885	8,955.0	52,803,698	1.61024E-08
10	123,181	8,950			1.59717E-08
11	131,434	9,000			1.58726E-08
12	140,564	11,345			1.22942E-08
13	151,352	12,275			1.12852E-08
14	146,926	12,400	12,377.5	77,280,167	1.11621E-08
15	130,963	12,525			1.10416E-08
16	144,630	12,310			1.12505E-08
17	147,041	13,700			1.00246E-08
18	179,021	15,000			9.09750E-09
19	166,200	15,175			8.98563E-09
20	180,732	14,995	15,095.0	120,571,040	9.10074E-09
21	178,187	15,050			9.06525E-09
22	185,304	15,200			8.96988E-09
23	155,931	15,150			9.00144E-09
24	172,579	16,800	16,650.0	132,388,990	8.06478E-09
25	188,851	16,500			8.22031E-09
26	192,424	17,830			7.57287E-09
27	203,112	19,500			6.89136E-09
28	192,482	19,200	19,262.5	138,856,867	7.00460E-09
29	218,715	19,000			7.08218E-09
30	214,317	19,350			6.94752E-09

advertising expense x is fitted by ordinary least squares, resulting in $\hat{y}= 49443.3838+8.0484x$. The residuals from this least squares fit are plotted against $\hat{y}_i$ and x_i in Figures 3.25 and 3.26, respectively. Both plots indicate violation of the constant variance assumption. Consequently, the ordinary least squares fit is inappropriate.

To correct this inequality of variance problem we must know the weights w_i. We note from examining the data in Table 3.8 that there are several sets of x values that are "near-neighbors," that is, have approximate repeat points on x. We will assume that these near-neighbors are close enough together to be considered repeat points and use the variance of the responses at those repeat points to investigate how $V(y)$ changes with x. Columns (c) and (d) of Table 3.8 show the average x value ($\bar{x}$) for each cluster of near-neighbors and the sample variance of the y's in each cluster. Plotting s_y^2 against the corresponding $\bar{x}$ implies that s_y^2 increases approximately linearly with $\bar{x}$. A least squares

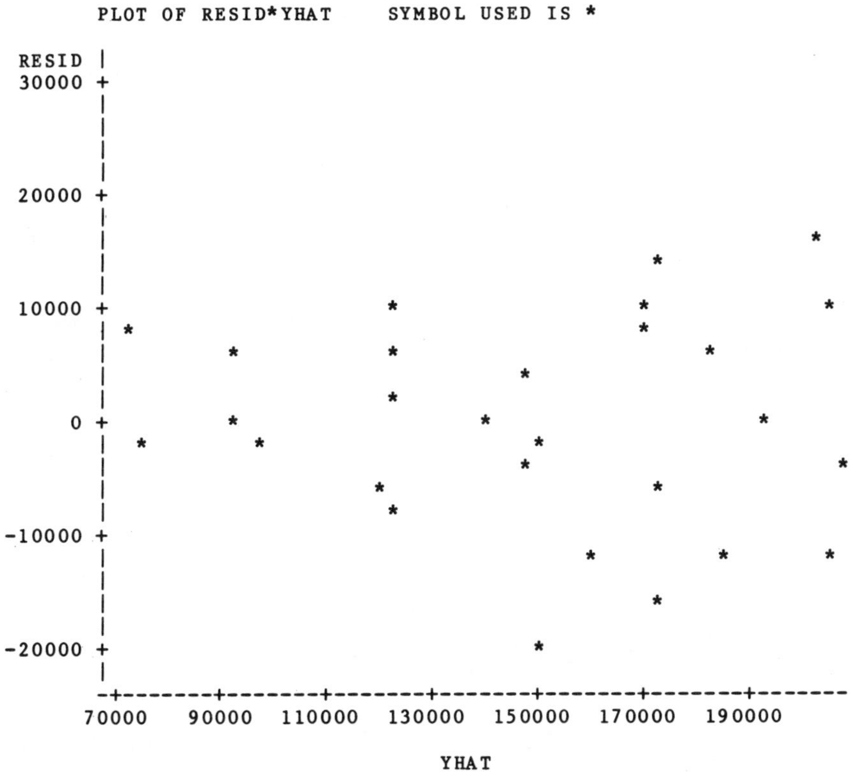

Figure 3.25 Plot (SAS) of ordinary least squares residuals versus fitted values, Example 3.11 (2 observations hidden).

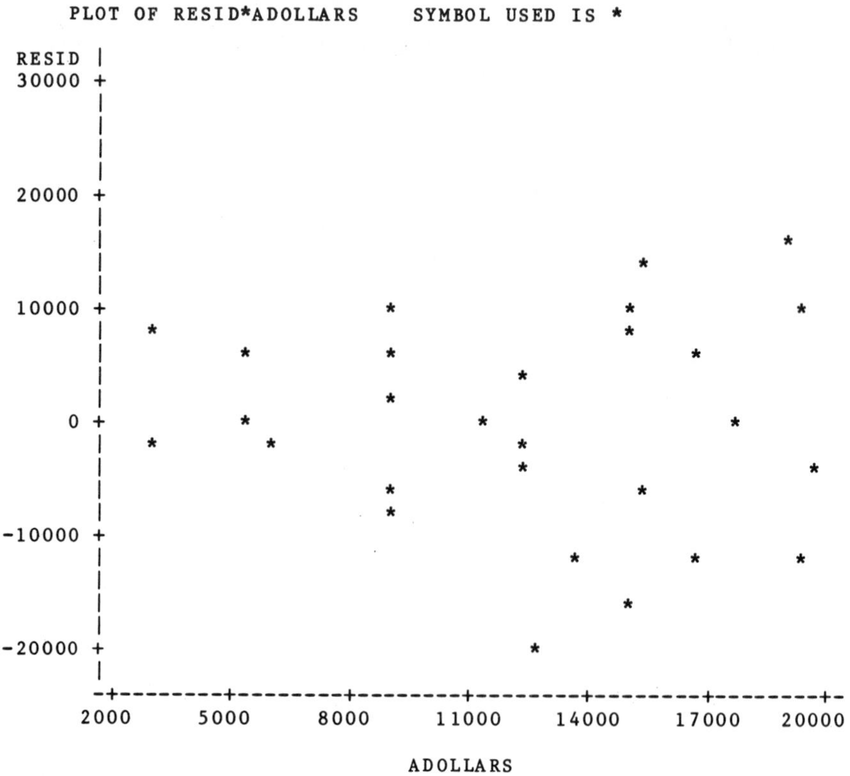

Figure 3.26 Plot (SAS) of ordinary least squares residuals versus advertising expenses, Example 3.11.

fit gives

$$\hat{s}_y^2 = -7376216.04 + 7819.77\bar{x}$$

Substituting each x_i value into this equation will give an estimate of the variance of the corresponding observation y_i. The inverse of these fitted values will be reasonable estimates of the weights w_i. These estimated weights are shown in column (e) of Table 3.8.

Applying weighted least squares to the data, using the weights in Table 3.8, gives the fitted model

$$\hat{y} = 50975.5667 + 7.9222x$$

We must now examine the residuals to determine if weighted least squares have improved the fit. To do this, plot the *weighted* residuals $w_i^{1/2}e_i = w_i^{1/2}(y_i - \hat{y}_i)$, where $\hat{y}_i$ comes from the weighted least squares fit, against $w_i^{1/2}\hat{y}_i$ and

```
      PLOT OF WRESID*WYHAT     SYMBOL USED IS *

WRESID  |
  3     +
        |
        |
        |
        |
  2     +
        |
        |                                                       *
        |                 *          *
        |   *
  1     +                *
        |           *    *            *
        |   *                 *
        |         *
        |   *
  0     +       *                 *
        |         *  *                 *
        |       * *       *
        |
        |   *                         *                   *  *
 -1     +   *                 *
        |             *
        |                 *
        |
        |
 -2     +          *
        |
        --+-------+-------+-------+-------+-------+-------+------
        15.25   15.75   16.25   16.75   17.25   17.75   18.25

                                   WYHAT
```

Figure 3.27 Plot (SAS) of weighted residuals $w_i^{1/2}e_i$ versus weighted fitted values $w_i^{1/2}\hat{y}_i$, Example 3.11 (1 observation hidden).

$w_i^{1/2}x_i$. These plots are shown in Figures 3.27 and 3.28, respectively, and are much improved when compared to the previous plots for the ordinary least squares fit. We conclude that weighted least squares have corrected the inequality of variance problem.

Two other points concerning this example should be made. First, we were fortunate to have several near-neighbors in the x-space. Furthermore, it was easy to identify these clusters of points by inspection of Table 3.8 because there was only one regressor involved. With several regressors, visual identification of these clusters would be more difficult. An analytical procedure for finding pairs of points that are close together in x-space will be presented in Chapter 4 (Section 4.7.3). The second point involves the use of a regression equation to estimate the weights. The analyst should carefully check the weights produced by the equation to be sure that they are reasonable. For example, in our problem a sufficiently small x value could result in a negative weight, which is clearly unreasonable.

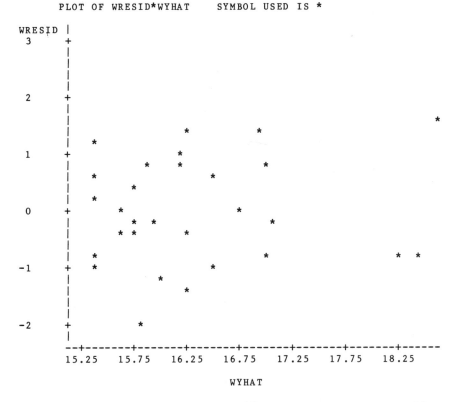

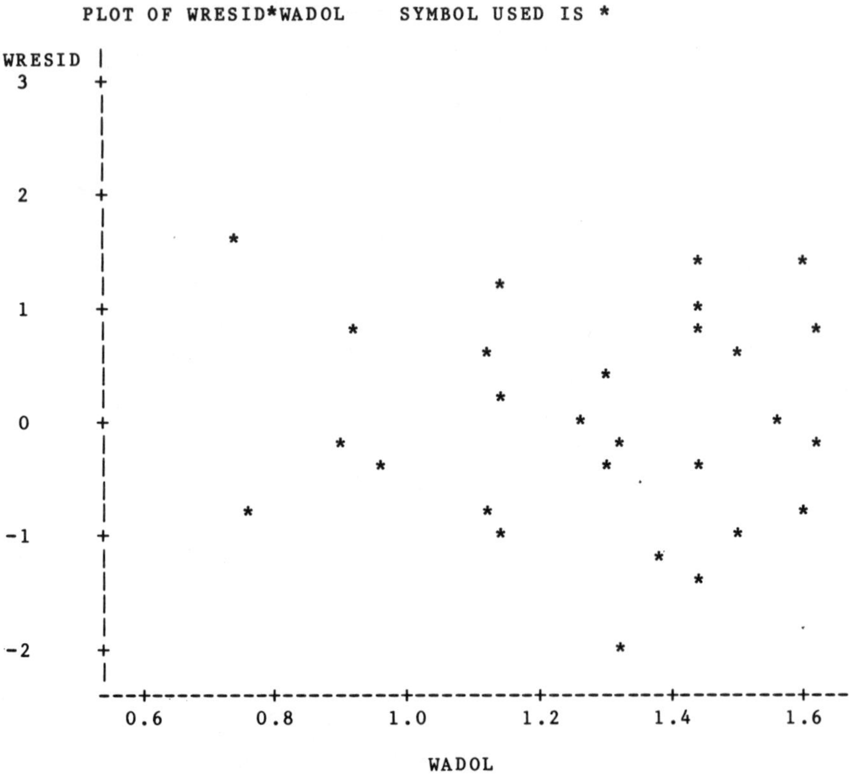

Figure 3.28 Plot (SAS) of weighted residuals $w_i^{1/2}e_i$ versus weighted advertising expenses $w_i^{1/2}x_i$, Example 3.11 (2 observations hidden).

Problems

3.1 Refer to Problem 2.1. Construct the following residual plots, and comment on model adequacy.

a. Normal probability plot.

b. Plot residuals against $\hat{y}_i$.

c. Plot residuals against x_{i8}.

3.2 Refer to Problem 2.1. Plot the residuals against the percent of rushing plays x_{i7}. Does this plot indicate that the model could be improved by adding this variable?

3.3 Refer to Problem 2.2. Construct the following residual plots, and draw conclusions about the adequacy of the model.

a. Normal probability plot.

b. Plot residuals against $\hat{y}_i$.
c. Plot residuals against x_{i4}.

3.4 Refer to Problem 2.4. Construct the following residual plots, and comment on model adequacy.
a. Normal probability plot.
b. Plot residuals against $\hat{y}_i$.
c. Plot residuals against x_{i1}.
d. Plot the residuals against the number of carburetor barrels, x_{i6}. Does this plot indicate that the model could be improved by adding the variable x_6?

3.5 Refer to Problem 2.5. Construct the following residual plots, and interpret the results.
a. Normal probability plot.
b. Plot residuals against $\hat{y}_i$.
c. Plot residuals against $x_{i,10}$.
d. Plot the residuals against x_{i8}, vehicle length. Does this plot indicate that the model could be improved by the addition of x_8?

3.6 Consider the house price model developed in Problem 2.6. Construct a normal probability plot and a plot of residuals versus $\hat{y}_i$. Comment on model adequacy.

3.7 Refer to Problem 2.7. Perform the following diagnostic checks and interpret the results.
a. Test for lack of fit.
b. Plot residuals against $\hat{y}_i$.
c. Plot residuals against x_i.

3.8 Consider the no-intercept model developed for the soft drink delivery time data in Problem 2.11. Construct the following residual plots:
a. Normal probability plot.
b. Plot residuals against $\hat{y}_i$.
c. Plot residuals against x_i.
What conclusions can you draw about the adequacy of the model?

3.9 Refer to Problem 2.12. Construct the following residual plots, and comment on model adequacy.
a. Normal probability plot.
b. Plot residuals against $\hat{y}_i$.
c. Plot residuals against x_i.

3.10 Consider the soft drink delivery time data in Example 2.8.
a. Compute the residuals and standardized residuals for this model.
b. Observation 9 has an unusually large residual. Assess the impact of

this observation on the model. Does it seem to control any key model properties?

3.11 Consider the two regression models

a. $y=\beta_0\beta_1^x+\varepsilon$

b. $y=\beta_0+\beta_1\sin x_1+\beta_2 e^{x_2}+\varepsilon$

where ε has mean zero and variance σ^2. Are these linear regression models? If so, write them in the linearized form.

3.12 The data shown below presents the average number of surviving bacteria in a canned food product and the minutes of exposure to 300°F heat.

Number of Bacteria	Minutes of Exposure
175	1
108	2
95	3
82	4
71	5
50	6
49	7
31	8
28	9
17	10
16	11
11	12

a. Plot a scatter diagram. Does it seem likely that a straight-line model will be adequate?

b. Fit the straight-line model. Compute the summary statistics and the residual plots. What are your conclusions regarding model adequacy?

c. Identify an appropriate transformed model for these data. Fit this model to the data and conduct the usual tests of model adequacy.

3.13 Consider the soft drink delivery time data in Example 2.8. Figures 3.7 and 3.8 indicate that there is some inequality of variance in these data. Suggest an appropriate transformation for this problem to stabilize the variance. Fit the transformed model and investigate the effect of this variance stabilizing transformation.

3.14 Consider the data shown below. Construct a scatter diagram and suggest an appropriate form for the regression model. Fit this model to the data and conduct the standard tests of model adequacy.

x	10	15	18	12	9	8	11	6
y	0.17	0.13	0.09	0.15	0.20	0.21	0.18	0.24

3.15 Consider the three models

a. $y=\beta_0+\beta_1\left(\frac{1}{x}\right)+\varepsilon$

b. $\frac{1}{y}=\beta_0+\beta_1 x+\varepsilon$

c. $y=\frac{x}{\beta_0-\beta_1 x+\varepsilon}$

All these models can be linearized by reciprocal transformations. Sketch the behavior of y as a function of x. What observed characteristics in the scatter diagram would lead you to choose one of these models?

3.16 A glass bottle manufacturing company has recorded data on the average number of defects per 10,000 bottles due to stones (small pieces of rock imbedded in the bottle wall) and the number of weeks since the last furnace overhaul. The data are shown below.

Defects/10,000	Weeks	Defects/10,000	Weeks
13.0	4	34.2	11
16.1	5	65.6	12
14.5	6	49.2	13
17.8	7	66.2	14
22.0	8	81.2	15
27.4	9	87.4	16
16.8	10	114.5	17

a. Fit a straight-line regression model to the data and perform the standard tests for model adequacy.

b. Suggest an appropriate transformation to eliminate the problems encountered in Part a. Fit the transformed model and check for adequacy.

3.17 Consider the electric utility data in Example 3.8.

a. Customer 16 is a potential outlier. What effect does removing this customer have on the least squares fit?

b. Customer 50 has an unusually large energy consumption. What effect does removing this customer have on the least squares fit?

c. Why does the removal of Customer 50 in Part b reduce the value of R^2?

3.18 Consider the simple linear regression model $y_i = \beta_0 + \beta_1 x_i + \varepsilon_i$, where the variance of ε_i is proportional to x_i^2; that is $V(\varepsilon_i) = \sigma^2 x_i^2$.

a. Suppose that we use the transformations $y' = y/x$ and $x' = 1/x$. Is this a variance stabilizing transformation?

b. What are the relationships between the parameters in the original and transformed models?

c. Suppose that we use the method of weighted least squares with $w_i = 1/x_i^2$. Is this equivalent to the transformation introduced in Part a?

3.19 Show that the slope of a least squares line relating e_i (ordinate) to y_i is $1 - R^2$, while the slope of a least squares line relating e_i to $\hat{y}_i$ is zero.

4

MULTIPLE LINEAR REGRESSION

A regression model that involves more than one regressor variable is called a multiple regression model. Fitting and analyzing these models is discussed in this chapter. The results are extensions of those in Chapter 2 for simple linear regression. We will also discuss measures of model adequacy that are useful in multiple regression.

4.1 Multiple regression models

Suppose that the effective life of a cutting tool depends on the cutting speed and the depth of cut. A multiple regression model that might describe this relationship is

$$y = \beta_0 + \beta_1 x_1 + \beta_2 x_2 + \varepsilon \tag{4.1}$$

where y denotes the effective tool life, x_1 denotes the cutting speed, and x_2 denotes the depth of cut. This is a multiple linear regression model with two regressor variables. The term "linear" is used because (4.1) is a linear function of the unknown parameters β_0, β_1, and β_2. The model describes a plane in the two-dimensional space of the regressor variables x_1 and x_2, as shown in Figure 4.1. The parameter β_0 is the intercept of the regression

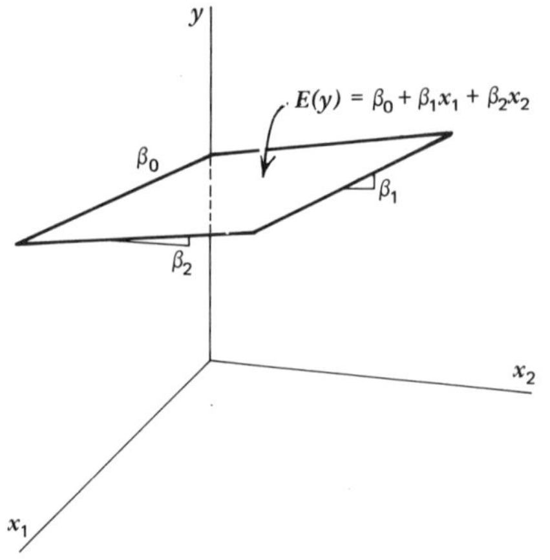

Figure 4.1 Example of a multiple regression model with two regressors.

plane. If the range of the data includes $x_1 = x_2 = 0$, then β_0 is the mean of y when $x_1 = x_2 = 0$. Otherwise β_0 has no physical interpretation. The parameter β_1 indicates the expected change in response (y) per unit change in x_1 when x_2 is held constant. Similarly, β_2 measures the expected change in y per unit change in x_2 when x_1 is held constant.

In general, the response y may be related to k regressor variables. The model

$$y = \beta_0 + \beta_1 x_1 + \beta_2 x_2 + \cdots + \beta_k x_k + \varepsilon \tag{4.2}$$

is called a multiple linear regression model with k regressors. The parameters β_j, $j = 0, 1, \ldots, k$ are called the regression coefficients. This model describes a hyperplane in the k-dimensional space of the regressor variables x_j. The parameter β_j represents the expected change in the response y per unit change in x_j when all the remaining regressor variables $x_i (i \neq j)$ are held constant. For this reason the parameters β_j, $j = 1, 2, \ldots, k$ are often called *partial* regression coefficients.

Multiple linear regression models are often used as approximating functions. That is, the true functional relationship between y and $x_1, x_2, \ldots, x_k$ is unknown, but over certain ranges of the regressor variables the linear regression model is an adequate approximation.

Models that are more complex in structure than (4.2) may often still be analyzed by multiple linear regression techniques. For example, consider the cubic polynomial model

$$y = \beta_0 + \beta_1 x + \beta_2 x^2 + \beta_3 x^3 + \varepsilon \tag{4.3}$$

If we let $x_1 = x$, $x_2 = x^2$, and $x_3 = x^3$, then (4.3) can be written as

$$y = \beta_0 + \beta_1 x_1 + \beta_2 x_2 + \beta_3 x_3 + \varepsilon \tag{4.4}$$

which is a multiple linear regression model with three regressor variables. Polynomial models will be discussed in more detail in Chapter 5. Models that include interaction effects may also be analyzed by multiple linear regression methods. For example, suppose that the model is

$$y = \beta_0 + \beta_1 x_1 + \beta_2 x_2 + \beta_{12} x_1 x_2 + \varepsilon \tag{4.5}$$

If we let $x_3 = x_1 x_2$ and $\beta_3 = \beta_{12}$ then (4.5) can be written as

$$y = \beta_0 + \beta_1 x_1 + \beta_2 x_2 + \beta_3 x_3 + \varepsilon \tag{4.6}$$

which is a linear regression model. In general, any regression model that is linear in the *parameters* (the β's) is a linear regression model, regardless of the shape of the surface that it generates.

4.2 Estimation of the model parameters

4.2.1 Least squares estimation of the regression coefficients

The method of least squares is used to estimate the regression coefficients in (4.2). Suppose that $n > k$ observations are available, and let y_i denote the ith observed response and x_{ij} denote the ith observation or level of regressor x_j. The data will appear as in Table 4.1. We assume that the error term ε in the model has $E(\varepsilon) = 0$, $V(\varepsilon) = \sigma^2$, and that the errors are uncorrelated.

Table 4.1 Data for multiple linear regression

Observation i	y	x_1	x_2	$\cdots$	x_k
1	y_1	x_{11}	x_{12}	$\cdots$	x_{1k}
2	y_2	x_{21}	x_{22}	$\cdots$	x_{2k}
$\vdots$	$\vdots$	$\vdots$	$\vdots$		$\vdots$
n	y_n	x_{n1}	x_{n2}	$\cdots$	x_{nk}

We may write the sample model corresponding to (4.2) as

$$\begin{aligned} y_i &= \beta_0 + \beta_1 x_{i1} + \beta_2 x_{i2} + \cdots + \beta_k x_{ik} + \varepsilon_i \\ &= \beta_0 + \sum_{j=1}^{k} \beta_j x_{ij} + \varepsilon_i, \qquad i = 1, 2, \ldots, n \end{aligned} \tag{4.7}$$

The least squares function is

$$\begin{aligned} S(\beta_0, \beta_1, \ldots, \beta_k) &= \sum_{i=1}^{n} \varepsilon_i^2 \\ &= \sum_{i=1}^{n} \left(y_i - \beta_0 - \sum_{j=1}^{k} \beta_j x_{ij} \right)^2 \end{aligned} \tag{4.8}$$

The function S is to be minimized with respect to $\beta_0, \beta_1, \ldots, \beta_k$. The least squares estimators of $\beta_0, \beta_1, \ldots, \beta_k$ must satisfy

$$\left.\frac{\partial S}{\partial \beta_0}\right|_{\hat{\beta}_0, \hat{\beta}_1, \ldots, \hat{\beta}_k} = -2 \sum_{i=1}^{n} \left(y_i - \hat{\beta}_0 - \sum_{j=1}^{k} \hat{\beta}_j x_{ij} \right) = 0 \tag{4.9a}$$

and

$$\left.\frac{\partial S}{\partial \beta_j}\right|_{\hat{\beta}_0, \hat{\beta}_1, \ldots, \hat{\beta}_k} = -2 \sum_{i=1}^{n} \left(y_i - \hat{\beta}_0 - \sum_{j=1}^{k} \hat{\beta}_j x_{ij} \right) x_{ij} = 0, \qquad j = 1, 2, \ldots, k \tag{4.9b}$$

Simplifying (4.9), we obtain the least squares normal equations

$$\begin{array}{l} n\hat{\beta}_0 + \hat{\beta}_1 \sum_{i=1}^{n} x_{i1} + \hat{\beta}_2 \sum_{i=1}^{n} x_{i2} + \cdots + \hat{\beta}_k \sum_{i=1}^{n} x_{ik} = \sum_{i=1}^{n} y_i \\ \hat{\beta}_0 \sum_{i=1}^{n} x_{i1} + \hat{\beta}_1 \sum_{i=1}^{n} x_{i1}^2 + \hat{\beta}_2 \sum_{i=1}^{n} x_{i1} x_{i2} + \cdots + \hat{\beta}_k \sum_{i=1}^{n} x_{i1} x_{ik} = \sum_{i=1}^{n} x_{i1} y_i \\ \quad \vdots \qquad\qquad \vdots \qquad\qquad \vdots \qquad\qquad \vdots \qquad\qquad \vdots \\ \hat{\beta}_0 \sum_{i=1}^{n} x_{ik} + \hat{\beta}_1 \sum_{i=1}^{n} x_{ik} x_{i1} + \hat{\beta}_2 \sum_{i=1}^{n} x_{ik} x_{i2} + \cdots + \hat{\beta}_k \sum_{i=1}^{n} x_{ik}^2 = \sum_{i=1}^{n} x_{ik} y_i \end{array} \tag{4.10}$$

Note that there are $p=k+1$ normal equations, one for each of the unknown regression coefficients. The solution to the normal equations will be the least squares estimators, $\hat{\beta}_0, \hat{\beta}_1, \ldots, \hat{\beta}_k$.

It is more convenient to deal with multiple regression models if they are expressed in matrix notation. This allows a very compact display of the model, data, and results. We now give a matrix development of the normal equations that parallels the development of (4.10). The model in terms of the observations (4.7), may be written in matrix notation as

$$\mathbf{y}=\mathbf{X}\boldsymbol{\beta}+\boldsymbol{\varepsilon}$$

where

$$\mathbf{y}=\begin{bmatrix} y_1 \\ y_2 \\ \vdots \\ y_n \end{bmatrix}, \qquad \mathbf{X}=\begin{bmatrix} 1 & x_{11} & x_{12} \cdots x_{1k} \\ 1 & x_{21} & x_{22} \cdots x_{2k} \\ \vdots & \vdots & \vdots \qquad \vdots \\ 1 & x_{n1} & x_{n2} \cdots x_{nk} \end{bmatrix}$$

$$\boldsymbol{\beta}=\begin{bmatrix} \beta_0 \\ \beta_1 \\ \vdots \\ \beta_k \end{bmatrix}, \qquad \text{and} \quad \boldsymbol{\varepsilon}=\begin{bmatrix} \varepsilon_1 \\ \varepsilon_2 \\ \vdots \\ \varepsilon_n \end{bmatrix}$$

In general, $\mathbf{y}$ is an $(n\times 1)$ vector of the observations, $\mathbf{X}$ is an $(n\times p)$ matrix of the levels of the regressor variables, $\boldsymbol{\beta}$ is a $(p\times 1)$ vector of the regression coefficients, and $\boldsymbol{\varepsilon}$ is an $(n\times 1)$ vector of random errors.

We wish to find the vector of least squares estimators, $\hat{\boldsymbol{\beta}}$, which minimizes

$$S(\boldsymbol{\beta})=\sum_{i=1}^{n}\varepsilon_i^2=\boldsymbol{\varepsilon}'\boldsymbol{\varepsilon}=(\mathbf{y}-\mathbf{X}\boldsymbol{\beta})'(\mathbf{y}-\mathbf{X}\boldsymbol{\beta})$$

Note that $S(\boldsymbol{\beta})$ may be expressed as

$$\begin{aligned} S(\boldsymbol{\beta})&=\mathbf{y}'\mathbf{y}-\boldsymbol{\beta}'\mathbf{X}'\mathbf{y}-\mathbf{y}'\mathbf{X}\boldsymbol{\beta}+\boldsymbol{\beta}'\mathbf{X}'\mathbf{X}\boldsymbol{\beta} \\ &=\mathbf{y}'\mathbf{y}-2\boldsymbol{\beta}'\mathbf{X}'\mathbf{y}+\boldsymbol{\beta}'\mathbf{X}'\mathbf{X}\boldsymbol{\beta} \end{aligned} \tag{4.11}$$

since $\boldsymbol{\beta}'\mathbf{X}'\mathbf{y}$ is a (1×1) matrix, or a scalar, and its transpose $(\boldsymbol{\beta}'\mathbf{X}'\mathbf{y})'=\mathbf{y}'\mathbf{X}\boldsymbol{\beta}$ is the same scalar. The least squares estimators must satisfy

$$\left.\frac{\partial S}{\partial \boldsymbol{\beta}}\right|_{\hat{\boldsymbol{\beta}}}=-2\mathbf{X}'\mathbf{y}+2\mathbf{X}'\mathbf{X}\hat{\boldsymbol{\beta}}=\mathbf{0}$$

which simplifies to

$$\mathbf{X}'\mathbf{X}\hat{\boldsymbol{\beta}}=\mathbf{X}'\mathbf{y} \tag{4.12}$$

Equations (4.12) are the least squares normal equations. They are identical to (4.10). To solve the normal equations multiply both sides of (4.12) by the inverse of $\mathbf{X}'\mathbf{X}$. Thus the least squares estimator of $\boldsymbol{\beta}$ is

$$\hat{\boldsymbol{\beta}}=(\mathbf{X}'\mathbf{X})^{-1}\mathbf{X}'\mathbf{y} \tag{4.13}$$

provided that $(\mathbf{X}'\mathbf{X})^{-1}$ exists. The $(\mathbf{X}'\mathbf{X})^{-1}$ matrix will always exist if the regressors are *linearly independent*; that is, if no column of the $\mathbf{X}$ matrix is a linear combination of the other columns.

It is easy to see that the matrix form of the normal equations (4.12) is identical to the scalar form (4.10). Writing out (4.12) in detail, we obtain

$$\begin{bmatrix} n & \sum_{i=1}^{n} x_{i1} & \sum_{i=1}^{n} x_{i2} & \cdots & \sum_{i=1}^{n} x_{ik} \\ \sum_{i=1}^{n} x_{i1} & \sum_{i=1}^{n} x_{i1}^2 & \sum_{i=1}^{n} x_{i1}x_{i2} & \cdots & \sum_{i=1}^{n} x_{i1}x_{ik} \\ \vdots & \vdots & \vdots & & \vdots \\ \sum_{i=1}^{n} x_{ik} & \sum_{i=1}^{n} x_{ik}x_{i1} & \sum_{i=1}^{n} x_{ik}x_{i2} & \cdots & \sum_{i=1}^{n} x_{ik}^2 \end{bmatrix} \begin{bmatrix} \hat{\beta}_0 \\ \hat{\beta}_1 \\ \vdots \\ \hat{\beta}_k \end{bmatrix} = \begin{bmatrix} \sum_{i=1}^{n} y_i \\ \sum_{i=1}^{n} x_{i1}y_i \\ \vdots \\ \sum_{i=1}^{n} x_{ik}y_i \end{bmatrix}$$

If the indicated matrix multiplication is performed, the scalar form of the normal equations (4.10) is obtained. In this display it is easy to see that $\mathbf{X}'\mathbf{X}$ is a $(p\times p)$ symmetric matrix and $\mathbf{X}'\mathbf{y}$ is a $(p\times 1)$ column vector. Note the special structure of the $\mathbf{X}'\mathbf{X}$ matrix. The diagonal elements of $\mathbf{X}'\mathbf{X}$ are the sums of squares of the elements in the columns of $\mathbf{X}$, and the off-diagonal elements are the sums of cross-products of the elements in the columns of $\mathbf{X}$. Furthermore note that the elements of $\mathbf{X}'\mathbf{y}$ are the sums of cross-products of the columns of $\mathbf{X}$ and the observations y_i.

The fitted regression model corresponding to the levels of the regressor variables $\mathbf{x}'=[1, x_1, x_2,\ldots, x_k]$ is

$$\hat{y}=\mathbf{x}'\hat{\boldsymbol{\beta}}$$

$$=\hat{\beta}_0+\sum_{j=1}^{k}\hat{\beta}_j x_j$$

The vector of fitted values $\hat{y}_i$ corresponding to the observed values y_i is

$$\hat{\mathbf{y}}=\mathbf{X}\hat{\boldsymbol{\beta}}$$

$$=\mathbf{X}(\mathbf{X}'\mathbf{X})^{-1}\mathbf{X}'\mathbf{y}$$

$$=\mathbf{H}\mathbf{y} \tag{4.14}$$

The $n\times n$ matrix $\mathbf{H}=\mathbf{X}(\mathbf{X}'\mathbf{X})^{-1}\mathbf{X}'$ is usually called the "hat" matrix, because it maps the vector of observed values into a vector of fitted values. The hat matrix and its properties play a central role in regression analysis.

The difference between the observed value y_i and the corresponding fitted value $\hat{y}_i$ is the residual $e_i=y_i-\hat{y}_i$. The n residuals may be conveniently written in matrix notation as

$$\mathbf{e}=\mathbf{y}-\hat{\mathbf{y}} \tag{4.15a}$$

There are several other ways to express the vector of residuals $\mathbf{e}$ that will prove useful, including:

$$\mathbf{e}=\mathbf{y}-\mathbf{X}\hat{\boldsymbol{\beta}} \tag{4.15b}$$

$$=\mathbf{y}-\mathbf{H}\mathbf{y} \tag{4.15c}$$

$$=(\mathbf{I}-\mathbf{H})\mathbf{y} \tag{4.15d}$$

- **Example 4.1** A soft drink bottler is analyzing the vending machine service routes in his distribution system. He is interested in predicting the amount of time required by the route driver to service the vending machines in an outlet.

This service activity includes stocking the machine with beverage products and minor maintenance or housekeeping. The industrial engineer responsible for the study has suggested that the two most important variables affecting the delivery time are the number of cases of product stocked and the distance walked by the route driver. The engineer has collected 25 observations on delivery time, which are shown in Table 4.2. (Note that this is an expansion of the data set used in Example 2.8.)

We shall fit the multiple linear regression model

$$y=\beta_0+\beta_1x_1+\beta_2x_2+\varepsilon$$

Table 4.2 Delivery time data for Example 4.1

Observation Number	Delivery Time (Minutes) y	Number of Cases x_1	Distance (Feet) x_2
1	16.68	7	560
2	11.50	3	220
3	12.03	3	340
4	14.88	4	80
5	13.75	6	150
6	18.11	7	330
7	8.00	2	110
8	17.83	7	210
9	79.24	30	1460
10	21.50	5	605
11	40.33	16	688
12	21.00	10	215
13	13.50	4	255
14	19.75	6	462
15	24.00	9	448
16	29.00	10	776
17	15.35	6	200
18	19.00	7	132
19	9.50	3	36
20	35.10	17	770
21	17.90	10	140
22	52.32	26	810
23	18.75	9	450
24	19.83	8	635
25	10.75	4	150

to these data. The **X** matrix and **y** vector are

$$\mathbf{X}=\begin{bmatrix} 1 & 7 & 560 \\ 1 & 3 & 220 \\ 1 & 3 & 340 \\ 1 & 4 & 80 \\ 1 & 6 & 150 \\ 1 & 7 & 330 \\ 1 & 2 & 110 \\ 1 & 7 & 210 \\ 1 & 30 & 1460 \\ 1 & 5 & 605 \\ 1 & 16 & 688 \\ 1 & 10 & 215 \\ 1 & 4 & 255 \\ 1 & 6 & 462 \\ 1 & 9 & 448 \\ 1 & 10 & 776 \\ 1 & 6 & 200 \\ 1 & 7 & 132 \\ 1 & 3 & 36 \\ 1 & 17 & 770 \\ 1 & 10 & 140 \\ 1 & 26 & 810 \\ 1 & 9 & 450 \\ 1 & 8 & 635 \\ 1 & 4 & 150 \end{bmatrix} \qquad \mathbf{y}=\begin{bmatrix} 16.68 \\ 11.50 \\ 12.03 \\ 14.88 \\ 13.75 \\ 18.11 \\ 8.00 \\ 17.83 \\ 79.24 \\ 21.50 \\ 40.33 \\ 21.00 \\ 13.50 \\ 19.75 \\ 24.00 \\ 29.00 \\ 15.35 \\ 19.00 \\ 9.50 \\ 35.10 \\ 17.90 \\ 52.32 \\ 18.75 \\ 19.83 \\ 10.75 \end{bmatrix}$$

The **X′X** matrix is

$$\mathbf{X}'\mathbf{X}=\begin{bmatrix} 1 & 1 & \cdots & 1 \\ 7 & 3 & \cdots & 4 \\ 560 & 220 & \cdots & 150 \end{bmatrix}\begin{bmatrix} 1 & 7 & 560 \\ 1 & 3 & 220 \\ \vdots & \vdots & \vdots \\ 1 & 4 & 150 \end{bmatrix}$$

$$=\begin{bmatrix} 25 & 219 & 10232 \\ 219 & 3055 & 133899 \\ 10232 & 133899 & 6725688 \end{bmatrix}$$

and the **X′y** vector is

$$\mathbf{X}'\mathbf{y}=\begin{bmatrix} 1 & 1 & \cdots & 1 \\ 7 & 3 & \cdots & 4 \\ 560 & 220 & \cdots & 150 \end{bmatrix}\begin{bmatrix} 16.68 \\ 11.50 \\ \vdots \\ 10.75 \end{bmatrix}=\begin{bmatrix} 559.60 \\ 7375.44 \\ 337072.00 \end{bmatrix}$$

The least squares estimator of $\boldsymbol{\beta}$ is

$$\hat{\boldsymbol{\beta}}=(\mathbf{X}'\mathbf{X})^{-1}\mathbf{X}'\mathbf{y}$$

or,

$$\begin{bmatrix}\hat{\beta}_0\\ \hat{\beta}_1\\ \hat{\beta}_2\end{bmatrix}=\begin{bmatrix}25 & 219 & 10232\\ 219 & 3055 & 133899\\ 10232 & 133899 & 6725688\end{bmatrix}^{-1}\begin{bmatrix}559.60\\ 7375.44\\ 337072.00\end{bmatrix}$$

$$=\begin{bmatrix}.11321518 & -.00444859 & -.00008367\\ -.00444859 & .00274378 & -.00004786\\ -.00008367 & -.00004786 & .00000123\end{bmatrix}\begin{bmatrix}559.60\\ 7375.44\\ 337072.00\end{bmatrix}$$

$$=\begin{bmatrix}2.34123115\\ 1.61590712\\ 0.01438483\end{bmatrix}$$

The least squares fit (with the regression coefficients reported to five decimals) is

$$\hat{y}=2.34123+1.61591x_1+0.01439x_2$$

Table 4.3 shows the observations y_i along with the corresponding fitted values
• $\hat{y}_i$ and the residuals e_i from this model.

4.2.2 A geometrical interpretation of least squares

An intuitive geometrical interpretation of least squares is sometimes helpful. We may think of the vector of observations $\mathbf{y}'=[y_1, y_2,\ldots, y_n]$ as defining a vector from the origin to the point A in Figure 4.2. Note that $y_1, y_2,\ldots, y_n$ form the coordinates of an n–dimensional sample space. The sample space in Figure 4.2 is three-dimensional.

The $\mathbf{X}$ matrix consists of $p(n\times 1)$ column vectors, for example $\mathbf{1}$ (a column vector of 1's), $\mathbf{x}_1,\mathbf{x}_2,\ldots,\mathbf{x}_k$. Each of these columns define a vector from the origin in the sample space. These p vectors form a p–dimensional subspace called the *estimation* space. The estimation space for $p=2$ is shown in Figure 4.2. We may represent any point in this subspace by a linear combination of the vectors $\mathbf{1},\mathbf{x}_1,\ldots,\mathbf{x}_k$. Thus any point in the estimation space is of the form $\mathbf{X}\boldsymbol{\beta}$. Let the vector $\mathbf{X}\boldsymbol{\beta}$ determine the point B in Figure

Table 4.3 Observations, estimated values, and residuals for Example 4.1

Observation Number	y_i	$\hat{y}_i$	$e_i = y_i - \hat{y}_i$
1	16.68	21.7081	−5.0281
2	11.50	10.3536	1.1464
3	12.03	12.0798	−0.0498
4	14.88	9.9556	4.9244
5	13.75	14.1944	−0.4444
6	18.11	18.3996	−0.2896
7	8.00	7.1554	0.8446
8	17.83	16.6734	1.1566
9	79.24	71.8203	7.4197
10	21.50	19.1236	2.3764
11	40.33	38.0925	2.2375
12	21.00	21.5930	−0.5930
13	13.50	12.4730	1.0270
14	19.75	18.6825	1.0675
15	24.00	23.3288	0.6712
16	29.00	29.6629	−0.6629
17	15.35	14.9136	0.4364
18	19.00	15.5514	3.4486
19	9.50	7.7068	1.7932
20	35.10	40.8880	−5.7880
21	17.90	20.5142	−2.6142
22	52.32	56.0065	−3.6865
23	18.75	23.3576	−4.6076
24	19.83	24.4028	−4.5728
25	10.75	10.9626	−0.2126

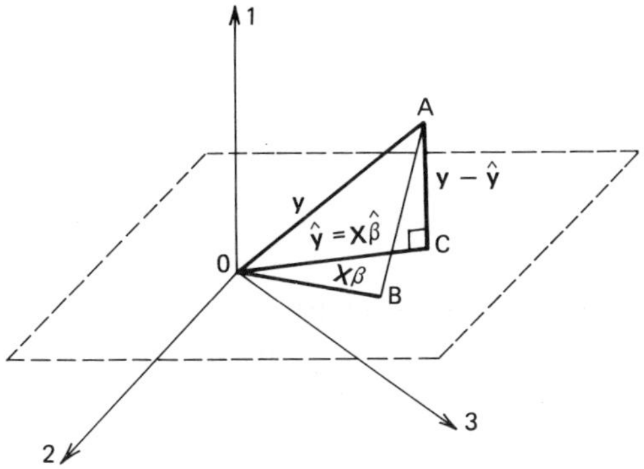

Figure 4.2 A geometric interpretation of least squares.

4.2. The squared distance from B to A is just

$$S(\boldsymbol{\beta})=(\mathbf{y}-\mathbf{X}\boldsymbol{\beta})'(\mathbf{y}-\mathbf{X}\boldsymbol{\beta})$$

Therefore minimizing the squared distance of the point A defined by the observation vector **y** to the estimation space requires finding the point in the estimation space that is closest to A. The squared distance will be a minimum when the point in the estimation space is the foot of the line from A normal (or perpendicular) to the estimation space. This is the Point C shown in Figure 4.2. This point is defined by the vector $\hat{\mathbf{y}}=\mathbf{X}\hat{\boldsymbol{\beta}}$. Therefore since $\mathbf{y}-\hat{\mathbf{y}}=\mathbf{y}-\mathbf{X}\hat{\boldsymbol{\beta}}$ is perpendicular to the estimation space, we may write

$$\mathbf{X}'(\mathbf{y}-\mathbf{X}\hat{\boldsymbol{\beta}})=\mathbf{0}$$

or

$$\mathbf{X}'\mathbf{X}\hat{\boldsymbol{\beta}}=\mathbf{X}'\mathbf{y}$$

which we recognize as the least squares normal equations.

4.2.3 Properties of the least squares estimators

The statistical properties of the least squares estimator $\hat{\boldsymbol{\beta}}$ may be easily demonstrated. Consider first bias:

$$\begin{aligned}
E(\hat{\boldsymbol{\beta}})&=E\left[(\mathbf{X}'\mathbf{X})^{-1}\mathbf{X}'\mathbf{y}\right]\\
&=E\left[(\mathbf{X}'\mathbf{X})^{-1}\mathbf{X}'(\mathbf{X}\boldsymbol{\beta}+\boldsymbol{\varepsilon})\right]\\
&=E\left[(\mathbf{X}'\mathbf{X})^{-1}\mathbf{X}'\mathbf{X}\boldsymbol{\beta}+(\mathbf{X}'\mathbf{X})^{-1}\mathbf{X}'\boldsymbol{\varepsilon}\right]\\
&=\boldsymbol{\beta}
\end{aligned}$$

since $E(\boldsymbol{\varepsilon})=\mathbf{0}$ and $(\mathbf{X}'\mathbf{X})^{-1}\mathbf{X}'\mathbf{X}=\mathbf{I}$. Thus $\hat{\boldsymbol{\beta}}$ is an unbiased estimator of $\boldsymbol{\beta}$.

The variance property of $\hat{\boldsymbol{\beta}}$ is expressed by the covariance matrix

$$\operatorname{Cov}(\hat{\boldsymbol{\beta}})=E\{[\hat{\boldsymbol{\beta}}-E(\hat{\boldsymbol{\beta}})][\hat{\boldsymbol{\beta}}-E(\hat{\boldsymbol{\beta}})]'\}$$

which is a $(p\times p)$ symmetric matrix whose jth diagonal element is the variance of $\hat{\beta}_j$ and whose ijth off-diagonal element is the covariance between $\hat{\beta}_i$ and $\hat{\beta}_j$. The covariance matrix of $\hat{\boldsymbol{\beta}}$ is

$$\operatorname{Cov}(\hat{\boldsymbol{\beta}})=\sigma^2(\mathbf{X}'\mathbf{X})^{-1}$$

Therefore if we let $\mathbf{C}=(\mathbf{X}'\mathbf{X})^{-1}$, the variance of $\hat{\beta}_j$ is $\sigma^2 C_{jj}$ and the covariance between $\hat{\beta}_i$ and $\hat{\beta}_j$ is $\sigma^2 C_{ij}$.

The least squares estimator $\hat{\boldsymbol{\beta}}$ is the best linear unbiased estimator of $\boldsymbol{\beta}$ (the Gauss–Markov theorem). If we further assume that the errors ε_i are normally distributed, then $\hat{\boldsymbol{\beta}}$ is also the maximum likelihood estimator of $\boldsymbol{\beta}$. The maximum likelihood estimator is the minimum variance unbiased estimator of $\boldsymbol{\beta}$.

4.2.4 Estimation of σ^2

As in simple linear regression, we may develop an estimator of σ^2 from the residual sum of squares

$$
\begin{aligned}
SS_E &= \sum_{i=1}^{n} (y_i - \hat{y}_i)^2 \\
&= \sum_{i=1}^{n} e_i^2 \\
&= \mathbf{e}'\mathbf{e}
\end{aligned}
$$

Substituting $\mathbf{e}=\mathbf{y}-\mathbf{X}\hat{\boldsymbol{\beta}}$, we have

$$
\begin{aligned}
SS_E &= (\mathbf{y}-\mathbf{X}\hat{\boldsymbol{\beta}})'(\mathbf{y}-\mathbf{X}\hat{\boldsymbol{\beta}}) \\
&= \mathbf{y}'\mathbf{y} - \hat{\boldsymbol{\beta}}'\mathbf{X}'\mathbf{y} - \mathbf{y}'\mathbf{X}\hat{\boldsymbol{\beta}} + \hat{\boldsymbol{\beta}}'\mathbf{X}'\mathbf{X}\hat{\boldsymbol{\beta}} \\
&= \mathbf{y}'\mathbf{y} - 2\hat{\boldsymbol{\beta}}'\mathbf{X}'\mathbf{y} + \hat{\boldsymbol{\beta}}'\mathbf{X}'\mathbf{X}\hat{\boldsymbol{\beta}}
\end{aligned}
$$

Since $\mathbf{X}'\mathbf{X}\hat{\boldsymbol{\beta}}=\mathbf{X}'\mathbf{y}$, this last equation becomes

$$SS_E = \mathbf{y}'\mathbf{y} - \hat{\boldsymbol{\beta}}'\mathbf{X}'\mathbf{y} \tag{4.16}$$

The residual sum of squares has $n-p$ degrees of freedom associated with it since p parameters are estimated in the regression model. The residual mean square is

$$MS_E = \frac{SS_E}{n-p} \tag{4.17}$$

We can show that the expected value of MS_E is σ^2, so an unbiased estimator of σ^2 is given by

$$\hat{\sigma}^2 = MS_E \tag{4.18}$$

As noted in the simple linear regression case, this estimator of σ^2 is *model-dependent*.

• **Example 4.2** We will estimate the error variance σ^2 for the multiple regression model fit to the soft drink delivery time data in Example 4.1. Since

$$\mathbf{y}'\mathbf{y} = \sum_{i=1}^{25} y_i^2 = 18{,}310.6290$$

and

$$\hat{\boldsymbol{\beta}}'\mathbf{X}'\mathbf{y} = [2.34123115 \quad 1.61590721 \quad 0.01438483]\begin{bmatrix} 559.60 \\ 7{,}375.44 \\ 337{,}072.00 \end{bmatrix}$$

$$= 18{,}076.90304$$

the residual sum of squares is

$$\begin{aligned} SS_E &= \mathbf{y}'\mathbf{y} - \hat{\boldsymbol{\beta}}'\mathbf{X}'\mathbf{y} \\ &= 18{,}310.6290 - 18{,}076.9030 \\ &= 233.7260 \end{aligned}$$

Therefore, the estimate of σ^2 is the residual mean square

$$\hat{\sigma}^2 = \frac{SS_E}{n-p} = \frac{233.7260}{25-3} = 10.6239$$

The model-dependent nature of this estimate σ^2 may be easily demonstrated. Figure 2.10 (page 51) displays the computer output from a least squares fit to the delivery time data using only one regressor, cases (x_1). The residual mean square for this model is 17.4841, which is considerably larger than the result obtained above for the two-regressor model. Which estimate is "correct"? Both estimates are in a sense correct, but they depend heavily on the choice of model. Perhaps a better question is which *model* is "correct"? Since σ^2 is the variance of the errors (the unexplained noise about the regression line), we would usually prefer a model with a small residual mean square to a model with a large one. •

4.2.5 Inadequacy of scatter diagrams in multiple regression

We saw in Chapter 2 that the scatter diagram is an important tool in analyzing the relationship between y and x in simple linear regression. It is tempting to conclude that this concept can be generalized to multiple regression, so that examining y versus x_1, y versus $x_2, \ldots$, y versus x_k plots would be useful in assessing the relationships between y and each of the regressors $x_1, x_2, \ldots, x_k$. Unfortunately, this is not true in general.

Following Daniel and Wood [1980], we illustrate the inadequacy of scatter diagrams for a problem with two regressors. Consider the data shown below:

y	x_1	x_2
10	2	1
17	3	2
48	4	5
27	1	2
55	5	6
26	6	4
9	7	3
16	8	4

These data were generated from the equation

$$y = 8 - 5x_1 + 12x_2$$

The scatter diagrams y versus x_1 and y versus x_2 are shown in Figure 4.3. The y versus x_1 plot does not exhibit any apparent relationship between the two variables. The y versus x_2 plot indicates that a linear relationship exists, with a slope of approximately 8. Note that both scatter diagrams convey erroneous information. Since in this data set there are two pairs of points that have the same x_2 values ($x_2 = 2$ and $x_2 = 4$), we could measure the x_1 effect at fixed x_2 from both pairs. This gives, for $x_2 = 2$, $\hat{\beta}_1 = (17 - 27)/(3 - 1) = -5$, and for $x_2 = 4$, $\hat{\beta}_1 = (26 - 16)/(6 - 8) = -5$, the correct results. Knowing $\hat{\beta}_1$, we could now estimate the x_2 effect. This procedure is not generally useful however, because many data sets do not have duplicate points.

This example illustrates that constructing scatter diagrams of y versus $x_j (j = 1, 2, \ldots, k)$ can be misleading, even in the case of only two regressors operating in a perfectly additive fashion, with no noise. A more realistic regression situation with several regressors and error in the y's would confuse the situation even further. If there is only one dominant regressor, the corresponding scatter diagram will generally reveal this. However, when several regressors are important, or when the regressors are themselves

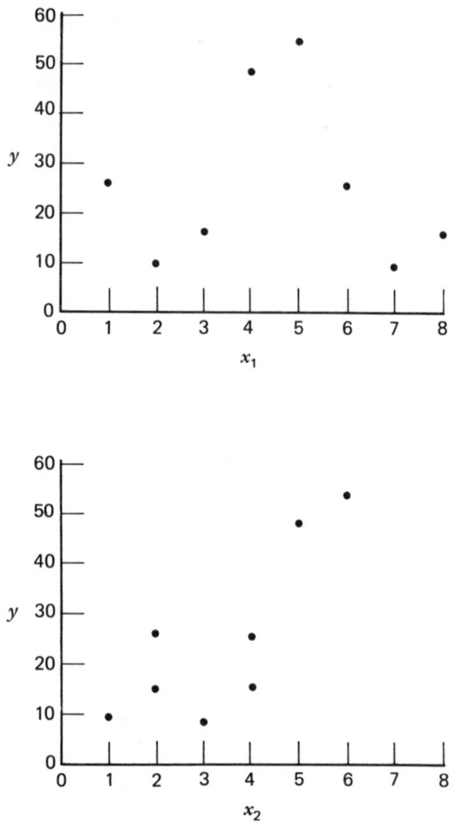

Figure 4.3 Scatter diagrams y versus x_1 and y versus x_2.

correlated, then these scatter diagrams are almost useless. Analytical methods for sorting out the relationships between several regressors and a response are discussed in Chapter 7.

4.3 Confidence intervals in multiple regression

4.3.1 Confidence intervals on the regression coefficients

To construct confidence interval estimates for the regression coefficients β_j, we must assume that the errors ε_i are normally and independently distributed with mean zero and variance σ^2. Therefore the observations y_i are normally and independently distributed with mean $\beta_0+\Sigma_{j=1}^{k}\beta_j x_{ij}$ and variance σ^2. Since the least squares estimator $\hat{\boldsymbol{\beta}}$ is a linear combination of the observations, it follows that $\hat{\boldsymbol{\beta}}$ is normally distributed with mean vector $\boldsymbol{\beta}$ and covariance matrix $\sigma^2(\mathbf{X}'\mathbf{X})^{-1}$. This implies that the marginal distribution of any regression coefficient $\hat{\beta}_j$ is normal with mean β_j and variance $\sigma^2 C_{jj}$, where C_{jj} is the jth diagonal element of the $(\mathbf{X}'\mathbf{X})^{-1}$ matrix. Consequently, each of the statistics

$$\frac{\hat{\beta}_j-\beta_j}{\sqrt{\hat{\sigma}^2 C_{jj}}}, \qquad j=0,1,\ldots,k \tag{4.19}$$

is distributed as t with $n-p$ degrees of freedom, where $\hat{\sigma}^2$ is the estimate of the error variance obtained from (4.18). Therefore, a $100(1-\alpha)$ percent confidence interval for the regression coefficient β_j, $j=0,1,\ldots,k$ is

$$\hat{\beta}_j-t_{\alpha/2,\,n-p}\sqrt{\hat{\sigma}^2 C_{jj}}\leqslant\beta_j\leqslant\hat{\beta}_j+t_{\alpha/2,\,n-p}\sqrt{\hat{\sigma}^2 C_{jj}} \tag{4.20}$$

We usually call the quantity

$$\operatorname{se}(\hat{\beta}_j)=\sqrt{\hat{\sigma}^2 C_{jj}} \tag{4.21}$$

the *standard error* of the regression coefficient $\hat{\beta}_j$.

- **Example 4.3** We will find a 95 percent confidence interval for the parameter β_1 in Example 4.1. The point estimate of β_1 is $\hat{\beta}_1=1.61591$, the diagonal element of $(\mathbf{X}'\mathbf{X})^{-1}$ corresponding to β_1 is $C_{11}=0.00274378$, and $\hat{\sigma}^2=10.6239$ (from Example 4.2). Using (4.20) we find that

$$\hat{\beta}_1-t_{.025,22}\sqrt{\hat{\sigma}^2 C_{11}}\leqslant\beta_1\leqslant\hat{\beta}_1+t_{.025,22}\sqrt{\hat{\sigma}^2 C_{11}}$$

$$1.61591-(2.074)\sqrt{(10.6239)(0.00274378)}\leqslant\beta_1$$
$$\leqslant 1.61591+(2.074)\sqrt{(10.6239)(0.00274378)}$$

$$1.61591-(2.074)(0.17073)\leqslant\beta_1\leqslant 1.61591+(2.074)(0.17073)$$

and the 95 percent confidence interval on β_1 is

$$1.26181\leqslant\beta_1\leqslant 1.97001$$

•

Although we can attach a confidence level of $100(1-\alpha)$ to each individual interval given by (4.20), the probability that all these statements are simultaneously true is not $1-\alpha$. It is possible to obtain a $100(1-\alpha)$ percent *joint confidence region* for all the parameters in $\boldsymbol{\beta}$ from

$$\frac{(\boldsymbol{\beta}-\hat{\boldsymbol{\beta}})'\mathbf{X}'\mathbf{X}(\boldsymbol{\beta}-\hat{\boldsymbol{\beta}})}{pMS_E} \leqslant F_{\alpha, p, n-p} \tag{4.22}$$

The inequality describes an elliptically shaped region. Figure 4.4 illustrates the joint confidence region and the individual confidence intervals obtained from (4.20) for $p=2$.

From Figure 4.4, we see the problem that arises when we attempt to interpret the individual confidence statements jointly. We may think that the point A provides reasonable values for β_0 and β_1, since A clearly falls within the rectangle formed by the two individual confidence statements. However, A is outside of the *joint* confidence region, so A cannot provide reasonable values for β_0 and β_1. Generally the individual confidence statements (4.20) are valid only for making inferences about one parameter without considering the values of the other parameters.

It is not difficult to construct the joint confidence region for $p=2$. If more than two parameters are involved however, the problem is somewhat harder. Joint inferences on regression models will be discussed further in Chapter 9.

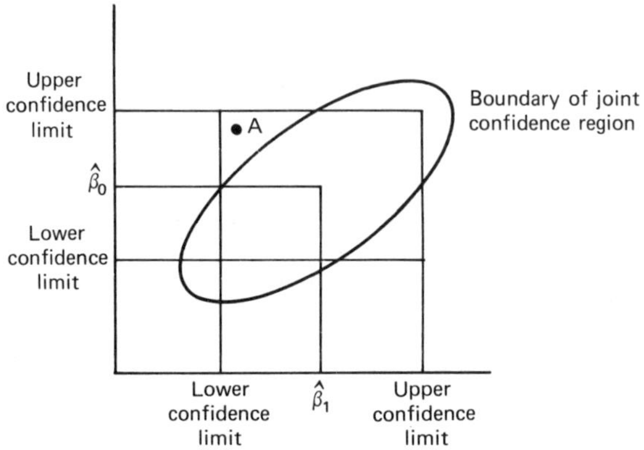

Figure 4.4 Comparison of the joint confidence region and individual confidence intervals for $p=2$.

4.3.2 Confidence interval estimation of the mean response

We may construct a confidence interval on the mean response at a particular point, such as $x_{01}, x_{02}, \ldots, x_{0k}$. Define the vector $\mathbf{x}_0$ as

$$\mathbf{x}_0 = \begin{bmatrix} 1 \\ x_{01} \\ x_{02} \\ \vdots \\ x_{0k} \end{bmatrix}$$

The fitted value at this point is

$$\hat{y}_0 = \mathbf{x}'_0 \hat{\boldsymbol{\beta}} \tag{4.23}$$

This is an unbiased estimator of y_0, since $E(\hat{y}_0) = \mathbf{x}'_0 \boldsymbol{\beta} = y_0$, and the variance of $\hat{y}_0$ is

$$V(\hat{y}_0) = \sigma^2 \mathbf{x}'_0 (\mathbf{X}'\mathbf{X})^{-1} \mathbf{x}_0 \tag{4.24}$$

Therefore a $100(1-\alpha)$ percent confidence interval on the mean response at the point $x_{01}, x_{02}, \ldots, x_{0k}$ is

$$\hat{y}_0 - t_{\alpha/2, n-p} \sqrt{\hat{\sigma}^2 \mathbf{x}'_0 (\mathbf{X}'\mathbf{X})^{-1} \mathbf{x}_0} \leqslant y_0 \leqslant \hat{y}_0 + t_{\alpha/2, n-p} \sqrt{\hat{\sigma}^2 \mathbf{x}'_0 (\mathbf{X}'\mathbf{X})^{-1} \mathbf{x}_0} \tag{4.25}$$

This is the multiple regression generalization of (2.42).

- **Example 4.4** The soft drink bottler in Example 4.1 would like to construct a 95 percent confidence interval on the mean delivery time for an outlet requiring $x_1 = 8$ cases and where the distance $x_2 = 275$ feet. Therefore

$$\mathbf{x}_0 = \begin{bmatrix} 1 \\ 8 \\ 275 \end{bmatrix}$$

The fitted value at this point is found from (4.23) as

$$\hat{y}_0 = \mathbf{x}'_0 \boldsymbol{\beta} = [1 \quad 8 \quad 275] \begin{bmatrix} 2.34123 \\ 1.61591 \\ 0.01438 \end{bmatrix} = 19.22 \text{ minutes}$$

The variance of $\hat{y}_0$ is estimated by

$$\hat{\sigma}^2 \mathbf{x}_0'(\mathbf{X}'\mathbf{X})^{-1}\mathbf{x}_0$$

$$=10.6239[1 \quad 8 \quad 275]\begin{bmatrix} 0.11321518 & -0.00444859 & -0.00008367 \\ -0.00444859 & 0.00274378 & -0.00004786 \\ -0.00008367 & -0.00004786 & 0.00000123 \end{bmatrix}\begin{bmatrix} 1 \\ 8 \\ 275 \end{bmatrix}$$

$$=10.6239(0.05346)=0.56794$$

Therefore a 95 percent confidence interval on the mean delivery time at this point is found from (4.25) as

$$19.22-2.074\sqrt{0.56794} \leq y_0 \leq 19.22+2.074\sqrt{0.56794}$$

which reduces to

$$17.66 \leq y_0 \leq 20.78$$

- Ninety-five percent of such intervals will contain the true delivery time.

4.4 Hypothesis testing in multiple linear regression

In multiple regression problems certain tests of hypotheses about the model parameters are useful in measuring model adequacy. In this section, we will describe several important hypothesis testing procedures. We will continue to require the normality assumption on the errors introduced in the previous section.

4.4.1 Test for significance of regression

The test for significance of regression is a test to determine if there is a linear relationship between the response y and any of the regressor variables $x_1, x_2, \ldots, x_k$. The appropriate hypotheses are

$$H_0\colon \beta_1=\beta_2=\cdots=\beta_k=0$$

$$H_1\colon \beta_j \neq 0 \text{ for at least one } j \tag{4.26}$$

Rejection of $H_0\colon \beta_j=0$ implies that at least one of the regressors $x_1, x_2, \ldots, x_k$ contributes significantly to the model. The test procedure is a generalization of that used in simple linear regression. The total sum of squares S_{yy} is partitioned into a sum of squares due to regression and a residual sum of squares, for example,

$$S_{yy} = SS_R + SS_E$$

and if H_0: $\beta_j = 0$ is true, then $SS_R/\sigma^2 \sim \chi^2_k$ where the number of degrees of freedom for χ^2 are equal to the number of regressor variables in the model. Also, we can show that $SS_E/\sigma^2 \sim \chi^2_{n-k-1}$, and that SS_E and SS_R are independent. The test procedure for H_0: $\beta_j = 0$ is to compute

$$F_0 = \frac{SS_R/k}{SS_E/(n-k-1)} = \frac{MS_R}{MS_E} \tag{4.27}$$

and to reject H_0 if $F_0 > F_{\alpha, k, n-k-1}$. The procedure is usually summarized in an analysis of variance table such as Table 4.4.

A computational formula for SS_R is found by starting with

$$SS_E = \mathbf{y}'\mathbf{y} - \hat{\boldsymbol{\beta}}\mathbf{X}'\mathbf{y} \tag{4.28}$$

and since

$$S_{yy} = \sum_{i=1}^{n} y_i^2 - \frac{\left(\sum_{i=1}^{n} y_i\right)^2}{n} = \mathbf{y}'\mathbf{y} - \frac{\left(\sum_{i=1}^{n} y_i\right)^2}{n}$$

we may rewrite the above equation as

$$SS_E = \mathbf{y}'\mathbf{y} - \frac{\left(\sum_{i=1}^{n} y_i\right)^2}{n} - \left[\hat{\boldsymbol{\beta}}'\mathbf{X}'\mathbf{y} - \frac{\left(\sum_{i=1}^{n} y_i\right)^2}{n}\right] \tag{4.29}$$

Table 4.4 Analysis of variance for significance of regression in multiple regression

Source of Variation	Sum of Squares	Degrees of Freedom	Mean Square	F_0
Regression	SS_R	k	MS_R	MS_R/MS_E
Residual	SS_E	$n-k-1$	MS_E	
Total	S_{yy}	$n-1$		

or

$$SS_E = S_{yy} - SS_R$$

Therefore the regression sum of squares is

$$SS_R = \hat{\boldsymbol{\beta}}'\mathbf{X}'\mathbf{y} - \frac{\left(\sum_{i=1}^{n} y_i\right)^2}{n} \tag{4.30}$$

the residual sum of squares is

$$SS_E = \mathbf{y}'\mathbf{y} - \hat{\boldsymbol{\beta}}'\mathbf{X}'\mathbf{y} \tag{4.31}$$

and the total sum of squares is

$$S_{yy} = \mathbf{y}'\mathbf{y} - \frac{\left(\sum_{i=1}^{n} y_i\right)^2}{n} \tag{4.32}$$

- **Example 4.5** We will test for significance of regression using the delivery time data from Example 4.1. Some of the numerical quantities required are calculated in Example 4.2. Note that

$$\begin{aligned} S_{yy} &= \mathbf{y}'\mathbf{y} - \frac{\left(\sum_{i=1}^{n} y_i\right)^2}{n} \\ &= 18{,}310.6290 - \frac{(559.60)^2}{25} \\ &= 5784.5426, \end{aligned}$$

$$\begin{aligned} SS_R &= \hat{\boldsymbol{\beta}}'\mathbf{X}'\mathbf{y} - \frac{\left(\sum_{i=1}^{n} y_i\right)^2}{n} \\ &= 18{,}076.9030 - \frac{(559.60)^2}{25} \\ &= 5550.8166 \end{aligned}$$

and

$$SS_E = S_{yy} - SS_R$$
$$= \mathbf{y}'\mathbf{y} - \hat{\boldsymbol{\beta}}'\mathbf{X}'\mathbf{y}$$
$$= 233.7260$$

The analysis of variance is shown in Table 4.5. To test H_0: $\beta_1 = \beta_2 = 0$, we calculate the statistic

$$F_0 = \frac{MS_R}{MS_E} = \frac{2775.4083}{10.6239} = 261.24$$

Since $F_0 > F_{.05,2,22} = 3.44$, we conclude that delivery time is related to delivery volume and/or distance. However this does not necessarily imply that the relationship found is an appropriate one for predicting delivery time as a function of volume and distance. Further tests of model adequacy are required. •

4.4.2 Tests on individual regression coefficients

We are frequently interested in testing hypotheses on the individual regression coefficients. These tests are helpful in determining the value of each of the regressors in the model. For example, the model might be more effective with the inclusion of additional regressors or perhaps with the deletion of one or more regressors presently in the model.

Adding a variable to a regression model always causes the sum of squares for regression to increase and the residual sum of squares to decrease. We must decide whether the increase in the regression sum of squares is sufficient to warrant using the additional regressor in the model. The addition of a regressor also increases the variance of the fitted value $\hat{y}$, so we must be careful to include only regressors that are of real value in explaining the response. Furthermore adding an unimportant regressor may increase the residual mean square, which may decrease the usefulness of the model.

Table 4.5 Test for significance of regression for Example 4.5

Source of Variation	Sum of Squares	Degrees of Freedom	Mean Square	F_0
Regression	5550.8166	2	2775.4083	261.24
Residual	233.7260	22	10.6239	
Total	5784.5426	24		

The hypotheses for testing the significance of any individual regression coefficient, such as β_j, are

$$H_0\colon \beta_j = 0$$

$$H_0\colon \beta_j \neq 0 \tag{4.33}$$

If $H_0\colon \beta_j = 0$ is not rejected, then this indicates that the regressor x_j can be deleted from the model. The test statistic for this hypothesis is

$$t_0 = \frac{\hat{\beta}_j}{\sqrt{\hat{\sigma}^2 C_{jj}}} = \frac{\hat{\beta}_j}{\operatorname{se}(\hat{\beta}_j)} \tag{4.34}$$

where C_{jj} is the diagonal element of $(\mathbf{X}'\mathbf{X})^{-1}$ corresponding to $\hat{\beta}_j$. The null hypothesis $H_0\colon \beta_j = 0$ is rejected if $|t_0| > t_{\alpha/2, n-k-1}$. Note that this is really a partial or marginal test, because the regression coefficient $\hat{\beta}_j$ depends on all the other regressor variables $x_i (i \neq j)$ that are in the model. Thus this is a test of the contribution of x_j given the other regressors in the model.

• **Example 4.6** To illustrate the procedure, consider Example 4.1. Suppose we wish to assess the value of the regressor variable "distance" (x_2) given that the regressor "cases" (x_1) is in the model. The hypotheses are

$$H_0\colon \beta_2 = 0$$

$$H_1\colon \beta_2 \neq 0$$

The main diagonal element of $(\mathbf{X}'\mathbf{X})^{-1}$ corresponding to β_2 is $C_{22} =$ 0.00000123, so the t-statistic (4.34) becomes

$$t_0 = \frac{\hat{\beta}_2}{\sqrt{\hat{\sigma}^2 C_{22}}} = \frac{0.01438}{\sqrt{(10.6239)(0.00000123)}} = 3.98$$

Since $t_{.025, 22} = 2.074$, we reject $H_0\colon \beta_2 = 0$ and conclude that the regressor "distance" or x_2 contributes significantly to the model, given that "cases" or x_1 is also in the model. •

We may also directly determine the contribution to the regression sum of squares of a regressor, for example x_j, given that other regressors $x_i (i \neq j)$ are included in the model by using the "extra sum of squares" method. This procedure can also be used to investigate the contribution of a *subset* of the regressor variables to the model. Consider the regression model with k

regressors

$$\mathbf{y}=\mathbf{X}\boldsymbol{\beta}+\boldsymbol{\varepsilon}$$

where $\mathbf{y}$ is $(n\times 1)$, $\mathbf{X}$ is $(n\times p)$, $\boldsymbol{\beta}$ is $(p\times 1)$, $\boldsymbol{\varepsilon}$ is $(n\times 1)$, and $p=k+1$. We would like to determine if some subset of $r<k$ regressors contributes significantly to the regression model. Let the vector of regression coefficients be partitioned as follows:

$$\boldsymbol{\beta}=\left[\frac{\boldsymbol{\beta}_1}{\boldsymbol{\beta}_2}\right]$$

where $\boldsymbol{\beta}_1$ is $(p-r)\times 1$ and $\boldsymbol{\beta}_2$ is $r\times 1$. We wish to test the hypotheses

$$H_0\colon \boldsymbol{\beta}_2=\mathbf{0}$$

$$H_0\colon \boldsymbol{\beta}_2\neq\mathbf{0} \qquad (4.35)$$

The model may be written as

$$\mathbf{y}=\mathbf{X}\boldsymbol{\beta}+\boldsymbol{\varepsilon}=\mathbf{X}_1\boldsymbol{\beta}_1+\mathbf{X}_2\boldsymbol{\beta}_2+\boldsymbol{\varepsilon} \qquad (4.36)$$

where the $n\times(p-r)$ matrix $\mathbf{X}_1$ represents the columns of $\mathbf{X}$ associated with $\boldsymbol{\beta}_1$ and the $n\times r$ matrix $\mathbf{X}_2$ represents the columns of $\mathbf{X}$ associated with $\boldsymbol{\beta}_2$. This is called the *full* model.

For the full model, we know that $\hat{\boldsymbol{\beta}}=(\mathbf{X}'\mathbf{X})^{-1}\mathbf{X}'\mathbf{y}$. The regression sum of squares for this model is

$$SS_R(\boldsymbol{\beta})=\hat{\boldsymbol{\beta}}'\mathbf{X}'\mathbf{y} \qquad (p \text{ degrees of freedom})$$

and

$$MS_E=\frac{\mathbf{y}'\mathbf{y}-\hat{\boldsymbol{\beta}}'\mathbf{X}'\mathbf{y}}{n-p}$$

To find the contribution of the terms in $\boldsymbol{\beta}_2$ to the regression, fit the model assuming that the null hypothesis H_0: $\boldsymbol{\beta}_2=\mathbf{0}$ is true. This *reduced* model is

$$\mathbf{y}=\mathbf{X}_1\boldsymbol{\beta}_1+\boldsymbol{\varepsilon} \qquad (4.37)$$

The least squares estimator of $\boldsymbol{\beta}_1$ in the reduced model is $\hat{\boldsymbol{\beta}}_1=(\mathbf{X}_1'\mathbf{X}_1)^{-1}\mathbf{X}_1'\mathbf{y}$. The regression sum of squares is

$$SS_R(\boldsymbol{\beta}_1)=\hat{\boldsymbol{\beta}}_1'\mathbf{X}_1'\mathbf{y} \qquad (p-r \text{ degrees of freedom}) \qquad (4.38)$$

The regression sum of squares due to $\boldsymbol{\beta}_2$ given that $\boldsymbol{\beta}_1$ is already in the model is

$$SS_R(\boldsymbol{\beta}_2|\boldsymbol{\beta}_1)=SS_R(\boldsymbol{\beta})-SS_R(\boldsymbol{\beta}_1) \tag{4.39}$$

with $p-(p-r)=r$ degrees of freedom. This sum of squares is called the "extra sum of squares" due to $\boldsymbol{\beta}_2$, because it measures the increase in the regression sum of squares that results from adding the regressors $x_{k-r+1}, x_{k-r+2}, \ldots, x_k$ to a model that already contains $x_1, x_2, \ldots, x_{k-r}$. Now $SS_R(\boldsymbol{\beta}_2|\boldsymbol{\beta}_1)$ is independent of MS_E and the null hypothesis $\boldsymbol{\beta}_2=\mathbf{0}$ may be tested by the statistic

$$F_0=\frac{SS_R(\boldsymbol{\beta}_2|\boldsymbol{\beta}_1)/r}{MS_E} \tag{4.40}$$

If $F_0>F_{\alpha, r, n-p}$ we reject H_0, concluding that at least one of the parameters in $\boldsymbol{\beta}_2$ is not zero, and consequently at least one of the regressors $x_{k-r+1}, x_{k-r+2}, \ldots, x_k$ in $\mathbf{X}_2$ contributes significantly to the regression model.

Some authors call the test in (4.40) a *partial* F-test, because it measures the contribution of the regressors in $\mathbf{X}_2$ given that the other regressors in $\mathbf{X}_1$ are in the model. To illustrate the usefulness of this procedure, consider the model

$$y=\beta_0+\beta_1 x_1+\beta_2 x_2+\beta_3 x_3+\varepsilon$$

The sums of squares

$$SS_R(\beta_1|\beta_0, \beta_2, \beta_3)$$
$$SS_R(\beta_2|\beta_0, \beta_1, \beta_3)$$

and

$$SS_R(\beta_3|\beta_0, \beta_1, \beta_2)$$

are single degree of freedom sums of squares that measure the contribution of each regressor x_j, $j=1,2,3$, to the model given that all the other regressors were already in the model. That is, we are assessing the value of adding x_j to a model that did not include this regressor. In general, we could find

$$SS_R(\beta_j|\beta_0, \beta_1, \ldots, \beta_{j-1}, \beta_{j+1}, \ldots, \beta_k), \qquad 1 \leqslant j \leqslant k$$

which is the increase in the regression sum of squares due to adding x_j to a model that already contains $x_1, \ldots, x_{j-1}, x_{j+1}, \ldots, x_k$. Some find it helpful to think of this as measuring the contribution of x_j as if it were the *last* variable added to the model.

We can show that the partial F-test on a single variable x_j is equivalent to the t-test in (4.34). However, the partial F-test is a more general procedure in that we can measure the effect of *sets* of variables. In Chapter 7 we will show how the partial F-test plays a major role in *model building*; that is, in searching for the best set of regressors to use in the model. .

The extra sum of squares method can be used to test hypotheses about any subset of regressor variables that seems reasonable for the particular problem under analysis. Sometimes we find that there is a natural hierarchy or ordering in the regressors, and this forms the basis of a test. For example, consider the quadratic polynomial

$$y=\beta_0+\beta_1 x+\beta_2 x^2+\varepsilon$$

Here we might be interested in finding

$$SS_R(\beta_1|\beta_0)$$

which would measure the linear effect of x, and

$$SS_R(\beta_2|\beta_0,\beta_1)$$

which would measure the contribution of adding a quadratic term to a model that already contained a linear term. When we think of adding regressors one at a time to a model and examining the contribution of the regressor added at each step given all regressors added previously, we can partition the regression sum of squares into marginal single degree of freedom components. For example, consider the model

$$y=\beta_0+\beta_1 x_1+\beta_2 x_2+\beta_3 x_3+\varepsilon$$

with the corresponding analysis of variance identity

$$S_{yy}=SS_R(\beta_1,\beta_2,\beta_3|\beta_0)+SS_E$$

We may decompose the three degree of freedom regression sum of squares as follows:

$$SS_R(\beta_1,\beta_2,\beta_3|\beta_0)=SS_R(\beta_1|\beta_0)+SS_R(\beta_2|\beta_1,\beta_0)+SS_R(\beta_3|\beta_1,\beta_2,\beta_0)$$

where each sum of squares on the right-hand side has one degree of freedom. Note that the order of the regressors in these marginal components is arbitrary. An alternate partitioning of $SS_R(\beta_1, \beta_2, \beta_3|\beta_0)$ is

$$SS_R(\beta_1,\beta_2,\beta_3|\beta_0)=SS_R(\beta_2|\beta_0)+SS_R(\beta_1|\beta_2,\beta_0)+SS_R(\beta_3|\beta_1,\beta_2,\beta_0)$$

However, the extra sum of squares method does not always produce a partitioning of the regression sum of squares, since in general

$$SS_R(\beta_1,\beta_2,\beta_3|\beta_0)\neq SS_R(\beta_1|\beta_2,\beta_3,\beta_0)+SS_R(\beta_2|\beta_1,\beta_3,\beta_0)$$
$$+SS_R(\beta_3|\beta_1,\beta_2,\beta_0)$$

• **Example 4.7** Consider the soft drink delivery time data in Example 4.1. Suppose that we wish to investigate the contribution of the variable distance (x_2) to the model. The appropriate hypotheses are

$$H_0\colon \beta_2=0$$
$$H_0\colon \beta_2\neq 0$$

To test these hypotheses, we need the extra sum of squares due to β_2, or

$$SS_R(\beta_2|\beta_1,\beta_0)=SS_R(\beta_1,\beta_2,\beta_0)-SS_R(\beta_1,\beta_0)$$
$$=SS_R(\beta_1,\beta_2|\beta_0)-SS_R(\beta_1|\beta_0)$$

From Example 4.5 we know that

$$SS_R(\beta_1,\beta_2|\beta_0)=\hat{\boldsymbol{\beta}}'\mathbf{X}'\mathbf{y}-\frac{\left(\sum_{i=1}^{n} y_i\right)^2}{n}=5550.8166 \qquad \text{(2 degrees of freedom)}$$

The reduced model $y=\beta_0+\beta_1 x_1+\varepsilon$ was fitted in Example 2.8, giving $\hat{y}=3.3208+2.1762x_1$. The regression sum of squares for this model is

$$SS_R(\beta_1|\beta_0)=\hat{\beta}_1 S_{xy}=(2.1762)(243.3440)=5382.4088$$
$$\text{(1 degree of freedom)}$$

Therefore we have

$$SS_R(\beta_2|\beta_1,\beta_0)=5550.8166-5382.4088=168.4078$$
$$\text{(1 degree of freedom)}$$

This is the increase in the regression sum of squares that results from adding x_2 to a model already containing x_1. To test H_0: $\beta_2=0$, form the test statistic

$$F_0=\frac{SS_R(\beta_2|\beta_1,\beta_0)/1}{MS_E}=\frac{168.4078/1}{10.6239}=15.85$$

Note that the MS_E from the *full* model using both x_1 and x_2 is used in the denominator of the test statistic. Since $F_{.05,1,22}=4.30$, we reject H_0: $\beta_2=0$ and conclude that distance (x_2) contributes significantly to the model.

Since this partial F-test involves a single variable, it is equivalent to the t-test. To see this, recall that the t-test on H_0: $\beta_2=0$ resulted in the test statistic $t_0=3.98$, and since the square of a t random variable with ν degrees of freedom is an F random variable with one and ν degrees of freedom, we have $t_0^2=(3.98)^2=15.84\simeq F_0$. •

4.4.3 Special case of orthogonal columns in X

Consider the model (4.36)

$$\begin{aligned}\mathbf{y}&=\mathbf{X}\boldsymbol{\beta}+\boldsymbol{\varepsilon}\\&=\mathbf{X}_1\boldsymbol{\beta}_1+\mathbf{X}_2\boldsymbol{\beta}_2+\boldsymbol{\varepsilon}\end{aligned}$$

The extra sum of squares method allows us to measure the effect of the regressors in $\mathbf{X}_2$ conditional on those in $\mathbf{X}_1$ by computing $SS_R(\boldsymbol{\beta}_2|\boldsymbol{\beta}_1)$. In general, we cannot talk about finding the sum of squares due to $\boldsymbol{\beta}_2$, $SS_R(\boldsymbol{\beta}_2)$, without accounting for the dependency of this quantity on the regressors in $\mathbf{X}_1$. However, if the columns in $\mathbf{X}_1$ are *orthogonal* to the columns in $\mathbf{X}_2$, we can determine a sum of squares due to $\boldsymbol{\beta}_2$ that is free of any dependency on the regressors in $\mathbf{X}_2$.

To demonstrate this form the normal equations $(\mathbf{X}'\mathbf{X})\hat{\boldsymbol{\beta}}=\mathbf{X}'\mathbf{y}$ for the model (4.36). The normal equations are

$$\begin{bmatrix}\mathbf{X}_1'\mathbf{X}_1 & \mathbf{X}_1'\mathbf{X}_2\\ \mathbf{X}_2'\mathbf{X}_1 & \mathbf{X}_2'\mathbf{X}_2\end{bmatrix}\begin{bmatrix}\hat{\boldsymbol{\beta}}_1\\ \hat{\boldsymbol{\beta}}_2\end{bmatrix}=\begin{bmatrix}\mathbf{X}_1'\mathbf{y}\\ \mathbf{X}_2'\mathbf{y}\end{bmatrix}$$

Now if the columns of $\mathbf{X}_1$ are orthogonal to the columns in $\mathbf{X}_2$, $\mathbf{X}_1'\mathbf{X}_2=\mathbf{0}$ and $\mathbf{X}_2'\mathbf{X}_1=\mathbf{0}$. Then the normal equations become

$$\begin{aligned}\mathbf{X}_1'\mathbf{X}_1\hat{\boldsymbol{\beta}}_1&=\mathbf{X}_1'\mathbf{y}\\ \mathbf{X}_2'\mathbf{X}_2\hat{\boldsymbol{\beta}}_2&=\mathbf{X}_2'\mathbf{y}\end{aligned}$$

with solution

$$\hat{\boldsymbol{\beta}}_1 = (\mathbf{X}_1'\mathbf{X}_1)^{-1}\mathbf{X}_1'\mathbf{y}$$

$$\hat{\boldsymbol{\beta}}_2 = (\mathbf{X}_2'\mathbf{X}_2)^{-1}\mathbf{X}_2'\mathbf{y}$$

Note that the least squares estimator of $\boldsymbol{\beta}_1$ is $\hat{\boldsymbol{\beta}}_1$, regardless of whether or not $\mathbf{X}_2$ is in the model, and the least squares estimator of $\boldsymbol{\beta}_2$ is $\hat{\boldsymbol{\beta}}_2$, regardless of whether or not $\mathbf{X}_1$ is in the model.

The regression sum of squares for the full model is

$$\begin{aligned} SS_R(\boldsymbol{\beta}) &= \hat{\boldsymbol{\beta}}'\mathbf{X}'\mathbf{y} \\ &= [\hat{\boldsymbol{\beta}}_1, \hat{\boldsymbol{\beta}}_2]\begin{bmatrix} \mathbf{X}_1'\mathbf{y} \\ \mathbf{X}_2'\mathbf{y} \end{bmatrix} \\ &= \hat{\boldsymbol{\beta}}'\mathbf{X}_1'\mathbf{y} + \boldsymbol{\beta}_2'\mathbf{X}_2'\mathbf{y} \\ &= \mathbf{y}'\mathbf{X}_1(\mathbf{X}_1'\mathbf{X}_1)^{-1}\mathbf{X}_1'\mathbf{y} + \mathbf{y}'\mathbf{X}_2(\mathbf{X}_2'\mathbf{X}_2)^{-1}\mathbf{X}_2'\mathbf{y} \end{aligned} \tag{4.41}$$

However, the normal equations form two sets, and for each set we note that

$$\begin{aligned} SS_R(\boldsymbol{\beta}_1) &= \hat{\boldsymbol{\beta}}_1'\mathbf{X}_1'\mathbf{y} = \mathbf{y}'\mathbf{X}_1(\mathbf{X}_1'\mathbf{X}_1)^{-1}\mathbf{X}_1'\mathbf{y} \\ SS_R(\boldsymbol{\beta}_2) &= \hat{\boldsymbol{\beta}}_2'\mathbf{X}_2'\mathbf{y} = \mathbf{y}'\mathbf{X}_2(\mathbf{X}_2'\mathbf{X}_2)^{-1}\mathbf{X}_2'\mathbf{y} \end{aligned} \tag{4.42}$$

Comparing (4.42) with (4.41), we see that

$$SS_R(\boldsymbol{\beta}) = SS_R(\boldsymbol{\beta}_1) + SS_R(\boldsymbol{\beta}_2) \tag{4.43}$$

Therefore

$$SS_R(\boldsymbol{\beta}_1|\boldsymbol{\beta}_2) = SS_R(\boldsymbol{\beta}) - SS_R(\boldsymbol{\beta}_2) \equiv SS_R(\boldsymbol{\beta}_1)$$

and

$$SS_R(\boldsymbol{\beta}_2|\boldsymbol{\beta}_1) = SS_R(\boldsymbol{\beta}) - SS_R(\boldsymbol{\beta}_1) \equiv SS_R(\boldsymbol{\beta}_2)$$

Consequently, $SS_R(\boldsymbol{\beta}_1)$ measures the contribution of the regressors in $\mathbf{X}_1$ to the model *unconditionally*, and $SS_R(\boldsymbol{\beta}_2)$ measures the contribution of the regressors in $\mathbf{X}_2$ to the model *unconditionally*. Because we can unambiguously determine the effect of each regressor when the regressors are orthogonal, data collection experiments are often designed to have orthogonal variables.

As an example of a regression model with orthogonal regressors, consider the model $y=\beta_0+\beta_1x_1+\beta_2x_2+\beta_3x_3+\varepsilon$, where the $\mathbf{X}$ matrix is

$$\mathbf{X}=\begin{array}{c}\begin{matrix}\beta_0 & \beta_1 & \beta_2 & \beta_3\end{matrix}\\ \begin{bmatrix}1 & -1 & -1 & -1\\ 1 & 1 & -1 & -1\\ 1 & -1 & 1 & -1\\ 1 & -1 & -1 & 1\\ 1 & 1 & 1 & -1\\ 1 & 1 & -1 & 1\\ 1 & -1 & 1 & 1\\ 1 & 1 & 1 & 1\end{bmatrix}\end{array}$$

The levels of the regressors correspond to the 2^3 factorial design. It is easy to see that the columns of $\mathbf{X}$ are orthogonal. Thus $SS_R(\beta_j)$, $j=1,2,3$, measures the contribution of the regressor x_j to the model regardless of whether any of the other regressors are included in the fit.

4.4.4 Testing the general linear hypothesis $\mathbf{T}\boldsymbol{\beta}=\mathbf{0}$

Many hypotheses about regression coefficients can be tested using a unified approach. The extra sum of squares method is a special case of this procedure. In the more general procedure, the sum of squares used to test the hypothesis is usually calculated as the difference between two residual sums of squares. We will now outline the procedure. For proofs and further discussion, refer to Graybill [1976], Searle [1971], or Seber [1977].

Suppose that the hypothesis of interest can be expressed as H_0: $\mathbf{T}\boldsymbol{\beta}=\mathbf{0}$, where $\mathbf{T}$ is an $m\times p$ matrix of constants, such that only r of the m equations in $\mathbf{T}\boldsymbol{\beta}=\mathbf{0}$ are independent. The full model is $\mathbf{y}=\mathbf{X}\boldsymbol{\beta}+\boldsymbol{\varepsilon}$, with $\hat{\boldsymbol{\beta}}=(\mathbf{X}'\mathbf{X})^{-1}\mathbf{X}'\mathbf{y}$, and the residual sum of squares for the full model is

$$SS_E(FM)=\mathbf{y}'\mathbf{y}-\hat{\boldsymbol{\beta}}'\mathbf{X}'\mathbf{y} \qquad (n-p \text{ degrees of freedom})$$

To obtain the reduced model, the r independent equations in $\mathbf{T}\boldsymbol{\beta}=\mathbf{0}$ are used to solve for r of the regression coefficients in the full model in terms of the remaining $p-r$ regression coefficients. This leads to the reduced model $\mathbf{y}=\mathbf{Z}\boldsymbol{\gamma}+\boldsymbol{\varepsilon}$, for example, where $\mathbf{Z}$ is an $n\times(p-r)$ matrix and $\boldsymbol{\gamma}$ is a $(p-r)\times 1$ vector of unknown regression coefficients. The estimate of $\boldsymbol{\gamma}$ is

$$\hat{\boldsymbol{\gamma}}=(\mathbf{Z}'\mathbf{Z})^{-1}\mathbf{Z}'\mathbf{y}$$

and the residual sum of squares for the reduced model is

$$SS_E(RM) = \mathbf{y}'\mathbf{y} - \hat{\boldsymbol{\gamma}}'\mathbf{Z}'\mathbf{y} \; (n-p+r \text{ degrees of freedom})$$

The reduced model contains fewer parameters than the full model, so consequently $SS_E(RM) \geq SS_E(FM)$. To test the hypothesis H_0: $\mathbf{T}\boldsymbol{\beta} = \mathbf{0}$, we use the difference in residual sums of squares

$$SS_H = SS_E(RM) - SS_E(FM) \tag{4.44}$$

with $n-p+r-(n-p) = r$ degrees of freedom. SS_H is called the sum of squares due to the hypothesis H_0: $\mathbf{T}\boldsymbol{\beta} = \mathbf{0}$. The test statistic for this hypothesis is

$$F_0 = \frac{SS_H/r}{SS_E(FM)/(n-p)} \tag{4.45}$$

We reject H_0: $\mathbf{T}\boldsymbol{\beta} = \mathbf{0}$ if $F_0 > F_{\alpha, r, n-p}$.

We now give two examples. Consider the model

$$y = \beta_0 + \beta_1 x_1 + \beta_2 x_2 + \beta_3 x_3 + \varepsilon \tag{4.46}$$

For the full model, $SS_E(FM)$ has $n-p = n-4$ degrees of freedom. We wish to test H_0: $\beta_1 = \beta_3$. This hypothesis may be stated as H_0: $\mathbf{T}\boldsymbol{\beta} = \mathbf{0}$, where

$$\mathbf{T} = [0, 1, 0, -1]$$

is a (1×4) row vector. There is only one equation in $\mathbf{T}\boldsymbol{\beta} = \mathbf{0}$, namely $\beta_1 - \beta_3 = 0$. Substituting this equation into the full model gives the reduced model

$$\begin{aligned} y &= \beta_0 + \beta_1 x_1 + \beta_2 x_2 + \beta_1 x_3 + \varepsilon \\ &= \beta_0 + \beta_1 (x_1 + x_3) + \beta_2 x_2 + \varepsilon \\ &= \gamma_0 + \gamma_1 z_1 + \gamma_2 z_2 + \varepsilon \end{aligned}$$

where $\gamma_0 = \beta_0$, $\gamma_1 = \beta_1 (= \beta_3)$, $z_1 = x_1 + x_3$, $\gamma_2 = \beta_2$, and $z_2 = x_2$. We would find $SS_E(RM)$ with $n-4+1 = n-3$ degrees of freedom by fitting the reduced model. The sum of squares due to hypothesis $SS_H = SS_E(RM) - SS_E(FM)$ has $n-3-(n-4) = 1$ degree of freedom. The F–ratio (4.45) is $F_0 = (SS_H/1)/[SS_E(FM)/(n-4)]$. Note that this hypothesis could also be tested by using the t-statistic

$$t_0=\frac{\hat{\beta}_1-\hat{\beta}_3}{\operatorname{se}(\hat{\beta}_1-\hat{\beta}_3)}=\frac{\hat{\beta}_1-\hat{\beta}_3}{\sqrt{\hat{\sigma}^2(C_{11}+C_{33}-2C_{13})}}$$

with $n-4$ degrees of freedom. This is equivalent to the F-test.

As a second example, suppose that for the model (4.46) we wish to test H_0: $\beta_1=\beta_3, \beta_2=0$. To state this in the form H_0: $\mathbf{T\beta}=\mathbf{0}$, write

$$\mathbf{T}=\begin{bmatrix}0 & 1 & 0 & -1\\ 0 & 0 & 1 & 0\end{bmatrix}$$

There are now two equations in $\mathbf{T\beta}=\mathbf{0}$, $\beta_1-\beta_3=0$ and $\beta_2=0$. These equations give the reduced model

$$\begin{aligned}y&=\beta_0+\beta_1x_1+\beta_1x_3+\varepsilon\\&=\beta_0+\beta_1(x_1+x_3)+\varepsilon\\&=\gamma_0+\gamma_1z_1+\varepsilon\end{aligned}$$

In this example, $SS_E(RM)$ has $n-2$ degrees of freedom, so SS_H has $n-2-(n-4)=2$ degrees of freedom. The F-ratio (4.45) is $F_0=(SS_H/2)/[SS_E(FM)/(n-4)]$.

4.5 Prediction of new observations

The regression model can be used to predict future observations on y corresponding to particular values of the regressor variables, for example, $x_{01}, x_{02},\ldots, x_{0k}$. If $\mathbf{x}'_0=[1, x_{01}, x_{02},\ldots, x_{0k}]$, then a point estimate of the future observation y_0 at the point $x_{01}, x_{02},\ldots, x_{0k}$ is

$$\hat{y}_0=\mathbf{x}'_0\hat{\boldsymbol{\beta}} \tag{4.47}$$

A $100(1-\alpha)$ percent prediction interval for this future observation is

$$\hat{y}_0-t_{\alpha/2,n-p}\sqrt{\hat{\sigma}^2\left(1+\mathbf{x}'_0(\mathbf{X}'\mathbf{X})^{-1}\mathbf{x}_0\right)}\leq y_0 \leq\hat{y}_0+t_{\alpha/2,n-p}\sqrt{\hat{\sigma}^2\left(1+\mathbf{x}'_0(\mathbf{X}'\mathbf{X})^{-1}\mathbf{x}_0\right)} \tag{4.48}$$

This is a generalization of the prediction interval for a future observation in simple linear regression, (2.44).

- **Example 4.8** Suppose that the soft drink bottler in Example 4.1 wishes to construct a 95 percent prediction interval on the delivery time at an outlet where $x_1 = 8$ cases are delivered and the distance walked by the deliveryman is $x_2 = 275$ feet. Note that $\mathbf{x}_0' = [1, \ 8, \ 275]$, and the point estimate of the delivery time is $\hat{y}_0 = \mathbf{x}_0'\hat{\boldsymbol{\beta}} = 19.22$ minutes. Also, in Example 4.4 we calculated $\mathbf{x}_0'(\mathbf{X}'\mathbf{X})^{-1}\mathbf{x}_0 = 0.05346$. Therefore from (4.48) we have

$$19.22 - 2.074\sqrt{10.6239(1+0.05346)} \leq y_0 \leq 19.22 + 2.074\sqrt{10.6239(1+0.05346)},$$

and the 95 percent prediction interval is

$$12.28 \leq y_0 \leq 26.16$$

4.6 Hidden extrapolation

In predicting new responses and in estimating the mean response at a given point $x_{01}, x_{02}, \ldots, x_{0k}$ one must be careful about extrapolating beyond the region containing the original observations. It is very possible that a model that fits well in the region of the original data will perform poorly outside that region. In multiple regression it is easy to inadvertently extrapolate, since the levels of the regressors $(x_{i1}, x_{i2}, \ldots, x_{ik})$, $i = 1, 2, \ldots, n$ jointly define the region containing the data. As an example, consider Figure 4.5, which

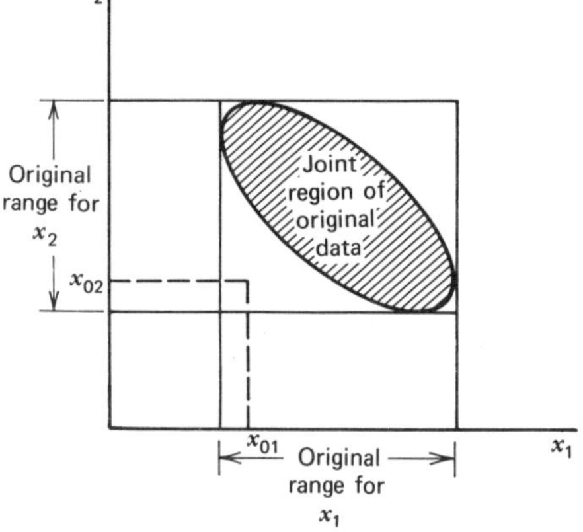

Figure 4.5 An example of extrapolation in multiple regression.

illustrates the region containing the original data for a two–regressor model. Note that the point (x_{01}, x_{02}) lies within the ranges of both regressors x_1 and x_2, but it is outside the region of the original data. Thus either predicting the value of a new observation or estimating the mean response at this point is an extrapolation of the original regression model.

Since simply comparing the levels of the x's for a new data point with the ranges of the original x's will not always detect a hidden extrapolation, it would be helpful to have a formal procedure to do so. We will define the smallest convex set containing all of the original n data points $(x_{i1}, x_{i2}, \ldots, x_{ik})$, $i=1,2,\ldots,n$ as the regressor variable hull (RVH). If a point $x_{01}, x_{02}, \ldots, x_{0k}$ lies inside or on the boundary of the RVH, then prediction or estimation involves interpolation, while if this point lies outside the RVH extrapolation is required.

The diagonal elements h_{ii} of the hat matrix $\mathbf{H}=\mathbf{X}(\mathbf{X}'\mathbf{X})^{-1}\mathbf{X}'$ are useful in detecting hidden extrapolation. The values of h_{ii} depend both on the Euclidean distance of the point $\mathbf{x}_i$ from the centroid and on the density of the points in the RVH. In general the point that has the largest value of h_{ii}, say $h_{\max}$, will lie on the boundary of the RVH in a region of the x–space where the density of the observations is relatively low. The set of points $\mathbf{x}$ (not necessarily data points used to fit the model) that satisfy

$$\mathbf{x}'(\mathbf{X}'\mathbf{X})^{-1}\mathbf{x} \leqslant h_{\max}$$

is an ellipsoid enclosing all points inside the RVH (see Cook [1979] and Weisberg [1980]). Thus if we are interested in prediction or estimation at the point $\mathbf{x}_0'=[1, x_{01}, x_{02}, \ldots, x_{0k}]$, the location of that point relative to the RVH is reflected by

$$h_{00} = \mathbf{x}_0'(\mathbf{X}'\mathbf{X})^{-1}\mathbf{x}_0$$

Points for which $h_{00} > h_{\max}$ are outside the ellipsoid enclosing the RVH, and are extrapolation points. However, if $h_{00} \leqslant h_{\max}$ then the point is inside the ellipsoid and possibly inside the RVH, and would be considered an interpolation point because it is close to the cloud of points used to fit the model. Generally the smaller the value of h_{00}, the closer the point $\mathbf{x}_0$ lies to the centroid of the x-space.*

*If $h_{\max}$ is much larger than the next largest value, the point is a severe outlier in x–space. The presence of such an outlier may make the ellipsoid much larger than desirable. In these cases one could use the second-largest value of h_{ii} as $h_{\max}$. This approach may be useful when the most remote point has been severely downweighted, say by the robust fitting techniques discussed in Chapter 9.

Weisberg [1980] notes that this procedure does not produce the smallest volume ellipsoid containing the RVH. This is called the minimum covering ellipsoid (MCE). He gives an iterative algorithm for generating the MCE. However, the test for extrapolation based on the MCE is still only an approximation, as there may still be regions inside the MCE where there are no sample points.

- **Example 4.9** We will illustrate detecting hidden extrapolation using the soft drink delivery time data in Example 4.1. The values of h_{ii} for the 25 data points are shown in Table 4.6. Note that Observation 9 has the largest value of h_{ii}. Since this data set has only two regressors, we may also graphically examine the disposition of the data in the x space. Figure 4.6 plots x_1 (cases)

Table 4.6 Values of h_{ii} for the delivery time data

Observation, i	Cases x_{i1}	Distance x_{i2}	h_{ii}
1	7	560	0.10180
2	3	220	0.07070
3	3	340	0.09874
4	4	80	0.08538
5	6	150	0.07501
6	7	330	0.04287
7	2	110	0.08180
8	7	210	0.06373
9	30	1460	$0.49829 = h_{max}$
10	5	605	0.19630
11	16	688	0.08613
12	10	215	0.11366
13	4	255	0.06113
14	6	462	0.07824
15	9	448	0.04111
16	10	776	0.16594
17	6	200	0.05943
18	7	132	0.09626
19	3	36	0.09645
20	17	770	0.10169
21	10	140	0.16528
22	26	810	0.39158
23	9	450	0.04126
24	8	635	0.12061
25	4	150	0.06664

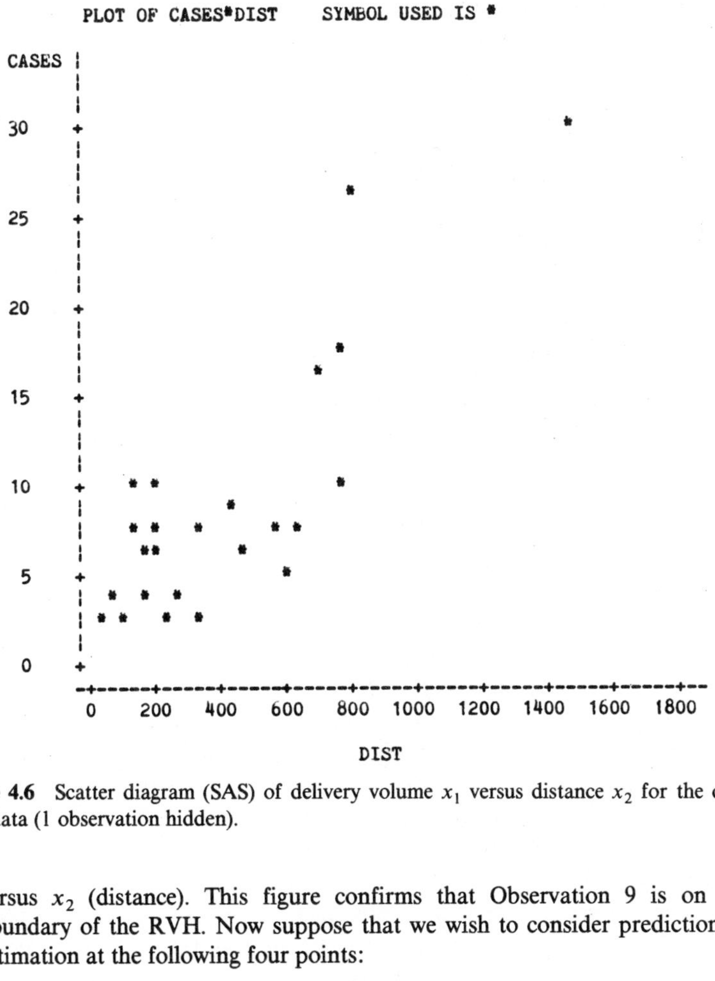

Figure 4.6 Scatter diagram (SAS) of delivery volume x_1 versus distance x_2 for the delivery time data (1 observation hidden).

versus x_2 (distance). This figure confirms that Observation 9 is on the boundary of the RVH. Now suppose that we wish to consider prediction or estimation at the following four points:

Point	x_{10}	x_{20}	h_{00}
a	8	275	0.05346
b	20	250	0.58917
c	28	500	0.89874
d	8	1200	0.86736

All these points lie within the ranges of the regressors x_1 and x_2. Point *a* (used in Examples 4.4 and 4.8 for estimation and prediction), for which $h_{00} =$

0.05346, is an interpolation point since $h_{00}=0.05346<h_{max}=0.49829$. The remaining points b, c, and d are all extrapolation points, since their values of h_{00} exceed h_{max}. This is readily confirmed by inspection of Figure 4.6. •

4.7 Measures of model adequacy in multiple regression

Evaluating model adequacy is an important part of a multiple regression problem. This section will present several methods for measuring model adequacy. Many of these techniques are extensions of those used in simple linear regression.

4.7.1 The coefficient of multiple determination

The coefficient of multiple determination R^2 is defined as

$$R^2=\frac{SS_R}{S_{yy}}=1-\frac{SS_E}{S_{yy}} \tag{4.49}$$

R^2 is a measure of the reduction in the variability of y obtained by using the regressor variables $x_1, x_2, \ldots, x_k$. As in the simple linear regression case, we must have $0 \leqslant R^2 \leqslant 1$. However, a large value of R^2 does not necessarily imply that the regression model is a good one. Adding a regressor to the model will always increase R^2, regardless of whether or not the additional regressor contributes to the model. Thus it is possible for models that have large values of R^2 to perform poorly in prediction or estimation.

The positive square root of R^2 is the multiple correlation coefficient between y and the set of regressor variables $x_1, x_2, \ldots, x_k$. That is, R is a measure of the linear association between y and $x_1, x_2, \ldots, x_k$. We may also show that R^2 is the square of the correlation between the vector of observations $\mathbf{y}$ and the vector of fitted values $\hat{\mathbf{y}}$ (see Problem 4.29).

• **Example 4.10** The coefficient of multiple determination for the regression model in Example 4.1 is

$$R^2=\frac{SS_R}{S_{yy}}=\frac{5550.8166}{5784.5426}=0.9596$$

That is, about 95.96 percent of the variability in delivery time y has been explained when the two regressor variables delivery volume (x_1) and distance (x_2) are used. In Example 2.8, a model relating y to x_1 only was developed. The value of R^2 for this model is $R^2=0.9305$. Therefore adding the variable

x_2 to the model has increased R^2 from 0.9305 to 0.9596. This increase in R^2 is relatively small.

4.7.2 Residual analysis

The residuals e_i from the multiple regression model play an important role in judging model adequacy, just as they do in simple linear regression. The residual plots introduced in Chapter 3 can be applied directly to multiple regression. Specifically, we often find it instructive to plot

1. Residuals on normal probability paper
2. Residuals versus each regressor x_j, $j=1,2,\ldots,k$
3. Residuals versus fitted $\hat{y}_i$, $i=1,2,\ldots,n$
4. Residuals in time sequence (if known)

These plots are used as described in Chapter 3 to detect departures from normality, outliers, inequality of variance, and the wrong functional specification for a regressor. Either the unscaled residuals e_i or the standardized residuals $d_i=e_i/\sqrt{MS_E}$ may be plotted. Inspection of the standardized residuals is often useful in detecting outliers.

There are several other residual analysis methods useful in multiple regression. These techniques will now be briefly described.

Plot residuals against regressors omitted from the model. If there are other candidate regressors that have not been included in the model, then plotting the residuals against the levels of these regressors (assuming that they are known) could reveal any dependency of the response y on the omitted factors. Any structure in the plot of residuals versus an omitted factor indicates that incorporation of that factor may improve the model.

Partial residual plots. These plots are designed to reveal more precisely the relationship between the residuals and the regressor variable x_j. Define the ith partial residual for regressor x_j as

$$\begin{aligned} e_{ij}^* &= y_i - \hat{\beta}_1 x_{i1} - \cdots - \hat{\beta}_{j-1} x_{i,j-1} - \hat{\beta}_{j+1} x_{i,j+1} - \cdots - \hat{\beta}_k x_{ik} \\ &= e_i + \hat{\beta}_j x_{ij}, \quad i=1,2,\ldots,n \end{aligned} \tag{4.50}$$

The plot of e_{ij}^* against x_{ij} is called a *partial residual plot*. These plots have been suggested by Ezekiel and Fox [1959] and Larsen and McCleary [1972].

Like the usual plot of residuals e_i versus x_{ij}, the partial residual plot is useful in detecting outliers and inequality of variance. However, since it displays the relationship between y and the regressor x_j after the effect of the other regressors $x_i(i \neq j)$ have been removed, the partial residual plot more clearly shows the influence of x_j on the response y in the presence of other regressors. Thus to some extent it is a substitute for y versus x_j plots in multiple regression.

Consider the linear regression through the origin of e_{ij}^* on x_{ij}. The slope of the least squares line for this regression will be $\hat{\beta}_j$, the same value obtained as an estimate of $\hat{\beta}_j$ in the full k–variable model. Therefore the partial residual plot will have a slope of $\hat{\beta}_j$ rather than zero as in the usual residual plot. This display allows the experimenter to easily evaluate the extent of departures from linearity, or the presence of outliers and inequality of variance. If the relationship between y and x_j is not linear, the partial residual plot usually indicates more precisely how to transform the data to achieve linearity than does the usual residual plot. Several excellent examples are in Larsen and McCleary [1972]. Some regression computer programs (such as BMD–P) will automatically produce partial residual plots.

Larsen and McCleary [1972] and Seber [1977] observe that because the variance of $\hat{\beta}_j$ in the regression of e_{ij}^* on x_{ij} is less than the variance of $\hat{\beta}_j$ from the full model, the experimenter tends to overestimate the stability of $\hat{\beta}_j$ by visually examining the partial residual plot. Consequently he overestimates the importance of x_j in explaining y. This is not serious unless x_j is highly correlated with the other regressors. Scaling the partial residuals would correct this deficiency, but scaling exaggerates any nonlinearity that is present. Therefore it is preferable to use unscaled residuals.

There are other ways to define partial residuals. For example, Daniel and Wood [1980] define

$$
\begin{aligned}
e'_{ij} &= y_i - \bar{y} - \sum_{\substack{u=1 \\ u \neq j}}^{k} \hat{\beta}_u (x_{iu} - \bar{x}_u) \\
&= y_i - \hat{y}_i + \hat{\beta}_j (x_{ij} - \bar{x}_j) \\
&= e_i + c_{ij}, \qquad i = 1, 2, \ldots, n
\end{aligned}
\tag{4.51}
$$

They refer to c_{ij} as the component effect of x_j on y. Wood [1973] recommends plotting e'_{ij} against x_{ij} $i = 1, 2, \ldots, n$. The resulting plot is called a component effect plus residual plot. Also see the references in Larsen and McCleary [1972].

Plot regressor x_j against regressor x_i. This plot may be useful in studying the relationship between regressor variables and the disposition of the data in x-space. Consider the plot of x_i versus x_j in Figure 4.7. This display indicates that x_i and x_j are highly correlated. Consequently it may not be necessary to include both regressors in the model. If two or more regressors are highly correlated, we say that *multicollinearity* is present in the data. Multicollinearity can seriously disturb the least squares fit, and in some situations render the regression model almost useless. This problem is discussed extensively in Chapter 8. Plots of x_i versus x_j may also be useful in discovering points that are remote from the rest of the data and which potentially influence key model properties. Anscombe [1973] presents several other types of plots between regressors.

Other methods for scaling residuals. We have noted that the standardized residuals d_i are useful in identifying outliers. We now explore other methods for scaling residuals. From (4.15), we may write the vector of residuals as

$$\mathbf{e}=(\mathbf{I}-\mathbf{H})\mathbf{y} \tag{4.52}$$

where $\mathbf{H}=\mathbf{X}(\mathbf{X}'\mathbf{X})^{-1}\mathbf{X}'$ is the hat matrix. The hat matrix has several useful properties. It is *symmetric* ($\mathbf{H}'=\mathbf{H}$) and idempotent ($\mathbf{HH}=\mathbf{H}$). Similarly, the matrix $\mathbf{I}-\mathbf{H}$ is symmetric and idempotent. Substituting $\mathbf{y}=\mathbf{X}\boldsymbol{\beta}+\boldsymbol{\varepsilon}$ into (4.52) yields

$$\begin{aligned}\mathbf{e}&=(\mathbf{I}-\mathbf{H})(\mathbf{X}\boldsymbol{\beta}+\boldsymbol{\varepsilon})\\&=\mathbf{X}\boldsymbol{\beta}-\mathbf{HX}\boldsymbol{\beta}+(\mathbf{I}-\mathbf{H})\boldsymbol{\varepsilon}\\&=\mathbf{X}\boldsymbol{\beta}-\mathbf{X}(\mathbf{X}'\mathbf{X})^{-1}\mathbf{X}'\mathbf{X}\boldsymbol{\beta}+(\mathbf{I}-\mathbf{H})\boldsymbol{\varepsilon}\\&=(\mathbf{I}-\mathbf{H})\boldsymbol{\varepsilon}\end{aligned} \tag{4.53}$$

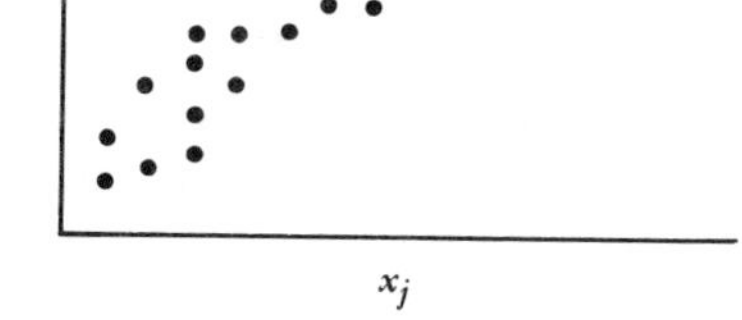

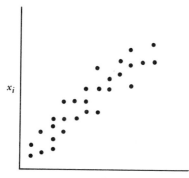

Figure 4.7 Plot of x_i versus x_j.

Thus the residuals are the same linear transformation of the observations $\mathbf{y}$ and the errors $\boldsymbol{\varepsilon}$.

The covariance matrix of the residuals is

$$\begin{aligned} V(\mathbf{e}) &= V[(\mathbf{I}-\mathbf{H})\boldsymbol{\varepsilon}] \\ &= (\mathbf{I}-\mathbf{H})V(\boldsymbol{\varepsilon})(\mathbf{I}-\mathbf{H})' \\ &= \sigma^2(\mathbf{I}-\mathbf{H}) \end{aligned} \tag{4.54}$$

since $V(\boldsymbol{\varepsilon})=\sigma^2\mathbf{I}$ and $\mathbf{I}-\mathbf{H}$ is symmetric and idempotent. The matrix $\mathbf{I}-\mathbf{H}$ is generally not diagonal, so the residuals have different variances and they are correlated.

The variance of the ith residual is

$$V(e_i)=\sigma^2(1-h_{ii}) \tag{4.55}$$

where h_{ii} is the ith diagonal element of $\mathbf{H}$. Since $0 \leqslant h_{ii} \leqslant 1$, using the residual mean square MS_E to estimate the variance of the residuals actually overestimates $V(e_i)$. Furthermore since h_{ii} is a measure of the *location* of the ith point in x-space, the variance of e_i depends upon where the point $\mathbf{x}_i$ lies. Generally points near the center of the x-space have larger variance (poorer least squares fit) than residuals at more remote locations. Violations of model assumptions are more likely at remote points, and these violations may be hard to detect from inspection of e_i (or d_i) because their residuals will usually be smaller.

Several authors (Behnken and Draper [1972], Davies and Hutton [1975], and Huber [1975]) suggest taking this inequality of variance into account when scaling the residuals. They recommend plotting the "studentized" residuals

$$f_i=\frac{e_i}{\sqrt{MS_E(1-h_{ii})}}, \qquad i=1,2,\ldots,n \tag{4.56}$$

instead of e_i (or d_i). The studentized residuals have constant variance $V(f_i)=1$ regardless of the location of $\mathbf{x}_i$ when the form of the model is correct. In many situations the variance of the residuals stabilizes, particularly for large data sets. In these cases, there may be little difference between the standardized and studentized residuals. Thus standardized and studentized residuals often convey equivalent information. However, since any point with a large residual *and* a large h_{ii} is potentially highly influential on the least squares fit, examination of the studentized residuals is generally recommended.

The covariance between e_i and e_j is

$$\text{Cov}(e_i, e_j) = -\sigma^2 h_{ij} \tag{4.57}$$

so another approach to scaling the residuals is to transform the n dependent residuals into $n-p$ orthogonal functions of the errors $\boldsymbol{\varepsilon}$. These transformed residuals are normally and independently distributed with constance variance σ^2. Several procedures have been proposed to investigate departures from the underlying assumptions using transformed residuals. A good review of this literature is in Seber [1977]. These procedures are not widely used in practice because it is difficult to make specific inferences about the transformed residuals, such as the interpretation of outliers. Furthermore, dependence between the residuals does not affect interpretation of the usual residual plots unless k is large relative to n.

- **Example 4.11** We will investigate the adequacy of the regression model for the soft drink delivery time data developed in Example 4.1. From examining Table 4.3, we note that one residual, $e_9 = 7.4197$, seems suspiciously large. The standardized residual is $d_9 = e_9/\sqrt{MS_E} = 7.4197/\sqrt{10.6239} = 2.28$. All other standardized residuals are inside the ± 2 limits. Column 2 of Table 4.7 shows the studentized residuals. The studentized residual at Point 9 is $f_9 = e_9/\sqrt{MS_E(1-h_{9,9})} = 7.4197/\sqrt{10.6239(1-0.49829)} = 3.2138$, which is substantially larger than the standardized residual. As we noted in Example 4.9, Point 9 has the largest value of x_1 (30 cases) and x_2 (1460 ft). If we take the remote location of Point 9 into account when scaling its residual, we conclude that the model does not fit this point well.

  A normal probability plot of the residuals is shown in Figure 4.8. Passing a straight line through the points falling roughly between the ± 0.5 expected normal values reveals two clusters of points in the tails of the distribution that fall off the line. This normal probability plot is characteristic of residuals from a distribution with heavier tails than the normal. The method of least squares often performs poorly in these situations. In Chapter 9 we will discuss alternatives to least squares that are more appropriate when the errors come from a heavy-tailed distribution. Several alternative fits to the delivery time data will be given.

  Figures 4.9, 4.10, and 4.11 present plots of the residuals versus $\hat{y}$, x_1 (cases), and x_2 (distance), respectively. These plots do not exhibit any strong unusual pattern, although the large residual e_9 shows up clearly. Figure 4.9 indicates that the model has a slight tendency to underpredict short delivery times and overpredict long delivery times.

  Figure 4.12 plots the residuals against the four data collection sites (Austin, Boston, San Diego, and Minneapolis). The tendency of the model to underpredict delivery times in Boston and overpredict delivery times in Austin

Table 4.7 Studentized residuals and partial residuals for Example 4.11

Observation Number	1. $e_i = y_i - \hat{y}_i$	2. $f_i = e_i/\sqrt{MS_E(1-h_{ii})}$	3. $e_{i1}^* = e_i + \hat{\beta}_1 x_{i1}$	4. $e_{i2}^* = e_i + \hat{\beta}_2 x_{i2}$
1	−5.0281	−1.6277	6.28327	3.03030
2	1.1464	0.3649	5.99413	4.31220
3	−0.0498	−0.0161	4.79793	4.8428
4	4.9244	1.5798	11.38804	6.0726
5	−0.4444	−0.1418	9.25106	1.7141
6	−0.2896	−0.0908	11.02177	4.4591
7	0.8446	0.2704	4.07642	2.4275
8	1.1566	0.3667	12.46797	4.1785
9	7.4197	3.2138	55.89700	28.4291
10	2.3764	0.8133	10.45595	11.08235
11	2.2375	0.7181	28.09206	12.13782
12	−0.5930	−0.1932	15.56610	2.50085
13	1.0270	0.3252	7.49064	4.69645
14	1.0675	0.3411	10.76296	7.71568
15	0.6712	0.2103	15.21439	7.11792
16	−0.6629	−0.2227	15.49620	10.50374
17	0.4364	0.1381	10.13186	3.3144
18	3.4486	1.1130	14.75997	5.34808
19	1.7932	0.5787	6.64093	2.31124
20	−5.7880	−1.8736	21.68247	5.2923
21	−2.6142	−0.8779	13.54490	−0.5996
22	−3.6865	−1.4500	38.32716	7.9694
23	−4.6076	−1.4437	9.93559	1.8679
24	−4.5728	−1.4961	8.35448	4.56485
25	−0.2126	−0.0675	6.25104	1.9459

noted previously in Section 3.2.5 is still evident, but the indication of inequality of variance observed in Figure 3.7 when only one regressor (x_1) was employed is no longer present.

The partial residuals are listed in columns 3 and 4 of Table 4.7. Figure 4.13 presents a plot of the partial residual $e_{i1}^* = e_i + \hat{\beta}_1 x_{i1}$ against x_1. The linear relationship between delivery time and the number of cases delivered is clearly evident in this plot. The large residual $e_{9,1}^*$ falls slightly off the straight line passing through the rest of the data. Figure 4.14 shows the partial residual plot of $e_{i2}^* = e_i + \hat{\beta}_2 x_{i2}$ against x_2. These partial residuals are listed in the last column of Table 4.7. The impression is that delivery time and distance are linearly related, and that residual $e_{9,2}^*$ is very far from the rest of the data. These plots indicate that Point 9 bears further investigation.

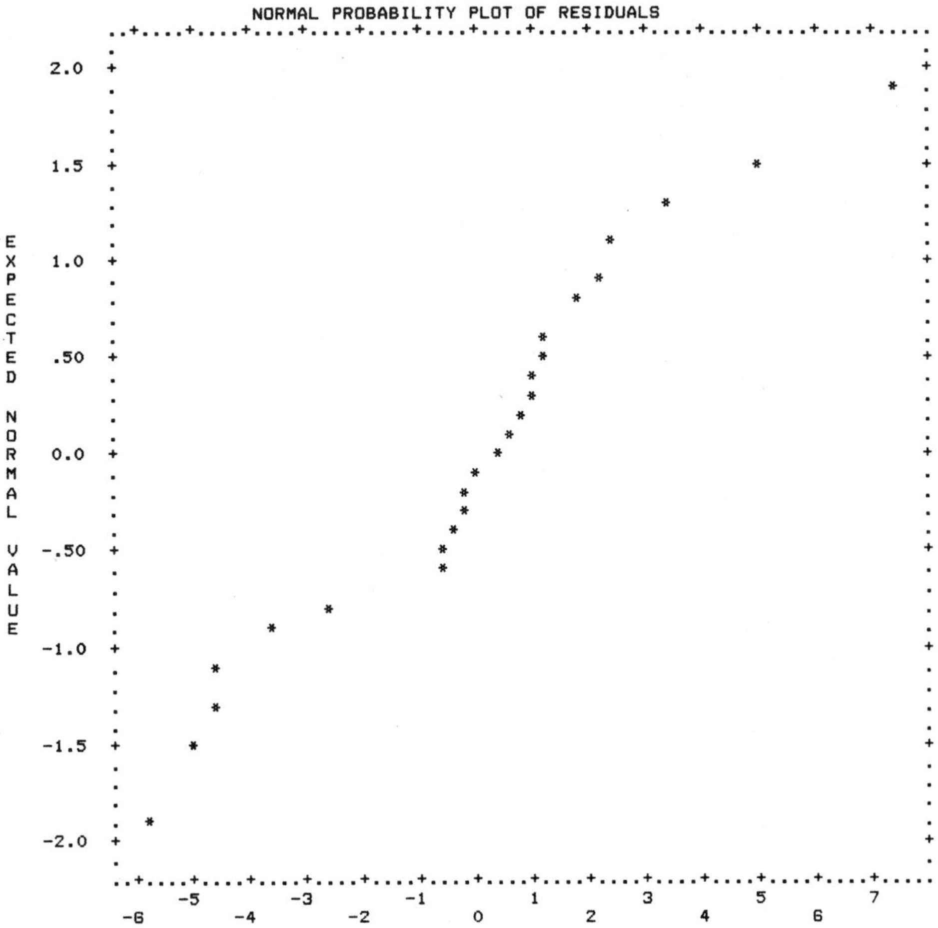

Figure 4.8 Normal probability plot (BMD–P) of residuals, Example 4.11.

Figure 4.6 (page 145) is a plot of x_1 (cases) versus x_2 (distance). Comparing Figure 4.6 and Figure 4.7, we see that cases and distance are positively correlated. In fact, the simple correlation between x_1 and x_2 is $r_{12}=0.82$. We have noted that highly correlated regressors can cause a number of serious problems in regression, although in this particular example there is no strong indication that such problems have occurred. We will discuss this subject in more detail in Chapter 8. As observed previously, Observation 9 ($x_1=30, x_2=1460$) is remote in x-space from the rest of the data. Observation 22 ($x_1=26, x_2=810$) is also quite far away from the rest of the data. It is possible that these points are controlling some of the properties of the regression model.

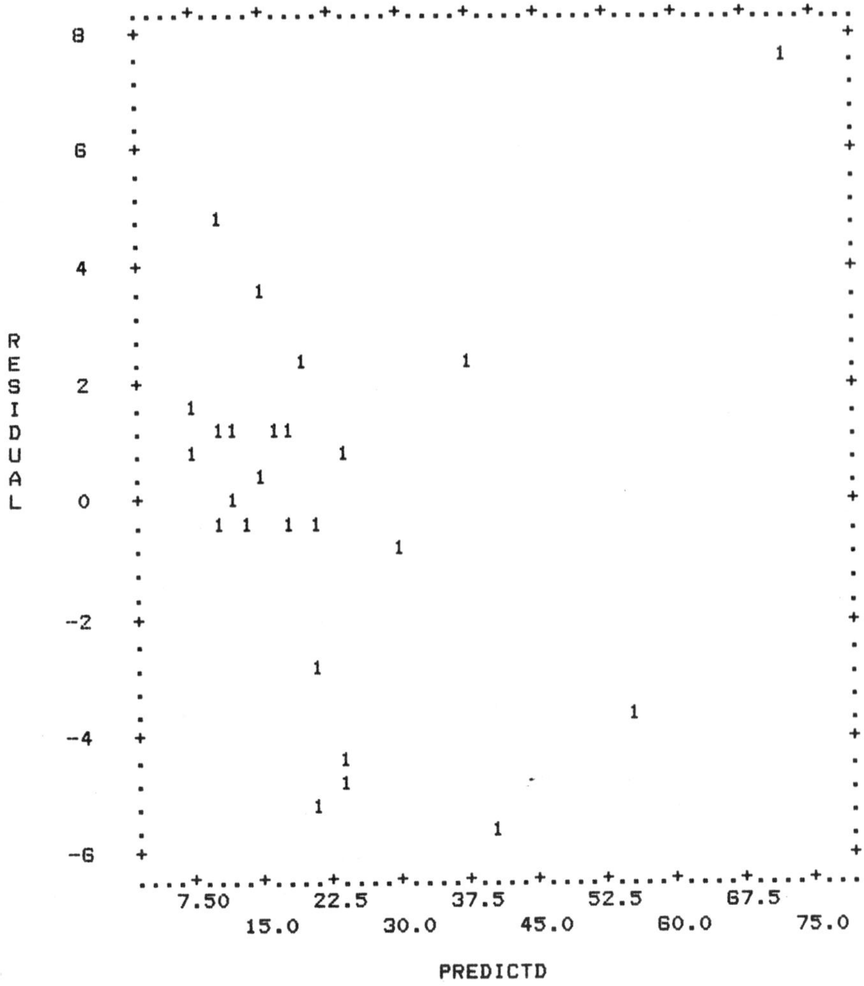

Figure 4.9 Plot (BMD–P) of residuals e_i versus fitted values $\hat{y}_i$, Example 4.11.

4.7.3 Estimation of pure error from near-neighbors

In Section 3.4 we described a test for lack of fit in simple linear regression. The procedure involved partitioning the error or residual sum of squares into a component due to "pure" error and a component due to lack of fit, such as

$$SS_E = SS_{PE} + SS_{LOF}$$

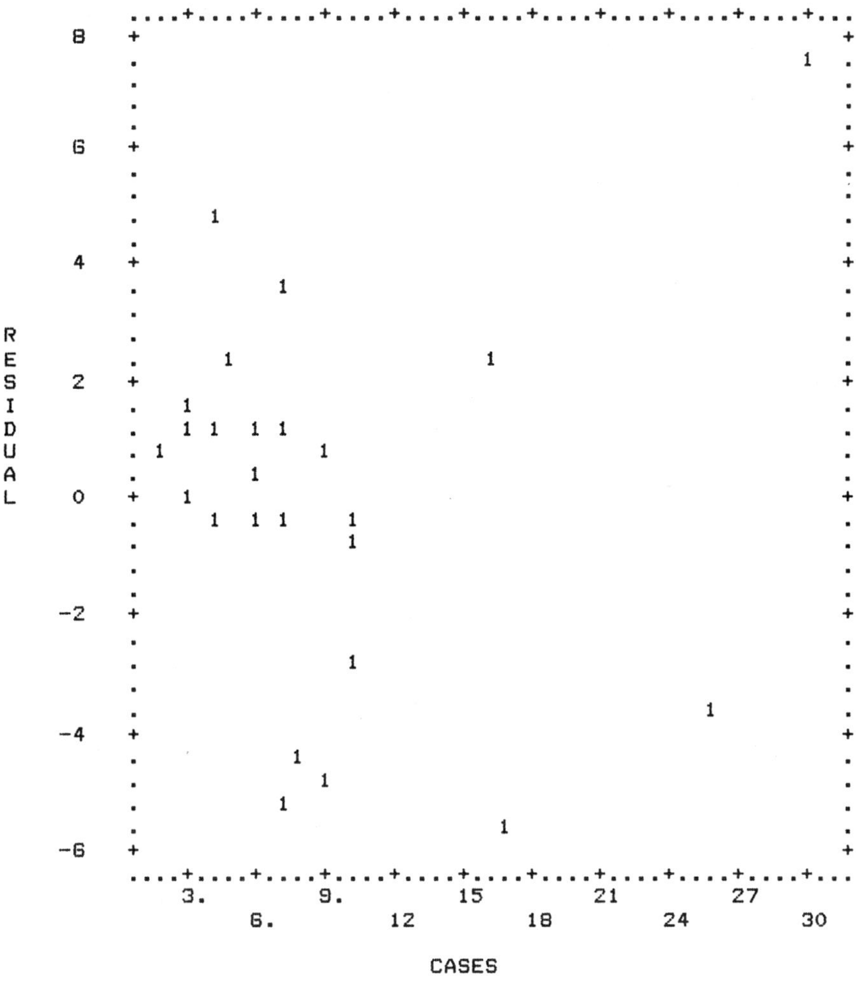

Figure 4.10 Plot (BMD–P) of residuals e_i versus delivery volume x_{i1}, Example 4.11.

The pure error sum of squares SS_{PE} is computed using responses at repeat observations at the same level of x. This is a model-independent estimate of σ^2.

This general procedure can in principle be extended to multiple regression. The calculation of SS_{PE} requires repeat observations on y at the same set of levels on the regressor variables $x_1, x_2, \ldots, x_k$. That is, some of the *rows* of the **X** matrix must be the same. However, repeat observations do not often occur in multiple regression, and the procedure described in Section 3.4 is not often useful.

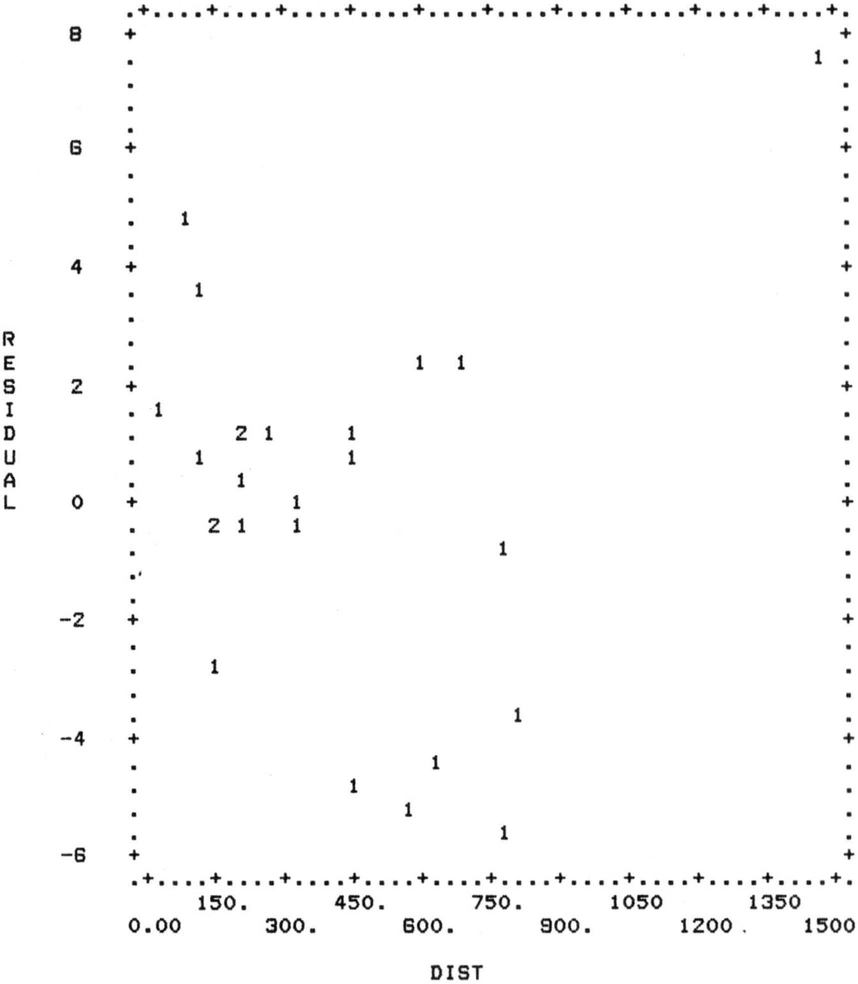

Figure 4.11 Plot (BMD–P) of residuals e_i versus distance x_{i2}, Example 4.11. The plotting symbol represents the number of observations at the point.

Daniel and Wood [1980] have suggested a method for obtaining a model-independent estimate of error when there are no exact repeat points. The procedure searches for points in the x space that are "near-neighbors," that is, sets of observations that have been taken with nearly identical levels of $x_1, x_2, \ldots, x_k$. The responses y_i from such near-neighbors can be considered as repeat points and used to obtain an estimate of pure error. As a measure of the distance between any two points, for example, $x_{i1}, x_{i2}, \ldots, x_{ik}$ and $x_{i'1}, x_{i'2}, \ldots, x_{i'k}$, Daniel and Wood propose the weighted sum of

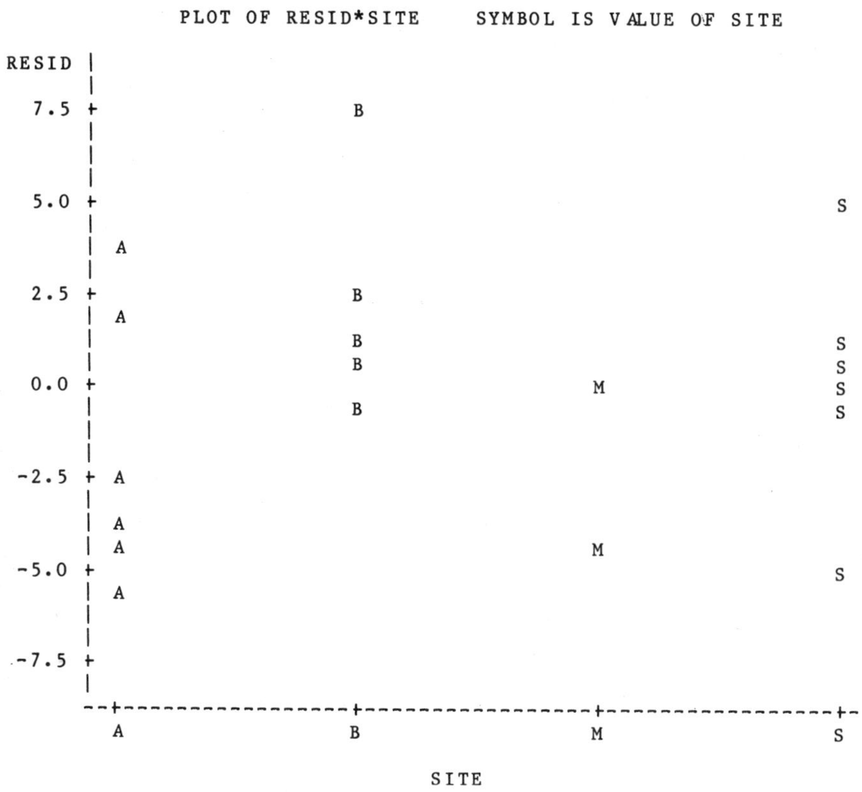

Figure 4.12 Plot (SAS) of residuals e_i versus site, Example 4.11.

squared distance

$$D_{ii'}^2 = \sum_{j=1}^{k} \left[\frac{\hat{\beta}_j (x_{ij} - x_{i'j})}{\sqrt{MS_E}} \right]^2 \tag{4.58}$$

Pairs of points that have small values of $D_{ii'}^2$, are "near-neighbors," that is, they are relatively close together in x-space. Pairs of points for which $D_{ii'}^2$ is large ($D_{ii'}^2 \gg 1$, for example) are widely separated in x-space. The residuals at two points with a small value of $D_{ii'}^2$ can be used to obtain an estimate of pure error. The estimate is obtained from the range of the residuals at the points i and i', as in

$$E_i = |e_i - e_{i'}|$$

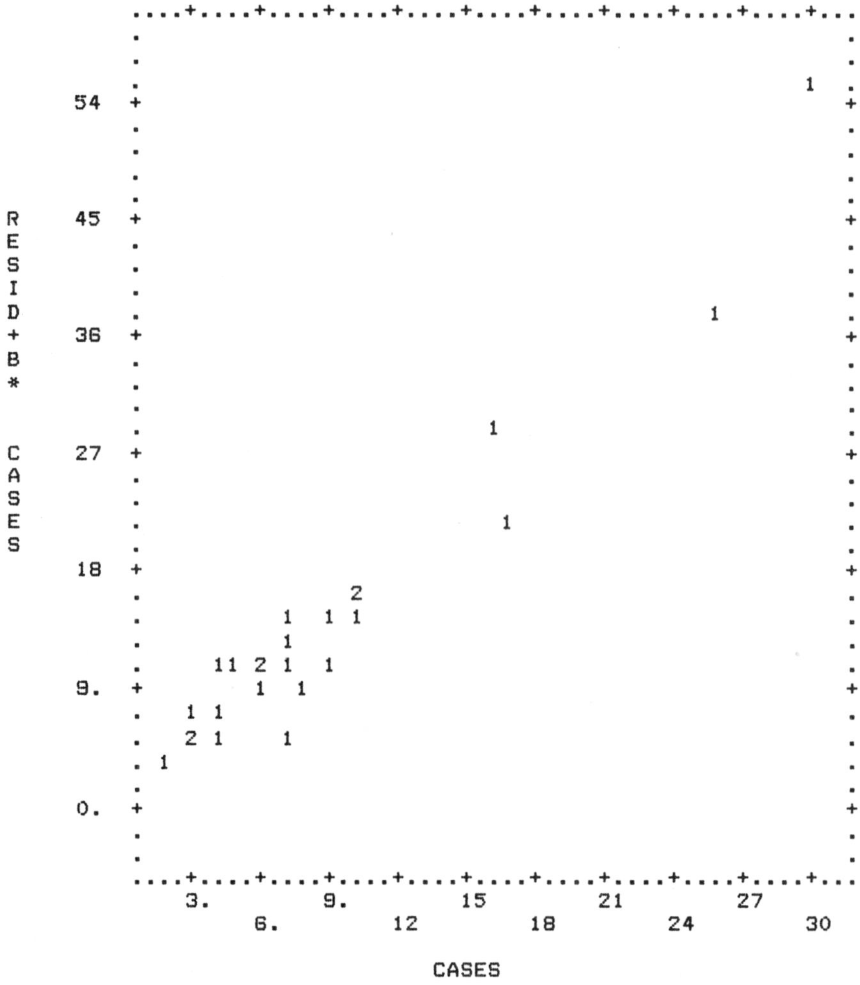

Figure 4.13 Plot (BMD–P) of partial residuals e_{i1}^* versus delivery volume x_{i1}, Example 4.11.

There is a relationship between the range of a sample from a normal population and the population standard deviation. For samples of size 2, this relationship is

$$\hat{\sigma}=(1.128)^{-1}E=0.886E$$

The quantity $\hat{\sigma}$ so obtained is an estimate of the standard deviation of pure error.

An efficient algorithm may be used to compute this estimate. A computer program for this algorithm is given in Montgomery, Martin, and Peck

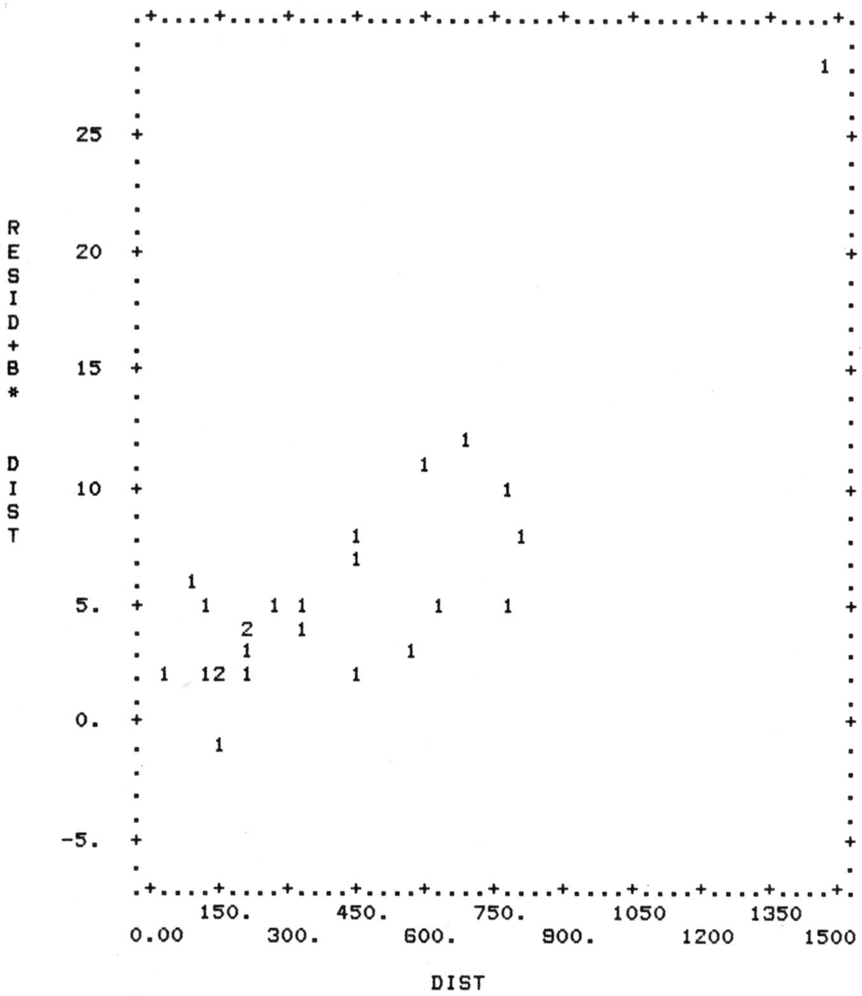

Figure 4.14 Plot (BMD–P) of partial residuals e_{i2}^* versus distance x_{i2}, Example 4.11.

[1980]. First arrange the data points $x_{i1}, x_{i2}, \ldots, x_{ik}$ in order of increasing $\hat{y}_i$. Note that points with very different values of $\hat{y}_i$ cannot be near-neighbors, but those with similar values of $\hat{y}_i$ could be neighbors (or they could be near the same contour of constant $\hat{y}$ but far apart in some x coordinates). Then

1. Compute the values of $D_{ii'}^2$ for all $n-1$ pairs of points with adjacent values of $\hat{y}$. Repeat this calculation for the pairs of points separated by 1, 2, and 3 intermediate $\hat{y}$ values. This will produce $4n-10$ values of $D_{ii'}^2$.

2. Arrange the $4n-10$ values of $D_{ii'}^2$ found in 1 above in ascending order. Let E_u, $u=1,2,\ldots,4n-10$ be the range of the residuals at these points.

3. For the first m values of E_u, calculate an estimate of the standard deviation of pure error as

$$\hat{\sigma}=\frac{0.886}{m}\sum_{u=1}^{m} E_u$$

Note that $\hat{\sigma}$ is based on the average range of the residuals associated with the m smallest values of $D_{ii'}^2$; m must be chosen after inspecting the values of $D_{ii'}^2$. One should not include values of E_u in the calculations for which the weighted sum of squared distance is too large.

• **Example 4.12** We shall use the procedure described above to calculate an estimate of the standard deviation of pure error for the soft drink delivery time data in Example 4.1. Table 4.8 displays the calculation of $D_{ii'}^2$ for pairs of points that, in terms of $\hat{y}$, are adjacent, one apart, two apart, and three apart. The columns labeled R in this table identify the 15 smallest values of $D_{ii'}^2$. The residuals at these 15 pairs of points are used to estimate σ. These calculations yield $\hat{\sigma}=1.969$, and are summarized in Table 4.9. From Table 4.5, we find that $\sqrt{MS_E}=\sqrt{10.6239}=3.259$. Now if there is no appreciable lack of fit, we would expect to find that $\hat{\sigma}\simeq\sqrt{MS_E}$. In this case $\sqrt{MS_E}$ is about 65 percent larger than $\hat{\sigma}$, indicating some lack of fit. This could be due to the effects of regressors not presently in the model, or the presence of one or more outliers. •

4.7.4 Detecting influential observations

We occasionally find that a small subset of the data exerts a disproportionate influence on the fitted regression model. That is, parameter estimates or predictions may depend more on the influential subset than on the majority of the data. We would like to locate these influential points and assess their impact on the model. If these influential points are "bad" values, then they should be eliminated. On the other hand, there may be nothing wrong with these points, but if they control key model properties we would like to know it, as it could affect the use of the model.

The disposition of points in the x-space is important in determining model properties. In particular, remote observations potentially influence the parameter estimates, predicted values, and the usual summary statistics. Daniel and Wood [1980] suggest that the weighted sum of squared distance

Table 4.8 Calculation of $D^2_{ii'}$ for Example 4.12

Near-Neighbor Calculations
Delta Residuals and the Weighted Standardized Squared Distances of Near-Neighbors

Observation	Ordered Fitted y	Residual	Adjacent Delta	Adjacent $D^2_{ii'}$	R[a]	1 Apart Delta	1 Apart $D^2_{ii'}$	R	2 Apart Delta	2 Apart $D^2_{ii'}$	R	3 Apart Delta	3 Apart $D^2_{ii'}$	R
7	7.155	.845	.949	.3524E+00		4.080	.1001E+01		.302	.4814E+00		1.057	.1014E+01	
19	7.707	1.793	3.131	.2835E+00	12	.647	.6594E+00		2.006	.4989E+00		1.843	.1800E+01	
4	9.956	4.924	3.778	.6275E+00		5.137	.9544E−01	3	4.974	.1562E+01		3.897	.5965E+00	
2	10.354	1.146	1.359	.3412E+00	15	1.196	.2805E+00	11	.119	.2696E+00	9	1.591	.2307E+01	
25	10.963	−.213	.163	.9489E+00		1.240	.2147E+00	6	.232	.9831E+00		.649	.1032E+01	
3	12.080	−.050	1.077	.3865E+00		.395	.2915E+01		.486	.2594E+01		3.498	.4775E+01	
13	12.473	1.027	1.471	.1198E+01		.591	.1042E+01		2.422	.2507E+01		.130	.2251E+01	
5	14.194	−.444	.881	.4869E−01	2	3.893	.2521E+00	8	1.601	.3159E+00	13	.155	.8768E+00	
17	14.914	.436	3.012	.3358E+00	14	.720	.2477E+00	7	.726	.5749E+00		.631	.1337E+01	
18	15.551	3.449	2.292	.1185E+00	5	3.738	.7636E+00		2.381	.2367E+01		1.072	.5341E+01	
8	16.673	1.157	1.446	.2805E+00	10	.089	.1483E+01		1.220	.4022E+01		3.771	.2307E+01	
6	18.400	−.290	1.357	.5851E+00		2.666	.2456E+01		2.325	.2915E+01		.303	.2470E+01	
14	18.682	1.068	1.309	.6441E+00		3.682	.5952E+01		1.661	.5121E+01		6.096	.4328E+00	
10	19.124	2.376	4.991	.1036E+02		2.969	.9107E+01		7.404	.1023E+01		1.705	.4412E+01	
21	20.514	−2.614	2.021	.1096E+00	4	2.414	.5648E+01		3.285	.2093E+01		1.993	.2117E+01	
12	21.593	−.593	4.435	.4530E+01		1.264	.1303E+01		4.015	.1321E+01		3.980	.4419E+01	
1	21.708	−5.028	5.699	.1227E+01		.421	.1219E+01		.455	.3553E+00		4.365	.3121E+01	
15	23.329	.671	5.279	.7791E−04	1	5.244	.9269E+00		1.334	.2341E+01		1.566	.1316E+02	
23	23.358	−4.608	.035	.9124E+00		3.945	.2316E+01		6.845	.1315E+02		1.180	.1772E+02	
24	24.403	−4.573	3.910	.1370E+01		6.810	.1578E+02		1.215	.2026E+02		.886	.8023E+02	
16	29.663	−.663	2.900	.8999E+01		5.125	.1204E+02		3.024	.6294E+02		8.083	.1074E+03	
11	38.093	2.237	8.025	.3767E+00		5.924	.2487E+02		5.182	.5978E+02				
20	40.888	−5.788	2.101	.1994E+02		13.208	.5081E+02							
22	56.007	−3.687	11.106	.1216E+02										
9	71.820	7.420												

[a]Column R gives the rank order of the 15 smallest $D^2_{ii'}$ values.

Table 4.9 Calculation of $\hat{\sigma}$ for Example 4.12

	Standard Deviation Estimated from Residuals of Neighboring Observations				
		Ordered by $D_{ii'}^2$			
Number	Cumulative Standard Deviation	$D_{ii'}^2$	Observation	Observation	Delta Residual
1	$.4677E+01$	$.7791E-04$	15	23	5.2788
2	$.2729E+01$	$.4869E-01$	5	17	.8807
3	$.3336E+01$	$.9544E-01$	4	25	5.1369
4	$.2950E+01$	$.1096E+00$	21	12	2.0211
5	$.2766E+01$	$.1185E+00$	18	8	2.2920
6	$.2488E+01$	$.2147E+00$	25	13	1.2396
7	$.2224E+01$	$.2477E+00$	17	8	.7203
8	$.2377E+01$	$.2521E+00$	5	18	3.8930
9	$.2125E+01$	$.2696E+00$	2	13	.1194
10	$.2040E+01$	$.2805E+00$	8	6	1.4462
11	$.1951E+01$	$.2805E+00$	2	3	1.1962
12	$.2020E+01$	$.2835E+00$	19	4	3.1312
13	$.1973E+01$	$.3159E+00$	5	8	1.6010
14	$.2023E+01$	$.3358E+00$	17	18	3.0123
15	$.1969E+01$	$.3412E+00$	2	25	1.3590
16	$.1898E+01$	$.3524E+00$	7	19	.9486
17	$.1810E+01$	$3553E+00$	1	24	.4552
18	$.2105E+01$	$.3767E+00$	11	20	8.0255
19	$.2044E+01$	$.3865E+00$	3	13	1.0768
20	$.2212E+01$	$.4328E+00$	14	1	6.0956
21	$.2119E+01$	$.4814E+00$	7	2	.3018
22	$.2104E+01$	$.4989E+00$	19	25	2.0058
23	$.2040E+01$	$.5749E+00$	17	6	.7259
24	$.2005E+01$	$.5851E+00$	6	14	1.3571
25	$.2063E+01$	$.5965E+00$	4	13	3.8973
26	$.2113E+01$	$.6275E+00$	4	2	3.7780
27	$.2077E+01$	$.6441E+00$	14	10	1.3089
28	$.2024E+01$	$.6594E+00$	19	2	.6468
29	$.2068E+01$	$.7636E+00$	18	6	3.7382
30	$.2004E+01$	$.8768E+00$	5	6	.1548
31	$.1940E+01$	$.9124E+00$	23	24	.0347
32	$.2025E+01$	$.9269E+00$	15	24	5.2441
33	$.1968E+01$	$.9489E+00$	25	3	.1628
34	$.1916E+01$	$.9831E+00$	25	5	.2318
35	$.1964E+01$	$.1001E+01$	7	4	4.0797
36	$.1936E+01$	$.1014E+01$	7	25	1.0572
37	$.2061E+01$	$.1023E+01$	10	1	7.4045
38	$.2022E+01$	$.1032E+01$	25	17	.6489
39	$.1983E+01$	$.1042E+01$	13	17	.5907
40	$.1966E+01$	$.1198E+01$	13	5	1.4714

of the ith point from the center of the data, say

$$WSSD_i = \sum_{j=1}^{k} \left[\frac{\hat{\beta}_j (x_{ij} - \bar{x}_j)}{\sqrt{MS_E}} \right]^2, \qquad i = 1, 2, \ldots, n \tag{4.59}$$

be used to locate points that are remote in x-space. The general procedure is to rank the points in increasing order of $WSSD_i$, and concentrate on points for which this statistic is large. It is difficult to give formal guidelines for identifying a "large" value of $WSSD_i$. Generally if the $WSSD_i$ values progress smoothly from small to large, there are probably no extremely remote points. However, if there is a sudden jump in the magnitude of $WSSD_i$, this often indicates that one or more extreme points are present.

Hoaglin and Welsch [1978] discuss the role of the hat matrix $\mathbf{H} = \mathbf{X}(\mathbf{X}'\mathbf{X})^{-1}\mathbf{X}'$ in identifying influential observations. As noted earlier, $\mathbf{H}$ determines the variances and covariances of $\hat{\mathbf{y}}$ and $\mathbf{e}$, since $V(\hat{\mathbf{y}}) = \sigma^2 \mathbf{H}$ and $V(\mathbf{e}) = \sigma^2(\mathbf{I} - \mathbf{H})$. The elements h_{ij} of $\mathbf{H}$ may be interpreted as the amount of leverage exerted by y_j on $\hat{y}_i$. Thus inspection of the elements of $\mathbf{H}$ can reveal points that are potentially influential by virtue of their location in x-space. Attention is usually focused on the diagonal elements h_{ii}. Since $\sum_{i=1}^{n} h_{ii} = \text{rank}(\mathbf{H}) = \text{rank}(\mathbf{X}) = p$, the average size of a diagonal element of the $\mathbf{H}$ matrix is p/n. As a rough guideline then, if a diagonal element $h_{ii} > 2p/n$, observation i is a high-leverage point. Further properties and uses of the elements of the hat matrix in regression diagnostics are discussed by Belsley, Kuh, and Welsch [1980].

Both these methods identify points that are potentially influential due to their location in x–space. It is desirable to consider both the location of the point and the response variable in measuring influence. Cook [1977, 1979] has suggested using a measure of the squared distance between the least squares estimate based on all n points $\hat{\boldsymbol{\beta}}$ and the estimate obtained by deleting the ith point, say $\hat{\boldsymbol{\beta}}_{(i)}$. This distance measure can be expressed in a general form as

$$D_i(\mathbf{M}, c) = \frac{(\hat{\boldsymbol{\beta}}_{(i)} - \hat{\boldsymbol{\beta}})' \mathbf{M} (\hat{\boldsymbol{\beta}}_{(i)} - \hat{\boldsymbol{\beta}})}{c}, \qquad i = 1, 2, \ldots, n \tag{4.60}$$

The usual choices of $\mathbf{M}$ and c are $\mathbf{M} = \mathbf{X}'\mathbf{X}$ and $c = pMS_E$, so that (4.60) becomes*

$$D_i(\mathbf{M}, c) \equiv D_i = \frac{(\hat{\boldsymbol{\beta}}_{(i)} - \hat{\boldsymbol{\beta}})' \mathbf{X}'\mathbf{X} (\boldsymbol{\beta}_{(i)} - \hat{\boldsymbol{\beta}})}{pMS_E}, \qquad i = 1, 2, \ldots, n$$

*An alternate version of Cook's distance measure is $D_i = (\hat{\mathbf{y}}_{(i)} - \hat{\mathbf{y}})'(\hat{\mathbf{y}}_{(i)} - \hat{\mathbf{y}})/pMS_E$, so that D_i may be interpreted as the squared Euclidean distance (apart from pMS_E) the vector of fitted values moves when the ith observation is deleted. Furthermore, alternate choices of $\mathbf{M}$ and c will produce other diagnostic statistics, including some of those in Belsley, Kuh and Welsch [1980].

Points with large values of D_i have considerable influence on the least squares estimate $\hat{\boldsymbol{\beta}}$. The magnitude of D_i may be assessed by comparing it to $F_{\alpha, p, n-p}$. If $D_i \simeq F_{.5, p, n-p}$, then deleting point i would move $\hat{\boldsymbol{\beta}}$ to the boundary of a 50 percent confidence region for $\boldsymbol{\beta}$ based on the complete data set. This is a large displacement, and indicates that the least squares estimate is sensitive to the ith data point. Since $F_{.5, p, n-p} \simeq 1$, we usually consider points for which $D_i > 1$ to be influential. Ideally we would like each estimate $\hat{\boldsymbol{\beta}}_{(i)}$ to stay within the boundary of a 10 or 20 percent confidence region.*

The D_i statistic may be rewritten as

$$D_i = \frac{f_i^2}{p} \frac{V(\hat{y}_i)}{V(e_i)} = \frac{f_i^2}{p} \frac{h_{ii}}{(1-h_{ii})}, \qquad i=1,2,\ldots,n \tag{4.61}$$

Thus we see that, apart from the constant p, D_i is the product of the square of the ith studentized residual and $h_{ii}/(1-h_{ii})$. This ratio can be shown to be the distance from the vector $\mathbf{x}_i$ to the centroid of the remaining data. Thus D_i is made up of a component that reflects how well the model fits the ith observation y_i and a component that measures how far that point is from the rest of the data. Either component (or both) may contribute to a large value of D_i.

Once a potentially influential observation has been identified, we may reanalyze the data with this point deleted to determine its actual impact on the model. If several points are potentially influential, several additional runs may have to be made.

All the diagnostic techniques we have discussed are *single-row* methods, since they are directed to a single row of the $\mathbf{X}$ matrix and the corresponding response y_i. It is also possible to devise *multiple-row* diagnostics for assessing the simultaneous influence of larger subsets of the data. For example let $\mathbf{i}$ be an $m \times 1$ vector of indices specifying the m points to be deleted. Then (4.60) becomes

$$D_{\mathbf{i}}(\mathbf{X}'\mathbf{X}, pMS_E) = D_{\mathbf{i}} = \frac{(\hat{\boldsymbol{\beta}}_{(\mathbf{i})} - \hat{\boldsymbol{\beta}})'\mathbf{X}'\mathbf{X}(\hat{\boldsymbol{\beta}}_{(\mathbf{i})} - \hat{\boldsymbol{\beta}})}{pMS_E}$$

The interpretation of $D_{\mathbf{i}}$ is identical to that of D_i. Selection of the subset of points to include in $\mathbf{i}$ is not obvious, however, because in some data sets, subsets of points are jointly influential but individual points are not. Cook

*The distance measure D_i is not an F-statistic, but is compared to an F-value because of the similarity of D_i to the normal theory confidence ellipsoid, (4.22).

and Weisberg [1980] discuss this problem in more detail and give some heuristic decision rules to assist in the selection of points to include. For more information on regression diagnostics, see Andrews and Pregibon [1978], Belsley, Kuh and Welsch [1980], Cook and Weisberg [1980], Draper and John [1981], Welsch and Kuh [1977], and Welsch and Peters [1978].

• **Example 4.13** We will investigate the soft drink delivery time data in Example 4.1 for high-leverage data points. Table 4.10 lists the values of $WSSD_i$ computed from (4.59), the diagonal elements of the hat matrix h_{ii} (computed previously in Example 4.9), and the values of D_i from (4.61). To illustrate the calculations, consider the first observation. The value of $WSSD_1$

Table 4.10 Statistics for detecting influential observations for the soft drink delivery time data

Observation i	$WSSD_i$	h_{ii}	D_i
1	1.204	0.10180	0.10009
2	8.852	0.07070	0.00338
3	8.248	0.09874	0.00001
4	7.680	0.08538	0.07766
5	3.182	0.07501	0.00054
6	0.884	0.04287	0.00012
7	12.980	0.08180	0.00217
8	1.535	0.06373	0.00305
9	132.400	0.49829	3.41835
10	4.221	0.19630	0.05385
11	14.400	0.08613	0.01620
12	1.113	0.11366	0.00160
13	6.032	0.06113	0.00229
14	1.926	0.07824	0.00329
15	0.043	0.04111	0.00063
16	2.997	0.16594	0.00329
17	2.725	0.05943	0.00040
18	2.259	0.09626	0.04398
19	10.870	0.09645	0.01192
20	19.220	0.10169	0.13246
21	1.790	0.16528	0.05086
22	76.180	0.39158	0.45106
23	0.046	0.04126	0.02990
24	1.135	0.12061	0.10232
25	6.878	0.06664	0.00011

is

$$WSSD_1 = \sum_{j=1}^{2} \left[\frac{\hat{\beta}_j (x_{1j} - \bar{x}_j)}{\sqrt{MS_E}} \right]^2$$

$$= \left[\frac{1.6159(7-8.76)}{\sqrt{10.6239}} \right]^2 + \left[\frac{0.01439(560-409.28)}{\sqrt{10.6239}} \right]^2$$

$$= 1.204$$

and D_1 is

$$D_1 = \frac{f_1^2}{p} \frac{h_{11}}{(1-h_{11})}$$

$$= \frac{(-1.6277)^2}{3} \frac{0.10180}{(1-0.10180)}$$

$$= 0.10009$$

The $WSSD_i$ statistic identifies Observations 9 and 22 as the two most remote points, and 9 is nearly twice as far from the rest of the data as is 22. This is confirmed by inspection of Figure 4.6, from which we see that for both Points 9 and 22, x_1 and x_2 are at or near the upper limits of their range. Note that Observation 9 has the largest residual (see Table 4.3 and Example 4.1). Since $p=3$ and $n=25$, any point for which $h_{ii} > 2p/n = 2(3)/25 = 0.24$ indicates an observation with unusually high leverage. Once again our attention is directed to Observations 9 and 22. The largest value of the D_i statistic is $D_9 = 3.41935$, which indicates that removal of Observation 9 would move the least squares estimate to approximately the boundary of a 96 percent confidence region around $\hat{\boldsymbol{\beta}}$. The next largest value is $D_{22} = 0.45106$, and removal of Point 22 will move the estimate of $\boldsymbol{\beta}$ to approximately the edge of a 35 percent confidence region.

Clearly Observations 9 and 22 have the greatest leverage on the least squares estimate $\hat{\boldsymbol{\beta}}$. To investigate the effect of these two points on the model, three additional analyses were performed; one deleting Observation 9, a second deleting Observation 22, and the third deleting both 9 and 22. The results of these additional runs are shown in the following table:

Run	$\hat{\beta}_0$	$\hat{\beta}_1$	$\hat{\beta}_2$	MS_E	R^2
9 and 22 in	2.341	1.616	0.014	10.624	0.9596
9 out	4.447	1.498	0.010	5.905	0.9487
22 out	1.916	1.786	0.012	10.066	0.9564
9 and 22 out	4.643	1.456	0.011	6.163	0.9072

Deleting Observation 9 produces only a minor change in $\hat{\beta}_1$, but results in approximately a 28 percent change in $\hat{\beta}_2$ and a 90 percent change in $\hat{\beta}_0$. This indicates that Observation 9 is off the plane passing through the other 24 points and it exerts a moderately strong influence on the regression coefficient associated with x_2 (distance). This is not surprising considering that the value of x_2 for this observation (1460 ft) is very different from the other observations. In effect, Observation 9 may be causing curvature in the x_2 direction. If Observation 9 were deleted, then MS_E would be reduced to 5.905. Note that $\sqrt{5.905} = 2.430$, which is not too different from the estimate of pure error $\hat{\sigma} = 1.969$ found in Example 4.12. It seems that most of the lack of fit noted in this model in Example 4.12 is due to Point 9's large residual. Deleting Point 22 produces relatively smaller changes in the regression coefficients and model summary statistics. Deleting both Points 9 and 22 produces changes similar to those observed when deleting only 9.

We conclude that Observations 9 and 22 have a moderately strong influence on the least squares fit. Subsequent investigation failed to reveal any legitimate reason for deleting these points, such as an error in recording the data, and so they are (at least for the present) retained. In Chapter 9 we will discuss estimation procedures that are less sensitive than least squares to influential observations with large residuals. •

4.8 Standardized regression coefficients

It is usually difficult to directly compare regression coefficients because the magnitude of $\hat{\beta}_j$ reflects the units of measurement of the regressor x_j. For example, suppose that the regression model is

$$\hat{y} = 5 + x_1 + 1000x_2$$

and y is measured in litres, x_1 is measured in millilitres, and x_2 is measured in litres. Note that although $\hat{\beta}_2$ is considerably larger than $\hat{\beta}_1$, the effect of both regressors on $\hat{y}$ is identical, since a one litre change in either x_1 or x_2 when the other variable is held constant produces the same change in $\hat{y}$. Generally the units of the regression coefficient β_j are (units of y/units of x_j). For this reason, it is sometimes helpful to work with scaled regressor and response variables that produce dimensionless regression coefficients.

There are two popular scaling techniques. The first of these is *unit normal scaling*

$$z_{ij} = \frac{x_{ij} - \bar{x}_j}{s_j}, \qquad \begin{cases} i = 1, 2, \ldots, n \\ j = 1, 2, \ldots, k \end{cases} \tag{4.62a}$$

and

$$y_i^* = \frac{y_i - \bar{y}}{s_y}, \qquad i=1,2,\ldots,n \tag{4.62b}$$

where

$$s_j^2 = \frac{\sum_{i=1}^{n} (x_{ij} - \bar{x}_j)^2}{n-1}$$

is the sample variance of regressor x_j and

$$s_y^2 = \frac{\sum_{i=1}^{n} (y_i - \bar{y})^2}{n-1}$$

is the sample variance of the response. Note the similarity of (4.62) to standardizing a normal random variable. All the scaled regressors and the scaled response have sample mean equal to zero and sample variance equal to one. Using these new variables, the regression model becomes

$$y_i^* = b_1 z_{i1} + b_2 z_{i2} + \cdots + b_k z_{ik} + \varepsilon_i, \qquad i=1,2,\ldots,n \tag{4.63}$$

Centering the regressor and response variables in (4.62) by subtracting $\bar{x}_j$ and $\bar{y}$ removes the intercept from the model (actually, the least squares estimate of b_0 is $\hat{b}_0 = \bar{y}^* = 0$). The least squares estimator of $\mathbf{b}$ is

$$\hat{\mathbf{b}} = (\mathbf{Z}'\mathbf{Z})^{-1}\mathbf{Z}'\mathbf{y}^* \tag{4.64}$$

The second popular scaling is *unit length scaling*

$$w_{ij} = \frac{x_{ij} - \bar{x}_j}{S_{jj}^{1/2}}, \qquad \begin{cases} i=1,2,\ldots,n \\ j=1,2,\ldots,k \end{cases} \tag{4.65a}$$

and

$$y_i^0 = \frac{y_i - \bar{y}}{S_{yy}^{1/2}}, \qquad i=1,2,\ldots,n \tag{4.65b}$$

where

$$S_{jj}=\sum_{i=1}^{n}(x_{ij}-\bar{x}_j)^2$$

is the corrected sum of squares for regressor x_j. In this scaling, each new regressor w_j has mean $\bar{w}_j=0$ and length $\sqrt{\sum_{i=1}^{n}(w_{ij}-\bar{w}_j)^2}=1$. In terms of these variables, the regression model is

$$y_i^0=b_1w_{i1}+b_2w_{i2}+\cdots+b_kw_{ik}+\varepsilon_i, \qquad i=1,2,\ldots,n \tag{4.66}$$

The vector of least squares regression coefficients is

$$\hat{\mathbf{b}}=(\mathbf{W}'\mathbf{W})^{-1}\mathbf{W}'\mathbf{y}^0 \tag{4.67}$$

In the unit length scaling, the $\mathbf{W}'\mathbf{W}$ matrix is in the form of a *correlation matrix*; that is

$$\mathbf{W}'\mathbf{W}=\begin{bmatrix} 1 & r_{12} & r_{13} & \cdots & r_{1k} \\ r_{12} & 1 & r_{23} & \cdots & r_{2k} \\ r_{13} & r_{23} & 1 & \cdots & r_{3k} \\ \vdots & \vdots & \vdots & & \vdots \\ r_{1k} & r_{2k} & r_{3k} & \cdots & 1 \end{bmatrix}$$

where

$$r_{ij}=\frac{\sum_{u=1}^{n}(x_{ui}-\bar{x}_i)(x_{uj}-\bar{x}_j)}{(S_{ii}S_{jj})^{1/2}}=\frac{S_{ij}}{(S_{ii}S_{jj})^{1/2}}$$

is the simple correlation between regressor x_i and x_j. Similarly, $\mathbf{W}'\mathbf{y}^0$ is

$$\mathbf{W}'\mathbf{y}^0=\begin{bmatrix} r_{1y} \\ r_{2y} \\ r_{3y} \\ \vdots \\ r_{ky} \end{bmatrix}$$

where

$$r_{jy}=\frac{\sum_{u=1}^{n}(x_{uj}-\bar{x}_j)(y_u-\bar{y})}{(S_{jj}S_{yy})^{1/2}}=\frac{S_{jy}}{(S_{jj}S_{yy})^{1/2}}$$

is the simple correlation* between the regressor x_j and the response y. If unit normal scaling is used, the $\mathbf{Z}'\mathbf{Z}$ matrix is closely related to $\mathbf{W}'\mathbf{W}$; in fact

$$\mathbf{Z}'\mathbf{Z}=(n-1)\mathbf{W}'\mathbf{W}$$

Consequently, the estimates of the regression coefficients in (4.64) and (4.67) are identical. That is, it doesn't matter which scaling we use; they both produce the same set of dimensionless regression coefficients $\hat{\mathbf{b}}$.

The regression coefficients $\hat{\mathbf{b}}$ are usually called *standardized regression coefficients*. The relationship between the original and standardized regression coefficients is

$$\hat{\beta}_j=\hat{b}_j\left(\frac{S_{yy}}{S_{jj}}\right)^{1/2}, \qquad j=1,2,\ldots,k \tag{4.68a}$$

and

$$\hat{\beta}_0=\bar{y}-\sum_{j=1}^{k}\hat{\beta}_j\bar{x}_j \tag{4.68b}$$

Many multiple regression computer programs use this scaling to reduce problems arising from round-off errors in the $(\mathbf{X}'\mathbf{X})^{-1}$ matrix. These round-off errors may be very serious if the original variables differ considerably in magnitude. Most computer programs also display both the original regression coefficients and the standardized regression coefficients, which are often referred to as "beta-coefficients." In interpreting standardized regression coefficients, we must remember that they are still *partial* regression coefficients (i.e., b_j measures the effect of x_j given that other regressors $x_i, i\neq j$ are in the model). Furthermore the $\hat{b}_j$ are affected by the range of values for the regressor variables. Consequently it may be dangerous to use the magnitude of the $\hat{b}_j$ as a measure of the relative importance of regressor x_j.

*It is customary to refer to r_{ij} and r_{jy} as correlations even though the regressors are not necessarily random variables.

• **Example 4.14** We will find the standardized regression coefficients for the delivery time data in Example 4.1. Since

$$S_{yy}=5784.5426 \qquad S_{11}=1136.5600$$
$$S_{1y}=2473.3440 \qquad S_{22}=2,537,935.0330$$
$$S_{2y}=108,038.6019 \qquad S_{12}=44266.6800$$

we find (assuming the unit length scaling) that

$$r_{12}=\frac{S_{12}}{(S_{11}S_{22})^{1/2}}=\frac{44266.6800}{\sqrt{(1136.5600)(2,537,935.0330)}}=0.824215$$

$$r_{1y}=\frac{S_{1y}}{(S_{11}S_{yy})^{1/2}}=\frac{2473.3440}{\sqrt{(1136.5600)(5784.5426)}}=0.964615$$

$$r_{2y}=\frac{S_{2y}}{(S_{22}S_{yy})^{1/2}}=\frac{108,038.6019}{\sqrt{2,537,935.0330)(5784.5426)}}=0.891670$$

and the correlation matrix for this problem is

$$\mathbf{W'W}=\begin{bmatrix}1 & 0.824215\\ 0.824215 & 1\end{bmatrix}$$

The normal equations in terms of the standardized regression coefficients are

$$\begin{bmatrix}1 & 0.824215\\ 0.824215 & 1\end{bmatrix}\begin{bmatrix}\hat{b}_1\\ \hat{b}_2\end{bmatrix}=\begin{bmatrix}0.964615\\ 0.891670\end{bmatrix}$$

Consequently, the standardized regression coefficients are

$$\begin{bmatrix}\hat{b}_1\\ \hat{b}_2\end{bmatrix}=\begin{bmatrix}1 & 0.824215\\ 0.824215 & 1\end{bmatrix}^{-1}\begin{bmatrix}0.964615\\ 0.891670\end{bmatrix}$$

$$=\begin{bmatrix}3.11841 & -2.57023\\ -2.57023 & 3.11841\end{bmatrix}\begin{bmatrix}0.964615\\ 0.891670\end{bmatrix}$$

$$=\begin{bmatrix}0.716267\\ 0.301311\end{bmatrix}$$

The fitted model is

$$\hat{y}^0=0.716267w_1+0.301311w_2$$

Thus increasing the standardized value of cases w_1 by one unit increases the standardized value of time $\hat{y}^0$ by 0.716267. Furthermore increasing the standardized value of distance w_2 by one unit increases $\hat{y}^0$ by 0.301311 units. Therefore it seems that the volume of product delivered is more important than the distance in that it has a larger effect on delivery time in terms of the standardized variables. However, we should be somewhat cautious in reaching this conclusion, as $\hat{b}_1$ and $\hat{b}_2$ are still *partial* regression coefficients, and $\hat{b}_1$ and $\hat{b}_2$ are affected by the spread in the regressors. That is, if we took another sample with a different range of values for cases and distance, we might draw different conclusions about the relative importance of these regressors. •

4.9 Sample computer output

Figure 4.15 presents the computer output from the SAS General Linear Models Procedure for the soft drink delivery time data in Example 4.1.

GENERAL LINEAR MODELS PROCEDURE

DEPENDENT VARIABLE: TIME

SOURCE	DF	SUM OF SQUARES	MEAN SQUARE	F VALUE
MODEL	2	5550.81092258	2775.40546129	261.24
ERROR	22	233.73167742	10.62416716	PR > F
CORRECTED TOTAL	24	5784.54260000		0.0001

R-SQUARE	C.V.	STD DEV	TIME MEAN
0.959594	14.5616	3.25947345	22.38400000

SOURCE	DF	TYPE I SS	F VALUE	PR > F
CASES	1	5382.40879702	506.62	0.0001
DIST	1	168.40212556	15.85	0.0006

SOURCE	DF	TYPE IV SS	F VALUE	PR > F
CASES	1	951.66266301	89.56	0.0001
DIST	1	168.40212556	15.85	0.0006

PARAMETER	ESTIMATE	T FOR H0: PARAMETER=0	PR > |T|	STD ERROR OF ESTIMATE
INTERCEPT	2.34123115	2.13	0.0442	1.09673017
CASES	1.61590721	9.46	0.0001	0.17073492
DIST	0.01438483	3.98	0.0006	0.00361309

Figure 4.15 Sample computer output (SAS) of two-variable model for delivery time data.

While the output format differs from one computer program to another, this display contains the information typically generated. We have labeled the response variable **TIME** and the regressor variables **CASES** and **DIST**. Most of the output in Figure 4.15 is self-explanatory, and is a straightforward extension to the multiple regression case of the computer output for simple linear regression discussed in Section 2.11. Two exceptions are the "Type I" and "Type IV" sums of squares. The Type I sum of squares result from a sequential single degree of freedom partition of the regression sum of squares. That is,

$$SS_R(\beta_1|\beta_0)=5382.40879702$$

$$SS_R(\beta_2|\beta_1,\beta_0)=168.40212556$$

and we note that

$$SS_R(\beta_1,\beta_2|\beta_0)=SS_R(\beta_1|\beta_0)+SS_R(\beta_2|\beta_1,\beta_0)$$

$$5550.81092258=5382.40879702+168.40212556$$

The Type IV sums of squares are the sums of squares associated with the usual partial F-tests, that is,

$$SS_R(\beta_1|\beta_2,\beta_0)=951.66266301$$

and

$$SS_R(\beta_2|\beta_1,\beta_0)=168.40212556$$

Both regressors **CASES** and **DIST** contribute significantly to the model.

The computer output does not exactly match the results given in previous examples. This is largely due to round-off errors, both in the hand calculations and in the computer program. Different computer programs could lead to different results for the same problem because the arithmetic operations are performed in different ways, and round-off errors are not dealt with equally well in all programs. Round-off errors can be particularly damaging when some of the regressors are highly correlated (the multicollinearity problem referred to in Section 4.7.2). It is a good idea to investigate the accuracy of a multiple regression computer program before using it on important problems. An obvious procedure is to analyze a set of data for which an accurate solution is known. Other error-checking procedures have been devised by Longley [1967] and Mullet and Murray [1971].

4.10 Computational aspects

In this section we briefly outline an important computational procedure for solving the least squares regression problem. The least squares criterion is

$$\min_{\boldsymbol{\beta}} S(\boldsymbol{\beta}) = (\mathbf{y} - \mathbf{X}\boldsymbol{\beta})'(\mathbf{y} - \mathbf{X}\boldsymbol{\beta})$$

and recall from Section 4.2.2 that the least squares solution vector is normal to the p–dimensional estimation space. Since the Euclidean norm is invariant under an orthogonal transformation, an equivalent formulation of the least squares problem is

$$\min_{\boldsymbol{\beta}} S(\boldsymbol{\beta}) = (\mathbf{Q}\mathbf{y} - \mathbf{Q}\mathbf{X}\boldsymbol{\beta})'(\mathbf{Q}\mathbf{y} - \mathbf{Q}\mathbf{X}\boldsymbol{\beta}) \tag{4.69}$$

where $\mathbf{Q}$ is an $n \times n$ orthogonal matrix. Now $\mathbf{Q}$ may be chosen so that

$$\mathbf{Q}\mathbf{X} = \begin{bmatrix} \mathbf{R} \\ \mathbf{0} \end{bmatrix}$$

where $\mathbf{R}$ is a $p \times p$ upper triangular matrix (i.e., a matrix with zeros below the main diagonal). If we let

$$\mathbf{Q}\mathbf{y} = \begin{bmatrix} \mathbf{q}_1 \\ \mathbf{q}_2 \end{bmatrix} = \begin{bmatrix} \mathbf{Q}_1'\mathbf{y} \\ \mathbf{Q}_2'\mathbf{y} \end{bmatrix}$$

where $\mathbf{Q}_1'$ is a $p \times n$ matrix consisting of the first p rows of $\mathbf{Q}$, $\mathbf{Q}_2'$ is an $(n-p) \times n$ matrix consisting of the last $n-p$ rows of $\mathbf{Q}$, and $\mathbf{q}_1$ is a $p \times 1$ column vector, then the solution to (4.69) satisfies

$$\mathbf{R}\hat{\boldsymbol{\beta}} = \mathbf{q}_1 \tag{4.70}$$

or

$$\hat{\boldsymbol{\beta}} = \mathbf{R}^{-1}\mathbf{q}_1 = \mathbf{R}^{-1}\mathbf{Q}_1'\mathbf{y} \tag{4.71}$$

One advantage of this approach is that we can obtain a numerically stable inverse of $\mathbf{R}$ by the method of back substitution. To illustrate, suppose that $\mathbf{R}$ and $\mathbf{q}_1 = \mathbf{Q}_1'\mathbf{y}$ are, for $p = 3$,

$$\mathbf{R} = \begin{bmatrix} 3 & 1 & 2 \\ 0 & 3 & 1 \\ 0 & 0 & 2 \end{bmatrix}, \qquad \mathbf{q}_1 = \mathbf{Q}_1'\mathbf{y} = \begin{bmatrix} 3 \\ -1 \\ 4 \end{bmatrix}$$

The equations (4.70) are

$$\begin{bmatrix} 3 & 1 & 2 \\ 0 & 3 & 1 \\ 0 & 0 & 2 \end{bmatrix} \begin{bmatrix} \hat{\beta}_0 \\ \hat{\beta}_1 \\ \hat{\beta}_2 \end{bmatrix} = \begin{bmatrix} 3 \\ -1 \\ 4 \end{bmatrix}$$

and the system of equations that must actually be solved is

$$3\hat{\beta}_0 + 1\hat{\beta}_1 + 2\hat{\beta}_2 = 3$$

$$3\hat{\beta}_1 + 1\hat{\beta}_2 = -1$$

$$2\hat{\beta}_2 = 4$$

From the bottom equation, we have $2\hat{\beta}_2 = 4$ or $\hat{\beta}_2 = 2$. Substituting in the equation directly above it yields $3\hat{\beta}_1 + 1(2) = -1$ or $\hat{\beta}_1 = -1$. Finally, the first equation gives $3\hat{\beta}_0 + 1\hat{\beta}_1 + 2\hat{\beta}_2 = 3$ or $\hat{\beta}_0 = 0$.

Algorithms for computing the **QR** decomposition are described by Golub [1969], Lawson and Hanson [1974], and Seber [1977]. FORTRAN subroutines for these computations are available for many types of computers in the LINPACK [1979] and ROSEPACK [1980] subroutine libraries.

The $(\mathbf{X}'\mathbf{X})^{-1}$ matrix can be found directly from the **QR** factorization. Since

$$\mathbf{QX} = \begin{bmatrix} \mathbf{R} \\ \mathbf{0} \end{bmatrix}$$

then

$$\mathbf{X} = \mathbf{Q}' \begin{bmatrix} \mathbf{R} \\ \mathbf{0} \end{bmatrix} = \mathbf{Q}_1 \mathbf{R}$$

Consequently, since $\mathbf{Q}_1'\mathbf{Q}_1 = \mathbf{I}$,

$$(\mathbf{X}'\mathbf{X})^{-1} = (\mathbf{R}'\mathbf{Q}_1'\mathbf{Q}_1\mathbf{R})^{-1} = (\mathbf{R}'\mathbf{R})^{-1} = \mathbf{R}^{-1}(\mathbf{R}')^{-1} \tag{4.72}$$

This decomposition also leads to efficient computation of the elements of the hat matrix, which we have seen to be useful in several respects. Note that

$$\begin{aligned} \mathbf{H} &= \mathbf{X}(\mathbf{X}'\mathbf{X})^{-1}\mathbf{X}' = \mathbf{Q}_1\mathbf{R}\mathbf{R}^{-1}(\mathbf{R}')^{-1}\mathbf{R}'\mathbf{Q}_1' \\ &= \mathbf{Q}_1\mathbf{Q}_1' \end{aligned} \tag{4.73}$$

Therefore the main diagonal elements of the hat matrix may be formed as the sums of squares of the rows of $\mathbf{Q}_1$. Thus we may easily compute many important regression diagnostic statistics, such as the studentized residuals and Cook's distance measure. Belsley, Kuh, and Welsch [1980] show how a number of regression diagnostics may be computed using these ideas.

Problems

4.1 Consider the National Football League data in Appendix Table B.1.

a. Fit a multiple linear regression model relating the number of games won to the team's passing yardage (x_2), the percent of rushing plays (x_7), and the opponents' yards rushing (x_8).

b. Construct the analysis of variance table and test for significance of regression.

c. Calculate R^2 for this model.

d. Construct the appropriate residual plots and discuss the adequacy of this model.

e. Using the partial F–test, determine the contribution of x_7 to the model.

4.2 Using the results of Problem 4.1, show that the square of the simple correlation coefficient between y_i and $\hat{y}_i$ equals R^2.

4.3 Refer to Problem 4.1. Find:

a. A 95 percent confidence interval on β_7.

b. A 95 percent confidence interval on the mean number of games won by a team when $x_2 = 2300$, $x_7 = 56.0$, and $x_8 = 2100$.

4.4 Consider the gasoline mileage data in Appendix Table B.3.

a. Fit a multiple linear regression model relating gasoline mileage y (MPG) to engine displacement x_1 and the number of carburetor barrels x_6.

b. Construct the analysis of variance table and test for significance of regression.

c. Calculate R^2 for this model. Compare this to the R^2 for the simple linear regression model relating mileage to engine displacement in Problem 2.4.

d. Find a 95 percent confidence interval for β_1.

e. Assess the importance of adding x_6 to a model that already contains x_1. Given that x_1 is in the model, is x_6 necessary?

f. Given that x_6 is in the model, is x_1 necessary?

4.5 Plot the residuals from the regression model in Problem 4.4 against $\hat{y}_i$, x_{i1}, and x_{i6}. Also construct a normal probability plot. What conclusions can you draw about model adequacy?

4.6 Consider the gasoline mileage data in Appendix Table B.3.

a. Fit a multiple linear regression model relating gasoline mileage (MPG) to vehicle length x_8 and the vehicle weight x_{10}.

b. Construct the analysis of variance table and test for significance of regression.

c. Calculate R^2 for this model. Compare this to the R^2 for the simple linear regression model relating mileage to vehicle weight x_{10} developed in Problem 2.5.

d. What is the value of adding x_8 to a model that already contains x_{10}?

e. Are both regressors necessary in this model?

4.7 Compute the residuals from the regression model in Problem 4.6.

a. Plot the residuals against $\hat{y}_i$, x_{i8}, and x_{i10}. Construct a normal probability plot. What can you say about model adequacy?

b. Plot the residuals against x_{i1} (engine displacement) and x_{i6} (number of carburetor barrels). Is there an indication that the model could be improved by the addition of either of these variables?

4.8 Refer to Problem 4.4.

a. Find a 95 percent confidence interval on the mean gasoline mileage when $x_1 = 250$ in^3 and $x_6 = 2$ barrels.

b. Find a 95 percent prediction interval for a new observation on gasoline mileage when $x_1 = 250$ in^3 and $x_6 = 2$ barrels.

4.9 Construct partial residual plots for the gasoline mileage regression model developed in Problem 4.4. What conclusions can you draw from these plots?

4.10 Construct partial residual plots for the gasoline mileage regression model developed in Problem 4.6. What conclusion can you draw from these plots?

4.11 Consider the house price data in Appendix Table B.4.

a. Fit a multiple regression model relating selling price to all nine regressors.

b. Test for significance of regression. What statements can you make about the contribution of each individual regressor to the model?

c. What is the contribution of lot size and living space to the fit, given that all other regressors are in the model.

d. Construct appropriate residual plots. Does the model seem to be adequate?

e. Discuss the practical aspects of using a model such as this to predict selling price for a house. Could this procedure be used to appraise property for tax purposes? What other factors do you think would have to be included in the model?

4.12 The data in Appendix Table B.5 presents the performance of a chemical process as a function of several controllable process variables.

a. Fit a multiple regression model relating CO_2 product (y) to total solvent (x_6) and hydrogen consumption (x_7).

b. Test for significance of regression.

c. Using the partial F-test, determine the contribution of x_6 to the model.

d. Using the partial F-test, determine the contribution of x_7 to the model.

e. Plot the residuals against $\hat{y}_i$, x_{i6}, and x_{i7}. Comment on these plots.

f. Construct a normal probability plot of residuals.

4.13 The concentration of $NbOCl_3$ in a tube-flow reactor as a function of several controllable variables is shown in Appendix Table B.6.

a. Fit a multiple regression model relating concentration of $NbOCl_3$ (y) to concentration of $COCl_2$ (x_1) and molar density (x_4).

b. Test for significance of regression.

c. Calculate R^2 for this model.

d. Using the partial F-test, determine the contribution of x_4 to the model. Are both regressors x_1 and x_4 necessary?

e. Construct appropriate residual plots for this model. Comment on the adequacy of the model.

4.14 The CPU time of a computer job is thought to be related to the number of lines of output generated, the number of program steps, and the number of mounted devices required by the job. Data for 38 jobs is given in Appendix Table B.7.

a. Fit a multiple regression model relating CPU time to these regressor variables.

b. Test for significance of regression.

c. What is the contribution of the number of lines of output to the model?

d. Construct appropriate residual plots for this model. Comment on model adequacy.

4.15 Refer to Problem 4.1. Using data points that are near-neighbors calculate an estimate of the standard deviation of pure error. Does this indicate any obvious lack of fit?

4.16 Refer to Problem 4.1. Are there any high-leverage points in the data?

4.17 Refer to Problem 4.4. Using data points that are near-neighbors, calculate an estimate of the standard deviation of pure error. Does this indicate any obvious lack of fit?

4.18 Refer to Problem 4.4. Are there any high-leverage points in the data?

4.19 Refer to Problem 4.1. Calculate the standardized regression coefficients. Can you interpret these coefficients?

4.20 Refer to Problem 4.4. Calculate the standardized regression coefficients. What conclusions can you draw?

4.21 Consider the multiple linear regression model

$$y=\beta_0+\beta_1 x_1+\beta_2 x_2+\beta_3 x_3+\beta_4 x_4+\varepsilon$$

Using the procedure for testing a general linear hypothesis, show how to test

a. H_0: $\beta_1=\beta_2=\beta_3=\beta_4=\beta$
b. H_0: $\beta_1=\beta_2, \beta_3=\beta_4$
c. H_0: $\beta_1-2\beta_2=4\beta_3$
$\beta_1+2\beta_2=0$

4.22 Suppose we fit the model $\mathbf{y}=\mathbf{X}\boldsymbol{\beta}+\boldsymbol{\varepsilon}$ and test the hypothesis H_0: $\mathbf{T}\boldsymbol{\beta}=\mathbf{0}$. If this hypothesis is not rejected, then we have found no strong evidence to contradict the statement $\mathbf{T}\boldsymbol{\beta}=\mathbf{0}$. Therefore it seems reasonable to require that $\mathbf{T}\hat{\boldsymbol{\beta}}=\mathbf{0}$ where $\hat{\boldsymbol{\beta}}$ is the estimator of $\boldsymbol{\beta}$. Now it is unlikely that the unconstrained least squares estimator $\hat{\boldsymbol{\beta}}$ will satisfy $\mathbf{T}\hat{\boldsymbol{\beta}}=\mathbf{0}$. Find an estimator $\tilde{\boldsymbol{\beta}}$ that minimizes the residual sum of squares such that the constraint $\mathbf{T}\tilde{\boldsymbol{\beta}}=\mathbf{0}$ is satisfied. Show how this estimator can be computed by modifying the elements in $\hat{\boldsymbol{\beta}}$.

4.23 Show that an alternate computing formula for the regression sum of squares is

$$SS_R=\sum_{i=1}^{n}\hat{y}_i^2-n\bar{y}^2$$

4.24 Show that $V(\hat{\mathbf{y}})=\sigma^2\mathbf{H}$.

4.25 Prove that the matrices $\mathbf{H}$ and $\mathbf{I}-\mathbf{H}$ are idempotent.

4.26 For the simple linear regression model, show that the elements of the hat matrix are

$$h_{ij}=\frac{1}{n}+\frac{(x_i-\bar{x})(x_j-\bar{x})}{S_{xx}} \quad \text{and} \quad h_{ii}=\frac{1}{n}+\frac{(x_i-\bar{x})^2}{S_{xx}}$$

Discuss the behavior of these quantities as x_i moves further from $\bar{x}$.

4.27 Show that the ith least squares residual can be written as

$$e_i = \varepsilon_i - \sum_{j=1}^{n} h_{ij}\varepsilon_j$$

Discuss the impact of this equation on the apparent distribution of e_i if n is large relative to p and if some of the h_{ij} are large. Does this explain why residual plots from nonnormal error distributions often fail to reveal any departure from normality?

4.28 For the multiple regression model, show that $SS_R(\boldsymbol{\beta}) = \mathbf{y}'\mathbf{H}\mathbf{y}$.

4.29 Prove that R^2 is the square of the correlation between $\mathbf{y}$ and $\hat{\mathbf{y}}$.

4.30 **Constrained Least Squares.** Suppose we wish to find the least squares estimator of $\boldsymbol{\beta}$ in the model $\mathbf{y} = \mathbf{X}\boldsymbol{\beta} + \boldsymbol{\varepsilon}$, subject to a set of equality contraints on $\boldsymbol{\beta}$, say $\mathbf{A}\boldsymbol{\beta} = \mathbf{a}$. Show that the estimator is

$$\tilde{\boldsymbol{\beta}} = \hat{\boldsymbol{\beta}} + (\mathbf{X}'\mathbf{X})^{-1}\mathbf{A}'\left[\mathbf{A}(\mathbf{X}'\mathbf{X})^{-1}\mathbf{A}'\right]^{-1}(\mathbf{a} - \mathbf{A}\hat{\boldsymbol{\beta}})$$

where $\hat{\boldsymbol{\beta}} = (\mathbf{X}'\mathbf{X})^{-1}\mathbf{X}'\mathbf{y}$. Discuss situations in which this constrained estimator might be appropriate. Find the residual sum of squares for the constrained estimator. Is it larger or smaller than the residual sum of squares in the unconstrained case?

5

POLYNOMIAL REGRESSION MODELS

5.1 Introduction

The linear regression model $\mathbf{y}=\mathbf{X}\boldsymbol{\beta}+\boldsymbol{\varepsilon}$ is a general model for fitting any relationship that is linear in the unknown parameters $\boldsymbol{\beta}$. This includes the important class of polynomial regression models. For example, the second-order polynomial in one variable

$$y=\beta_0+\beta_1 x+\beta_2 x^2+\varepsilon$$

and the second-order polynomial in two variables

$$y=\beta_0+\beta_1 x_1+\beta_2 x_2+\beta_{11} x_1^2+\beta_{22} x_2^2+\beta_{12} x_1 x_2+\varepsilon$$

are linear regression models.

Polynomials are widely used in situations where the response is curvilinear, because even complex nonlinear relationships can be adequately modeled by polynomials over reasonably small ranges of the x's. This chapter will survey several major types of problems associated with fitting polynomials.

5.2 Polynomial models in one variable

5.2.1 Basic principles

As an example of a polynomial regression model in one variable, consider

$$y=\beta_0+\beta_1x+\beta_2x^2+\varepsilon \tag{5.1}$$

This model is called a second-order model in one variable. It is also sometimes called the *quadratic model*, since the expected value of y is

$$E(y)=\beta_0+\beta_1x+\beta_2x^2$$

which describes a quadratic function. A typical example is shown in Figure 5.1. We often call β_1 the linear effect parameter and β_2 the quadratic effect parameter. The parameter β_0 is the mean of y when $x=0$ if the range of the data includes $x=0$. Otherwise β_0 has no physical interpretation.

In general the kth–order polynomial model in one variable is

$$y=\beta_0+\beta_1x+\beta_2x^2+\cdots+\beta_kx^k+\varepsilon \tag{5.2}$$

If we set $x_j=x^j$, $j=1,2,\ldots,k$, then (5.2) becomes a multiple linear regres-

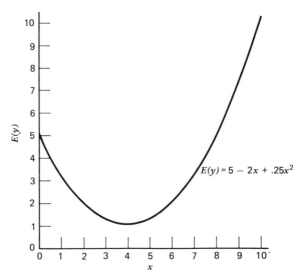

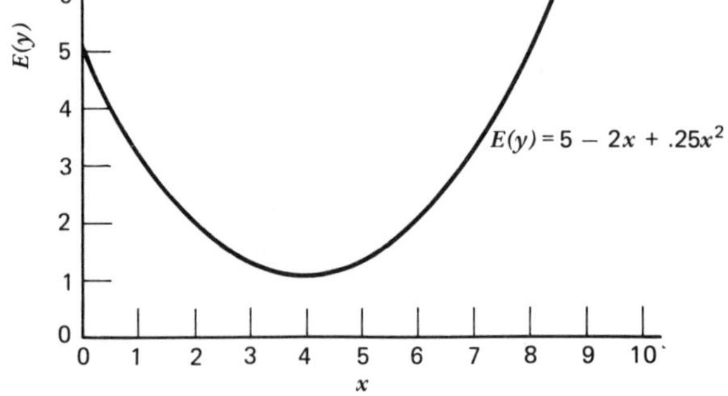

Figure 5.1 An example of a quadratic polynomial.

sion model in the k regressors $x_1, x_2, \ldots, x_k$. Thus a polynomial model of order k may be fitted using the techniques studied previously.

Polynomial models are useful in situations where the analyst knows that curvilinear effects are present in the true response function. They are also useful as approximating functions to unknown and possibly very complex nonlinear relationships. In this sense, the polynomial model is just the Taylor series expansion of the unknown function. This type of application seems to occur most often in practice.

There are seveal special problems that arise when fitting a polynomial in one variable. Some of these are discussed below.

1. It is important to keep the order of the model as low as possible. When the response function appears to be curvilinear, transformations should be tried to keep the model first-order. The methods discussed in Chapter 3 are useful in this regard. If this fails, a second-order polynomial should be tried. As a general rule the use of high-order polynomials ($k>2$) should be avoided unless they can be justified for reasons outside the data. A low-order model in a transformed variable is almost always preferable to a high-order model in the original metric. Arbitrary fitting of high-order polynomials is a serious abuse of regression analysis. One should always maintain a sense of *parsimony*—that is, use the simplest possible model that is consistent with the data and knowledge of the problem environment. Remember that in an extreme case it is always possible to pass a polynomial of order $n-1$ through n points, so that a polynomial of sufficiently high degree can always be found that provides a "good" fit to the data. Such a model would do nothing to enhance understanding of the unknown function, nor will it likely be a good predictor.

2. Various strategies for choosing the order of an approximating polynomial have been suggested. One approach is to successively fit models of increasing order until the t-test for the highest-order term is nonsignificant. An alternate procedure is to fit the highest-order model appropriate and then delete terms one at a time, starting with the highest-order, until the highest-order remaining term has a significant t-statistic. These two procedures are called *forward selection* and *backward elimination*, respectively. They do not necessarily lead to the same model. In light of the comment in (1) above, these procedures should be used carefully. In most situations, we should restrict our attention to first and second-order polynomials.

3. Extrapolation with polynomial models can be extremely hazardous. For example, consider the second-order model in Figure 5.2. If we extrapolate beyond the range of the original data, the predicted response turns downward. This may be at odds with the true behavior of the system. In

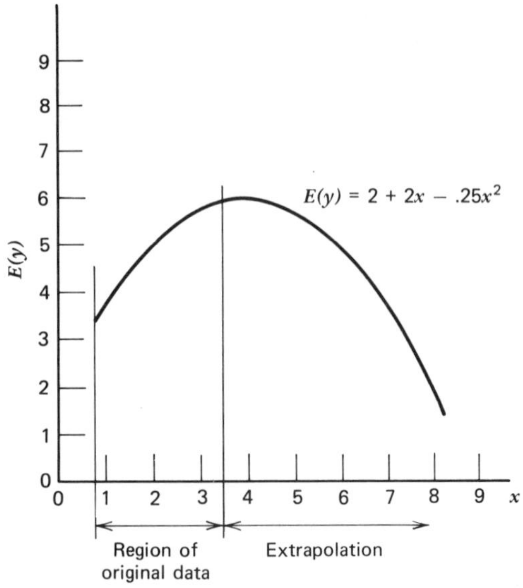

Figure 5.2 Danger of extrapolation.

general polynomial models may turn in unanticipated and inappropriate directions, both in interpolation and extrapolation.

4. As the order of the polynomial increases, the **X′X** matrix becomes ill-conditioned. This means that the matrix inversion calculations will be inaccurate, and considerable error may be introduced into the parameter estimates. For example, see Forsythe [1957]. Nonessential ill-conditioning caused by the arbitrary choice of origin can be removed by first centering the regressor variables (that is, correcting x for its average $\bar{x}$), but as Bradley and Srivastava [1979] point out, even centering the data can still result in large sample correlations between certain regression coefficients. A method for dealing with this problem will be discussed in Section 5.4.

5. If the values of x are limited to a narrow range, there can be significant ill-conditioning or multicollinearity in the columns of the **X** matrix. For example, if x varies between 1 and 2, x^2 varies between 1 and 4, which could create strong multicollinearity between x and x^2.

We will now illustrate some of the analyses typically associated with fitting a polynomial model in one variable.

• **Example 5.1** Table 5.1 presents data concerning the strength of Kraft paper and the percentage of hardwood in the batch of pulp from which the paper

Table 5.1 Hardwood concentration in pulp and tensile strength of Kraft paper, Example 5.1

x_i, Hardwood Concentration (%)	y_i, Tensile Strength (Psi)
1	6.3
1.5	11.1
2	20.0
3	24.0
4	26.1
4.5	30.0
5	33.8
5.5	34.0
6	38.1
6.5	39.9
7	42.0
8	46.1
9	53.1
10	52.0
11	52.5
12	48.0
13	42.8
14	27.8
15	21.9

was produced. A scatter diagram of this data is shown in Figure 5.3. This display and knowledge of the production process suggests that a quadratic model may adequately describe the relationship between tensile strength and hardwood concentration. Following the suggestion that centering the data may remove nonessential ill-conditioning, we will fit the model

$$y=\beta_0+\beta_1(x-\bar{x})+\beta_2(x-\bar{x})^2+\varepsilon$$

Since fitting this model is equivalent to fitting a two-variable regression model, we can use the general approach in Chapter 4. The fitted model is

$$\hat{y}=45.295+2.546(x-7.2632)-0.635(x-7.2632)^2$$

The analysis of variance for this model is shown in Table 5.2. The observed value of $F_0=79.434>F_{.01,2,16}=6.23$, so the hypothesis $H_0: \beta_1=\beta_2=0$ is rejected. We conclude that either the linear or the quadratic term (or both) contribute significantly to the model. The other summary statistics for this model are $R^2=0.9085$, se $(\hat{\beta}_1)=0.254$, and se $(\hat{\beta}_2)=0.062$.

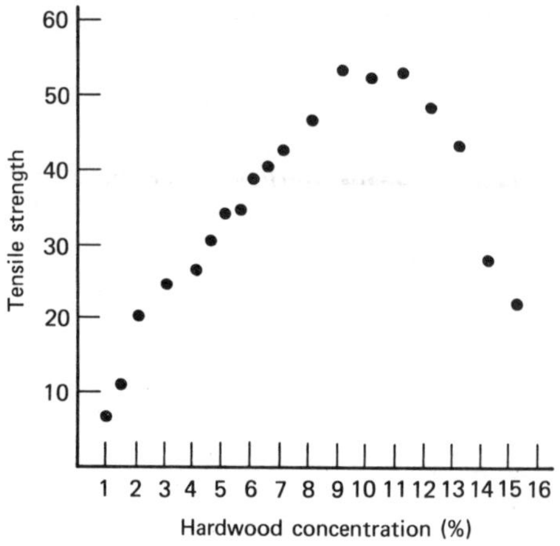

Figure 5.3 Scatter plot of data, Example 5.1.

The plots of the residuals versus $\hat{y}_i$ and $(x_i - \bar{x})$ are shown in Figures 5.4 and 5.5, respectively. These plots do not reveal any serious model inadequacy. The normal probability plot of the residuals, shown in Figure 5.6, is mildly disturbing, indicating that the error distribution has heavier tails than the normal. However, at this point we do not seriously question the normality assumption.

Now suppose that we wish to investigate the contribution of the quadratic term to the model. That is, we wish to test

$$H_0: \beta_2 = 0$$

$$H_1: \beta_2 \neq 0$$

We will test this hypothesis using the extra sum of squares method. If $\beta_2 = 0$, then the reduced model is the straight line $y = \beta_0 + \beta_1(x - \bar{x}) + \varepsilon$. The least squares fit is

$$\hat{y} = 34.184 + 1.771(x - 7.2632)$$

Table 5.2 Analysis of variance for the quadratic model for Example 5.1

Source of Variation	Sum of Squares	Degrees of Freedom	Mean Square	F_0
Regression	3104.247	2	1552.123	79.434
Residual	312.638	16	19.540	
Total	3416.885	18		

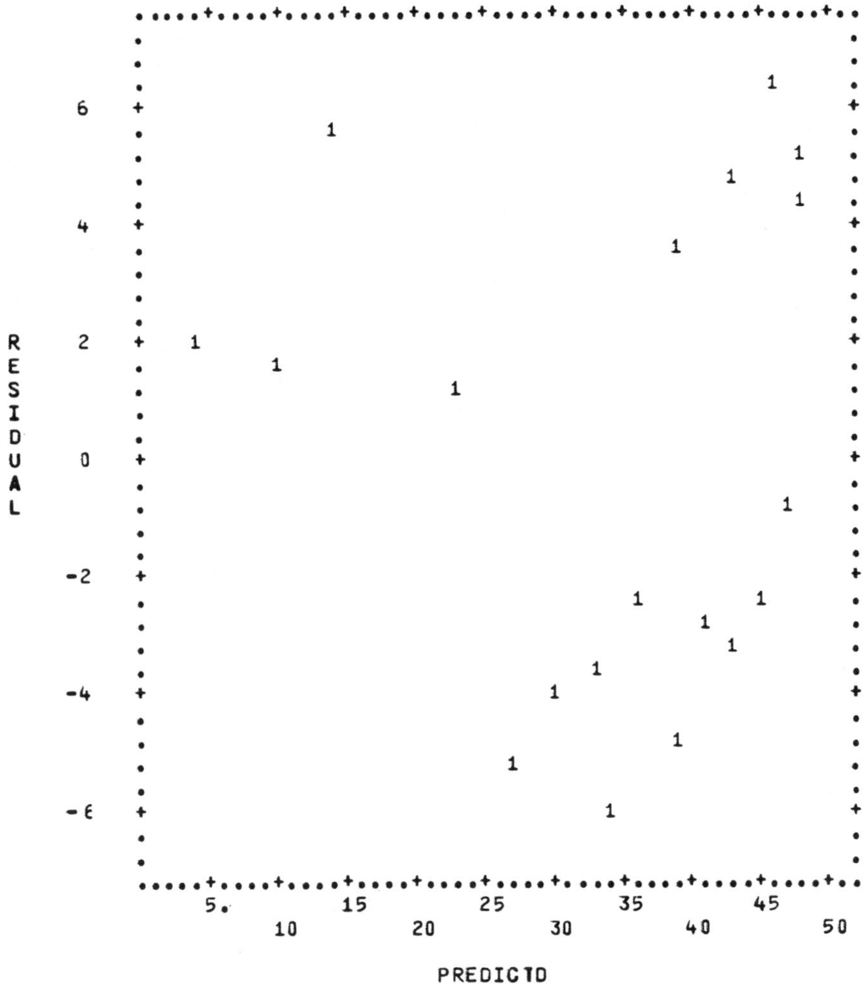

Figure 5.4 Plot (BMD–P) of residuals e_i versus fitted values $\hat{y}_i$, Example 5.1.

The summary statistics for this model are $MS_E = 139.615$, $R^2 = 0.3054$, se($\hat{\beta}_1$) $= 0.648$, and $SS_R(\beta_1|\beta_0) = 1043.427$. We note that deleting the quadratic term has substantially affected R^2, MS_E, and se($\hat{\beta}_1$). These summary statistics are much worse than they were for the quadratic model. The extra sum of squares for testing H_0: $\beta_2 = 0$ is

$$\begin{aligned} SS_R(\beta_2|\beta_1, \beta_0) &= SS_R(\beta_1, \beta_2|\beta_0) - SS_R(\beta_1|\beta_0) \\ &= 3104.247 - 1043.427 \\ &= 2060.820 \end{aligned}$$

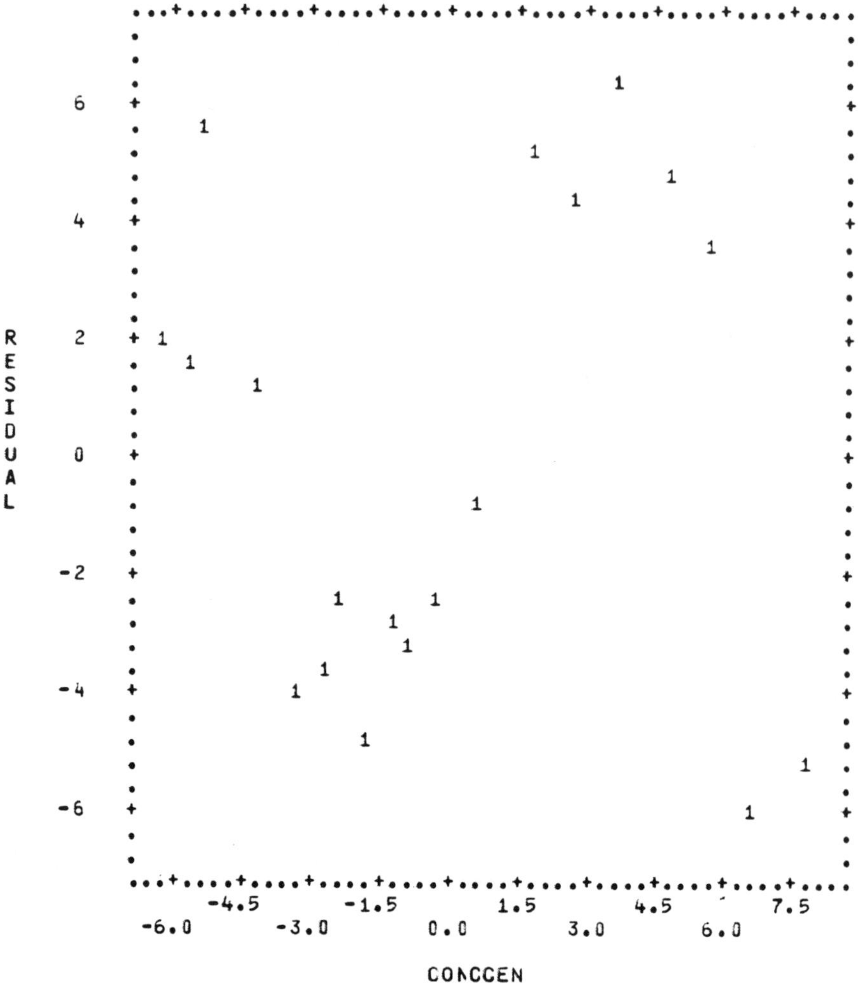

Figure 5.5 Plot (BMD–P) of residuals e_i versus $(x_i - \bar{x})$, Example 5.1.

with one degree of freedom. The F-statistic is

$$F_0 = \frac{SS_R(\beta_2|\beta_1, \beta_0)/1}{MS_E} = \frac{2060.820/1}{19.540} = 105.47$$

and since $F_{.01,1,16} = 8.53$, we conclude that $\beta_2 \neq 0$. Thus the quadratic term contributes significantly to the model.

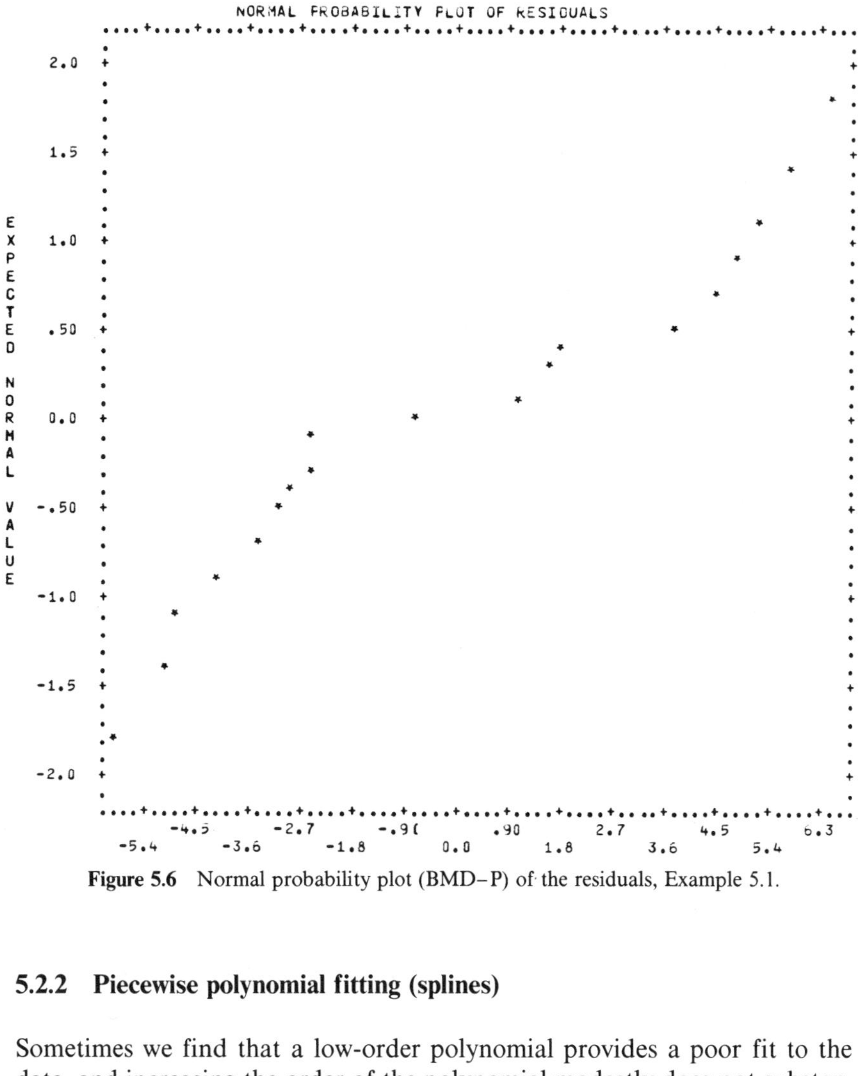

Figure 5.6 Normal probability plot (BMD–P) of the residuals, Example 5.1.

5.2.2 Piecewise polynomial fitting (splines)

Sometimes we find that a low-order polynomial provides a poor fit to the data, and increasing the order of the polynomial modestly does not substantially improve the situation. Symptoms of this are the failure of the residual sum of squares to stabilize, or residual plots that exhibit remaining unexplained structure. This problem may occur when the function behaves differently in different parts of the range of x. Occasionally transformations on x and/or y will eliminate this problem. The usual approach however, is to divide the range of x into segments and fit an appropriate curve in each segment. Spline functions offer a useful way to perform this type of piecewise polynomial fitting.

Splines are piecewise polynomials of order k. The joint points of the pieces are usually called *knots*. Generally we require the function values and the first $k-1$ derivatives to agree at the knots, so that the spline is a continuous function with $k-1$ continuous derivatives. The *cubic* spline ($k=3$) is usually adequate for most practical problems.

A cubic spline with h knots, $t_1<t_2<\cdots<t_h$, with continuous first and second derivatives, can be written as

$$E(y)=S(x)=\sum_{j=0}^{3}\beta_{0j}x^j+\sum_{i=1}^{h}\beta_i(x-t_i)_+^3 \tag{5.3}$$

where

$$(x-t_i)_+=\begin{cases}(x-t_i) & \text{if } x-t_i>0\\ 0 & \text{if } x-t_i\leq 0\end{cases}$$

We assume that the positions of the knots are known. If the knot positions are parameters to be estimated, the resulting problem is a nonlinear regression problem. When the knot positions are known however, fitting (5.3) can be accomplished by a straightfoward application of linear least squares.

Deciding on the number and position of the knots, and the order of the polynomial in each segment is not simple. Wold [1974] suggests that there should be as few knots as possible, with at least 4 or 5 data points per segment. Considerable caution should be exercised here, because the great flexibility of spline functions makes it very easy to "overfit" the data. Wold also suggests that there should be no more than one extreme point (maximum or minimum) and one point of inflexion per segment. Insofar as possible, the extreme points should be centered in the segment and the points of inflexion should be near the knots. When prior information about the data generating process is available, this can sometimes aid in knot positioning.

The basic cubic spline model (5.3) can be easily modified to fit polynomials of different order in each segment, and to impose different continuity restrictions at the knots. If all $h+1$ polynomial pieces are of order 3, then a cubic spline model with no continuity restrictions is

$$E(y)=S(x)=\sum_{j=0}^{3}\beta_{0j}x^j+\sum_{i=1}^{h}\sum_{j=0}^{3}\beta_{ij}(x-t_i)_+^j \tag{5.4}$$

where $(x-t)_+^0=1$ if $x>t$ and 0 if $x\leq t$. Thus if a term $\beta_{ij}(x-t_i)_+^j$ is in the model, this forces a discontinuity at t_i in the jth derivative of $S(x)$. If this term is absent, the jth derivative of $S(x)$ is continuous at t_i. The fewer continuity restrictions required, the better the fit because more parameters are in the model, while the more continuity restrictions required, the worse the fit but the smoother the final curve will be. Determining both the order of the polynomial segments and the continuity restrictions which do not substantially degrade the fit can be done using standard multiple regression hypothesis testing methods.

As an illustration consider a cubic spline with a single knot at t and no continuity restrictions, for example,

$$E(y)=S(x)=\beta_{00}+\beta_{01}x+\beta_{02}x^2+\beta_{03}x^3+\beta_{10}(x-t)_+^0$$
$$+\beta_{11}(x-t)_+^1+\beta_{12}(x-t)_+^2+\beta_{13}(x-t)_+^3$$

Note that neither $S(x)$, $S'(x)$, nor $S''(x)$ is necessarily continuous at t because of the presence of the terms involving β_{10}, β_{11}, and β_{12} in the model. Testing whether imposing continuity restrictions reduces the quality of the fit is done by testing the hypotheses H_0: $\beta_{10}=0$ [continuity of $S(x)$], H_0: $\beta_{10}=\beta_{11}=0$ [continuity of $S(x)$ and $S'(x)$], and H_0: $\beta_{10}=\beta_{11}=\beta_{12}=0$ [continuity of $S(x)$, $S'(x)$, and $S''(x)$]. To determine whether the cubic spline fits the data better than a single cubic polynomial over the range of x, simply test H_0: $\beta_{10}=\beta_{11}=\beta_{12}=\beta_{13}=0$.

An excellent description of this approach to fitting splines is in Smith [1979]. A potential disadvantage of this method is that the $\mathbf{X}'\mathbf{X}$ matrix becomes ill-conditioned if there are a large number of knots. This problem can be overcome by using a different representation of the spline called the *cubic B-spline*. The cubic B-splines are defined in terms of divided differences

$$B_i(x)=\sum_{j=i-4}^{i}\left[\frac{(x-t_j)_+^3}{\prod_{\substack{m=i-4\\ m\neq j}}^{i}(t_j-t_m)}\right], \qquad i=1,2,\ldots,h+4 \tag{5.5}$$

and

$$E(y)=S(x)=\sum_{i=1}^{h+4}\gamma_i B_i(x) \tag{5.6}$$

where γ_i, $i=1,2,\ldots,h+4$, are parameters to be estimated. In (5.5) there are

Table 5.3 Voltage drop data

Observation i	Time (seconds) x_i	Voltage Drop y_i
1	0.0	8.33
2	0.5	8.23
3	1.0	7.17
4	1.5	7.14
5	2.0	7.31
6	2.5	7.60
7	3.0	7.94
8	3.5	8.30
9	4.0	8.76
10	4.5	8.71
11	5.0	9.71
12	5.5	10.26
13	6.0	10.91
14	6.5	11.67
15	7.0	11.76
16	7.5	12.81
17	8.0	13.30
18	8.5	13.88
19	9.0	14.59
20	9.5	14.05
21	10.0	14.48
22	10.5	14.92
23	11.0	14.37
24	11.5	14.63
25	12.0	15.18
26	12.5	14.51
27	13.0	14.34
28	13.5	13.81
29	14.0	13.79
30	14.5	13.05
31	15.0	13.04
32	15.5	12.60
33	16.0	12.05
34	16.5	11.15
35	17.0	11.15
36	17.5	10.14
37	18.0	10.08
38	18.5	9.78
39	19.0	9.80
40	19.5	9.95
41	20.0	9.51

eight additional knots $t_{-3}<t_{-2}<t_{-1}<t_0$ and $t_{h+1}<t_{h+2}<t_{h+3}<t_{h+4}$. We usually take $t_0=x_{\min}$ and $t_{h+1}=x_{\max}$; the other knots are arbitrary. For further reading on splines, see Buse and Lim [1977], Curry and Schoenberg [1966], Gallant and Fuller [1973], Hayes [1970, 1974], Poirier [1973, 1975], and Wold [1974].

• **Example 5.2** The battery voltage drop in a guided missile motor observed over the time of missile flight is shown in Table 5.3. The scatter plot in Figure 5.7 suggests that voltage drop behaves differently in different segments of time, and so we will model the data with a cubic spline, using two knots at $t_1=6.5$ and $t_2=13$ seconds after launch, respectively. This placement of knots roughly agrees with course changes by the missile (with associated changes in power requirements), which are known from trajectory data. The voltage drop model is intended for use in a digital-analog simulation model of the missile.

The cubic spline model is

$$y=\beta_{00}+\beta_{01}x+\beta_{02}x^2+\beta_{03}x^3+\beta_1(x-6.5)_+^3+\beta_2(x-13)_+^3+\varepsilon$$

and the least squares fit is

$$\hat{y}=8.4657-1.4531x+0.4899x^2-0.0295x^3$$
$$+0.0247(x-6.5)_+^3+0.0271(x-13)_+^3$$

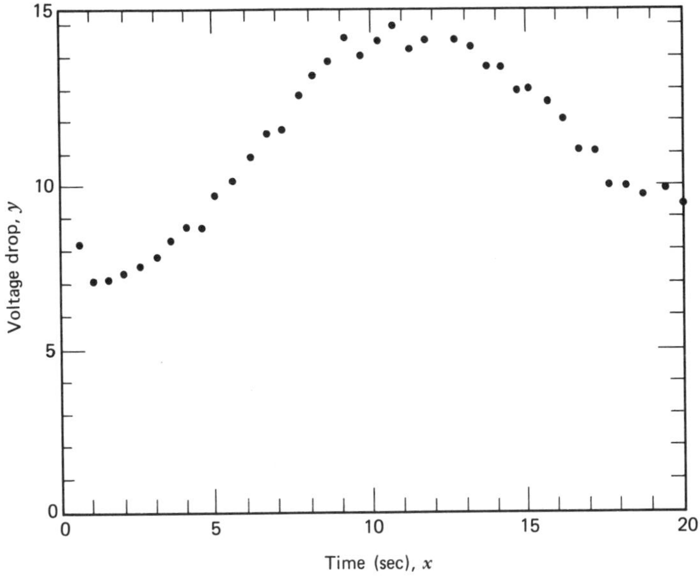

Figure 5.7 Scatter plot of voltage drop data.

The model summary statistics are displayed in Table 5.4. A plot of the residuals versus $\hat{y}$ and a normal probability plot of the residuals are shown in Figures 5.8 and 5.9, respectively. These plots do not reveal any serious departures from assumptions, so we conclude that the cubic spline model is an adequate fit to the voltage drop data.

We may easily compare the cubic spline model fit with a single cubic polynomial over the entire time of missile flight, for example

$$\hat{y}=6.4910+0.7032x+0.0340x^2-0.0033x^3$$

This is a simpler model containing fewer parameters, and would be preferable to the cubic spline model if it provides a satisfactory fit. The residuals from this cubic polynomial are plotted versus $\hat{y}$ in Figure 5.10. This plot exhibits strong indication of curvature, and on the basis of this remaining unexplained structure we conclude that the simple cubic polynomial is an inadequate model for the voltage drop data.

We may also investigate whether the cubic spline model improves the fit by testing the hypothesis H_0: $\beta_1=\beta_2=0$ using the extra sum of squares method. The regression sum of squares for the cubic polynomial is

$$SS_R(\beta_{01}, \beta_{02}, \beta_{03}|\beta_{00})=230.4444$$

with 3 degrees of freedom. The extra sum of squares for testing H_0: $\beta_1=\beta_2=0$

Table 5.4 Summary statistics for the cubic spline model of the voltage drop data

Source of Variation	Sum of Squares	Degrees of Freedom	Mean Square	F_0
Regression	260.1784	5	52.0357	725.52
Residual	2.5102	35	0.0717	
Total	262.6886	40		

Parameter	Estimate	Standard Error	t–value for H_0: $\beta=0$
β_{00}	8.4657	0.2005	42.22
β_{01}	−1.4531	0.1816	−8.00
β_{02}	0.4899	0.0430	11.39
β_{03}	−0.0295	0.0028	−10.35
β_1	0.0247	0.0040	6.12
β_2	0.0271	0.0036	7.58

$R^2=0.9904$

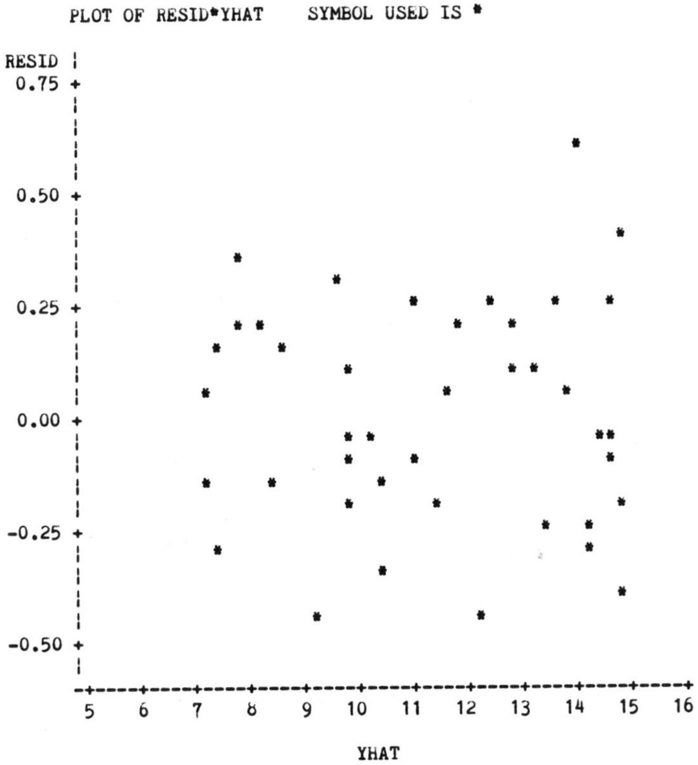

Figure 5.8 Plot (SAS) of residuals e_i versus fitted values $\hat{y}_i$ for the cubic spline model.

is

$$\begin{aligned} SS_R(\beta_1, \beta_2 | \beta_{00}, \beta_{01}, \beta_{02}, \beta_{03}) &= SS_R(\beta_{01}, \beta_{02}, \beta_{03}, \beta_1, \beta_2 | \beta_{00}) \\ &\quad - SS_R(\beta_{01}, \beta_{02}, \beta_{03} | \beta_{00}) \\ &= 260.1784 - 230.4444 \\ &= 29.7340 \end{aligned}$$

with 2 degrees of freedom. Since

$$F_0 = \frac{SS_R(\beta_1, \beta_2 | \beta_{00}, \beta_{01}, \beta_{02}, \beta_{03})/2}{MS_E} = \frac{29.7340/2}{0.0717} = 207.35$$

which would be referred to the $F_{2,35}$ distribution, we reject the hypothesis that H_0: $\beta_1 = \beta_2 = 0$. We conclude that the cubic spline model provides a better fit.

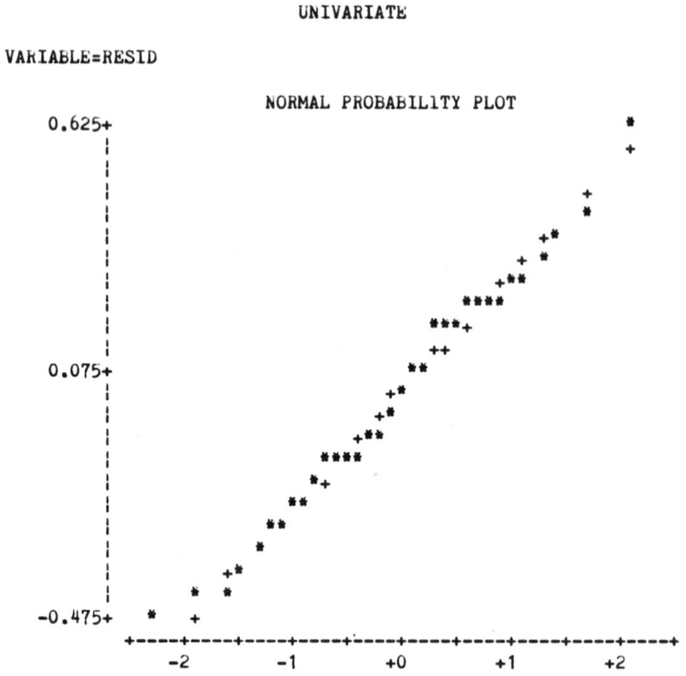

Figure 5.9 Normal probability plot (SAS) of the residuals for the cubic spline model.

- **Example 5.3 Piecewise linear regression** An important special case of practical interest involves fitting piecewise linear regression models. This can be treated easily using *linear* splines. For example, suppose that there is a single knot at t, and that there could be both a slope change and a discontinuity at the knot. The resulting linear spline model is

$$E(y)=S(x)=\beta_{00}+\beta_{01}x+\beta_{10}(x-t)_+^0+\beta_{11}(x-t)_+^1$$

Now if $x \leqslant t$, the straight line model is

$$E(y)=\beta_{00}+\beta_{01}x$$

and if $x>t$ the model is

$$\begin{aligned}E(y)&=\beta_{00}+\beta_{01}x+\beta_{10}(1)+\beta_{11}(x-t)\\&=(\beta_{00}+\beta_{10}-\beta_{11}t)+(\beta_{01}+\beta_{11})x\end{aligned}$$

That is, if $x \leqslant t$ the model has intercept β_{00} and slope β_{01}, while if $x>t$, the intercept is $\beta_{00}+\beta_{10}-\beta_{11}t$ and the slope is $\beta_{01}+\beta_{11}$. The regression function

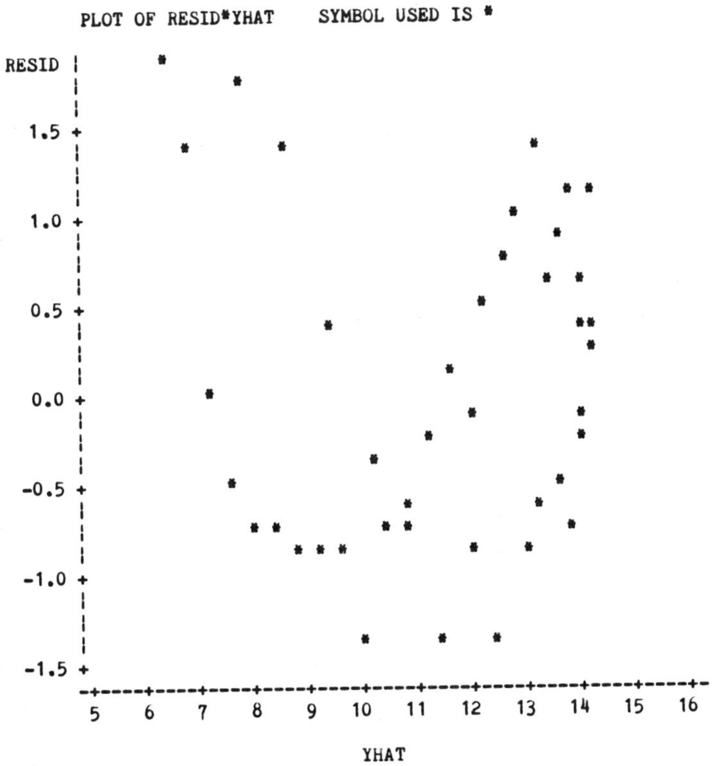

Figure 5.10 Plot (SAS) of residuals e_i versus fitted values $\hat{y}_i$ for the cubic polynomial model.

is shown in Figure 5.11*a*. Note that the parameter β_{10} represents the difference in mean response at the knot t.

A smoother function would result if we required the regression function to be continuous at the knot. This is easily accomplished by deleting the term $\beta_{10}(x-t)_+^0$ from the original model, giving

$$E(y)=S(x)=\beta_{00}+\beta_{01}+\beta_{11}(x-t)_+^1$$

Now if $x \leq t$ the model is

$$E(y)=\beta_{00}+\beta_{01}x$$

and if $x>t$ the model is

$$\begin{aligned} E(y)&=\beta_{00}+\beta_{01}x+\beta_{11}(x-t) \\ &=(\beta_{00}-\beta_{11}t)+(\beta_{01}+\beta_{11})x \end{aligned}$$

• The two regression functions are shown in Figure 5.11*b*.

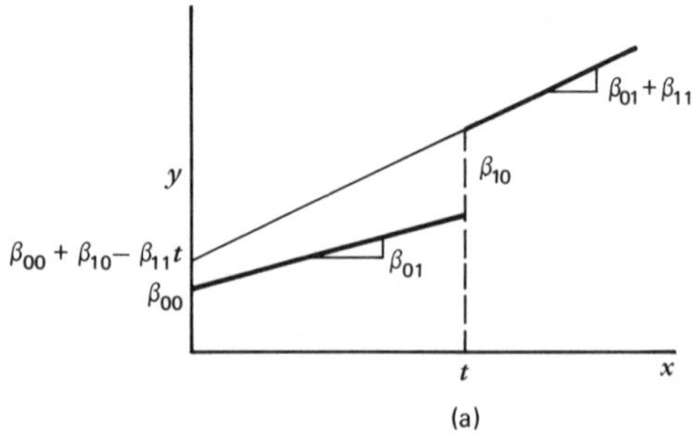

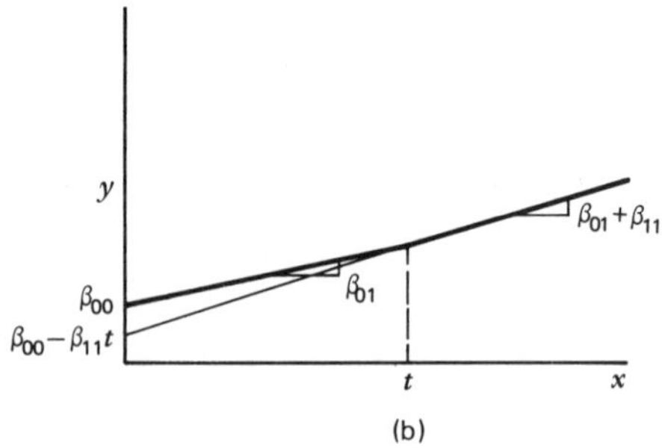

Figure 5.11 Piecewise linear regression. (**a**) Discontinuity at the knot. (**b**) Continuous Piecewise linear regression model.

5.3 Polynomial models in two or more variables

Fitting a polynomial regression model in two or more regressor variables is a straightforward extension of the approach in Section 5.2.1. For example, a second-order polynomial model in two variables would be

$$y=\beta_0+\beta_1x_1+\beta_2x_2+\beta_{11}x_1^2+\beta_{22}x_2^2+\beta_{12}x_1x_2+\varepsilon$$

Note that this model contains two linear effect parameters β_1 and β_2, two quadratic effect parameters β_{11} and β_{22}, and an interaction effect parameter β_{12}. Fitting and analyzing this model has received considerable attention in the literature. We usually call the regression function

$$E(y)=\beta_0+\beta_1x_1+\beta_2x_2+\beta_{11}x_1^2+\beta_{22}x_2^2+\beta_{12}x_1x_2$$

a *response surface*. We may represent the two-dimensional response surface graphically by drawing the x_1 and x_2 axes in the plane of the paper and

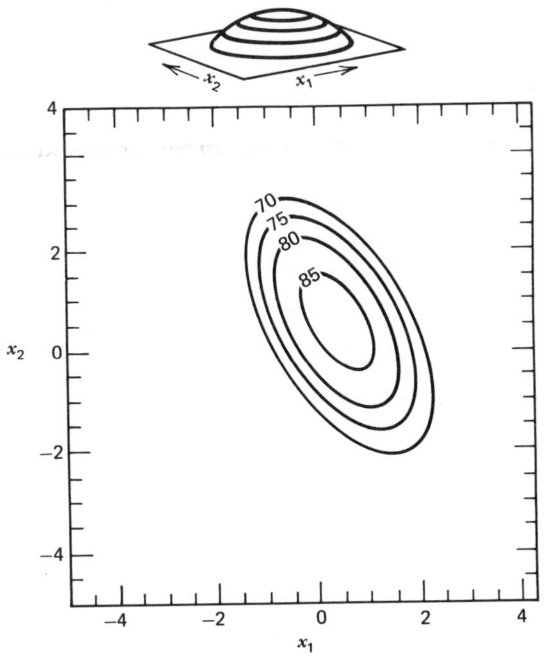

Figure 5.12 A second-order response surface, $E(y)=83.57+9.39x_1+7.12x_2-7.44x_1^2-3.71x_2^2-5.80x_1x_2$. (Adapted with permission from *Statistics for Experimenters*, by G. E. P. Box, W. G. Hunter and J. S. Hunter, Wiley, New York 1978.)

visualizing the $E(y)$ axis perpendicular to the plane of the paper. Plotting contours of constant expected response $E(y)$ produces the response surface. Figure 5.12 shows the response surface

$$E(y)=83.57+9.39x_1+7.12x_2-7.44x_1^2-3.71x_2^2-5.80x_1x_2$$

Note that this response surface is a hill, containing a point of maximum response. Other possibilities include a valley containing a point of minimum response and a saddle system. For a detailed treatment of response surfaces see Box, Hunter and Hunter [1978], Montgomery [1976, Ch. 14], and Myers [1971].

We will now illustrate fitting a second-order response surface in two variables.

- **Example 5.4** A chemical engineer is investigating the influence of two variables, reaction time and temperature, on process yield. Twenty-four observations on yield and the corresponding reaction times and temperatures were collected using a designed experiment and are shown in Table 5.5. Since the relationship between yield, temperature, and time is known to be nonlinear over the ranges of temperature and time observed in the data, we will fit the second-order polynomial

$$y=\beta_0+\beta_1(x_1-\bar{x}_1)+\beta_2(x_2-\bar{x}_2)+\beta_{11}(x_1-\bar{x}_1)^2+\beta_{22}(x_2-\bar{x}_2)^2$$
$$+\beta_{12}(x_1-\bar{x}_1)(x_2-\bar{x}_2)+\varepsilon$$

Notice that we have centered the data to minimize problems with ill-conditioning, as discussed in Section 5.2.1. The fitted model is

$$\hat{y}=50.4171-0.7198(x_1-85.467)+0.1053(x_2-175.792)$$
$$-0.0597(x_1-85.467)^2-0.0377(x_2-175.792)^2$$
$$+0.0126(x_1-85.467)(x_2-175.792)$$

Table 5.5 Process yield data for Example 5.4

Observation Number	Yield y	Reaction Time, x_1	Temperature, x_2
1	50.95	76.0	170
2	47.35	80.5	165
3	50.99	78.0	182
4	44.96	89.0	185
5	41.89	93.0	180
6	41.44	92.1	172
7	51.79	77.8	170
8	50.78	84.0	180
9	42.48	87.3	165
10	49.80	75.0	172
11	48.74	85.0	185
12	46.20	90.0	176
13	50.49	85.0	178
14	52.78	79.2	174
15	49.71	83.0	168
16	52.75	82.0	179
17	39.41	94.0	181
18	43.63	91.4	184
19	38.19	95.0	173
20	50.92	81.1	169
21	46.55	88.8	183
22	44.28	91.0	178
23	48.72	87.0	175
24	49.13	86.0	175

Table 5.6 Analysis of variance for the quadratic model Example 5.4

Source of Variation	Degrees of Freedom	Sum of Squares	Mean Square	F_0
Regression	5	416.31112514	83.26222503	206.28
Residual	18	7.26537069	0.40363170	
Total	23	423.57649583		

The analysis of variance for the quadratic model is shown in Table 5.6. Since the observed value of F_0 exceeds $F_{.01,5,18}=4.25$, we conclude that at least one of the regressors is useful in explaining the relationship of yield to reaction time and temperature.

The residuals from this model are plotted against $\hat{y}_i$ in Figure 5.13, and against the levels of $(x_{i1}-\bar{x}_1)$ and $(x_{i2}-\bar{x}_2)$ in Figures 5.14 and 5.15,

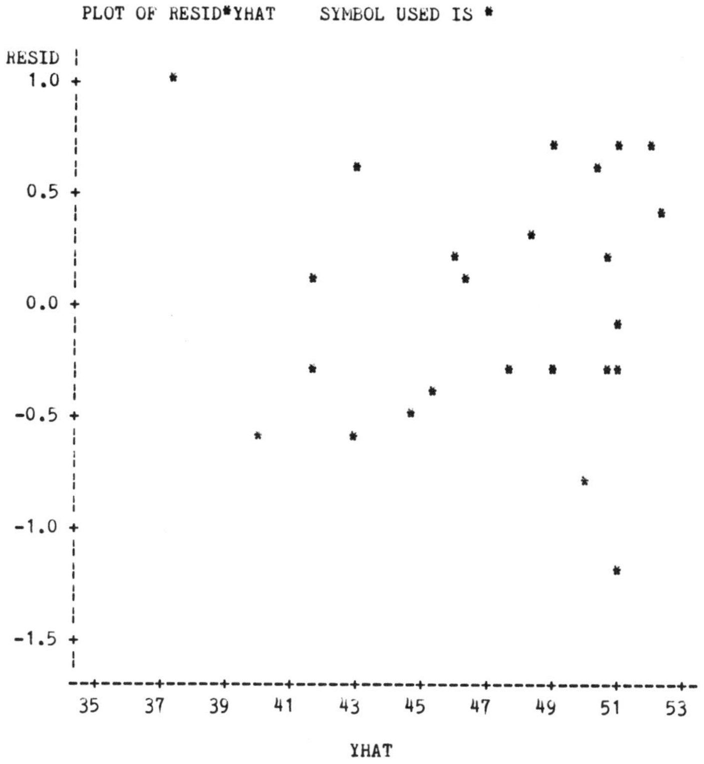

Figure 5.13 Plot (SAS) of residuals e_i versus fitted values $\hat{y}_i$, Example 5.4.

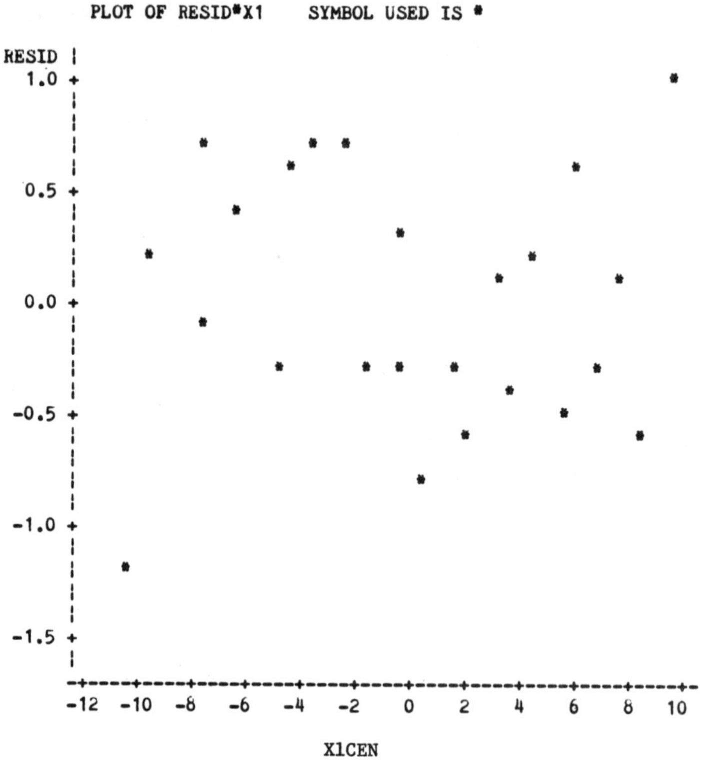

Figure 5.14 Plot (SAS) of residuals e_i versus $(x_{1i}-\bar{x}_1)$, Example 5.4.

respectively. These plots do not reveal any serious violation of the underlying assumptions.

The goodness of fit of this quadratic model may be analyzed using the procedure described in Section 4.7.3 to obtain a model-independent estimate of error. Table 5.7 displays the calculation of the weighted standardized squared distances $D^2_{ii'}$, from (4.58). Table 5.8 summarizes the calculation of the standard deviation of pure error, which stabilizes at about $\hat{\sigma}=0.52$. The square root of the residual mean square from Table 5.6 is $\sqrt{0.4036}=0.6353$, which is about one-fifth larger than $\hat{\sigma}$. This difference is not considered large enough to cast doubt on the adequacy of the second-order model.

We may also use the extra sum of squares method to test the contribution of the quadratic terms to the model. That is, we wish to test

$$H_0\colon \beta_{11}=\beta_{22}=\beta_{12}=0$$

$$H_1\colon \text{at least one } \beta_{ij}\neq 0$$

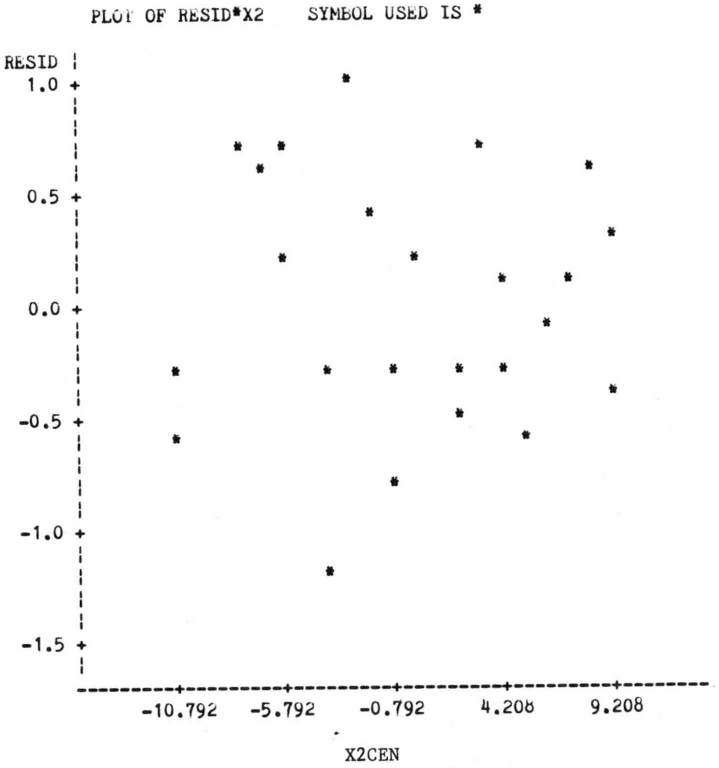

Figure 5.15 Plot (SAS) of Residuals e_i versus $(x_{2i}-\bar{x}_2)$. Example 5.4.

The reduced model for testing this hypothesis is

$$y=\beta_0+\beta_1(x_1-\bar{x}_1)+\beta_2(x_2-\bar{x}_2)+\varepsilon$$

The fitted first order model is

$$\hat{y}=47.2471-.6898(x_1-85.467)+.1686(x_2-175.792)$$

and

$$SS_R(\beta_1,\beta_2|\beta_0)=333.08805020.$$

The extra sum of squares for testing H_0: $\beta_{11}=\beta_{22}=\beta_{12}=0$ is

$$\begin{aligned}SS_R(\beta_{11},\beta_{22},\beta_{12}|\beta_1,\beta_2,\beta_0)&=SS_R(\beta_1,\beta_2,\beta_{11},\beta_{22},\beta_{12}|\beta_0)\\&\quad-SS_R(\beta_1,\beta_2|\beta_0)\\&=416.31112514-333.08805020\\&=83.22307490\end{aligned}$$

Table 5.7 Calculation of weighted sum of squared distances $D_{ii'}^2$ for Example 5.4

Near-Neighbor Calculations
Delta Residuals and the Weighted Standardized Squared Distances of Near-neighbors

	Ordered Fitted		Adjacent			1 Apart			2 Apart			3 Apart		
Observation	y	Residual	Delta	$D_{ii'}^2$	R[a]	Delta	$D_{ii'}^2$	R	Delta	$D_{ii'}^2$	R	Delta	D_{ii}^2	R
19	37.211	.979	1.585	.5118E+00	14	1.299	.1686E+01		.873	.1024E+01		.333	.3444E+01	
17	40.016	−.606	.286	.9249E+00		.713	.2201E+00	5	1.252	.1506E+01		.038	.7604E+01	
6	41.760	−.320	.427	.3076E+00	9	.966	.9601E+00		.324	.4563E+01		.210	.2770E+00	7
5	41.784	.106	.539	.9226E+00		.751	.6019E+01		.636	.6663E+00		.459	.3018E+01	
18	42.984	.646	1.290	.2819E+01		1.176	.8688E+00		.999	.7301E+00		.406	.1278E+01	
9	43.124	−.644	.114	.4053E+01		.291	.1113E+01		.884	.3506E+01		.736	.1539E+01	
22	44.810	−.530	.177	.1778E+01		.770	.1344E+00	4	.622	.1006E+01		.209	.1061E+02	
4	45.313	−.353	.593	.1652E+01		.445	.2221E+00	6	.032	.6041E+01		.632	.1244E+01	
12	45.960	.240	.148	.7593E+00		.561	.9337E+01		.039	.3513E+01		.506	.4421E+01	
21	46.458	.092	.412	.6329E+01		.188	.1322E+01		.655	.2770E+01		.423	.8950E+00	
2	47.671	−.321	.600	.2616E+01		1.067	.1262E+01		.010	.6096E+01		.453	.5286E+01	
11	48.461	.279	.467	.8684E+00		.610	.1824E+01		1.053	.1608E+01		.298	.1966E+01	
15	48.964	.746	1.077	.1937E+01		1.520	.1447E+01		.169	.3855E+00	12	.498	.7090E+01	
23	49.051	−.331	.443	.7342E−01	1	.908	.3099E+01		.580	.1268E+02		.045	.3047E+00	8
24	49.904	−.774	1.351	.2364E+01		1.023	.1136E+02		.488	.8858E+01	2	.400	.1456E+02	
20	50.343	.577	.328	.4339E+01		.863	.1735E+01		1.751	.6877E+01		.840	.1110E+01	
1	50.701	.249	.534	.1002E+02		1.423	.3488E+00	11	.511	.8594E+01		.343	.1294E+01	
13	50.776	−.286	.888	.1309E+02		.023	.1120E+00	3	.191	.5295E+01		.962	.5680E+01	
10	50.974	−1.174	.912	.1157E+02		1.080	.2487E+01		1.850	.1900E+01		1.919	.8288E+01	
8	51.043	−.263	.168	.4101E+01		.938	.4569E+01		1.007	.3457E+00	10	.626	.2421E+01	
3	51.084	−.094	.770	.4112E+00	13	.839	.2274E+01		.458	.6495E+00				
7	51.114	.676	.069	.2625E+01		.313	.5543E+00	15						
16	52.006	.744	.381	.9817E+00										
14	52.417	.363												

[a] Column R gives the rank order of the 15 smallest D values.

Table 5.8 Calculation of the standard deviation of pure error for Example 5.4

Linear Least Squares Curve Fitting
Standard Deviation Estimated from Residuals of Neighboring Observations

	Cumulative Standard Deviation	WSSD	Ordered by $D_{ii'}^2$ Observation	Observation	Delta Residual
1	.3925E+00	.7342E−01	23	24	.4430
2	.4125E+00	.8858E−01	24	13	.4881
3	.2818E+00	.1120E+00	13	8	.0231
4	.3819E+00	.1344E+00	22	12	.7700
5	.4318E+00	.2201E+00	17	5	.7125
6	.4255E+00	.2212E+00	4	21	.4447
7	.3913E+00	.2770E+00	6	22	.2098
8	.3473E+00	.3047E+00	23	13	.0451
9	.3508E+00	.3076E+00	6	5	.4267
10	.4049E+00	.3457E+00	8	16	1.0071
11	.4827E+00	.3488E+00	1	10	1.4230
12	.4550E+00	.3855E+00	15	20	.1691
13	.4725E+00	.4112E+00	3	7	.7701
14	.5390E+00	.5118E+00	19	17	1.5851
15	.5216E+00	.5543E+00	7	14	.3126
16	.5143E+00	.6495E+00	3	14	.4575
17	.5172E+00	.6663E+00	5	22	.6364
18	.5376E+00	.7301E+00	18	4	.9986
19	.5162E+00	.7593E+00	12	21	.1482
20	.5111E+00	.8684E+00	11	15	.4670
21	.5364E+00	.8688E+00	18	22	1.1757
22	.5290E+00	.8950E+00	21	23	.4225
23	.5268E+00	.9226E+00	5	18	.5393
24	.5154E+00	.9249E+00	17	6	.2858
25	.5290E+00	.9601E+00	6	18	.9660
26	.5217E+00	.9817E+00	16	14	.3813
27	.5227E+00	.1006E+01	22	21	.6218
28	.5317E+00	.1024E+01	19	5	.8726
29	.5390E+00	.1110E+01	20	8	.8398
30	.5296E+00	.1131E+01	9	4	.2912
31	.5306E+00	.1244E+01	4	11	.6323
32	.5436E+00	.1262E+01	2	15	1.0671
33	.5380E+00	.1278E+01	18	12	.4058
34	.5311E+00	.1294E+01	1	3	.3431
35	.5207E+00	.1322E+01	21	11	.1876
36	.5437E+00	.1447E+01	15	24	1.5202
37	.5589E+00	.1506E+01	17	18	1.2518
38	.5614E+00	.1539E+01	9	21	.7358
39	.5709E+00	.1608E+01	11	24	.1.0532
40	.5698E+00	.1652E+01	4	12	.5929

Table 5.9 Regression coefficients, standard errors, and *t*-statistics for Example 5.4

Parameter	Estimate	*t*-Statistic	Standard Error
β_0	50.41707498	192.85	0.26143227
β_1	−0.71980624	−29.36	0.02451463
β_2	0.10528134	4.43	0.02378950
β_{11}	−0.05965306	−13.09	0.00455567
β_{22}	−0.03767571	−9.19	0.00410124
β_{12}	0.01257698	2.40	0.00523189

The *F*-statistic is

$$F_0=\frac{SS_R(\beta_{11},\beta_{22},\beta_{12}|\beta_1,\beta_2,\beta_0)/3}{MS_E}=\frac{83.22307490/3}{0.40363170}=68.7286$$

Now $F_{.01,3,18}=5.09$, and since the observed value F_0 exceeds this limit, we reject H_0: $\beta_{11}=\beta_{22}=\beta_{12}=0$ and conclude that at least one of the quadratic terms is necessary.

Table 5.9 presents the estimates of the regression coefficients, their standard errors, and the value of the *t*-statistic for testing whether or not that parameter equals zero. Referring each of these *t*-statistics to $t_{.01,1,18}=2.552$ indicates that all the regression coefficients differ significantly from zero. Even if some of the individual *t*-statistics are small, it is debatable whether those terms should be deleted from the model. We can adopt the viewpoint that it is the *order* of the model that is important, not the individual terms. From this perspective the model is an interpolation equation, and all terms are needed. •

5.4 Orthogonal polynomials

We have noted that in fitting polynomial models in one variable, even if nonessential ill-conditioning is removed by centering, we may still have high correlation between some regression coefficients. Some of these difficulties can be eliminated by using orthogonal polynomials to fit the model.

Suppose that the model is

$$y_i=\beta_0+\beta_1x_i+\beta_2x_i^2+\cdots+\beta_kx_i^k+\varepsilon_i,\qquad i=1,2,\ldots,n \qquad (5.7)$$

Generally the columns of the **X** matrix will not be orthogonal. Furthermore if we increase the order of the polynomial by adding a term $\beta_{k+1}x^{k+1}$ we must recompute $(\mathbf{X}'\mathbf{X})^{-1}$ and the estimates of the lower-order parameters $\hat{\beta}_0, \hat{\beta}_1, \ldots, \hat{\beta}_k$ will change.

Now suppose that we fit the model

$$y_i = \alpha_0 P_0(x_i) + \alpha_1 P_1(x_i) + \alpha_2 P_2(x_i) + \cdots + \alpha_k P_k(x_i) + \varepsilon_i, \qquad i = 1, 2, \ldots, n \tag{5.8}$$

where $P_u(x_i)$ is a uth order orthogonal polynomial defined such that

$$\sum_{i=1}^{n} P_r(x_i) P_s(x_i) = 0, \qquad r \neq s, r, s = 0, 1, 2, \ldots, k$$

$$P_0(x_i) = 1$$

Then the model becomes $\mathbf{y} = \mathbf{X}\boldsymbol{\alpha} + \boldsymbol{\varepsilon}$, where the $\mathbf{X}$ matrix is

$$\mathbf{X} = \begin{bmatrix} P_0(x_1) & P_1(x_1) & \cdots & P_k(x_1) \\ P_0(x_2) & P_1(x_2) & \cdots & P_k(x_2) \\ \vdots & \vdots & & \vdots \\ P_0(x_n) & P_1(x_n) & \cdots & P_k(x_n) \end{bmatrix}$$

Since this matrix has orthogonal columns, the $\mathbf{X}'\mathbf{X}$ matrix is

$$\mathbf{X}'\mathbf{X} = \begin{bmatrix} \sum_{i=1}^{n} P_0^2(x_i) & 0 & \cdots & 0 \\ 0 & \sum_{i=1}^{n} P_1^2(x_i) & \cdots & 0 \\ \vdots & \vdots & & \vdots \\ 0 & 0 & \cdots & \sum_{i=1}^{n} P_k^2(x_i) \end{bmatrix}$$

The least squares estimators of the $\boldsymbol{\alpha}$ are found from $(\mathbf{X}'\mathbf{X})^{-1}\mathbf{X}'\mathbf{y}$ as

$$\hat{\alpha}_j = \frac{\sum_{i=1}^{n} P_j(x_i) y_i}{\sum_{i=1}^{n} P_j^2(x_i)}, \qquad j = 0, 1, \ldots, k \tag{5.9}$$

Since $P_0(x_i)$ is a polynomial of degree zero, we can set $P_0(x_i)=1$, and consequently

$$\hat{\alpha}_0=\bar{y}$$

The residual sum of squares is

$$SS_E(k)=S_{yy}-\sum_{j=1}^{k}\hat{\alpha}_j\left[\sum_{i=1}^{n}P_j(x_i)y_i\right] \tag{5.10}$$

The regression sum of squares for any model parameter does not depend on the other parameters in the model. This regression sum of squares is

$$SS_R(\alpha_j)=\hat{\alpha}_j\sum_{i=1}^{n}P_j(x_i)y_i \tag{5.11}$$

If we wish to assess the significance of the highest-order term, we should test H_0: $\alpha_k=0$ [this is equivalent to testing H_0: $\beta_k=0$ in (5.4)]; we would use

$$F_0=\frac{SS_R(\alpha_k)}{SS_E(k)/(n-k-1)}=\frac{\hat{\alpha}_k\sum_{i=1}^{n}P_k(x_i)y_i}{SS_E(k)/(n-k-1)} \tag{5.12}$$

as the F-statistic. Furthermore note that if the order of the model is changed to $k+r$, only the r new coefficients must be computed. The coefficients $\hat{\alpha}_0, \hat{\alpha}_1, \ldots, \hat{\alpha}_k$ do not change due to the orthogonality property of the polynomials. Thus sequential fitting of the model is computationally easy.

The orthogonal polynomials $P_j(x_i)$ are easily constructed for the case where the levels of x are equally spaced. The first five orthogonal polynomials are

$$P_0(x_i)=1$$

$$P_1(x_i)=\lambda_1\left[\frac{x_i-\bar{x}}{d}\right]$$

$$P_2(x_i)=\lambda_2\left[\left(\frac{x_i-\bar{x}}{d}\right)^2-\left(\frac{n^2-1}{12}\right)\right]$$

$$P_3(x_i)=\lambda_3\left[\left(\frac{x_i-\bar{x}}{d}\right)^3-\left(\frac{x_i-\bar{x}}{d}\right)\left(\frac{3n^2-7}{20}\right)\right]$$

$$P_4(x_i)=\lambda_4\left[\left(\frac{x_i-\bar{x}}{d}\right)^4-\left(\frac{x_i-\bar{x}}{d}\right)^2\left(\frac{3n^2-13}{14}\right)+\frac{3(n^2-1)(n^2-9)}{560}\right]$$

where d is the spacing between the levels of x and the $\{\lambda_j\}$ are constants chosen so that the polynomials will have integer values. A brief table of the numerical values of these orthogonal polynomials is in Appendix Table A.5. More extensive tables are in Delury [1960], and Pearson and Hartley [1966]. Orthogonal polynomials can also be constructed and used in cases where the x's are not equally spaced. A survey of methods for generating orthogonal polynomials is in Seber [1977, Ch. 8].

- **Example 5.5** An operations research analyst has developed a computer simulation model of a single item inventory system. He has experimented with the simulation model to investigate the effect of various reorder quantities on the average annual cost of the inventory. The data are shown in Table 5.10.

Since we know that average annual inventory cost is a convex function of the reorder quantity, we suspect that a second-order polynomial is the highest-order model that must be considered. Therefore we will fit

$$y_i = \alpha_0 P_0(x_i) + \alpha_1 P_1(x_i) + \alpha_2 P_2(x_i)\varepsilon_i, \qquad i = 1, 2, \ldots, 10$$

The coefficients of the orthogonal polynomials $P_0(x_i)$, $P_1(x_i)$, and $P_2(x_i)$, obtained from Appendix Table A.5, are shown in Table 5.11.

Table 5.10 Inventory simulation output for Example 5.5

Reorder Quantity, x_i	Average Annual Cost, y_i
50	\$335
75	326
100	316
125	313
150	311
175	314
200	318
225	328
250	337
275	345

Table 5.11 Coefficients of orthogonal polynomials for Example 5.3

i	$P_0(x_i)$	$P_1(x_i)$	$P_2(x_i)$
1	1	−9	6
2	1	−7	2
3	1	−5	−1
4	1	−3	−3
5	1	−1	−4
6	1	1	−4
7	1	3	−3
8	1	5	−1
9	1	7	2
10	1	9	6
	$\sum_{i=1}^{10} P_0^2(x_i)=10$	$\sum_{i=1}^{10} P_1^2(x_i)=330$	$\sum_{i=1}^{10} P_2^2(x_i)=132$
		$\lambda_1=2$	$\lambda_2=1/2$

Thus

$$\mathbf{X'X}=\begin{bmatrix}\sum_{i=1}^{10} P_0^2(x_i) & 0 & 0\\ 0 & \sum_{i=1}^{10} P_1^2(x_i) & 0\\ 0 & 0 & \sum_{i=1}^{10} P_2^2(x_i)\end{bmatrix}=\begin{bmatrix}10 & 0 & 0\\ 0 & 330 & 0\\ 0 & 0 & 132\end{bmatrix}$$

$$\mathbf{X'y}=\begin{bmatrix}\sum_{i=1}^{10} P_0(x_i)y_i\\ \sum_{i=1}^{10} P_1(x_i)y_i\\ \sum_{i=1}^{10} P_2(x_i)y_i\end{bmatrix}=\begin{bmatrix}3243\\ 245\\ 369\end{bmatrix}$$

and

$$\hat{\boldsymbol{\beta}}=(\mathbf{X'X})^{-1}\mathbf{X'y}$$

$$=\begin{bmatrix}1/10 & 0 & 0\\ 0 & 1/330 & 0\\ 0 & 0 & 1/132\end{bmatrix}\begin{bmatrix}3243\\ 245\\ 369\end{bmatrix}=\begin{bmatrix}324.3000\\ 0.7424\\ 2.7955\end{bmatrix}$$

Table 5.12 Analysis of variance for the quadratic model in Example 5.5

Source of Variation	Sum of Squares	Degrees of Freedom	Mean Square	F_0
Regression	1213.43	2	606.72	159.24
Linear, α_1	(181.89)	1	181.89	47.74
Quadratic, α_2	(1031.54)	1	1031.54	270.75
Residual	26.67	7	3.81	
Total	1240.10	9		

The fitted model is

$$\hat{y}=324.30+0.7424P_1(x)+2.7955P_2(x)$$

The regression sum of squares is

$$SS_R(\alpha_1,\alpha_2)=\sum_{j=1}^{2}\hat{\alpha}_j\left[\sum_{i=1}^{10}P_j(x_i)y_i\right]$$

$$=0.7424(245)+2.7955(369)$$

$$=181.89+1031.54=1213.43$$

The analysis of variance is shown in Table 5.12. Both the linear and quadratic terms contribute significantly to the model. Since these terms account for most of the variation in the data, we tentatively adopt the quadratic model, subject to a satisfactory residual analysis.

We may obtain a fitted equation in terms of the original regressor by substituting for $P_j(x_i)$ as follows:

$$\hat{y}=324.30+0.7424P_1(x)+2.7955P_2(x)$$

$$=324.30+0.7424(2)\left(\frac{x-162.5}{25}\right)+2.7955\frac{1}{2}\left[\left(\frac{x-162.5}{25}\right)^2-\frac{(10)^2-1}{12}\right]$$

$$=312.7686+0.0594(x-162.5)+0.0022(x-162.5)^2$$

- This form of the model should be reported to the user.

Problems

5.1 Consider the values of x shown below:

$x=1.00 \quad 1.70 \quad 1.25 \quad 1.20 \quad 1.45 \quad 1.85 \quad 1.60 \quad 1.50 \quad 1.95 \quad 2.00$

Suppose that we wish to fit a second-order model using these levels for the regressor variable x. Calculate the correlation between x and x^2. Do you see any potential difficulties in fitting the model?

5.2 A solid-fuel rocket propellant loses weight after it is produced. The following data are available:

x(Months Since Production)	y(Weight Loss, kgm)
0.25	1.42
0.50	1.39
0.75	1.55
1.00	1.89
1.25	2.43
1.50	3.15
1.75	4.05
2.00	5.15
2.25	6.43
2.50	7.89

a. Fit a second-order polynomial that expresses weight loss as a function of the number of months since production.

b. Test for significance of regression.

c. Test the hypothesis H_0: $\beta_2 = 0$. Comment on the need for the quadratic term in this model.

d. Are there any potential hazards in extrapolating with this model?

5.3 Refer to Problem 5.2. Compute the residuals for the second-order model. Analyze the residuals and comment on the adequacy of the model.

5.4 Consider the data shown below:

x	y
4.00	24.60
4.00	24.71
4.00	23.90
5.00	39.50
5.00	39.60
6.00	57.12
6.50	67.11
6.50	67.24
6.75	67.15
7.00	77.87
7.10	80.11
7.30	84.67

a. Fit a second-order polynomial model to these data.
b. Test for significance of regression.
c. Test for lack of fit and comment on the adequacy of the second-order model.
d. Test the hypothesis H_0: $\beta_2=0$. Can the quadratic term be deleted from this equation?

5.5 Refer to Problem 5.4. Compute the residuals from the second-order model. Analyze the residuals and draw conclusions about the adequacy of the model.

5.6 The carbonation level of a soft drink beverage is affected by the temperature of the product and the filler operating pressure. Twelve observations were obtained and the resulting data are shown below.

y(Carbonation)	x_1(Temperature)	x_2(Pressure)
2.60	31.0	21.0
2.40	31.0	21.0
17.32	31.5	24.0
15.60	31.5	24.0
16.12	31.5	24.0
5.36	30.5	22.0
6.19	31.5	22.0
10.17	30.5	23.0
2.62	31.0	21.5
2.98	30.5	21.5
6.92	31.0	22.5
7.06	30.5	22.5

a. Fit a second-order polynomial.
b. Test for significance of regression.
c. Test for lack of fit and draw conclusions.
d. Does the interaction term contribute significantly to the model?
e. Do the second-order terms contribute significantly to the model?

5.7 Refer to Problem 5.6. Compute the residuals from the second–order model. Analyze the residuals and comment on the adequacy of the model.

5.8 An entomologist is studying the effect of several factors on the activity of cholinestrase in cricket eggs. The data resulting from the study are in Appendix Table B.8.
a. Fit a regression model of the form

$$y=\beta_0+\beta_1x_1+\beta_2x_2+\beta_3x_3+\beta_{11}x_1^2+\beta_{33}x_3^2+\beta_{13}x_1x_3+\varepsilon$$

where y is the activity of cholinestrase as a percent of normal, x_1 is the age of the eggs at the time of treatment, x_2 is the age of the eggs at the time of observation, and x_3 is the dosage level of the eggs.

b. Test for significance of regression.

c. Do the quadratic terms contribute significantly to the model?

d. Should the terms in x_2^2, x_1x_2, and x_2x_3 be added to the model?

5.9 Refer to Problem 5.8. Compute the residuals for the model

$$\hat{y}=\hat{\beta}_0+\hat{\beta}_1x_1+\hat{\beta}_2x_2+\hat{\beta}_3x_3+\hat{\beta}_{11}x_1^2+\hat{\beta}_{33}x_3^2+\hat{\beta}_{13}x_1x_3$$

a. Analyze the residuals and draw conclusions about the adequacy of the model.

b. Delete the three outliers identified in Part (a). Refit the equation and comment on the validity of the model.

5.10 Consider the data in Problem 5.2.

a. Fit a second-order model to these data using orthogonal polynomials.

b. Suppose that we wish to investigate the addition of a third-order term to this model. Comment on the necessity of this additional term. Support your conclusions with an appropriate statistical analysis.

5.11 Suppose we wish to fit the piecewise quadratic polynomial with a knot at $x=t$:

$$E(y)=S(x)=\beta_{00}+\beta_{01}x+\beta_{02}x^2+\beta_{10}(x-t)_+^0+\beta_{11}(x-t)_+^1$$
$$+\beta_{12}(x-t)_+^2$$

a. Show how to test the hypothesis that this quadratic spline model fits the data significantly better than an ordinary quadratic polynomial.

b. The quadratic spline polynomial model is not continuous at the knot t. How can the model be modified so that continuity at $x=t$ is obtained?

c. Show how the model can be modified so that both $E(y)$ and $dE(y)/dx$ are continuous at $x=t$.

d. Discuss the significance of the continuity restrictions on the model in Parts b and c. In practice, how would you select the type of continuity restrictions to impose?

5.12 Suppose that we wish to fit a piecewise polynomial model with three segments: if $x<t_1$ the polynomial is linear, if $t_1 \leqslant x<t_2$ the polynomial is

quadratic, and if $x \geq t_2$ the polynomial is linear. Consider the model

$$E(y) = S(x) = \beta_{00} + \beta_{01}x + \beta_{02}x^2 + \beta_{10}(x-t_1)_+^0 + \beta_{11}(x-t_1)_+^1 + \beta_{12}(x-t_1)_+^2 + \beta_{20}(x-t_2)_+^0 + \beta_{21}(x-t_2)_+^1 + \beta_{22}(x-t_2)_+^2$$

a. Does this segmented polynomial satisfy our requirements? If not, show how it can be modified to do so.

b. Show how the segmented model would be modified to insure that $E(y)$ is continuous at the knots t_1 and t_2.

c. Show how the segmented model would be modified to insure that both $E(y)$ and $dE(y)/dx$ are continuous at the knots t_1 and t_2.

5.13 An operations research analyst is investigating the relationship between production lot size x and the average production cost per unit y. A study of recent operations provides the following data:

x	100	120	140	160	180	200	220	240	260	280	300
y	\$9.73	9.61	8.15	6.98	5.87	4.98	5.09	4.79	4.02	4.46	3.82

The analyst suspects that a piecewise linear regression model should be fit to these data. Estimate the parameters in such a model assuming that the slope of the line changes at $x = 200$ units. Do the data support the use of this model?

5.14 Modify the model in Problem 5.13 to investigate the possibility that a discontinuity exists in the regression function at $x = 200$ units. Estimate the parameters in this model. Test appropriate hypotheses to determine if the regression function has a change in both the slope and the intercept at $x = 200$ units.

6

INDICATOR VARIABLES

6.1 The general concept of indicator variables

The variables employed in regression analysis are usually *quantitative* variables; that is, the variables have a well defined scale of measurement. Variables such as temperature, distance, pressure, and income are quantitative variables. Occasionally it is necessary to use *qualitative* or categorical variables as predictor variables in regression. Examples of qualitative or categorical variables are operators, employment status (employed or unemployed), shifts (day, evening, or night), and sex (male or female). In general a qualitative variable has no natural scale of measurement. We must assign a set of levels to a qualitative variable to account for the effect that the variable may have on the response. This is done through the use of indicator variables. Sometimes indicator variables are called "dummy" variables.

Suppose that a mechanical engineer wishes to relate the effective life of a cutting tool (y) used on a lathe to the lathe speed in RPM (x_1) and the type of cutting tool used. The second regressor variable, tool type, is qualitative, and has two levels (tool types A and B, for example). We will use an indicator variable that takes on the values 0 and 1 to identify the classes of the regressor variable "tool type." Let

$$x_2 = \begin{cases} 0 & \text{if the observation is from tool type A} \\ 1 & \text{if the observation is from tool type B} \end{cases}$$

The choice of 0 and 1 to identify the levels of a qualitative variable is arbitrary. Any two distinct values for x_2 would be satisfactory, although 0 and 1 are usually best.

Assuming that a first-order model is appropriate, we have

$$y=\beta_0+\beta_1x_1+\beta_2x_2+\varepsilon \tag{6.1}$$

To interpret the parameters in this model, consider first tool type A, for which $x_2=0$. The regression model becomes

$$\begin{aligned} y&=\beta_0+\beta_1x_1+\beta_2(0)+\varepsilon \\ &=\beta_0+\beta_1x_1+\varepsilon \end{aligned} \tag{6.2}$$

Thus the relationship between tool life and lathe speed for tool type A is a straight line with intercept β_0 and slope β_1. For tool type B, we have $x_2=1$, and

$$\begin{aligned} y&=\beta_0+\beta_1x_1+\beta_2(1)+\varepsilon \\ &=(\beta_0+\beta_2)+\beta_1x_1+\varepsilon \end{aligned} \tag{6.3}$$

That is, for tool type B the relationship between tool life and lathe speed is also a straight line with slope β_1 but intercept $\beta_0+\beta_2$.

The two response functions are shown in Figure 6.1. The models (6.2) and (6.3) describe two parallel regression lines, that is, two lines with a common slope β_1 and different intercepts. Also, the variance of the errors ε is assumed to be the same for both tool types A and B. The parameter β_2

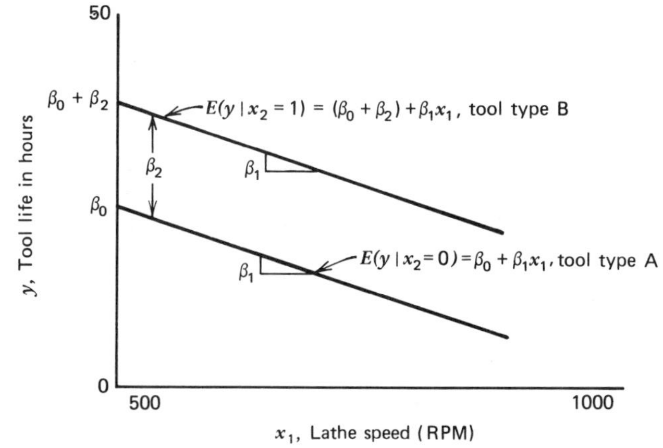

Figure 6.1 Response functions for the tool life example.

expresses the difference in heights between the two regression lines; that is, β_2 is a measure of the difference in mean tool life resulting from changing from tool type A to tool type B.

We may generalize this approach to qualitative factors with any number of levels. For example, suppose that three tool types, A, B, and C, are of interest. Two indicator variables, such as x_2 and x_3, will be required to incorporate the three levels of tool type into the model. The levels of the indicator variables are

x_2	x_3	
0	0	if the observation is from tool type A
1	0	if the observation is from tool type B
0	1	if the observation is from tool type C

and the regression model is

$$y=\beta_0+\beta_1x_1+\beta_2x_2+\beta_3x_3+\varepsilon$$

In general a qualitative variable with a levels is represented by $a-1$ indicator variables, each taking on the values 0 and 1.

Table 6.1 Data, fitted values, and residuals for Example 6.1

i	y_i(Hours)	x_{i1} (RPM)	Tool Type	$\hat{y}_i$	e_i
1	18.73	610	A	20.7552	−2.0252
2	14.52	950	A	11.7087	2.8113
3	17.43	720	A	17.8284	−0.3984
4	14.54	840	A	14.6355	−0.0955
5	13.44	980	A	10.9105	2.5295
6	24.39	530	A	22.8838	1.5062
7	13.34	680	A	18.8927	−5.5527
8	22.71	540	A	22.6177	0.0923
9	12.68	890	A	13.3052	−0.6252
10	19.32	730	A	17.5623	1.7577
11	30.16	670	B	34.1630	−4.0030
12	27.09	770	B	31.5023	−4.4123
13	25.40	880	B	28.5755	−3.1755
14	26.05	1000	B	25.3826	0.6674
15	33.49	760	B	31.7684	1.7216
16	35.62	590	B	36.2916	−0.6716
17	26.07	910	B	27.7773	−1.7073
18	36.78	650	B	34.6952	2.0848
19	34.95	810	B	30.4380	4.5120
20	43.67	500	B	38.6862	4.9838

• **Example 6.1** Twenty observations on tool life and lathe speed are presented in Table 6.1, and the scatter diagram is shown in Figure 6.2. Inspection of this scatter diagram indicates that two different regression lines are required to adequately model this data, with the intercept depending on the type of tool used. Therefore we will fit the model

$$y=\beta_0+\beta_1 x_1+\beta_2 x_2+\varepsilon$$

where the indicator variable $x_2=0$ if the observation is from tool type A and $x_2=1$ if the observation is from tool type B. The $\mathbf{X}$ matrix and $\mathbf{y}$ vector for fitting this model are

$$\mathbf{X}=\begin{bmatrix} 1 & 610 & 0 \\ 1 & 950 & 0 \\ 1 & 720 & 0 \\ 1 & 840 & 0 \\ 1 & 980 & 0 \\ 1 & 530 & 0 \\ 1 & 680 & 0 \\ 1 & 540 & 0 \\ 1 & 890 & 0 \\ 1 & 730 & 0 \\ 1 & 670 & 1 \\ 1 & 770 & 1 \\ 1 & 880 & 1 \\ 1 & 1000 & 1 \\ 1 & 760 & 1 \\ 1 & 590 & 1 \\ 1 & 910 & 1 \\ 1 & 650 & 1 \\ 1 & 810 & 1 \\ 1 & 500 & 1 \end{bmatrix} \quad \text{and} \quad \mathbf{y}=\begin{bmatrix} 18.73 \\ 14.52 \\ 17.43 \\ 14.54 \\ 13.44 \\ 24.39 \\ 13.34 \\ 22.71 \\ 12.68 \\ 19.32 \\ 30.16 \\ 27.09 \\ 25.40 \\ 26.05 \\ 33.49 \\ 35.62 \\ 26.07 \\ 36.78 \\ 34.95 \\ 43.67 \end{bmatrix}$$

The least squares fit is

$$\hat{y}=36.986-0.027x_1+15.004x_2$$

The analysis of variance and other summary statistics for this model are shown in Table 6.2. Since the observed value of F_0 exceeds $F_{.01,2,17}=6.11$, the hypothesis of significance of regression is rejected, and since the t–statistics for β_1 and β_2 exceed $t_{.01,17}=2.567$, we conclude that both regressors x_1 (RPM) and x_2 (tool type) contribute to the model. The parameter β_2 is the change in mean tool life resulting from a change from tool type A to tool type B. Using (4.20), we may find a 95 percent confidence interval on β_2 as follows:

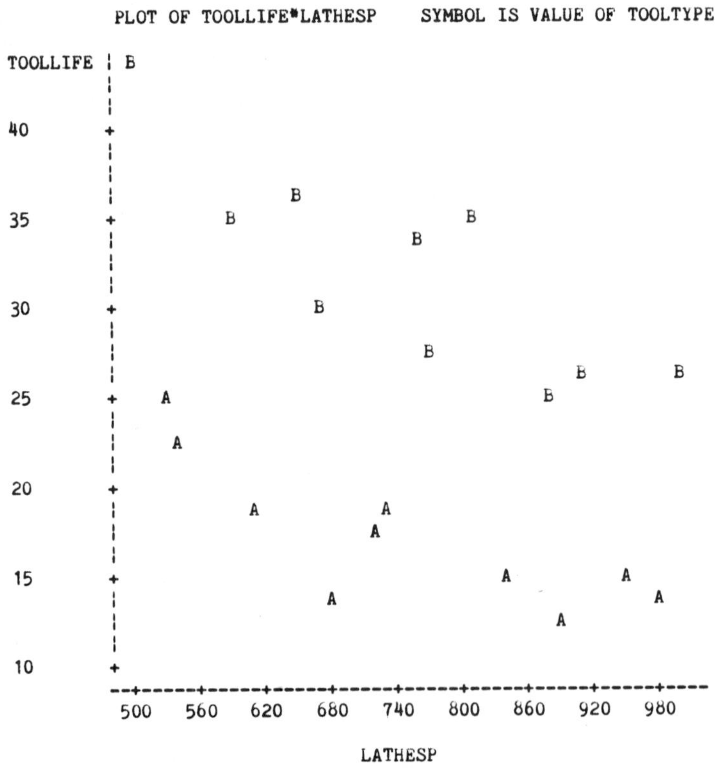

Figure 6.2 Plot (SAS) of tool life y versus lathe speed x_1 for tool types A and B.

Table 6.2 Summary statistics for the regression model in Example 6.1

Source of Variation	Sum of Squares	Degree of Freedom	Mean Square	F_0
Regression	1418.034	2	709.017	76.75
Error	157.055	17	9.239	
Total	1575.089	19		
Coefficient	**Estimate**	**Standard Error**		t_0
β_0	36.986			
β_1	−0.027	0.005		−5.887
β_2	15.004	1.360		11.035
		$R^2 = 0.9003$		

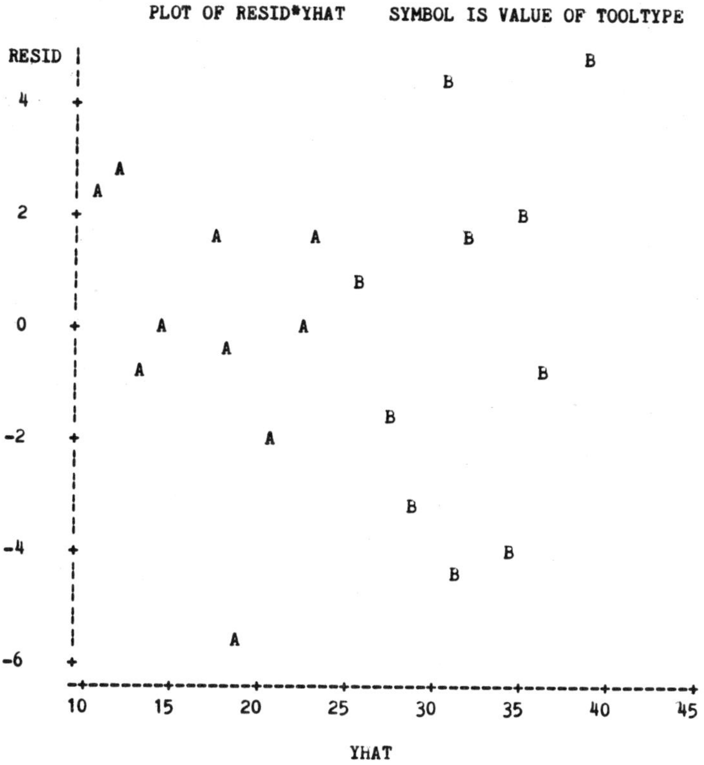

Figure 6.3 Plot (SAS) of residuals e_i versus fitted values $\hat{y}_i$, Example 6.1.

$$\hat{\beta}_2 - t_{.025,17}\,\mathrm{se}(\hat{\beta}_2) \leq \beta_2 \leq \hat{\beta}_2 + t_{.025,17}\,\mathrm{se}(\hat{\beta}_2)$$

$$15.004 - 2.110(1.360) \leq \beta_2 \leq 15.004 + 2.110(1.360)$$

or

$$12.135 \leq \beta_2 \leq 17.873$$

Therefore we are 95 percent confident that changing from tool type A to tool type B increases the mean tool life by between 12.135 hours and 17.873 hours.

The fitted values $\hat{y}_i$ and the residuals e_i from this model are shown in the last two columns of Table 6.1. A plot of the residuals versus $\hat{y}_i$ is shown in Figure 6.3. The residuals in this plot are identified by tool type (A or B). If the variance of the errors is not the same for both tool types, this should show up in the plot. Note that the "B" residuals in Figure 6.3 exhibit slightly more scatter than the "A" residuals, implying that there may be a mild inequality of variance problem. Figures 6.4 and 6.5 give plots of the residuals versus x_{i1} and the normal probability plot, respectively. Inspection of these plots does not reveal any serious model inadequacies. •

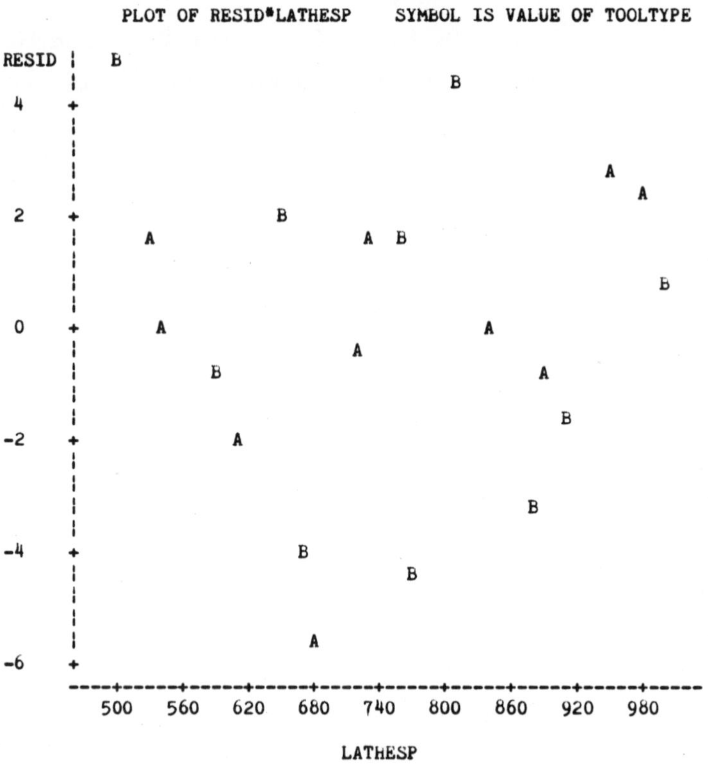

Figure 6.4 Plot (SAS) of residuals e_i versus lathe speed x_{i1}, Example 6.1.

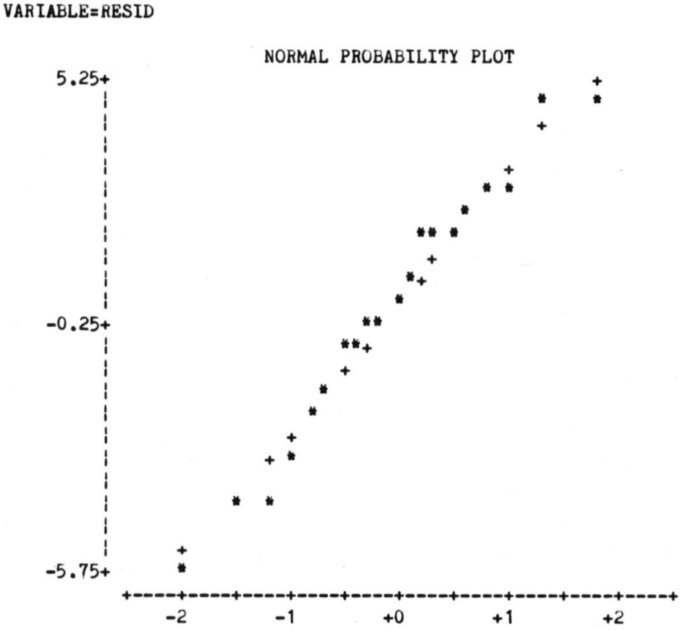

Figure 6.5 Normal probability plot (SAS) of residuals, Example 6.1.

Since two different regression lines are employed to model the relationship between tool life and lathe speed, we could have initially fit two separate straight-line models instead of a single model with an indicator variable. However, the single-model approach is preferred because the analyst has only one final equation to work with instead of two, a much simpler practical result. Furthermore since both straight lines are assumed to have the same slope, it makes sense to combine the data from both tool types to produce a single estimate of this common parameter. This approach also gives one estimate of the common error variance σ^2 and more residual degrees of freedom than would result from fitting two separate regression lines.

Now suppose that we expect the regression lines relating tool life to lathe speed to differ in both intercept and slope. It is possible to model this situation with a single regression equation by using indicator variables. The model is

$$y=\beta_0+\beta_1x_1+\beta_2x_2+\beta_3x_1x_2+\varepsilon \tag{6.4}$$

Comparing (6.4) with (6.1) we observe that a cross-product between lathe speed x_1 and the indicator variable denoting tool type x_2 has been added to the model. To interpret the parameters in this model, first consider tool type A, for which $x_2=0$. The model (6.4) becomes

$$\begin{aligned} y&=\beta_0+\beta_1x_1+\beta_2(0)+\beta_3x_1(0)+\varepsilon \\ &=\beta_0+\beta_1x_1+\varepsilon \end{aligned} \tag{6.5}$$

which is a straight line with intercept β_0 and slope β_1. For tool type B, we have $x_2=1$, and

$$\begin{aligned} y&=\beta_0+\beta_1x_1+\beta_2(1)+\beta_3x_1(1)+\varepsilon \\ &=(\beta_0+\beta_2)+(\beta_1+\beta_3)x_1+\varepsilon \end{aligned} \tag{6.6}$$

This is a straight-line model with intercept $\beta_0+\beta_2$ and slope $\beta_1+\beta_3$. Both regression functions are plotted in Figure 6.6. Note that (6.4) defines two regression lines with different slopes and intercepts. Therefore the parameter β_2 reflects the change in the intercept associated with changing from tool type A to tool type B (the classes 0 and 1 for the indicator variable x_2), and β_3 indicates the change in the slope associated with changing from tool type A to tool type B.

Fitting the model (6.4) is equivalent to fitting two separate regression equations. An advantage to the use of indicator variables is that tests of

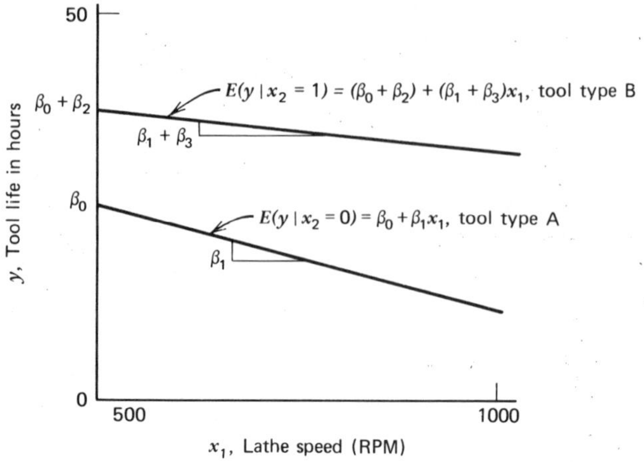

Figure 6.6 Response functions for Equation (6.4).

hypotheses can be performed directly using the extra sum of squares method. For example, to test whether or not the two regression models are identical, we would test

$$H_0: \beta_2 = \beta_3 = 0$$

$$H_1: \beta_2 \neq 0 \quad \text{and/or} \quad \beta_3 \neq 0$$

If H_0: $\beta_2 = \beta_3 = 0$ is not rejected, this would imply that a single regression model can explain the relationship between tool life and lathe speed. To test that the two regression lines have a common slope but possibly different intercepts, the hypotheses are

$$H_0: \beta_3 = 0$$

$$H_1: \beta_3 \neq 0$$

By using the model (6.4), both regression lines can be fitted and these tests performed with one computer run, provided the program produces the sums of squares $SS_R(\beta_1|\beta_0)$, $SS_R(\beta_2|\beta_0, \beta_1)$, and $SS_R(\beta_3|\beta_0, \beta_1, \beta_2)$.

- **Example 6.2.** We will fit the regression model

$$y = \beta_0 + \beta_1 x_1 + \beta_2 x_2 + \beta_3 x_1 x_2 + \varepsilon$$

to the tool life data in Table 6.1. The **X** matrix and **y** vector for this model are

$$
\mathbf{X} = \begin{array}{c} \begin{array}{ccc} x_1 & x_2 & x_1 x_2 \end{array} \\ \begin{bmatrix} 1 & 610 & 0 & 0 \\ 1 & 950 & 0 & 0 \\ 1 & 720 & 0 & 0 \\ 1 & 840 & 0 & 0 \\ 1 & 980 & 0 & 0 \\ 1 & 530 & 0 & 0 \\ 1 & 680 & 0 & 0 \\ 1 & 540 & 0 & 0 \\ 1 & 890 & 0 & 0 \\ 1 & 730 & 0 & 0 \\ 1 & 670 & 1 & 670 \\ 1 & 770 & 1 & 770 \\ 1 & 880 & 1 & 880 \\ 1 & 1000 & 1 & 1000 \\ 1 & 760 & 1 & 760 \\ 1 & 590 & 1 & 590 \\ 1 & 910 & 1 & 910 \\ 1 & 650 & 1 & 650 \\ 1 & 810 & 1 & 810 \\ 1 & 500 & 1 & 500 \end{bmatrix} \end{array} \quad \text{and} \quad \mathbf{y} = \begin{bmatrix} 18.73 \\ 14.52 \\ 17.43 \\ 14.54 \\ 13.44 \\ 24.39 \\ 13.34 \\ 22.71 \\ 12.68 \\ 19.32 \\ 30.16 \\ 27.09 \\ 25.40 \\ 26.05 \\ 33.49 \\ 35.62 \\ 26.07 \\ 36.78 \\ 34.95 \\ 43.67 \end{bmatrix}
$$

The fitted regression model is

$$\hat{y} = 32.775 - 0.021x_1 + 23.971x_2 - 0.012x_1 x_2$$

The summary statistics for this model are presented in Table 6.3. To test the hypothesis that the two regression lines are identical ($H_0: \beta_2 = \beta_3 = 0$), use the statistic

$$F_0 = \frac{SS_R(\beta_2, \beta_3 | \beta_1, \beta_0)/2}{MS_E}$$

Since

$$SS_R(\beta_2, \beta_3 | \beta_1, \beta_0) = SS_R(\beta_1, \beta_2, \beta_3 | \beta_0) - SS_R(\beta_1 | \beta_0)$$

$$= 1434.112 - 293.005$$

$$= 1141.007$$

the test statistic is

$$F_0 = \frac{SS_R(\beta_2, \beta_3 | \beta_1, \beta_0)/2}{MS_E} = \frac{1141.007/2}{8.811} = 64.75$$

Table 6.3 Summary statistics for the tool life regression model in Example 6.2.

Source of Variation	Sum of Squares	Degrees of Freedom	Mean Square	F_0
Regression	1434.112	3	478.037	54.25
Error	140.976	16	8.811	
Total	1575.088	19		

Coefficient	Estimate	Standard Error	t_0	Sum of Squares
β_0	32.775			
β_1	−0.021	0.0061	−3.45	$SS_R(\beta_1 \mid \beta_0) = 293.005$
β_2	23.971	6.7690	3.54	$SS_R(\beta_2 \mid \beta_1, \beta_0) = 1125.029$
β_3	−0.012	0.0880	−1.35	$SS_R(\beta_3 \mid \beta_2, \beta_1, \beta_0) = 16.078$
		$R^2 = 0.9105$		

and since $F_{.05,2,16} = 3.63$ we conclude that the two regression lines are not identical. To test the hypothesis that the two lines have different intercepts and a common slope (H_0: $\beta_3 = 0$), use the statistic

$$F_0 = \frac{SS_R(\beta_3 \mid \beta_2, \beta_1, \beta_0)/1}{MS_E} = \frac{16.078}{8.11} = 1.98$$

and since $F_{.05,1,16} = 4.49$, we conclude that the slopes of the two straight lines are the same. This can also be determined by using the t–statistics for β_2 and
• β_3 in Table 6.3.

Indicator variables are useful in a variety of regression situations. We will now present three further typical applications of indicator variables.

• **Example 6.3 An Indicator Variable with More than Two Levels** An electric utility is investigating the effect of the size of a single-family house and the type of air conditioning used in the house on the total electricity consumption during warm weather months. Let y be the total electricity consumption (KWH) during the period June through September and x_1 be the size of the house (square feet of floor space). There are four types of air conditioning systems: (1) no air conditioning, (2) window units, (3) heat pump, and (4) central air conditioning. The four levels of this factor can be modeled by three indicator variables x_2, x_3, and x_4, defined as follows:

Type of Air Conditioning	x_2	x_3	x_4
No air conditioning	0	0	0
Window units	1	0	0
Heat pump	0	1	0
Central air conditioning	0	0	1

The regression model is

$$y=\beta_0+\beta_1x_1+\beta_2x_2+\beta_3x_3+\beta_4x_4+\varepsilon \tag{6.7}$$

If the house has no air conditioning, (6.7) becomes

$$y=\beta_0+\beta_1x_1+\varepsilon$$

If the house has window units, then

$$y=(\beta_0+\beta_2)+\beta_1x_1+\varepsilon$$

If the house has a heat pump, the regression model is

$$y=(\beta_0+\beta_3)+\beta_1x_1+\varepsilon$$

while if the house has central air conditioning, then

$$y=(\beta_0+\beta_4)+\beta_1x_1+\varepsilon$$

Thus the model (6.7) assumes that the relationship between warm weather electricity consumption, and the size of the house is linear, and that the slope does not depend on the type of air conditioning system employed. The parameters β_2, β_3, and β_4 modify the height (or intercept) of the regression model for the different types of air conditioning systems. That is, β_2, β_3, and β_4 measure the effect of window units, a heat pump, and a central air conditioning system respectively, compared to no air conditioning. Furthermore other effects can be determined by directly comparing the appropriate regression coefficients. For example, $\beta_3-\beta_4$ reflects the relative efficiency of a heat pump compared to central air conditioning. Note also the assumption that the variance of energy consumption does not depend on the type of air conditioning system used. This assumption may be inappropriate.

In this problem it would seem unrealistic to assume that the slope of the regression function relating mean electricity consumption to the size of the house does not depend on the type of air conditioning system. For example, we would expect the mean electricity consumption to increase with the size of the house, but the rate of increase should be different for a central air conditioning system than for window units, because central air conditioning should be more efficient than window units for larger houses. That is, there should be an *interaction* between the size of the house and the type of air conditioning system. This can be incorporated into the model by expanding (6.7) to include interaction terms. The resulting model is

$$y=\beta_0+\beta_1x_1+\beta_2x_2+\beta_3x_3+\beta_4x_4+\beta_5x_1x_2+\beta_6x_1x_3+\beta_7x_1x_4+\varepsilon \quad (6.8)$$

The four regression models corresponding to the four types of air conditioning systems are as follows:

$$\begin{aligned}
&y=\beta_0+\beta_1x_1+\varepsilon_1 && \text{(no air conditioning)}\\
&y=(\beta_0+\beta_2)+(\beta_1+\beta_5)x_1+\varepsilon && \text{(window units)}\\
&y=(\beta_0+\beta_3)+(\beta_1+\beta_6)x_1+\varepsilon && \text{(heat pump)}\\
&y=(\beta_0+\beta_4)+(\beta_1+\beta_7)x_1+\varepsilon && \text{(central air conditioning)}
\end{aligned}$$

Note that model (6.8) implies that each type of air conditioning system can have a separate regression line, with a unique slope and intercept. •

• **Example 6.4 More Than One Indicator Variable** Frequently there are several different qualitative variables that must be incorporated into the model. To illustrate, suppose that in Example 6.1 a second qualitative factor, the type of cutting oil used, must be considered. Assuming that this factor has two levels, we may define a second indicator variable x_3 as follows:

$$x_3=\begin{cases}0 & \text{if low viscosity oil used}\\ 1 & \text{if medium viscosity oil used}\end{cases}$$

A regression model relating tool life (y) to cutting speed (x_1), tool type (x_2), and type of cutting oil (x_3) is

$$y=\beta_0+\beta_1x_1+\beta_2x_2+\beta_3x_3+\varepsilon \quad (6.9)$$

Clearly the slope β_1 of the regression model relating tool life to cutting speed does not depend on either the type of tool or the type of cutting oil. The intercept of the regression line depends on these factors in an additive fashion.

Various types of interaction effects may be added to the model. For example, suppose that we consider interactions between cutting speed and the two qualitative factors, so that the model (6.9) becomes

$$y=\beta_0+\beta_1x_1+\beta_2x_2+\beta_3x_3+\beta_4x_1x_2+\beta_5x_1x_3+\varepsilon \quad (6.10)$$

This implies the following situation:

Tool Type	Cutting Oil	Regression Model
A	low viscosity	$y=\beta_0+\beta_1x_1+\varepsilon$
B	low viscosity	$y=(\beta_0+\beta_2)+(\beta_1+\beta_4)x_1+\varepsilon$
A	medium viscosity	$y=(\beta_0+\beta_3)+(\beta_1+\beta_5)x_1+\varepsilon$
B	medium viscosity	$y=(\beta_0+\beta_2+\beta_3)+(\beta_1+\beta_4+\beta_5)x_1+\varepsilon$

Notice that each combination of tool type and cutting oil results in a separate regression line, with different slopes and intercepts. However, the model is still additive with respect to the levels of the indicator variables. That is, changing from low viscosity to medium viscosity cutting oil changes the intercept by β_3 and the slope by β_5, regardless of the type of tool used.

Suppose that we add a cross-product term involving the two indicator variables x_2 and x_3 to the model, resulting in

$$y=\beta_0+\beta_1x_1+\beta_2x_2+\beta_3x_3+\beta_4x_1x_2+\beta_5x_1x_3+\beta_6x_2x_3+\varepsilon \quad (6.11)$$

We then have the following:

Tool Type	Cutting Oil	Regression Model
A	low viscosity	$y=\beta_0+\beta_1x_1+\varepsilon$
B	low viscosity	$y=(\beta_0+\beta_2)+(\beta_1+\beta_4)x_1+\varepsilon$
A	medium viscosity	$y=(\beta_0+\beta_3)+(\beta_1+\beta_5)x_1+\varepsilon$
B	medium viscosity	$y=(\beta_0+\beta_2+\beta_3+\beta_6)+(\beta_1+\beta_4+\beta_5)x_1+\varepsilon$

The addition of the cross-product term $\beta_6x_2x_3$ in (6.11) results in the effect of one indicator variable on the intercept depending on the level of the other indicator variable. That is, changing from low viscosity to medium viscosity cutting oil changes the intercept by β_3 if tool type A is used, but the same change in cutting oil changes the intercept by $\beta_3+\beta_6$ if tool type B is used. If an interaction term $\beta_7x_1x_2x_3$ were added to the model in (6.11), then changing from low viscosity to medium viscosity cutting oil would have an effect on *both* the intercept *and* the slope that depends on the type of tool used.

Unless prior information is available concerning the anticipated effect of tool type and cutting oil viscosity on tool life, we will have to let the data guide us in selecting the correct form of the model. This may generally be done by testing hypotheses about individual regression coefficients using the partial F-test. For example, testing H_0: $\beta_6=0$ for model (6.11) would allow us to discriminate between the two candidate models (6.11) and (6.10).

Example 6.5 Comparing Regression Models Consider the case of simple linear regression where the n observations can be formed into M groups, with the *mth* group having n_m observations. The most general model consists of M separate equations, such as

$$y=\beta_{0m}+\beta_{1m}x+\varepsilon, \qquad m=1,2,\ldots,M \tag{6.12}$$

It is often of interest to compare this general model to a more restrictive one. Indicator variables are helpful in this regard. We consider the following cases:

a. Parallel lines In this situation all M slopes are identical, $\beta_{11}=\beta_{12}=\cdots=\beta_{1M}$, but the intercepts may differ. Note that this is the type problem encountered in Example 6.1 (where $M=2$), leading to the use of an additive indicator variable. More generally, we may use the extra sum of squares method to test the hypothesis H_0: $\beta_{11}=\beta_{12}=\cdots=\beta_{1M}$. Recall that this procedure involves fitting a *full model* (FM) and a *reduced model* (RM) restricted to the null hypothesis and computing the F-statistic

$$F_0=\frac{[SS_E(RM)-SS_E(FM)]/(df_{RM}-df_{FM})}{SS_E(FM)/df_{FM}} \tag{6.13}$$

If the reduced model is as satisfactory as the full model, then F_0 will be small compared to $F_{\alpha,\,df_{RM}-df_{FM},\,df_{FM}}$. Large values of F_0 imply that the reduced model is inadequate.

To fit the full model (6.12) simply fit M separate regression equations. Then $SS_E(FM)$ is found by adding the residual sums of squares from each separate regression. The degrees of freedom for $SS_E(FM)$ is $df_{FM}=\sum_{m=1}^{M}(n_m-2)=n-2M$. To fit the reduced model, define $M-1$ indicator variables $D_1, D_2,\ldots, D_{M-1}$ corresponding to the M groups and fit

$$y=\beta_0+\beta_1x+\beta_2D_1+\beta_3D_2+\cdots+\beta_MD_{M-1}+\varepsilon$$

The residual sum of squares from this model is $SS_E(RM)$ with $df_{RM}=n-(M+1)$ degrees of freedom.

If the F-test (6.13) indicates that the M regression models have a common slope, then $\hat{\beta}_1$ from the reduced model is an estimate of this parameter found by pooling or combining all of the data. This was illustrated in Example 6.1. More generally, *analysis of covariance* is used to pool the data to estimate the common slope. The analysis of covariance is a special type of linear model that is a combination of a regression model (with quantitative factors) and an analysis of variance model (with qualitative factors). For an introduction to analysis of covariance, see Montgomery [1976, Ch. 15] or Neter and Wasserman [1974, Ch. 22].

b. Concurrent lines In this section, all M intercepts are equal, $\beta_{01}=\beta_{02}=\cdots=\beta_{0M}$, but the slopes may differ. The reduced model is

$$y=\beta_0+\beta_1x+\beta_2Z_1+\beta_2Z_2+\cdots+\beta_MZ_{M-1}+\varepsilon$$

where $Z_k=xD_k$, $k=1,2,\ldots,M-1$. The residual sum of squares from this model is $SS_E(RM)$ with $df_{RM}=n-(M+1)$ degrees of freedom. Note that we are assuming concurrence at the origin. The more general case of concurrence at an arbitrary point x_0 is treated by Graybill [1976] and Seber [1977].

c. Coincident lines In this case both the M slopes and the M intercepts are the same, $\beta_{01}=\beta_{02}=\cdots=\beta_{0M}$, and $\beta_{11}=\beta_{12}=\cdots=\beta_{1M}$. The reduced model is simply

$$y=\beta_0+\beta_1x+\varepsilon$$

and the residual sum of squares $SS_E(RM)$ has $df_{RM}=n-2$ degrees of freedom. Indicator variables are not necessary in the test of coincidence, but we include this case for completeness.

6.2 Comments on the use of indicator variables

6.2.1 Indicator variables versus regression on allocated codes

Another approach to the treatment of a qualitative variable in regression is to "measure" the levels of the variable by an allocated code. Recall Example 6.3, where an electric utility is investigating the effect of size of house and type of air conditioning system on residential electricity consumption. Instead of using three indicator variables to represent the four levels of the qualitative factor "type of air conditioning system," we could use one quantitative factor x_2 with the following allocated code:

Type of Air Conditioning System	x_2
No air conditioning	1
Window units	2
Heat pumps	3
Central air conditioning	4

We may now fit the regression model

$$y=\beta_0+\beta_1x_1+\beta_2x_2+\varepsilon \qquad (6.14)$$

where x_1 is the size of the house. This model implies that

$$\begin{aligned} E(y|x_1, \text{no air conditioning}) &= \beta_0 + \beta_1 x_1 + \beta_2 \\ E(y|x_1, \text{window units}) &= \beta_0 + \beta_1 x_1 + 2\beta_2 \\ E(y|x_1, \text{heat pump}) &= \beta_0 + \beta_1 x_1 + 3\beta_2 \\ E(y|x_1, \text{central air conditioning}) &= \beta_0 + \beta_1 x_1 + 4\beta_2 \end{aligned}$$

A direct consequence of this is that

$$\begin{aligned} &E(y|x_1, \text{central air conditioning}) - E(y|x_1, \text{heat pump}) \\ &\quad = E(y|x_1, \text{heat pump}) - E(y|x_1, \text{window units}) \\ &\quad = E(y|x_1, \text{window units}) - E(y|x_1, \text{no air conditioning}) \\ &\quad = \beta_2 \end{aligned}$$

which may be quite unrealistic. The allocated codes impose a particular metric on the levels of the qualitative factor. Other choices of the allocated code would imply different distances between the levels of the qualitative factor, but there is no guarantee that any particular allocated code leads to a spacing that is appropriate.

Indicator variables are more informative for this type problem because they do not force any particular metric on the levels of the qualitative factor. Furthermore regression using indicator variables always leads to a larger R^2 than does regression on allocated codes (for example, see Searle and Udell [1970]).

6.2.2 Indicator variables as a substitute for a quantitative regressor

Quantitative regressors can also be represented by indicator variables. Sometimes this is necessary because it is difficult to collect accurate information on the quantitative regressor. Consider the electric power usage study in Example 6.3 and suppose that a second quantitative regressor, household income, is included in the analysis. Because it is difficult to obtain this information precisely, the quantitative regressor "income" may be collected by grouping income into classes such as

$0 to $4,999
$5,000 to $9,999
$10,000 to $14,999
$15,000 to $19,999
$20,000 and over

We may now represent the factor "income" in the model by using four indicator variables.

One disadvantage of this approach is that more parameters are required to represent the information content of the quantitative factor. In general if the quantitative regressor is grouped into a classes, $a-1$ parameters will be required, while only one parameter would be required if the original quantitative regressor is used. Thus treating a quantitative factor as a qualitative one increases the complexity of the model. This approach also reduces the degrees of freedom for error, although if the data are numerous this is not a serious problem. An advantage of the indicator variable approach is that it does not require the analyst to make any prior assumptions about the functional form of the relationship between the response and the regressor variable.

6.2.3 Models with only indicator variables

Sometimes all the regressor variables are qualitative. This would lead to a regression model with only 0–1 indicator variables as regressors. Models of this type are called analysis of variance models, and are discussed briefly in Chapter 9.

6.3 Regression models with an indicator response variable

Occasionally the response variable in a regression problem is *binary*; that is, it can assume only two values. Thus the response is an indicator variable with value either 0 or 1. For example, suppose that an aeronautical engineer is investigating whether a shoulder-fired antiaircraft missile hits the target as a function of target speed. The response variable in this problem is binary; either the missile misses the target, or it hits the target. We may assign the values 0 and 1 to these outcomes, respectively.

The expected response has a special interpretation in this situation. Consider a model with only one regressor, say

$$y_i=\beta_0+\beta_1 x_i+\varepsilon, \qquad i=1,2,\ldots,n$$

If $E(\varepsilon_i)=0$, then

$$E(y_i|x_i)=\beta_0+\beta_1 x_i$$

Since y_i can take on only the values 0 and 1, a reasonable probability model for the response is the Bernoulli distribution. Recall that in the Bernoulli distribution the random variable y_i takes on the value 1 with probability

$P(y_i=1)=p_i$ and the value 0 with probability $P(y_i=0)=(1-p_i)$. Since the mean of the Bernoulli distribution is $E(y_i|x_i)=p_i$, and

$$E(y_i|x_i)=\beta_0+\beta_1 x_i=p_i \tag{6.15}$$

the mean response is interpreted as the probability that $y_i=1$ when the regressor variable takes on the value x_i.

Fitting a model with an indicator response variable is not straightforward. To illustrate one of the difficulties encountered, consider the variance of the errors. We may show that

$$\begin{aligned} V(\varepsilon_i|x_i) &= V(y_i|x_i) \\ &= p_i(1-p_i) \\ &= (\beta_0+\beta_1 x_i)(1-\beta_0-\beta_1 x_i) \end{aligned} \tag{6.16}$$

since the variance of the Bernoulli distribution is $V(y_i|x_i)=p_i(1-p_i)$. Equation (6.16) implies that the variance of the errors is not constant and in fact depends on the value of the regressor variable x_i. This is a violation of the basic regression assumptions. The use of weighted least squares with the weights chosen inversely proportional to the variance of y_i will eliminate this problem. We also note that the errors cannot be normally distributed, since at each possible level of the regressor there are only two possible values of ε_i. Finally, because $E(y_i|x_i)$ is the probability that $y_i=1$ when the value of the regressor is x_i, it seems logical to require that the predicted responses lie between 0 and 1 for all x_i within the range of the original data. There is no assurance that the fitted model will have this property.

We will now illustrate fitting regression models using an indicator response variable. We begin with linear regression models, and then present a nonlinear regression model based on the logistic function.

6.3.1 A linear model

Suppose that

$$y=\beta_0+\beta_1 x_1+\beta_2 x_2+\cdots\beta_k x_k+\varepsilon$$

where $y=0$ or 1 is an indicator variable. As noted previously, weighted least squares should be used to estimate the parameters in this model, since the variance of the errors is not constant.

- **Example 6.6** Table 6.4 presents the results of test-firing 25 ground-to-air antiaircraft missiles at targets of varying speed. The result of each test-firing is either a hit ($y=1$) or a miss ($y=0$). A scatter diagram is shown in Figure 6.7.

Table 6.4 Missile test data for Example 6.6

Test-Firing i	Target Speed (knots) x_i	Hit or Miss y_i
1	400	0
2	220	1
3	490	0
4	410	1
5	500	0
6	270	0
7	200	1
8	470	0
9	480	0
10	310	1
11	240	1
12	490	0
13	420	0
14	330	1
15	280	1
16	210	1
17	300	1
18	470	1
19	230	0
20	430	0
21	460	0
22	220	1
23	250	1
24	200	1
25	390	0

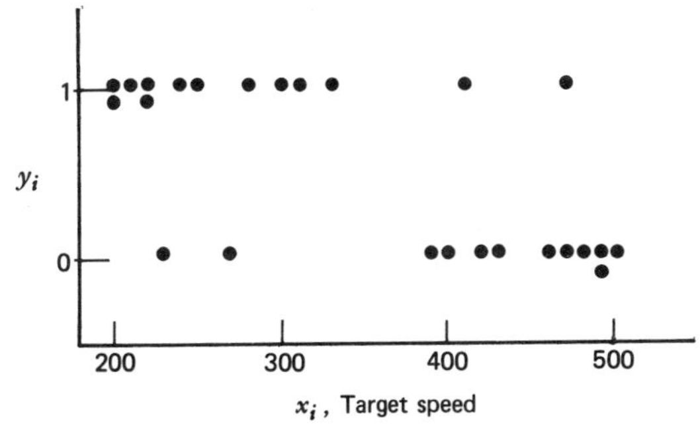

Figure 6.7 Plot of y_i versus target speed x_i (knots).

This diagram indicates that the probability of hitting the target seems to decrease as target speed increases. We will fit the straight-line model $y=\beta_0+\beta_1 x+\varepsilon$ to these data, using weighted least squares.

An immediate difficulty is that the weights required for weighted least squares are unknown. That is, the weights w_i should be

$$w_i=\frac{1}{V(y_i|x_i)}=\frac{1}{p_i(1-p_i)}$$
$$=\frac{1}{(\beta_0+\beta_1 x_i)(1-\beta_0-\beta_1 x_i)}$$

and w_i is a function of unknown parameters β_0 and β_1. This problem can be overcome by initially estimating the model parameters using ordinary (unweighted) least squares, and then calculating the weights using $\hat{\beta}_0$ and $\hat{\beta}_1$, the ordinary least squares parameter estimates, as follows:

$$\hat{w}_i=\frac{1}{(\hat{\beta}_0+\hat{\beta}_1 x_i)(1-\hat{\beta}_0-\hat{\beta}_1 x_i)}=\frac{1}{\hat{y}_i(1-\hat{y}_i)} \tag{6.17}$$

If necessary, additional iterations could be performed, revising the weights at each step. Usually however, further iterations do not produce much improvement in the estimates.

The ordinary least squares estimates of β_0 and β_1 and their standard errors are shown below:

Coefficient	Estimate	Standard Error
β_0	1.56228	0.26834
β_1	−0.00301	0.00074

Table 6.5 shows the fitted values $\hat{y}_i$ (which may be interpreted as the probability of a hit when the target speed is x_i knots) and the estimated weights $\hat{w}_i$. For example, consider the first test-firing, with a target speed of $x_1=400$ knots. We have

$$\hat{y}_1=\hat{\beta}_0+\hat{\beta}_1 x_1$$
$$=1.56228-0.00301(400)=0.3601$$

and

$$\hat{w}_1=\frac{1}{\hat{y}_1(1-\hat{y}_1)}=\frac{1}{(0.3601)(1-0.3601)}=4.3397$$

Table 6.5 Fitted values and estimated weights for Example 6.6

Test-Firing, i	$\hat{y}_i$	$\hat{w}_i = 1/[\hat{y}_i(1-\hat{y}_i)]$
1	.3601	4.3397
2	.9011	11.2198
3	.0896	12.2563
4	.3301	4.5224
5	.0596	17.8507
6	.7508	5.3450
7	.9612	26.8113
8	.1497	7.8547
9	.1197	9.4918
10	.6306	4.2929
11	.8410	7.4776
12	.0896	12.2563
13	.3000	4.7619
14	.5705	4.0811
15	.7208	4.9686
16	.9311	15.5967
17	.6607	4.4605
18	.1497	7.8547
19	.8710	8.9020
20	.2699	5.0742
21	.1798	6.7814
22	.9011	11.2198
23	.8109	6.5221
24	.9612	16.8113
25	.3902	4.2028

The fitted model produces values of $\hat{y}_i$ that fall between 0 and 1 if the target speed is in the range of the data. If it did not, this would be a symptom of model inadequacy, and a nonlinear model such as the one in Section 6.3.2, must be used.

Using the estimated weights and the method of weighted least squares (see Section 3.8 for details), we obtain the revised estimates of the model parameters as

Coefficient	Estimate	Standard Error
β_0	1.58669	0.07344
β_1	-0.00309	0.00022

Apart from smaller standard errors, these parameter estimates are not dramatically different from those found using ordinary least squares. The final fitted model is

•
$$\hat{y}=1.58669-0.00309x$$

6.3.2 A nonlinear model

In some problems with an indicator response variable, the relationship between y and x is nonlinear. Very frequently we find that the response function is S-shaped, as in Figure 6.8. There are several approaches to fitting such a function. One method involves modeling the response function in Figure 6.8 with the normal cumulative distribution function. This approach is called *probit analysis* (see Finney [1952]). A second method of analysis is to model the response using the logistic function

$$E(y|x)=\frac{\exp(\beta_0+\beta_1 x)}{1+\exp(\beta_0+\beta_1 x)} \tag{6.18}$$

The logistic function (6.18) has the characteristic S-shape shown in Figure 6.8. It also has asymptotes at 0 and 1, guaranteeing that the estimated response function lies between 0 and 1. Fitting the logistic function is usually called *logit analysis*.

Both the probit analysis and logit analysis models arise from a consideration of threshold values. For example, in determining the structural reliability of a metal fastener, a fastener is assumed to have a threshold strength s such that if the fastener is tested under a load less than or equal to s it will not fail but if tested at a load greater than s it will fail. Thus if the applied load is x, and if $y=1$ when $s\leqslant x$ and $y=0$ when $s>x$, then $E(y)=P(y=1|x)=P(s\leqslant x)$. Note that $P(s\leqslant x)$ is the cumulative distribution of the threshold strength of the population of fasteners. If this cumulative distribution is normal, then the probit analysis approach should be used, while if

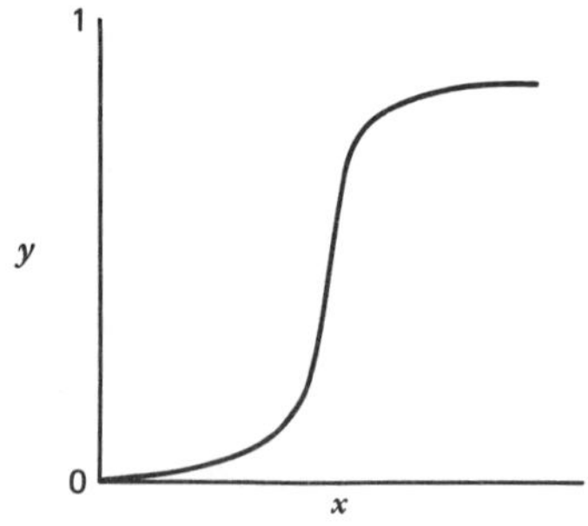

Figure 6.8 A nonlinear response function.

the cumulative distribution is logistic, then the logit analysis approach is appropriate. As the logistic function is somewhat easier to fit, we will illustrate its use.

The logistic function (6.18) is intrinsically linear. Recall that the mean response $E(y|x)=p$ when y is an indicator variable. We may linearize (6.18) by using the transformation

$$p^*=\ln\left[\frac{E(y|x)}{1-E(y|x)}\right]=\ln\left(\frac{p}{1-p}\right) \tag{6.19}$$

The resulting linearized model is

$$p^*=\beta_0+\beta_1 x \tag{6.20}$$

To fit the logistic response function, we will assume that there are repeat observations on y at each level of x. Denote the levels of x by $x_1, x_2, \ldots, x_m$ and let there be n_i observations at level x_i. Let c_i be the number of 1's at level x_i. The observed proportion of 1's at each level of x is then

$$\bar{p}_i=\frac{c_i}{n_i}, \qquad i=1,2,\ldots,m \tag{6.21}$$

The transformed logistic response function (6.20) may be fit by linear least squares, using the transformed observed proportions

$$\bar{p}_i^*=\ln\left(\frac{\bar{p}_i}{1-\bar{p}_i}\right) \tag{6.22}$$

as the response variable. However, the error terms in the linearized model have unequal variance. In fact, if the number of observations at each level of x is large, the variance of the observed proportion $\bar{p}_i^*$ is approximately

$$V(\bar{p}_i^*)=\frac{1}{n_i p_i(1-p_i)}, \qquad i=1,2,\ldots,m$$

where p_i is the true probability that $y_i=1$ when the level of $x=x_i$. If this probability is estimated by $\bar{p}_i$, then an estimate of the variance of $\bar{p}_i^*$ is

$$V(\bar{p}_i^*)=\frac{1}{n_i \bar{p}_i(1-\bar{p}_i)}, \qquad i=1,2,\ldots,m$$

Because the variance of the error term is not constant, weighted least squares should be used to estimate β_0 and β_1. The appropriate choice of weights is

$$w_i = n_i \bar{p}_i(1-\bar{p}_i)$$

The fitted response model is

$$\hat{p}^* = \hat{\beta}_0 + \hat{\beta}_1 x$$

where the parameters $\hat{\beta}_0$ and $\hat{\beta}_1$ are estimated by weighted least squares. Expressed in terms of the original units, the fitted response model is

$$\hat{p} = \frac{\exp(\hat{\beta}_0 + \hat{\beta}_1 x)}{1+\exp(\hat{\beta}_0 + \hat{\beta}_1 x)}$$

This method of fitting the logistic function requires that none of the $p_i = 0$ or 1. If this assumption is violated, and if some of the $\bar{p}_i$ are very close to 0 or 1, refer to the methods described in Cox [1970].

• **Example 6.7** The compressive strength of an alloy fastener used in aircraft construction is being studied. Ten loads were selected over the range 2500 psi to 4300 psi, and a number of fasteners tested at those loads. The resulting data is shown in Table 6.6. We will assume that the logistic function is an appropriate model for this data. The weights for the weighted least squares procedure are shown in the last column of this table.

Applying the method of weighted least squares to this data yields the fitted model

$$\hat{p}^* = -5.3361 + 0.001547x$$

If we wished to find the probability of failure for a fastener that is subjected

Table 6.6 Aircraft fastener failure data for Example 6.7

Load psi x_i	Sample Size n_i	Number Failing c_i	Percent $\bar{p}_i^*$	$\frac{\bar{p}_i}{1-\bar{p}_i}$	$\bar{p}_i^* = \ln\left(\frac{\bar{p}_i}{1-\bar{p}_i}\right)$	Weight $w_i = n_i \bar{p}_i(1-\bar{p}_i)$
2500	50	10	.20	.2500	−1.3864	8.0000
2700	70	17	.2429	.3208	−1.1369	12.8714
2900	100	30	.30	.4286	−.8473	21.0000
3100	60	21	.35	.5386	−.6190	13.6500
3300	40	18	.45	.8182	−.2007	9.9000
3500	85	43	.5059	1.0239	.0236	24.2471
3700	90	54	.60	1.5000	.4055	21.6000
3900	50	33	.66	1.9412	.6633	11.2200
4100	80	60	.75	3.0000	1.0986	15.0000
4300	65	51	.7846	3.6425	1.2927	10.9846

to a load of 2800 psi, we would first evaluate the fitted model as

$$\begin{aligned}\hat{p}^* &= -5.3361 + 0.001547(2800)\\ &= -1.0045\end{aligned}$$

and then transform $\hat{p}^*$ to the original units as follows:

$$\hat{p} = \frac{\exp(\hat{\beta}_0 + \hat{\beta}_1 x)}{1+\exp(\hat{\beta}_0 + \hat{\beta}_1 x)} = \frac{e^{\hat{p}*}}{1+e^{\hat{p}*}} = \frac{e^{-1.0045}}{1+e^{-1.0045}} = 0.2681$$

That is, the estimated failure probability of a fastener subjected to a load of 2800 psi is 0.2681. Of course before adopting this equation as an adequate model of the fastener failure data the adequacy of the model should be investigated through residual analysis. •

Problems

6.1 Consider the regression model (6.8) described in Example 6.3. Graph the response function for this model and indicate the role the model parameters play in determining the shape of this function.

6.2 Consider the regression models described in Example 6.4.

a. Graph the response function associated with (6.10).

b. Graph the response function associated with (6.11).

6.3 Consider the delivery time data in Example 2.8. In Section 3.2.5 we noted that these observations were collected in four cities, San Diego, Boston, Austin, and Minneapolis.

a. Develop a model that relates delivery time y to delivery volume x_1 and the city in which the delivery was made. Estimate the parameters of the model.

b. Is there an indication that delivery site is an important variable?

c. Analyze the residuals from this model. What conclusions can you draw regarding model adequacy?

6.4 Consider the automobile gasoline mileage data in Appendix Table B.3.

a. Build a linear regression model relating gasoline mileage y to engine displacement x_1 and the type of transmission x_{11}. Does the type of transmission significantly affect the mileage performance?

b. Modify the model developed in Part a to include an interaction between engine displacement and the type of transmission. What conclusions can you draw about the effect of the type of transmission on gasoline mileage? Interpret the parameters in this model.

6.5 Consider the automobile gasoline mileage data in Appendix Table B.3.

a. Build a linear regression model relating gasoline mileage y to vehicle weight x_{10} and the type of transmission x_{11}. Does the type of transmission significantly affect the mileage performance?

b. Modify the model developed in Part a to include an interaction between vehicle weight and the type of transmission. What conclusions can you draw about the effect of the type of transmission on gasoline mileage? Interpret the parameters in this model.

6.6 Consider the National Football League data in Appendix Table B.1. Build a linear regression model relating the number of games won to the yards gained rushing by opponents x_8, the percent of rushing plays x_7, and a modification of the turnover differential x_5. Specifically, let the turnover differential be an indicator variable whose value is determined by whether the actual turnover differential is positive, negative, or zero. What conclusions can you draw about the effect of turnovers on the number of games won?

6.7 A study was conducted attempting to relate home ownership to family income. Twenty households were selected, and family income along with whether the home was owned ($y=1$) or rented ($y=0$) recorded. The data are shown below. Fit a linear regression model to these data. Does it appear that the linear model is reasonable?

Household Number	Income	Home Ownership Status
1	8,300	0
2	21,200	1
3	9,100	0
4	13,400	1
5	17,700	0
6	23,000	0
7	11,500	1
8	10,800	0
9	15,400	1
10	22,400	1
11	18,700	1
12	10,100	0
13	19,500	1
14	8,000	0
15	12,000	1
16	24,000	1
17	21,700	1
18	9,400	0
19	10,900	0
20	22,800	1

6.8 The market research department of a soft drink manufacturer is investigating the effectiveness of a price discount coupon on a 6-pack of a 2-litre beverage product. A sample of 5500 consumers was selected, and a coupon given to each. The coupons offered different price discounts varying from 5 cents off to 25 cents off in increments of 2 cents, and 500 consumers were assigned to one of the eleven price discount categories. The response variable was whether or not the coupon was redeemed after one month. The data are shown below. Fit a logistic response function to these data. Does this model seem to adequately describe the data?

Price Discount (x_j)	Sample Size (n_j)	Number of Coupons Redeemed (c_j)
5	500	100
7	500	122
9	500	147
11	500	176
13	500	211
15	500	244
17	500	277
19	500	310
21	500	343
23	500	372
25	500	391

6.9 Piecewise Linear Regression. In Example 5.3 we showed how a linear regression model with a change in slope at some point t ($x_{\min}<t<x_{\max}$) could be fitted using splines. Develop a formulation of the piecewise linear regression model using indicator variables. Assume that the function is continuous at the point t.

6.10 Continuation of Problem 6.9. Show how indicator variables can be used to develop a piecewise linear regression model with a discontinuity at the join point t.

7

VARIABLE SELECTION AND MODEL BUILDING

7.1 Introduction

7.1.1 The model building problem

In the preceding chapters, we have assumed that the regressor variables included in the model are known to be influential. Our focus was on techniques to insure that the functional form of the model was correct and that the underlying assumptions were not violated. In some applications, theoretical considerations or prior experience can be helpful in selecting the regressors to be used in the model. However, in most practical problems the analyst has a pool of *candidate* regressors that should include all the influential factors, but the actual subset of regressors that should be used in the model needs to be determined. Finding an appropriate subset of regressors for the model is called the *variable selection problem*.

Building a regression model that includes only a subset of the available regressors involves two conflicting objectives. (1) We would like the model to include as many regressors as possible so that the "information content" in these factors can influence the predicted value of y. (2) We want the model to include as few regressors as possible because the variance of the prediction $\hat{y}$ increases as the number of regressors increases. Also, the more regressors there are in a model, the greater the costs of data collection and model maintenance. The process of finding a model that is a compromise between these two objectives is called selecting the "best" regression equation. Unfortunately, as we will see in this chapter, there is no unique definition of "best." Furthermore there are several algorithms that can be used for variable selection, and these procedures frequently specify different subsets of the candidate regressors as "best."

The variable selection problem is often discussed in an idealized setting. It is usually assumed that the correct functional specification of the regressors is known (such as $1/x_1$, $\ln x_2$, etc.), and that no outliers or influential observations are present. In practice, these assumptions are rarely met. Residual analysis, such as described in Chapters 3 and 4, is useful in revealing functional forms for regressors that might be investigated, in pointing out new candidate regressors, and for identifying defects in the data such as outliers. The effect of influential or high-leverage observations should also be determined. Investigation of model adequacy is linked to the variable selection problem. Although ideally these problems should be solved simultaneously, an iterative approach is often employed, in which (1) a particular variable selection strategy is employed, and then (2) the resulting subset model is checked for correct functional specification, outliers, and influential observations. This may indicate that Step (1) must be repeated. Several iterations may be required to produce an adequate model.

None of the variable selection procedures described in this chapter are guaranteed to produce the "best" regression equation for a given data set. In fact there usually is not a single "best" equation but rather several equally good ones. Because variable selection algorithms are heavily computer-dependent, the analyst is sometimes tempted to place too much reliance on the results of a particular procedure. Such temptation is to be avoided. Experience, professional judgement in the subject matter field, and subjective considerations all enter into the variable selection problem. Variable selection procedures should be used by the analyst as methods to explore the structure of the data. Good general discussions of variable selection in regression include Cox and Snell [1974], Hocking [1976], and Thompson [1978a, b].

7.1.2 Consequences of model misspecification

To provide motivation for variable selection, we will briefly review the consequences of incorrect model specification. Assume that there are K candidate regressors $x_1, x_2, \ldots, x_K$ and $n \geqslant K+1$ observations on these regressors and the response y. The *full* model, containing all K regressors, is

$$y_i = \beta_0 + \sum_{j=1}^{K} \beta_j x_{ij} + \varepsilon_i, \qquad i = 1, 2, \ldots, n \tag{7.1a}$$

or equivalently

$$\mathbf{y} = \mathbf{X}\boldsymbol{\beta} + \boldsymbol{\varepsilon} \tag{7.1b}$$

We assume that the list of candidate regressors contains all the influential variables. Note that (7.1) contains an intercept term β_0. While β_0 could also be a candidate for selection, it is typically forced into the model. We assume that all equations include an intercept term. Let r be the number of regressors that are deleted from (7.1). Then the number of variables that are retained is $p = K + 1 - r$. Since the intercept is included, the subset model contains $p - 1 = K - r$ of the original regressors.

The model (7.1) may be written as

$$\mathbf{y} = \mathbf{X}_p\boldsymbol{\beta}_p + \mathbf{X}_r\boldsymbol{\beta}_r + \boldsymbol{\varepsilon} \tag{7.2}$$

where the $\mathbf{X}$ matrix has been partitioned into $\mathbf{X}_p$, an $n \times p$ matrix whose columns represent the intercept and the $p-1$ regressors to be retained in the subset model, and $\mathbf{X}_r$, an $n \times r$ matrix whose columns represent the regressors to be deleted from the full model. Let $\boldsymbol{\beta}$ be partitioned conformably into $\boldsymbol{\beta}_p$ and $\boldsymbol{\beta}_r$. For the full model the least squares estimate of $\boldsymbol{\beta}$ is

$$\hat{\boldsymbol{\beta}}^* = (\mathbf{X}'\mathbf{X})^{-1}\mathbf{X}'\mathbf{y} \tag{7.3}$$

and an estimate of the residual variance σ^2 is

$$\hat{\sigma}_*^2 = \frac{\mathbf{y}'\mathbf{y} - \hat{\boldsymbol{\beta}}^{*\prime}\mathbf{X}'\mathbf{y}}{n-K-1} = \frac{\mathbf{y}'\left[\mathbf{I} - \mathbf{X}(\mathbf{X}'\mathbf{X})^{-1}\mathbf{X}'\right]\mathbf{y}}{n-K-1} \tag{7.4}$$

The components of $\hat{\boldsymbol{\beta}}^*$ are denoted by $\hat{\boldsymbol{\beta}}_p^*$ and $\hat{\boldsymbol{\beta}}_r^*$, and $\hat{y}_i^*$ are the fitted values. For the subset model

$$\mathbf{y} = \mathbf{X}_p\boldsymbol{\beta}_p + \boldsymbol{\varepsilon} \tag{7.5}$$

the least squares estimate of $\boldsymbol{\beta}_p$ is

$$\hat{\boldsymbol{\beta}}_p = (\mathbf{X}_p'\mathbf{X}_p)^{-1}\mathbf{X}_p'\mathbf{y} \tag{7.6}$$

the estimate of the residual variance is

$$\hat{\sigma}^2 = \frac{\mathbf{y}'\mathbf{y} - \hat{\boldsymbol{\beta}}_p'\mathbf{X}_p'\mathbf{y}}{n-p} = \frac{\mathbf{y}'\left[\mathbf{I} - \mathbf{X}_p(\mathbf{X}_p'\mathbf{X}_p)^{-1}\mathbf{X}_p'\right]\mathbf{y}}{n-p} \tag{7.7}$$

and the fitted values are $\hat{y}_i$.

The properties of the estimates $\hat{\boldsymbol{\beta}}_p$ and $\hat{\sigma}^2$ from the subset model have been investigated by several authors, including Hocking [1974, 1976], Narula and Ramberg [1972], Rao [1971], Rosenberg and Levy [1972], and Walls and Weeks [1969]. The results can be summarized as follows:

1. The expected value of $\hat{\boldsymbol{\beta}}_p$ is

$$E(\hat{\boldsymbol{\beta}}_p) = \boldsymbol{\beta}_p + (\mathbf{X}_p'\mathbf{X}_p)^{-1}\mathbf{X}_p'\mathbf{X}_r\boldsymbol{\beta}_r$$

$$= \boldsymbol{\beta}_p + \mathbf{A}\mathbf{B}_r$$

where $\mathbf{A} = (\mathbf{X}_p'\mathbf{X}_p)^{-1}\mathbf{X}_p'\mathbf{X}_r$. $\mathbf{A}$ is sometimes called the alias matrix. Thus $\hat{\boldsymbol{\beta}}_p$ is a biased estimate of $\boldsymbol{\beta}_p$ unless the regression coefficients corresponding to the deleted variables ($\boldsymbol{\beta}_r$) are zero or the retained variables are orthogonal to the deleted variables ($\mathbf{X}_p'\mathbf{X}_r = \mathbf{0}$).

2. The variances of $\hat{\boldsymbol{\beta}}_p$ and $\hat{\boldsymbol{\beta}}^*$ are $V(\hat{\boldsymbol{\beta}}_p) = \sigma^2(\mathbf{X}_p'\mathbf{X}_p)^{-1}$ and $V(\hat{\boldsymbol{\beta}}^*) = \sigma^2(\mathbf{X}'\mathbf{X})^{-1}$, respectively. Also the matrix $V(\hat{\boldsymbol{\beta}}_p^*) - V(\hat{\boldsymbol{\beta}}_p)$ is positive semidefinite; that is, the variances of the least squares estimates of the parameters in the full model are greater than or equal to the variances of the corresponding parameters in the subset model. Consequently, deleting variables never increases the variances of the estimates of the remaining parameters.

3. Since $\hat{\boldsymbol{\beta}}_p$ is a biased estimate of $\boldsymbol{\beta}_p$ and $\hat{\boldsymbol{\beta}}_p^*$ is not, it is more reasonable to compare the precision of the parameter estimates from the full and subset models in terms of mean square error. Recall that if $\hat{\theta}$ is an estimate of the parameter θ, the mean square error of $\hat{\theta}$ is

$$\text{MSE}(\hat{\theta}) = V(\hat{\theta}) + \left[E(\hat{\theta}) - \theta\right]^2$$

The mean square error of $\hat{\boldsymbol{\beta}}_p$ is

$$\text{MSE}(\hat{\boldsymbol{\beta}}_p) = \sigma^2(\mathbf{X}_p'\mathbf{X}_p)^{-1} + \mathbf{A}\boldsymbol{\beta}_r\boldsymbol{\beta}_r'\mathbf{A}'$$

If the matrix $V(\hat{\boldsymbol{\beta}}_r^*) - \boldsymbol{\beta}_r\boldsymbol{\beta}_r'$ is positive semidefinite, the matrix $V(\hat{\boldsymbol{\beta}}_p^*) - \text{MSE}(\hat{\boldsymbol{\beta}}_p)$ is positive semidefinite. This means that the least squares estimates of the parameters in the subset model have smaller mean square error than the corresponding parameter estimates from the full model when the deleted variables have regression coefficients that are smaller than the standard errors of their estimates in the full model.

4. $\hat{\sigma}_*^2$ from the full model is an unbiased estimate of σ^2. However for the subset model

$$E(\hat{\sigma}^2)=\frac{\sigma^2+\boldsymbol{\beta}_r'\mathbf{X}_r'\left[\mathbf{I}-\mathbf{X}_p(\mathbf{X}_p'\mathbf{X}_p)^{-1}\mathbf{X}_p'\right]\mathbf{X}_r\boldsymbol{\beta}_r}{(n-p)}$$

That is, $\hat{\sigma}^2$ is generally biased upward as an estimate of σ^2.

5. Suppose that we wish to predict the response at the point $\mathbf{x}'=[\mathbf{x}_p', \mathbf{x}_r']$. If we use the full model, the predicted value is $\hat{y}^*=\mathbf{x}'\hat{\boldsymbol{\beta}}^*$, with mean $\mathbf{x}'\boldsymbol{\beta}$ and prediction variance

$$V(\hat{y}^*)=\sigma^2\left[1+\mathbf{x}'(\mathbf{X}'\mathbf{X})^{-1}\mathbf{x}\right]$$

However if the subset model is used, $\hat{y}=\mathbf{x}_p'\hat{\boldsymbol{\beta}}_p$, with mean

$$E(\hat{y})=\mathbf{x}_p'\boldsymbol{\beta}_p+\mathbf{x}_p'\mathbf{A}\boldsymbol{\beta}_r$$

and prediction mean square error

$$\text{MSE}(\hat{y})=\sigma^2\left[1+\mathbf{x}_p'(\mathbf{X}_p'\mathbf{X}_p)^{-1}\mathbf{x}_p\right]+\left(\mathbf{x}_p'\mathbf{A}\boldsymbol{\beta}_r-\mathbf{x}_r'\boldsymbol{\beta}_r\right)^2$$

Note that $\hat{y}$ is a biased estimate of y unless $\mathbf{x}_p'\mathbf{A}\boldsymbol{\beta}_r=0$ which is only true in general if $\mathbf{X}_p'\mathbf{X}_r\boldsymbol{\beta}_r=\mathbf{0}$. Furthermore the variance of $\hat{y}^*$ from the full model is not less than the variance of $\hat{y}$ from the subset model. In terms of mean square error we can show that

$$V(\hat{y}^*)\geq \text{MSE}(\hat{y})$$

provided that the matrix $V(\hat{\boldsymbol{\beta}}_r^*)-\boldsymbol{\beta}_r\boldsymbol{\beta}_r'$ is positive semidefinite.

Our motivation for variable selection can be summarized as follows. By deleting variables from the model we may improve the precision of the parameter estimates of the retained variables even though some of the deleted variables are not negligible. This is also true for the variance of a predicted response. Deleting variables potentially introduces bias into the estimates of the coefficients of retained variables and the response. However if the deleted variables have small effects, the MSE of the biased estimates will be less than the variance of the unbiased estimates. That is, the amount of bias introduced is less than the reduction in the variance. There is danger in retaining negligible variables, that is, variables with zero coefficients or coefficients less than their corresponding standard errors from the full model. This danger is that the variances of the estimates of the parameters and the predicted response are increased.

A final point to remember is that regression models are frequently built from "happenstance" data (Box, Hunter and Hunter [1978]); that is, data that have been extracted from historical records. Happenstance data are often saturated with defects including outliers, "wild" points, and inconsistencies resulting from changes in the organization's data collection and information processing system over time. These data defects can have great impact on the variable selection process and lead to model misspecification. A very common problem in happenstance data is to find that some candidate regressors have been controlled so that they vary over a very limited range. These are often the most influential variables, and so they were tightly controlled to keep the response within acceptable limits. Yet because of the limited range of the data, the regressor may seem unimportant in the least squares fit. This is a serious model misspecification that only the model builder's nonstatistical knowledge of the problem environment may prevent. When the range of variables thought to be important is tightly controlled, the analyst may have to collect new data specifically for the model building effort. Designed experiments are helpful in this regard.

7.1.3 Criteria for evaluating subset regression models

Two key aspects of the variable selection problem are generating the subset models and deciding if one subset is better than another. In this section we discuss criteria for evaluating and comparing subset regression models. Section 7.2 will present computational methods for variable selection.

The coefficient of multiple determination. A measure of the adequacy of a regression model that has been widely used is the coefficient of multiple determination, R^2. Let R_p^2 denote the coefficient of multiple determination for a subset regression model with p terms; that is, $p-1$ regressors and an intercept term β_0. Computationally

$$R_p^2 = \frac{SS_R(p)}{S_{yy}} = 1 - \frac{SS_E(p)}{S_{yy}} \tag{7.8}$$

where $SS_R(p)$ and $SS_E(p)$ denote the regression sum of squares and the residual sum of squares respectively, for a p-term subset model. Note that there are $\binom{K}{p-1}$ values of R_p^2 for each value of p; one for each possible subset model of size p. Now R_p^2 increases as p increases, and is a maximum when $p = K+1$. Therefore the analyst uses this criterion by adding regres-

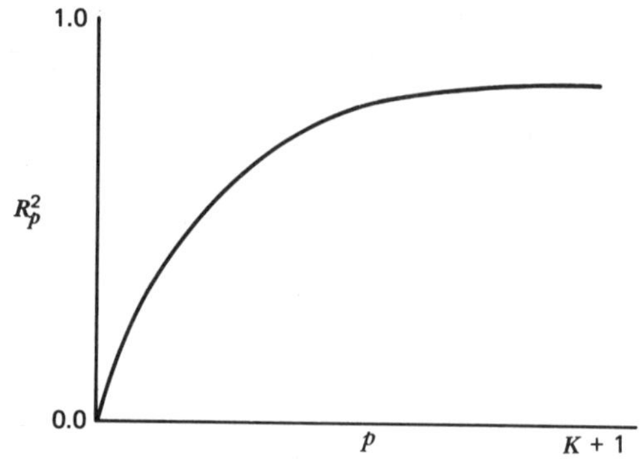

Figure 7.1 Plot of R_p^2 versus p.

sors to the model up to the point where an additional variable is not useful in that it provides only a small increase in R_p^2. The general approach is illustrated in Figure 7.1, which presents a hypothetical plot of the maximum value of R_p^2 for each subset of size p against p. Typically one examines a display such as this and then specifies the number of regressors for the final model as the point at which the "knee" in the curve becomes apparent. Clearly this requires judgement on the part of the analyst.

Since we cannot find an "optimum" value of R^2 for a subset regression model, we must look for a "satisfactory" value. Aitken [1974] has proposed one solution to this problem by providing a test by which all subset regression models that have an R^2 not significantly different from the R^2 for the full model can be identified. Let

$$R_0^2 = 1 - (1 - R_{K+1}^2)(1 + d_{\alpha, n, K}) \tag{7.9}$$

where

$$d_{\alpha, n, K} = \frac{KF_{\alpha, n, n-K-1}}{n-K-1}$$

and R_{K+1}^2 is the value of R^2 for the full model. Aitken calls any subset of regressor variables producing an R^2 greater than R_0^2 an R^2*-adequate* (α) *subset*.

Generally it is not straightforward to use R^2 as a criterion for choosing the number of regressors to include in the model. However for a fixed number of variables p, R_p^2 can be used to compare the $\binom{K}{p-1}$ subset models so generated. Models having large values of R_p^2 are preferred.

Adjusted R^2. To avoid the difficulties of interpreting R^2, some analysts prefer to use an adjusted R^2 statistic (for example, see Ezekiel [1930]), defined as

$$\overline{R}_p^2 = 1 - \left(\frac{n-1}{n-p}\right)\left(1 - R_p^2\right) \tag{7.10}$$

The $\overline{R}_p^2$-statistic does not necessarily increase as additional regressors are introduced into the model. In fact it can be shown (Edwards [1969], Haitovski [1969], and Seber [1977]) that if s regressors are added to the model, $\overline{R}_{p+s}^2$ will exceed $\overline{R}_p^2$ if and only if the partial F-statistic for testing the significance of the s additional regressors exceeds one. Consequently, one criterion for selection of an optimum subset model is to choose the model that has a maximum $\overline{R}_p^2$. However, this is equivalent to another criterion that we now present.

Residual mean square. The residual mean square for a subset regression model, for example

$$MS_E(p) = \frac{SS_E(p)}{n-p} \tag{7.11}$$

may be used as a model evaluation criterion. The general behavior of $MS_E(p)$ as p increases is illustrated in Figure 7.2. Because $SS_E(p)$ always decreases as p increases, $MS_E(p)$ initially decreases, then stabilizes, and eventually may increase. The eventual increase in $MS_E(p)$ occurs when the reduction in $SS_E(p)$ from adding a regressor to the model is not sufficient to compensate for the loss of one degree of freedom in the denominator of (7.11). That is, adding a regressor to a p-term model will cause $MS_E(p+1)$ to be greater than $MS_E(p)$ if the decrease in the residual sum of squares is less than $MS_E(p)$. Advocates of the $MS_E(p)$ criterion will plot $MS_E(p)$

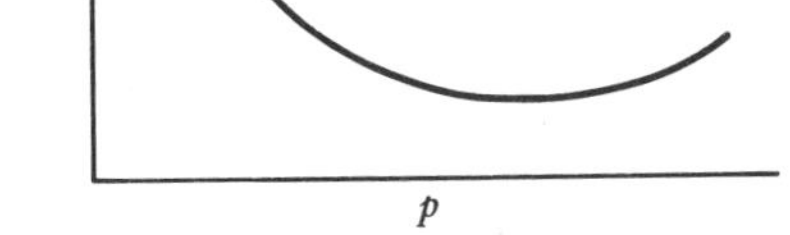

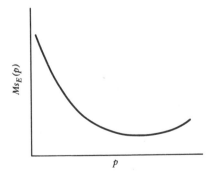

Figure 7.2 Plot of $MS_E(p)$ versus p.

versus p and base the choice of p on either

1. The minimum $MS_E(p)$,

2. The value of p such that $MS_E(p)$ is approximately equal to MS_E for the full model, or

3. A value of p near the point where the smallest $MS_E(p)$ turns upward.

The subset regression model that minimizes $MS_E(p)$ will also maximize $\bar{R}_p^2$. To see this, note that

$$\begin{aligned}\bar{R}_p^2 &= 1-\frac{n-1}{n-p}\left(1-R_p^2\right)\\ &= 1-\frac{n-1}{n-p}\,\frac{SS_E(p)}{S_{yy}}\\ &= 1-\frac{n-1}{S_{yy}}\,\frac{SS_E(p)}{n-p}\\ &= 1-\frac{n-1}{S_{yy}}MS_E(p)\end{aligned}$$

Thus the criteria minimum $MS_E(p)$ and maximum $\bar{R}_p^2$ are equivalent.

Mallows' C_p statistic. Mallows [1964, 1966, 1973] has proposed a criterion that is related to the mean square error of a fitted value, that is,

$$E[\hat{y}_i-E(\hat{y}_i)]^2=[E(y_i)-E(\hat{y}_i)]^2+V(\hat{y}_i) \tag{7.12}$$

Note that $E(y_i)$ is the expected response from the true regression equation and $E(\hat{y}_i)$ is the expected response from the p-term subset model. Thus $E(y_i)-E(\hat{y}_i)$ is the bias at the ith data point. Consequently, the two terms on the right-hand side of (7.12) are the squared bias and variance components respectively, of the mean square error. Let the total squared bias for a p-term equation be

$$SS_B(p)=\sum_{i=1}^{n}[E(y_i)-E(\hat{y}_i)]^2$$

and define the standardized total mean square error as

$$\Gamma_p = \frac{1}{\sigma^2}\left\{ \sum_{i=1}^{n} [E(y_i) - E(\hat{y}_i)]^2 + \sum_{i=1}^{n} V(\hat{y}_i) \right\}$$

$$= \frac{SS_B(p)}{\sigma^2} + \frac{1}{\sigma^2} \sum_{i=1}^{n} V(\hat{y}_i) \tag{7.13}$$

It can be shown that

$$\sum_{i=1}^{n} V(\hat{y}_i) = p\sigma^2$$

and that the expected value of the residual sum of squares from a p-term equation is

$$E[SS_E(p)] = SS_B(p) + (n-p)\sigma^2$$

Substituting for $\sum_{i=1}^{n} V(\hat{y}_i)$ and $SS_B(p)$ in (7.13) gives

$$\Gamma_p = \frac{1}{\sigma^2}\{E[SS_E(p)] - (n-p)\sigma^2 + p\sigma^2\}$$

$$= \frac{E[SS_E(p)]}{\sigma^2} - n + 2p \tag{7.14}$$

Suppose that $\hat{\sigma}^2$ is a good estimate of σ^2. Then replacing $E[SS_E(p)]$ by the observed value $SS_E(p)$ produces an estimate of Γ_p, say

$$C_p = \frac{SS_E(p)}{\hat{\sigma}^2} - n + 2p \tag{7.15}$$

If the p-term model has negligible bias, then $SS_B(p) = 0$. Consequently $E[SS_E(p)] = (n-p)\sigma^2$, and

$$E[C_p \mid \text{Bias}=0] = \frac{(n-p)\sigma^2}{\sigma^2} - n + 2p = p$$

When using the C_p criterion it is helpful to construct a plot of C_p as a function of p for each regression equation, such as shown in Figure 7.3. Regression equations with little bias will have values of C_p that fall near the

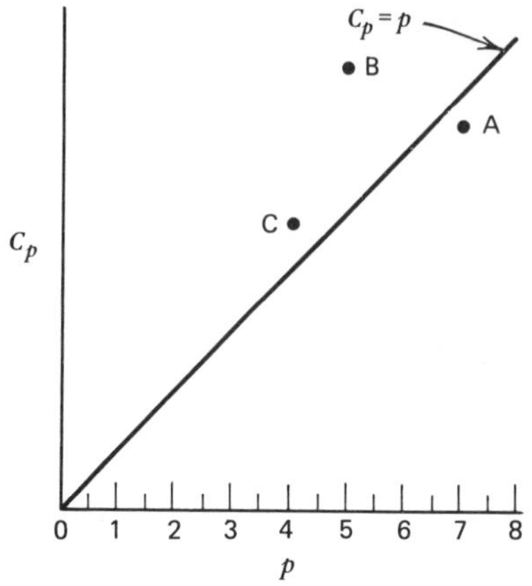

Figure 7.3 C_p plot.

line $C_p = p$ (Point A in Figure 7.3) while those equations with substantial bias will fall above this line (Point B in Figure 7.3). Generally, small values of C_p are desirable. For example, although Point C in Figure 7.3 is above the line $C_p = p$, it is below Point A, and thus represents a model with lower total error. It may be preferable to accept some bias in the equation to reduce the average error of prediction.

To calculate C_p, we need an unbiased estimate of σ^2. Frequently we use the residual mean square for the full equation for this purpose. However, this forces $C_p = p = K+1$ for the full equation. Using $MS_E(K+1)$ from the full model as an estimate of σ^2 assumes that the full model has negligible bias. If the full model has several regressors that do not contribute significantly to the model (zero regression coefficients), then $MS_E(K+1)$ will often overestimate σ^2, and consequently the values of C_p will be small. If the C_p statistic is to work properly, a good estimate of σ^2 must be used. As an alternative to $MS_E(K+1)$, we could base our estimate of σ^2 on pairs of points that are "near-neighbors" in the x-space, as illustrated in Section 4.7.3.

Uses of regression and model evaluation criteria. As we have seen, there are several criteria that can be used to evaluate subset regression models. The criterion that we use for model selection should certainly be related to the intended use of the model. There are several possible uses of regression including (1) data description, (2) prediction and estimation, (3) parameter estimation, and (4) control.

If the objective is to obtain a good description of a given process or to model a complex system, a search for regression equations with small residual sums of squares is indicated. Since SS_E is minimized by using all K candidate regressors, we usually prefer to eliminate some variables if only a small increase in SS_E results. In general we would like to describe the system with as few regressors as possible while simultaneously explaining the substantial portion of the variability in y.

Frequently regression equations are used for prediction of future observations or estimation of the mean response. In general we would like to select the regressors such that the mean square error of prediction is minimized. This usually implies that regressors with small effects should be deleted from the model. Allen (1971b, 1974) has suggested using all but the ith observation to obtain a p-term predictor of y_i, for example $\hat{y}_{(i)}$ and forming the sum of squared differences between the observed and predicted value for $i=1,2,\ldots,n$. Allen calls this the prediction error sum of squares, as in

$$\text{PRESS}_p = \sum_{i=1}^{n} \left[y_i - \hat{y}_{(i)} \right]^2 \tag{7.16}$$

One then selects the subset regression model based on small values of PRESS_p. While PRESS_p has intuitive appeal particularly for the prediction problem, it is not a simple function of the residual sum of squares and developing an algorithm for variable selection based on this criterion is not straightforward. This statistic is however, quite useful for discriminating between alternative models (for an example, see Chapter 10).

If we are interested in parameter estimation, then clearly we should consider both the bias that results from deleting variables and the variances of the estimated coefficients. When the regressors are highly intercorrelated the least squares estimates of the individual regression coefficients may be extremely poor. In Chapter 8 we will discuss several methods that may improve parameter estimates when the $\mathbf{X}$ matrix is ill-conditioned. These methods often lead to a predictive equation that is more effective in extrapolation than one based on least squares, for in extrapolation good estimates of the parameters are essential.

When a regression model is used for control, accurate estimates of the parameters are important. This implies that the standard errors of the regression coefficients should be small. Furthermore since the adjustments made on the x's to control y will be proportional to the $\hat{\beta}$'s, the regression coefficients should closely represent the effects of the regressors. If the regressors are highly intercorrelated, the $\hat{\beta}$'s may be very poor estimates of the effects of individual regressors.

7.2 Computational techniques for variable selection

We have seen that it is desirable to consider regression models that employ a subset of the candidate regressor variables. To find the subset of variables to use in the final equation, it is natural to consider fitting models with various combinations of the candidate regressors. In this section we will discuss several computational techniques for generating subset regression models and illustrate criteria for evaluation of these models.

7.2.1 All possible regressions

This procedure requires that the analyst fit all the regression equations involving one candidate regressor, two candidate regressors, and so on. These equations are evaluated according to some suitable criterion and the "best" regression model selected. If we assume that the intercept term β_0 is included in all equations, then if there are K candidate regressors, there are 2^K total equations to be estimated and examined. For example, if $K=4$, then there are $2^4=16$ possible equations, while if $K=10$, there are $2^{10}=1024$ possible regression equations. Clearly the number of equations to be examined increases rapidly as the number of candidate regressors increases. Prior to the development of efficient computer codes, generating all possible regressions was impractical for problems involving more than a few regressors. The availability of highspeed computers has motivated the development of several very efficient algorithms for all possible regressions. Some of these algorithms will be discussed subsequently.

• **Example 7.1** Hald [1952]* presents data concerning the heat evolved in calories per gram of cement (y) as a function of the amount of each of four ingredients in the mix; tricalcium aluminate (x_1), tricalcium silicate (x_2), tetracalcium alumino ferrite (x_3) and dicalcium silicate (x_4). The data are shown in Table 7.1. We will use this data to illustrate the all possible regressions approach to variable selection.

Since there are $K=4$ candidate regressors, there are $2^4=16$ possible regression equations if we always include the intercept β_0. The results of fitting these 16 equations is displayed in Table 7.2. The R_p^2, $\bar{R}_p^2$, $MS_E(p)$, and C_p statistics are also given in this table.

Table 7.3 displays the least squares estimates of the regression coefficients. The partial nature of regression coefficients is readily apparent from examination of this table. For example, consider x_2. When the model contains only x_2, the least squares estimate of the x_2 effect is .789. If x_4 is added to the model the x_2 effect is .311, a reduction of over 50 percent. Further addition of x_3 changes the x_2 effect to $-.923$. Clearly the least squares estimate of an individual regression coefficient depends heavily on the *other* regressors in the model. The large changes in the regression coefficients observed in the Hald

*These are "classical" data for illustrating the problems inherent in variable selection. For other analyses, see Daniel and Wood [1980], Draper and Smith [1981], and Seber [1977].

Table 7.1 Data for Example 7.1

Observation, i	y_i	x_{i1}	x_{i2}	x_{i3}	x_{i4}
1	78.5	7	26	6	60
2	74.3	1	29	15	52
3	104.3	11	56	8	20
4	87.6	11	31	8	47
5	95.9	7	52	6	33
6	109.2	11	55	9	22
7	102.7	3	71	17	6
8	72.5	1	31	22	44
9	93.1	2	54	18	22
10	115.9	21	47	4	26
11	83.8	1	40	23	34
12	113.3	11	66	9	12
13	109.4	10	68	8	12

Table 7.2 Summary of all possible regression for Example 7.1

Number of Regressors in Model	p	Regressors in Model	$SS_E(p)$	R_p^2	$\bar{R}_p^2$	$MS_E(p)$	C_p
None	1	None	2715.7635	0	0	226.3136	442.92
1	2	x_1	1265.6867	0.53395	0.49158	115.0624	202.55
1	2	x_2	906.3363	0.66627	0.63593	82.3942	142.49
1	2	x_3	1939.4005	0.28587	0.22095	176.3092	315.16
1	2	x_4	883.8669	0.67459	0.64495	80.3515	138.73
2	3	x_1x_2	57.9045	0.97868	0.97441	5.7904	2.68
2	3	x_1x_3	1227.0721	0.54817	0.45780	122.7073	198.10
2	3	x_1x_4	74.7621	0.97247	0.96697	7.4762	5.50
2	3	x_2x_3	415.4427	0.84703	0.81644	41.5443	62.44
2	3	x_2x_4	868.8801	0.68006	0.61607	86.8880	138.23
2	3	x_3x_4	175.7380	0.93529	0.92235	17.5738	22.37
3	4	$x_1x_2x_3$	48.1106	0.98228	0.97638	5.3456	3.04
3	4	$x_1x_2x_4$	47.9727	0.98234	0.97645	5.3303	3.02
3	4	$x_1x_3x_4$	50.8361	0.98128	0.97504	5.6485	3.50
3	4	$x_2x_3x_4$	73.8145	0.97282	0.96376	8.2017	7.34
4	5	$x_1x_2x_3x_4$	47.8636	0.98238	0.97356	5.9829	5.00

cement data when variables are added or removed indicates that there is substantial correlation between the four regressors (the multicollinearity problem discussed previously in Chapter 4).

Consider evaluating the subset models by the R_p^2 criterion. A plot of R_p^2 versus p is shown in Figure 7.4. From examining this display it is clear that after two regressors are in the model there is little to be gained in terms of R^2 by introducing additional variables. Both of the two-regressor models (x_1, x_2) and (x_1, x_4) have essentially the same R^2 values, and in terms of this criteria, it would make little difference which model is selected as the final regression equation. Draper and Smith [1981, p. 297] suggest that it may be preferable to use (x_1, x_4) because x_4 provides the best one-regressor model. From Equation (7.9) we find that if we take $\alpha=.05$,

$$R_0^2 = 1-\left(1-R_5^2\right)\left(1+\frac{4F_{.05,4,8}}{8}\right)$$

$$=1-0.01762\left[1+\frac{4(3.84)}{8}\right]$$

$$=0.94855$$

Table 7.3 Least squares estimates for all possible regressions (Hald cement data)

Variables in Model	$\hat{\beta}_0$	$\hat{\beta}_1$	$\hat{\beta}_2$	$\hat{\beta}_3$	$\hat{\beta}_4$
x_1	81.479	1.869			
x_2	57.424		.789		
x_3	110.203			−1.256	
x_4	117.568				−.738
x_1x_2	52.577	1.468	.662		
x_1x_3	72.349	2.312		.494	
x_1x_4	103.097	1.440			−.614
x_2x_3	72.075		.731	−1.008	
x_2x_4	94.160		.311		−.457
x_3x_4	131.282			−1.200	−.724
$x_1x_2x_3$	48.194	1.696	.657	.250	
$x_1x_2x_4$	71.648	1.452	.416		−.237
$x_2x_3x_4$	203.642		−.923	−1.448	−1.557
$x_1x_3x_4$	111.684	1.052		−.410	−.643
$x_1x_2x_3x_4$	62.405	1.551	.510	.102	−.144

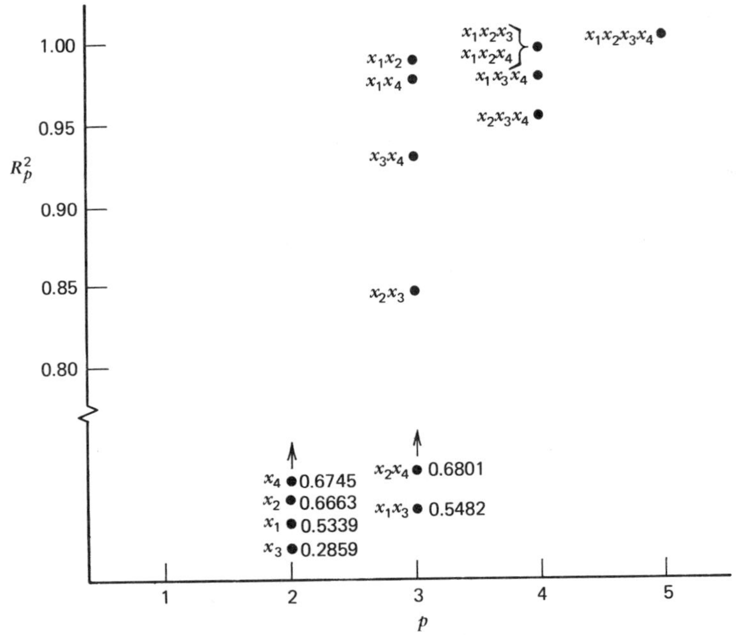

Figure 7.4 Plot of R_p^2 versus p, Example 7.1.

Therefore any subset regression model for which $R_p^2 > R_0^2 = 0.94855$ is R^2-adequate (.05); that is, its R^2 is not significantly different from R_{K+1}^2. Clearly several models in Table 7.2 satisfy this criterion, and so the choice of the final model is still not clear.

It is instructive to examine the pairwise correlations between x_i and x_j, and between x_i and y. These simple correlations are shown in Table 7.4. Note that the pairs of regressors (x_1, x_3) and (x_2, x_4) are highly correlated, since

$$r_{13} = -0.824 \quad \text{and} \quad r_{24} = -0.973$$

Consequently, adding further regressors when x_1 and x_2 or when x_1 and x_4

Table 7.4 Matrix of simple correlations for Hald's data in Example 7.1

	x_1	x_2	x_3	x_4	y
x_1	1.0				
x_2	0.229	1.0			
x_3	−0.824	−0.139	1.0		
x_4	−0.245	−0.973	0.030		
y	0.731	0.816	−0.535	−0.821	1.0

are already in the model will be of little use since the information content in the excluded regressors is essentially present in the regressors that are in the model. This correlative structure is partially responsible for the large changes in the regression coefficients noted in Table 7.3.

A plot of $MS_E(p)$ versus p is shown in Figure 7.5. The minimum residual mean square model is (x_1, x_2, x_4), with $MS_E(4)=5.3303$. Note that as expected, the model that minimizes $MS_E(p)$ also maximizes the adjusted $\bar{R}_p^2$. However, two of the other three-regressor models $[(x_1, x_2, x_3)$ and $(x_1, x_3, x_4)]$ and the two-regressor models $[(x_1, x_2)$ and $(x_1, x_4)]$ have comparable values

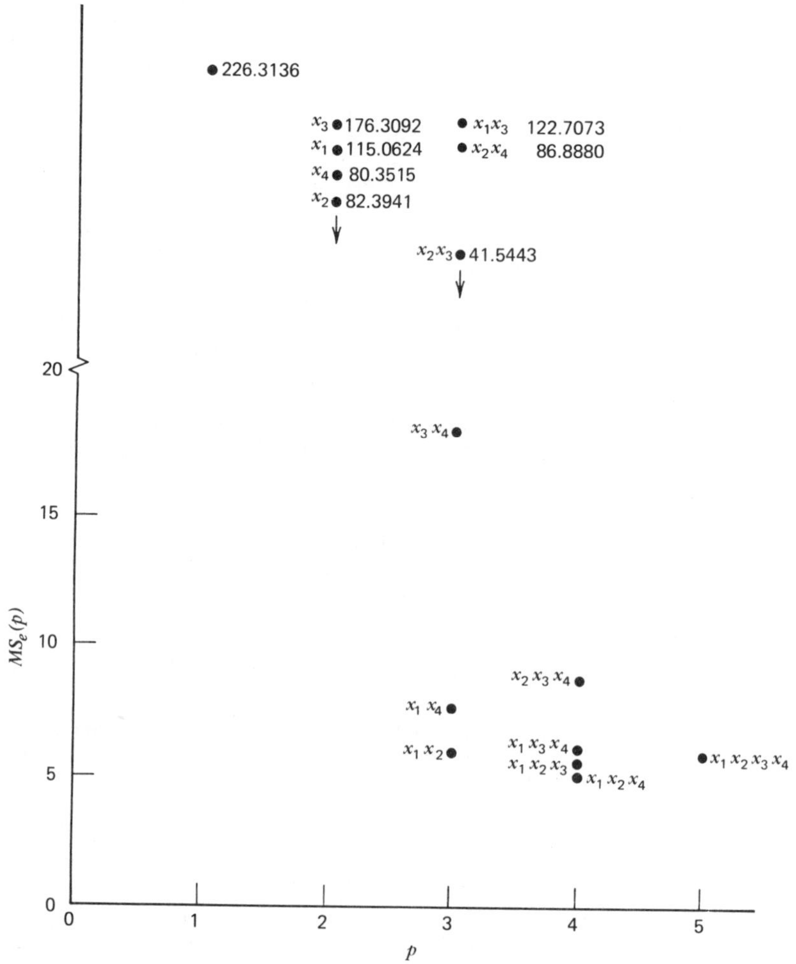

Figure 7.5 Plot of $MS_E(p)$ versus p, Example 7.1.

of the residual mean square. If either (x_1, x_2) or (x_1, x_4) are in the model, there is little reduction in residual mean square by adding further regressors. The subset model (x_1, x_2) may be more appropriate than (x_1, x_4) because it has a smaller value of the residual mean square.

A C_p plot is shown in Figure 7.6. To illustrate the calculations, suppose we take $\hat{\sigma}^2 = 5.9829$ (MS_E from the full model), and calculate C_3 for the model

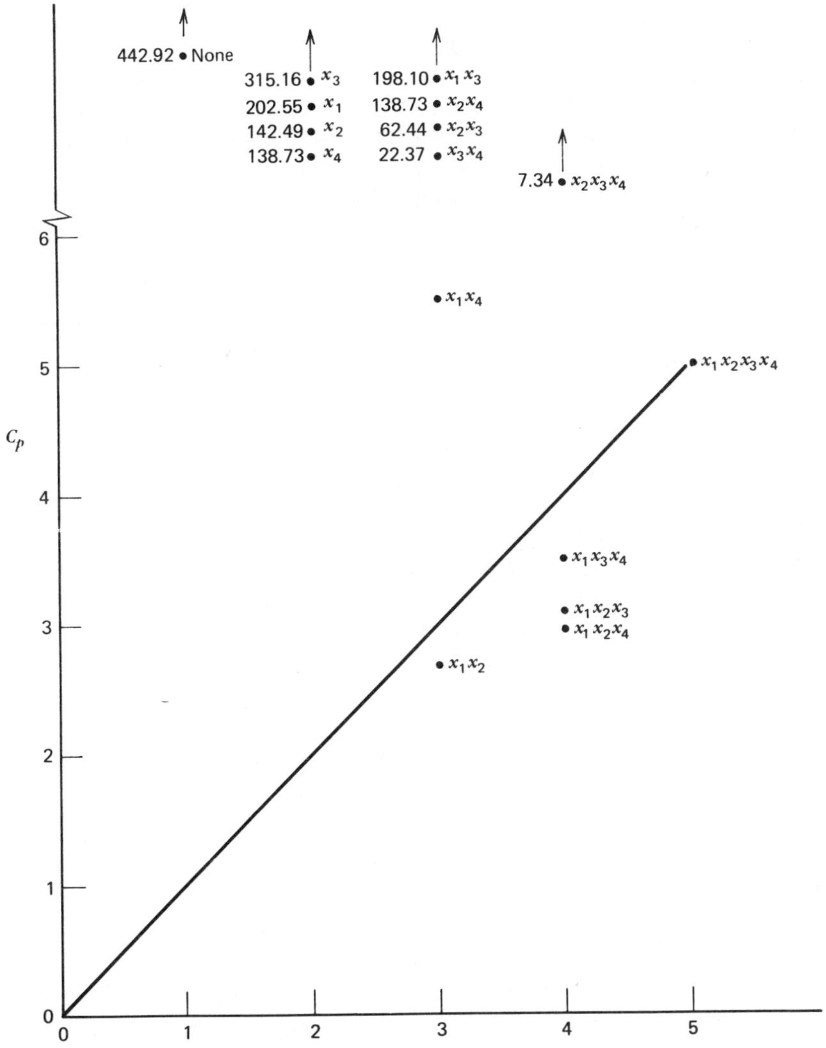

Figure 7.6 C_p plot, Example 7.1.

(x_1, x_4). From (7.15) we find that

$$\begin{aligned} C_3 &= \frac{SS_E(3)}{\hat{\sigma}^2} - n + 2p \\ &= \frac{74.7621}{5.9829} - 13 + 2(3) \\ &= 5.50 \end{aligned}$$

From examination of this plot we find that there are four models that could be acceptable: (x_1, x_2), (x_1, x_2, x_3), (x_1, x_2, x_4), and (x_1, x_3, x_4). Without considering additional factors such as technical information about the regressors or the costs of data collection, it may be appropriate to choose the simplest model (x_1, x_2) as the final model because it has the smallest C_p.

In summary, we have found that the model (x_1, x_2) has good values of the R^2, $MS_E(p)$, and C_p statistics. Note however that there is no clear cut choice of the "best" regression equation. Very often we find that different criteria suggest different subsets of the regressors as "best." At this point the model (x_1, x_2) should be subjected to the usual checks for adequacy. If it is going to be used for prediction, it would be highly desirable to test the ability of the model to predict fresh data. These subsequent investigations may lead us to reconsider one or more of the other models identified above. •

Efficient generation of all possible regressions. There are presently several algorithms available for efficiently generating all possible regressions. For example see Furnival [1971], Furnival and Wilson [1974], Garside [1965, 1971], Morgan and Tatar [1972], and Schatzoff et al. [1978]. The basic idea underlying all these algorithms is to perform the calculations for the 2^K possible subset models in such a way that sequential subset models differ by only one variable. This allows very efficient numerical methods to be used in performing the calculations. These methods are usually based on either Gauss–Jordan reduction or the sweep operator (see Beaton [1964], or Seber [1977]). Some of these algorithms are available commercially. For example, the Furnival and Wilson [1974] algorithm is an option in the BMD-P library (see Dixon [1977]).

A sample computer output for the BMD-P Furnival and Wilson algorithm applied to the Hald cement data is shown in Figure 7.7. This program allows the user to select the best subset regression model of each size for $1 \leqslant p \leqslant K+1$ using either the C_p, R_p^2, or $\overline{R}_p^2$ criteria. It also displays the values of the C_p, R_p^2 and $\overline{R}_p^2$ statistics for several (but not all) models for each value of p. The program has the capability of identifying the m-best (for $m \leqslant 10$) subset regression models in addition to the single best subset. For example, in Figure 7.7, we see that the best five subsets in terms of C_p have been found for the cement data. For these "best" and "nearly best"

UNIVARIATE SUMMARY STATISTICS

VARIABLE	MEAN	STANDARD DEVIATION	COEFFICIENT OF VARIATION
2 X1	7.46154	5.88239	.788362
3 X2	48.15385	15.56088	.323149
4 X3	11.76923	6.40513	.544226
5 X4	30.00000	16.73818	.557939
1 Y	95.42308	15.04372	.157653

SMALLEST VALUE	LARGEST VALUE	SMALLEST STANDARD SCORE	LARGEST STANDARD SCORE	SKEWNESS	KURTOSIS
1.00000	21.00000	-1.10	2.30	.61	-.38
26.00000	71.00000	-1.42	1.47	-.04	-1.57
4.00000	23.00000	-1.21	1.75	.54	-1.36
6.00000	60.00000	-1.43	1.79	.29	-1.31
72.50000	115.90000	-1.52	1.36	-.17	-1.59

VALUES FOR KURTOSIS GREATER THAN ZERO INDICATE DISTRIBUTIONS WITH HEAVIER TAILS THAN THE NORMAL DISTRIBUTION.

CORRELATIONS

		X1 2	X2 3	X3 4	X4 5	Y 1
X1	2	1.000				
X2	3	.229	1.000			
X3	4	-.824	-.139	1.000		
X4	5	-.245	-.973	.030	1.000	
Y	1	.731	.816	-.535	-.821	1.000

Figure 7.7 Computer output (BMD–P) for Furnival and Wilson all possible regressions algorithm.

```
FOR EACH SUBSET SELECTED BY YOUR CRITERION, THE R-SQUARED,
ADJUSTED R-SQUARED, MALLOWS" CP, AND THE VARIABLE NAMES ARE
PRINTED.  THE REGRESSION COEFFICIENTS AND T-STATISTICS ARE
PRINTED TO THE RIGHT OF THE VARIABLE NAMES.

MANY OTHER SUBSETS MAY ALSO BE REPORTED THAT ARE NOT
ACCOMPANIED BY REGRESSION COEFFICIENTS AND T-STATISTICS.
SOME OF THESE SUBSETS MAY BE QUITE GOOD ALTHOUGH THEY ARE
NOT NECESSARILY BETTER THAN ANY SUBSET THAT HAS NOT BEEN
PRINTED.


                                          ****  SUBSETS WITH    1 VARIABLES  ****
              ADJUSTED
 R-SQUARED   R-SQUARED         CP

   .674542     .644955     138.73  X4

   .666268     .635929     142.49  X2

   .533948     .491580     202.55  X1

   .285873     .220952     315.15  X3


                                          ****  SUBSETS WITH    2 VARIABLES  ****
              ADJUSTED
 R-SQUARED   R-SQUARED         CP

   .978678     .974414       2.68  VARIABLE          COEFFICIENT   T-STATISTIC
                                   2 X1                  1.46831      12.10
                                   3 X2                  .662250      14.44
                                    INTERCEPT            52.5773

   .972471     .966965       5.50  X1         X4

   .935290     .922348      22.37  X3         X4

   .680060     .616072     138.23  X2         X4

   .548167     .457800     198.09  X1         X3
```

Figure 7.7 *Continued*

subsets more detailed information is reported, including values of the regression coefficient and t-statistics.

Early versions of all possible regressions algorithms were capable of handling problems with ten to twelve candidate regressors. More recent efforts have expanded this capability considerably. The BMD-P version of Furnival and Wilson's algorithm will very efficiently process up to about 30 candidate regressors with computing times that are comparable to the usual stepwise-type regression algorithms that are discussed in Section 7.2.3. Our experience indicates that problems with 30 or less candidate regressors can usually be solved relatively easily with one of the all possible regressions algorithms.

An alternative to all possible regressions is to generate only the *better* regressions. Computational methods have been suggested by Furnival and

```
                                         ****  SUBSETS WITH    3 VARIABLES   ****
                 ADJUSTED
  R-SQUARED     R-SQUARED        CP

   .982335       .976447       3.02  VARIABLE          COEFFICIENT   T-STATISTIC
                                     2 X1                  1.45194       12.41
                                     3 X2                  .416110        2.24
                                     5 X4                 -.236540       -1.37
                                      INTERCEPT            71.6483

   .982265       .976360       3.04  VARIABLE          COEFFICIENT   T-STATISTIC
                                     2 X1                  1.69569        8.29
                                     3 X2                  .656915       14.85
                                     4 X3                  .250018        1.35
                                      INTERCEPT            48.1936

   .981281       .975041       3.50  VARIABLE          COEFFICIENT   T-STATISTIC
                                     2 X1                  1.05185        4.70
                                     4 X3                 -.410043       -2.06
                                     5 X4                 -.642796      -14.43


                                      INTERCEPT            111.684

   .972820       .963760       7.34  X2          X3          X4


                                         ****  SUBSETS WITH    4 VARIABLES   ****
                 ADJUSTED
  R-SQUARED     R-SQUARED        CP

   .982376       .973563       5.00  VARIABLE          COEFFICIENT   T-STATISTIC
                                     2 X1                  1.55110        2.08
                                     3 X2                  .510168         .70
                                     4 X3                  .101909         .14
                                     5 X4                 -.144061        -.20
                                      INTERCEPT            62.4054


STATISTICS FOR "BEST" SUBSET
MALLOWS" CP                              2.68
SQUARED MULTIPLE CORRELATION           .97868
MULTIPLE CORRELATION                   .98928
ADJUSTED SQUARED MULT. CORR.           .97441
RESIDUAL MEAN SQUARE                 5.790448
STANDARD ERROR OF EST.               2.406335
F-STATISTIC                            229.50
NUMERATOR DEGREES OF FREEDOM                2
DENOMINATOR DEGREES OF FREEDOM             10
SIGNIFICANCE                            .0000


  VARIABLE       REGRESSION       STANDARD  STAND.     T-   2TAIL      TOL-   CONTRIBUTION
NO.   NAME       COEFFICIENT         ERROR   COEF.   STAT.   SIG.    ERANCE   TO R-SQUARED

   INTERCEPT         52.5773       2.28617   3.495   23.00   .000
 2 X1                1.46831       .121301    .574   12.10   .000   .947751        .312410
 3 X2                .662250      .0458547    .685   14.44   .000   .947751        .444730

THE CONTRIBUTION TO R-SQUARED FOR EACH VARIABLE IS THE AMOUNT
BY WHICH R-SQUARED WOULD BE REDUCED IF THAT VARIABLE WERE
REMOVED FROM THE REGRESSION EQUATION.
```

Figure 7.7 *Continued*

Wilson [1974], Beale et al. [1967], and Hocking and Leslie [1967]. These algorithms use a branch-and-bound type procedure that avoids searching in unfavorable directions. Some extensions of the basic Hocking–Leslie procedure have been reported by LaMotte and Hocking [1970]. LaMotte [1972] has developed a computer program called SELECT for the algorithm that allows relatively large problems to be treated. Hocking [1976] reports that a 70-regressor problem has been analyzed with only a moderate amount of computation.

7.2.2 Directed search on *t*

The test statistic for testing H_0: $\beta_j=0$ for the full model with $p=K+1$ regressors is

$$t_{K,j}=\frac{\hat{\beta}_j}{\text{se}(\hat{\beta}_j)}$$

Regressors that contribute significantly to the full model will have a large $|t_{K,j}|$ and will tend to be included in the "best" p-regressor subset, where "best" implies minimum residual sum of squares or C_p. Consequently, ranking the regressors according to decreasing order of magnitude of the $|t_{K,j}|$ $j=1,2,\ldots,K$, and then introducing the regressors into the model one at a time in this order should lead to the best (or one of the best) subset models for each p. Daniel and Wood [1980] call this variable selection procedure the directed search on t. It is often a very effective variable selection strategy when the number of candidate regressors is relatively large, for example $K>20$ or 30.

- **Example 7.2** We will apply the directed search on t to the Hald cement data given in Example 7.1. The values of the $|t_{K,j}|$ for the full model are given in Figure 7.7, under the heading "Subsets with 4 variables." Table 7.5 shows the results of applying the procedure. We see that the directed search on t has produced the best (in terms of C_p) two-regressor subset (x_1, x_2) and the best

Table 7.5 A directed search on *t* applied to Hald's cement data

Regressor x_j	$\lvert t_{K,j}\rvert$	Regressors in Subset	p	C_p
x_1	2.08	(x_1)	2	202.55
x_2	0.70	(x_1, x_2)	3	2.68
x_4	0.20	(x_1, x_2, x_4)	4	3.02
x_3	0.14	(x_1, x_2, x_3, x_4)	5	5.00

three-regressor subset (x_1, x_2, x_4). Note that these two subsets have the smallest and next-smallest values of C_p. Table 7.5 implies that x_1 and x_2 form a basic set of regressors to be included in any well-fitting model. Further variable selection efforts would be confined to models containing x_1 and x_2 (since in this example $K=4$, the other "interesting" subsets would be (x_1, x_2, x_3) and (x_1, x_2, x_4)).

Example 7.3 Gorman and Toman [1966] present data concerning the rut depth of 31 asphalt pavements prepared under different conditions specified

Table 7.6 Data for Example 7.3

Observation, i	y_i	x_{i1}	x_{i2}	x_{i3}	x_{i4}	x_{i5}	x_{i6}
1	0.82930	0.44716	4.68	4.87	−1.00	8.4	4.916
2	1.11394	0.14613	5.19	4.50	−1.00	6.5	4.563
3	1.16879	0.14613	4.82	4.73	−1.00	7.9	5.321
4	1.10037	0.51851	4.85	4.76	−1.00	8.3	4.865
5	0.91645	0.23045	4.86	4.95	−1.00	8.4	3.776
6	1.02816	0.46240	5.16	4.45	−1.00	7.4	4.397
7	0.86213	0.56820	4.82	5.05	−1.00	6.8	4.867
8	1.10278	0.23045	4.86	4.70	−1.00	8.6	4.828
9	1.09968	−0.03621	4.78	4.84	−1.00	6.7	4.865
10	1.31387	−0.16749	5.16	4.76	−1.00	7.7	4.034
11	0.55388	0.77815	4.57	4.82	−1.00	7.4	5.450
12	0.84510	0.63347	4.61	4.65	−1.00	6.7	4.853
13	1.41830	−0.22185	5.07	5.10	−1.00	7.5	4.257
14	1.06707	0.25537	4.66	5.09	−1.00	8.2	5.144
15	0.88479	0.77815	5.42	4.41	−1.00	5.8	3.718
16	1.08814	0.64345	5.01	4.74	−1.00	7.1	4.715
17	−0.11919	1.94448	4.97	4.66	1.00	6.5	4.625
18	0.13033	1.79239	4.01	4.72	1.00	8.0	4.977
19	0.15836	1.69897	4.96	4.90	1.00	6.8	4.322
20	0.20412	1.76343	5.20	4.70	1.00	8.2	5.087
21	0.04139	1.95424	4.80	4.60	1.00	6.6	5.971
22	0.07058	1.81954	4.98	4.69	1.00	6.4	4.647
23	0.07918	2.14613	5.35	4.76	1.00	7.3	5.115
24	0.25181	2.38021	5.04	4.80	1.00	7.8	5.939
25	0.14267	2.62325	4.80	4.80	1.00	7.4	5.916
26	0.32790	2.69897	4.83	4.60	1.00	6.7	5.471
27	0.48149	2.25527	4.66	4.72	1.00	7.2	4.602
28	0.58503	2.43136	4.67	4.50	1.00	6.3	5.043
29	0.11919	2.23045	4.72	4.70	1.00	6.8	5.075
30	0.09691	1.99123	5.00	5.07	1.00	7.2	4.334
31	0.30102	1.54407	4.70	4.80	1.00	7.7	5.705

by five regressors. A sixth regressor is used as an indicator variable to separate the data into two sets of runs. The data are shown in Table 7.6. The full model is

$$\hat{y}=\hat{\beta}_0+\hat{\beta}_1x_1+\hat{\beta}_2x_2+\hat{\beta}_3x_3+\hat{\beta}_4x_4+\hat{\beta}_5x_5+\hat{\beta}_6x_6$$

where $y=\log$ (change in rut depth/million wheel passes)

$x_1=\log$ (viscosity of asphalt)

$x_2=$percentage of asphalt in surface course

$x_3=$percentage of asphalt in base course

$x_4=$indicator variable

$x_5=$percentage of fines in surface course

$x_6=$percentage of voids in surface course

Summary statistics for the full model are shown in Table 7.7, along with the values of C_p for all $2^6=64$ equations.

The results of a directed search on t are displayed in Table 7.8. By comparing the C_p statistics in Table 7.8 with those in Table 7.7, we observe that the directed search on t has produced the equation with the smallest value of $C_p[(x_1, x_2, x_4, x_6)$, with $C_5=4.3]$ along with two other subsets with small values of C_p. Note that the value of C_p first decreases as regressors are added to the model, then it reaches a minimum and begins to increase. Daniel

Table 7.7 Summary statistics for Example 7.3

	Full Equation			Matrix of Correlation Coefficients							
Term	$\hat{\beta}_j$	se($\hat{\beta}_j$)	$\lvert t_{K,j}\rvert$		x_1	x_2	x_3	x_4	x_5	x_6	y
Constant	−2.64	-	-	x_1	1.0						
x_1	−0.51	0.073	7.0	x_2	−0.06	1.0					
x_2	0.50	0.12	4.3	x_3	−0.22	−0.26	1.0				
x_3	0.10	0.14	0.7	x_4	0.93	0.01	−0.12	1.0			
x_4	−0.13	0.064	2.1	x_5	−0.30	−0.14	0.44	−0.23	1.0		
x_5	0.02	0.034	0.6	x_6	0.46	−0.43	−0.03	0.40	0.12	1.0	
x_6	0.14	0.047	2.9	y	−0.97	0.15	0.19	−0.93	0.32	−0.41	1.0
F-value	140										
R^2	0.972										
SS_E	0.3070										
MS_E	0.01279										

Table 7.7 Summary statistics for Example 7.3 (*Continued*)

C_p's for All Equations

Variables in Equation	p	C_p	Variables in Equation	p	C_p	Variables in Equation	p	C_p	Variables in Equation	p	C_p
none	1	835.6	x_5	2	748.8	x_6	2	692	x_5x_6	3	575.5
x_1	2	20.43	x_1x_5	3	21.68	x_1x_6	3	20.00	$x_1x_5x_6$	4	21.87
x_2	2	817.0	x_2x_5	3	716.1	x_2x_6	3	693.7	$x_2x_5x_6$	4	577.4
x_1x_2	3	13.98	$x_1x_2x_5$	4	14.29	$x_1x_2x_6$	4	5.67	$x_1x_2x_5x_6$	5	7.45
x_3	2	806.7	x_3x_5	3	748.3	x_3x_6	3	666.8	$x_3x_5x_6$	4	577.3
x_1x_3	3	21.90	$x_1x_2x_5$	4	22.53	$x_1x_3x_6$	4	21.37	$x_1x_3x_5x_6$	5	22.84
x_2x_3	3	770.0	$x_2x_3x_5$	4	709.0	$x_2x_3x_6$	4	668.0	$x_2x_3x_5$	5	579.1
$x_1x_2x_3$	4	15.95	$x_1x_2x_3x_5$	5	16.17	$x_1x_2x_3x_6$	5	7.56	$x_1x_2x_3x_5x_6$	6	9.43
x_4	2	92.0	x_4x_5	3	84.2	x_4x_6	3	92.8	$x_4x_5x_6$	4	82.7
x_1x_4	3	20.53	$x_1x_4x_5$	4	21.46	$x_1x_4x_6$	4	20.61	$x_1x_4x_5x_6$	5	22.24
x_2x_4	3	70.23	$x_2x_4x_5$	4	57.56	$x_2x_4x_6$	4	70.21	$x_2x_4x_5x_6$	5	59.09
$x_1x_2x_4$	4	12.22	$x_1x_2x_4x_5$	5	11.32	$x_1x_2x_4x_6$	5	4.26	$x_1x_2x_4x_5x_6$	6	5.51
x_3x_4	3	88.6	$x_3x_4x_5$	4	85.1	$x_3x_4x_6$	4	89.36	$x_3x_4x_5x_6$	5	83.9
$x_1x_2x_4$	4	22.42	$x_1x_3x_4x_5$	5	22.86	$x_1x_3x_4x_6$	5	22.36	$x_1x_3x_4x_5x_6$	6	23.65
$x_2x_3x_4$	4	58.38	$x_2x_3x_4x_5$	5	53.92	$x_2x_3x_4x_6$	5	56.83	$x_2x_3x_4x_5x_6$	6	54.38
$x_1x_2x_3x_4$	5	13.51	$x_1x_2x_3x_4x_5$	6	13.24	$x_1x_2x_3x_4x_6$	6	5.31	$x_1x_2x_3x_4x_5x_6$	7	7.00

Table 7.8 A directed search on t for Example 7.3

Regressor, x_j	$\lvert t_{K,j} \rvert$	Regressors in Subset	p	C_p
x_1	7.0	(x_1)	2	20.4
x_2	4.3	(x_1, x_2)	3	14.0
x_6	2.9	(x_1, x_2, x_6)	4	5.7
x_4	2.1	(x_1, x_2, x_6, x_4)	5	4.3
x_3	0.7	$(x_1, x_2, x_6, x_4, x_3)$	6	5.3
x_5	0.6	$(x_1, x_2, x_6, x_4, x_3, x_5)$	7	7.0

and Wood [1980] recommend using the turning point as a guide to determining the basic set of regressors for the model. If the smallest C_p in the listing is less than or equal to p, then we can safely include all regressors down to and including the turning point in the basic set of variables. However, if the minimum C_p exceeds p the basic set of regressors should include only the variables down to two less than the turning point.

Once a basic set of K_B regressors has been determined, we can perform all possible regressions on the remaining $K-K_B$ candidate regressors with the regressors in K_B forced into all models. If $K-K_B$ is large this may still require considerable computation. In such cases one can look at fractions of the 2^{K-K_B} runs containing the basic set of regressors. For further details see Daniel and Wood [1980] or Gorman and Toman [1966].

In this example the basic set of regressors is (x_1, x_2, x_4, x_6). The directed search on t leads immediately to the equation with the smallest C_p. However, this equation has a serious defect—x_4 is an indicator variable used to separate two sets of runs, and hence it is not possible to set x_4 at either of its former levels. Therefore if prediction or extrapolation is important, this equation is seriously deficient. One possible solution to this problem would be to attempt to identify the reasons that the two sets of runs reacted differently and incorporate this into the model via new regressors. Another possibility is to try to avoid the use of x_4 with the present data. Inspection of Table 7.6 reveals that deleting x_4 leads to a poorer fit for (x_1, x_2, x_6) as well as for all combinations of these variables with x_3 and x_5. Thus there seems to be no alternative but to include the indicator variable x_4 for the present.

7.2.3 Stepwise regression methods

Because evaluating all possible regressions can be burdensome computationally, various methods have been developed for evaluating only a small number of subset regression models by either adding or deleting regressors one at a time. These methods are generally referred to as stepwise-type procedures. They can be classified into three broad categories: (1) forward selection, (2) backward elimination, and (3) stepwise regression which is a popular combination of procedures (1) and (2). We now briefly describe and illustrate these procedures.

Forward Selection. This procedure begins with the assumption that there are no regressors in the model other than the intercept. An effort is made to find an optimal subset by inserting regressors into the model one at a time. The first regressor selected for entry into the equation is the one that has the largest simple correlation with the response variable y. Suppose that this regressor is x_1. This is also the regressor that will produce the largest value of the F-statistic for testing significance of regression. This regressor is entered if the F-statistic exceeds a preselected F-value, say F_{IN} (or F-to-enter). The second regressor chosen for entry is the one that now has the largest correlation with y after adjusting for the effect of the first regressor entered (x_1) on y. We refer to these correlations as partial correlations. They are the simple correlations between the residuals from the regression $\hat{y}=\hat{\beta}_0+\hat{\beta}_1 x_1$ and the residuals from the regressions of the other candidate regressors on x_1, say $\hat{x}_j=\hat{\alpha}_{0j}+\hat{\alpha}_{1j}x_1, j=2,3,\ldots, K$.

Suppose that at step two the regressor with the highest partial correlation with y is x_2. This implies that the largest partial F-statistic is

$$F=\frac{SS_R(x_2|x_1)}{MS_E(x_1,x_2)}$$

If this F value exceeds F_{IN}, then x_2 is added to the model. In general at each step the regressor having the highest partial correlation with y (or equivalently the largest partial F-statistic given the other regressors already in the model) is added to the model if its partial F-statistic exceeds the preselected entry level F_{IN}. The procedure terminates either when the partial F-statistic at a particular step does not exceed F_{IN}, or when the last candidate regressor is added to the model.

- **Example 7.4** We will apply the forward selection procedure to the Hald cement data given in Example 7.1. Figure 7.8 shows the results obtained when a particular computer program, the SAS forward selection algorithm, was applied to these data. In this program the user specifies the cutoff value F_{IN} by choosing a type I error rate α so that the regressor with the highest partial correlation with y is added to the model if its partial F-statistic exceeds $F_{\alpha,1,n-p}$. In this example we will use $\alpha=.10$ to determine F_{IN}. Some computer codes require that a numerical value be selected for F_{IN}, popular choices being between 2.0 and 4.0.

FORWARD SELECTION PROCEDURE FOR DEPENDENT VARIABLE Y

STEP 1 VARIABLE X4 ENTERED R SQUARE = 0.67454196
C(P) = 138.73063349

	DF	SUM OF SQUARES	MEAN SQUARE	F	PROB>F
REGRESSION	1	1831.89616002	1831.89616002	22.80	0.0006
ERROR	11	883.86691690	80.35153790		
TOTAL	12	2715.76307692			

	B VALUE	STD ERROR	TYPE II SS	F	PROB>F
INTERCEPT	117.56793118				
X4	-0.73816181	0.15459600	1831.89616002	22.80	0.0006

STEP 2 VARIABLE X1 ENTERED R SQUARE = 0.97247105
C(P) = 5.49585082

	DF	SUM OF SQUARES	MEAN SQUARE	F	PROB>F
REGRESSION	2	2641.00096477	1320.50048238	176.63	0.0001
ERROR	10	74.76211216	7.47621122		
TOTAL	12	2715.76307692			

	B VALUE	STD ERROR	TYPE II SS	F	PROB>F
INTERCEPT	103.09738164				
X1	1.43995828	0.13841664	809.10480474	108.22	0.0001
X4	-0.61395363	0.04864455	1190.92463664	159.30	0.0001

STEP 3 VARIABLE X2 ENTERED R SQUARE = 0.98233545
C(P) = 3.01823347

	DF	SUM OF SQUARES	MEAN SQUARE	F	PROB>F
REGRESSION	3	2667.79034752	889.26344917	166.83	0.0001
ERROR	9	47.97272940	5.33030327		
TOTAL	12	2715.76307692			

	B VALUE	STD ERROR	TYPE II SS	F	PROB>F
INTERCEPT	71.64830697				
X1	1.45193796	0.11699759	820.90740153	154.01	0.0001
X2	0.41610976	0.18561049	26.78938276	5.03	0.0517
X4	-0.23654022	0.17328779	9.93175378	1.86	0.2054

NO OTHER VARIABLES MET THE 0.1000 SIGNIFICANCE LEVEL FOR ENTRY INTO THE MODEL.

Figure 7.8 Computer output (SAS) of forward selection applied to Hald's cement data.

From Table 7.3, we see that the regressor most highly correlated with y is $x_4(r_{4y}=-0.821)$, and since the F-statistic associated with the model using x_4 is $F=22.80>F_{.10,1,11}=3.23$, x_4 is added to the equation. At Step 2 the regressor having the largest partial correlation with y (or the largest partial F-statistic given that x_4 is in the model) is x_1, and since the partial F-statistic for this regressor is

$$F=\frac{SS_R(x_1|x_4)}{MS_E(x_1,x_4)}=\frac{809.1048}{7.4762}=108.22$$

which exceeds $F_{\text{IN}}=F_{.10,1,10}=3.29$, x_1 is added to the model. In the third step, x_2 exhibits the highest partial correlation with y. The partial F-statistic is

$$\frac{SS_R(x_2|x_1,x_4)}{MS_E(x_1,x_2,x_4)}=\frac{27.7894}{5.3303}=5.03$$

which is larger than $F_{\text{IN}}=F_{.10,1,9}=3.36$, and so x_2 is added to the model. At this point the only remaining candidate regressor is x_3, for which the partial F-statistic does not exceed $F_{\text{IN}}=F_{.10,1,8}=3.46$, so the forward selection procedure terminates with

$$\hat{y}=71.6483+1.4519x_1+0.4161x_2-0.2365x_4$$

• as the final model.

Backward elimination. Forward selection begins with no regressors in the model and attempts to insert variables until a suitable model is obtained. Backward elimination attempts to find a good model by working in the opposite direction. That is, we begin with a model that includes all K candidate regressors. Then the partial F-statistic is computed for each regressor as if it were the last variable to enter the model. The smallest of these partial F-statistics is compared with a preselected value, F_{OUT} (or F-to-remove), for example, and if the smallest partial F-value is less than F_{OUT} that regressor is removed from the model. Now a regression model with $K-1$ regressors is fit, the partial F-statistics for this new model calculated, and the procedure repeated. The backward elimination algorithm terminates when the smallest partial F-value is not less than the preselected cutoff value F_{OUT}.

Backward elimination is often a very good variable selection procedure. It is particularly favored by analysts who like to see the effect of including all the candidate regressors, just so that nothing "obvious" will be missed.

• **Example 7.5** We will illustrate backward elimination using the Hald cement data from Example 7.1. Figure 7.9 presents the results of using the SAS

```
                 BACKWARD ELIMINATION PROCEDURE FOR DEPENDENT VARIABLE Y

STEP 0     ALL VARIABLES ENTERED          R SQUARE = 0.98237562
                                          C(P) =     5.00000000

                   DF         SUM OF SQUARES     MEAN SQUARE            F      PROB>F

REGRESSION          4          2667.89943757    666.97485939       111.48      0.0001
ERROR               8            47.86363935      5.98295492
TOTAL              12          2715.76307692

                      B VALUE      STD ERROR     TYPE II SS            F      PROB>F

INTERCEPT     62.40536930
X1             1.55110265     0.74476987     25.95091138         4.34      0.0708
X2             0.51016758     0.72378800      2.97247824         0.50      0.5009
X3             0.10190940     0.75470905      0.10909005         0.02      0.8959
X4            -0.14406103     0.70905206      0.24697472         0.04      0.8441
--------------------------------------------------------------------------------------

STEP 1     VARIABLE X3 REMOVED            R SQUARE = 0.98233545
                                          C(P) =     3.01823347

                   DF         SUM OF SQUARES     MEAN SQUARE            F      PROB>F

REGRESSION          3          2667.79034752    889.26344917       166.83      0.0001
ERROR               9            47.97272940      5.33030327
TOTAL              12          2715.76307692

                      B VALUE      STD ERROR     TYPE II SS            F      PROB>F

INTERCEPT     71.64830697
X1             1.45193796     0.11699759    820.90740153       154.01      0.0001
X2             0.41610976     0.18561049     26.78938276         5.03      0.0517
X4            -0.23654022     0.17328779      9.93175378         1.86      0.2054
--------------------------------------------------------------------------------------

STEP 2     VARIABLE X4 REMOVED            R SQUARE = 0.97867837
                                          C(P) =     2.67824160

                   DF         SUM OF SQUARES     MEAN SQUARE            F      PROB>F

REGRESSION          2          2657.85859375   1328.92929687       229.50      0.0001
ERROR              10            57.90448318      5.79044832
TOTAL              12          2715.76307692

                      B VALUE      STD ERROR     TYPE II SS            F      PROB>F

INTERCEPT     52.57734888
X1             1.46830574     0.12130092    848.43186034       146.52      0.0001
X2             0.66225049     0.04585472   1207.78226562       208.58      0.0001
--------------------------------------------------------------------------------------

ALL VARIABLES IN THE MODEL ARE SIGNIFICANT AT THE 0.1000 LEVEL.
```

Figure 7.9 Computer output (SAS) of backward elimination applied to Hald's cement data.

version of backward elimination on those data. In this run, we have selected the cutoff value F_{OUT} by choosing $\alpha=.10$; thus a regressor is dropped if its partial F-statistic is less than $F_{.10,1,n-p}$. Step 0 shows the results of fitting the full model. The smallest partial F-value is $F=0.02$ and it is associated with x_3. Thus since $F=0.02<F_{OUT}=F_{.10,1,8}=3.46$, x_3 is removed from the model. At Step 1 in Figure 7.9, we see the results of fitting the three-variable model involving (x_1, x_2, x_4). The smallest partial F-value in this model, $F=1.86$, is associated with x_4. Since $F=1.86<F_{OUT}=F_{.10,1,9}=3.36$, x_4 is removed from the model. At Step 2, we see the results of fitting the two-variable model involving (x_1, x_2). The smallest partial F-statistic in this model is $F=146.52$, associated with x_1, and since this exceeds $F_{OUT}=F_{.10,1,10}=3.29$, no further regressors can be removed from the model. Therefore backward elimination terminates, yielding the final model

$$\hat{y}=52.5773+1.4683x_1+0.6623x_2$$

Note that this is a different model from that found by forward selection. Furthermore it is the model tentatively identified as "best" by the all possible regressions procedure. •

Stepwise regression. The two procedures described above suggest a number of possible combinations. One of the most popular is the stepwise regression algorithm of Efroymson [1960]. Stepwise regression is a modification of forward selection in which at each step all regressors entered into the model previously are reassessed via their partial F-statistics. A regressor added at an earlier step may now be redundant because of the relationships between it and regressors now in the equation. If the partial F-statistic for a variable is less than F_{OUT}, that variable is dropped from the model.

Stepwise regression requires two cutoff values, F_{IN} and F_{OUT}. Some analysts prefer to choose $F_{IN}=F_{OUT}$, although this is not necessary. Frequently we choose $F_{IN}>F_{OUT}$, making it relatively more difficult to add a regressor than to delete one.

• **Example 7.6** Figure 7.10 presents the results of using the SAS stepwise regression algorithm on the Hald cement data. We have specified the α-level for either adding or removing a regressor as 0.10. At Step 1, the procedure begins with no variables in the model and tries to add x_4. Since the partial F-statistic at this step exceeds $F_{IN}=F_{.10,1,11}=3.23$, x_4 is added to the model. At Step 2, x_1 is added to the model. If the partial F-value for x_4 is less than $F_{OUT}=F_{.10,1,10}=3.29$, x_4 would be deleted. However, the partial F-value for x_4 at Step 2 is $F=159.30$, so x_4 is retained. In Step 3, the stepwise regression algorithm adds x_2 to the model. Then the partial F-statistics for x_1 and x_4 are compared to $F_{OUT}=F_{.10,1,9}=3.36$. Since for x_4 we find a partial F-value of 1.86 which is less than $F_{OUT}=3.36$, x_4 is deleted. Step 4 shows the results of

```
          STEPWISE REGRESSION PROCEDURE FOR DEPENDENT VARIABLE Y

STEP 1    VARIABLE X4 ENTERED            R SQUARE = 0.67454196
                                         C(P) =   138.73083349

                 DF         SUM OF SQUARES    MEAN SQUARE          F      PROB>F

REGRESSION        1         1831.89616002   1831.89616002      22.80      0.0006
ERROR            11          883.86691690     80.35153790
TOTAL            12         2715.76307692

                    B VALUE       STD ERROR     TYPE II SS         F      PROB>F

INTERCEPT    117.56793118
X4            -0.73816181      0.15459600   1831.89616002      22.80      0.0006
--------------------------------------------------------------------------------

STEP 2    VARIABLE X1 ENTERED            R SQUARE = 0.97247105
                                         C(P) =     5.49585082

                 DF         SUM OF SQUARES    MEAN SQUARE          F      PROB>F

REGRESSION        2         2641.00096477   1320.50048238     176.63      0.0001
ERROR            10           74.76211216      7.47621122
TOTAL            12         2715.76307692

                    B VALUE       STD ERROR     TYPE II SS         F      PROB>F

INTERCEPT    103.09738164
X1             1.43995828      0.13841664    809.10480474     108.22      0.0001
X4            -0.61395363      0.04864455   1190.92463664     159.30      0.0001
--------------------------------------------------------------------------------

STEP 3    VARIABLE X2 ENTERED            R SQUARE = 0.98233545
                                         C(P) =     3.01823347

                 DF         SUM OF SQUARES    MEAN SQUARE          F      PROB>F

REGRESSION        3         2667.79034752    889.26344917     166.83      0.0001
ERROR             9           47.97272940      5.33030327
TOTAL            12         2715.76307692

                    B VALUE       STD ERROR     TYPE II SS         F      PROB>F

INTERCEPT     71.64830697
X1             1.45193796      0.11699759    820.90740153     154.01      0.0001
X2             0.41610976      0.18561049     26.78938276       5.03      0.0517
X4            -0.23654022      0.17328779      9.93175378       1.86      0.2054
--------------------------------------------------------------------------------
```

Figure 7.10 Computer output (SAS) of stepwise regression algorithm applied to Hald's cement data.

removing x_4 from the model. At this point the only remaining candidate regressor is x_3, which cannot be added because its partial F-value does not exceed F_{IN}. Therefore stepwise regression terminates with the model

$$\hat{y}=52.5773+1.4683x_1+0.6623x_2$$

This is the same equation identified by the all possible regressions and backward elimination procedures.

STEPWISE REGRESSION PROCEDURE FOR DEPENDENT VARIABLE Y

STEP 4 VARIABLE X4 REMOVED R SQUARE = 0.97867837
C(P) = 2.67824160

	DF	SUM OF SQUARES	MEAN SQUARE	F	PROB>F
REGRESSION	2	2657.85859375	1328.92929687	229.50	0.0001
ERROR	10	57.90448318	5.79044832		
TOTAL	12	2715.76307692			

	B VALUE	STD ERROR	TYPE II SS	F	PROB>F
INTERCEPT	52.57734888				
X1	1.46830574	0.12130092	848.43186034	146.52	0.0001
X2	0.66225049	0.04585472	1207.78226562	208.58	0.0001

NO OTHER VARIABLES MET THE 0.1000 SIGNIFICANCE LEVEL FOR ENTRY INTO THE MODEL.

Figure 7.10 *Continued*

General comments on stepwise type procedures. The stepwise regression algorithms described above have been criticized on various grounds, the most common being that none of the procedures generally guarantees that the "best" subset regression model of any size will be identified. Furthermore since all the stepwise-type procedures terminate with one final equation, the inexperienced analyst may conclude that he has found a model that is in some sense optimal. Part of the problem is that it is unlikely that there is one "best" subset model, but probably several equally good ones.

The analyst should also keep in mind that the order in which the regressors enter or leave the model does not necessarily imply an order of importance to the regressors. It is not unusual to find that a regressor inserted into the model early in the procedure becomes negligible at a subsequent step. This is evident in the Hald cement data, for which forward selection chooses x_4 as the first regressor to enter. However, when x_2 is added at a subsequent step x_4 is no longer required because of the high intercorrelation between x_2 and x_4. This is in fact a *general* problem with the forward selection procedure. Once a regressor has been added it cannot be removed at a later step.

Note that forward selection, backward elimination, and stepwise regression do not necessarily lead to the *same* choice of final model. The intercorrelation between the regressors affect the order of entry and removal. For example, using the Hald cement data we found that the regressors selected by each procedure were as follows:

Forward Selection	x_1 x_2 x_4
Backward Elimination	x_1 x_2
Stepwise Regression	x_1 x_2

Some users have recommended that all the procedures be applied in the hopes of either seeing some agreement or learning something about the structure of the data that might be overlooked by using only one selection procedure. Furthermore there is not necessarily any agreement between any of the stepwise-type procedures and all possible regressions. However, Berk [1978] has noted that forward selection tends to agree with all possible regressions for small subset sizes but not for large ones, while backward elimination tends to agree with all possible regressions for large subset sizes but not for small ones.

For these reasons stepwise-type variable selection procedures should be used with caution. Our own preference is for the stepwise regression algorithm, followed by backward elimination. The backward elimination algorithm is often less adversely affected by the correlative structure of the regressors than is forward selection (see Mantel [1970]).

Stopping rules for stepwise procedures. Choosing the cutoff values F_{IN} and/or F_{OUT} in stepwise-type procedures can be thought of as specifying a "stopping rule" for these algorithms. Some computer programs allow the analyst to specify these numbers directly, while others require the choice of a Type I error rate α to generate F_{IN} and/or F_{OUT}. However, because the partial F-value examined at each stage is the maximum of several correlated partial F variables, thinking of α as a level of significance or Type I error rate is misleading. Several authors (e.g., Draper et al. [1971], and Pope and Webster [1972]) have investigated this problem, and little progress has been made towards either finding conditions under which the "advertised" level of significance on F is meaningful or developing the exact distribution of the F-to-enter and F-to-remove statistics.

Some users prefer to choose relatively small values of F_{IN} and F_{OUT} so that several additional regressors that would ordinarily be rejected by more conservative F-values may be investigated. In the extreme, we may choose F_{IN} and F_{OUT} so that all regressors are entered by forward selection or removed by backward elimination revealing one subset model of each size for $p=2,3,\ldots, K+1$. These subset models may then be evaluated by criteria such as C_p or MS_E to determine the final model. We do not recommend this extreme strategy because the analyst may think that the subsets so determined are in some sense optimal, when it is likely that the "best" subset model was overlooked. A very popular procedure is to set $F_{\text{IN}}=F_{\text{OUT}}=4$, as this corresponds roughly to the upper five percent point of the F distribution. Still another possibility is to make several runs using different values for F_{IN} and F_{OUT} and observe the effect of the choice of criteria on the subsets obtained.

There have been several studies directed towards providing practical guidelines in the choice of stopping rules. Bendel and Afifi [1974] recommend $\alpha=0.25$ for forward selection. This would typically result in a numerical value of F_{IN} of between 1.3 and 2. Kennedy and Bancroft [1971] also suggest $\alpha=0.25$ for forward selection and recommend $\alpha=0.10$ for backward elimination. The choice of values for F_{IN} and F_{OUT} is largely a matter of the personal preference of the analyst, and considerable latitude is often taken in this area.

7.2.4 Other procedures

Several variable selection procedures have been suggested as compromises between the single-final-model output of the stepwise-type methods and all possible regressions. Two techniques implemented in SAS are maximum R^2 improvement and minimum R^2 improvement. Both procedures are extensions of the stepwise concept. The maximum R^2 improvement procedure attempts to find the best p-term equation, starting with a $p-1$ term equation, by adding the regressor that provides the greatest increase in R^2. Given this subset a check is made to determine if switching a regressor currently in the model with one that is currently excluded will increase R^2. If so, the switch that provides the maximum increase in R^2 is made. This process continues until no further switching will increase R^2. The resulting model is called the "best" p-term model. In this manner the procedure produces a "best" subset model for each $p=2,3,\ldots, K+1$. However, these models are not necessarily the optimal set of p-term models because not all subsets of size p have been evaluated. For example, suppose that $K=4$, and that x_1 gives the best one-regressor equation. Now suppose that adding x_2 to this model gives the largest increase in R^2. The procedure will now attempt to increase R^2 for the model (x_1, x_2) by swapping x_3 or x_4 for x_1. If a swap can be made, then the procedure will try to improve R^2 by swapping x_2 for another regressor. The result of these comparisons will be called the "best two–variable model found." Note that this is not guaranteed to be the best two–variable model, as (x_3, x_4) may give a higher R^2 than any other two–variable model, but this combination of regressors may never be considered. The minimum R^2 improvement techniques operates in a similar fashion, except that adding and switching regressors is performed so that R^2 is increased as *little* as possible at each stage. This has the effect of showing the analyst a relatively large number of candidate regression equations.

These two procedures will allow the analyst to examine the summary statistics for a larger number of models than would ordinarily be generated

MAXIMUM R-SQUARE IMPROVEMENT FOR DEPENDENT VARIABLE Y

STEP 1 VARIABLE X4 ENTERED R SQUARE = 0.67454196
C(P) = 138.73083349

	DF	SUM OF SQUARES	MEAN SQUARE	F	PROB>F
REGRESSION	1	1831.89616002	1831.89616002	22.80	0.0006
ERROR	11	883.86691690	80.35153790		
TOTAL	12	2715.76307692			

	B VALUE	STD ERROR	TYPE II SS	F	PROB>F
INTERCEPT	117.56793118				
X4	-0.73816181	0.15459600	1831.89616002	22.80	0.0006

THE ABOVE MODEL IS THE BEST 1 VARIABLE MODEL FOUND.

STEP 2 VARIABLE X1 ENTERED R SQUARE = 0.97247105
C(P) = 5.49585082

	DF	SUM OF SQUARES	MEAN SQUARE	F	PROB>F
REGRESSION	2	2641.00096477	1320.50048238	176.63	0.0001
ERROR	10	74.76211216	7.47621122		
TOTAL	12	2715.76307692			

	B VALUE	STD ERROR	TYPE II SS	F	PROB>F
INTERCEPT	103.09738164				
X1	1.43995828	0.13841664	809.10480474	108.22	0.0001
X4	-0.61395363	0.04864455	1190.92463664	159.30	0.0001

STEP 2 X4 REPLACED BY X2 R SQUARE = 0.97867837
C(P) = 2.67824160

	DF	SUM OF SQUARES	MEAN SQUARE	F	PROB>F
REGRESSION	2	2657.85859375	1328.92929687	229.50	0.0001
ERROR	10	57.90448318	5.79044832		
TOTAL	12	2715.76307692			

	B VALUE	STD ERROR	TYPE II SS	F	PROB>F
INTERCEPT	52.57734888				
X1	1.46830574	0.12130092	848.43186034	146.52	0.0001
X2	0.66225049	0.04585472	1207.78226562	208.58	0.0001

THE ABOVE MODEL IS THE BEST 2 VARIABLE MODEL FOUND.

Figure 7.11 Computer output (SAS) of maximum R^2 improvement for Hald's cement data.

via stepwise methods. Note that minimum R^2 improvement usually produces many more equations than does maximum R^2 improvement. Both procedures still fall short of all possible regressions, and there is no guarantee that the optimal model will not be overlooked. Furthermore the analyst is still confronted with the problem of choosing the final equation to use.

- **Example 7.7** Figure 7.11 shows the SAS maximum R^2 improvement procedure applied to the Hald cement data. For this particular data the procedure correctly identifies the two-variable and three-variable models with the largest values of R^2. There is of course no guarantee that this will always happen. •

```
               MAXIMUM R-SQUARE IMPROVEMENT FOR DEPENDENT VARIABLE Y

STEP 3     VARIABLE X4 ENTERED                  R SQUARE = 0.98233545
                                                C(P) =      3.01823347

                   DF          SUM OF SQUARES       MEAN SQUARE             F       PROB>F

REGRESSION          3           2667.79034752      889.26344917        166.83       0.0001
ERROR               9             47.97272940        5.33030327
TOTAL              12           2715.76307692

                      B VALUE       STD ERROR       TYPE II SS              F       PROB>F

INTERCEPT     71.64830697
X1             1.45193796      0.11699759      820.90740153        154.01       0.0001
X2             0.41610976      0.18561049       26.78938276          5.03       0.0517
X4            -0.23654022      0.17328779        9.93175378          1.66       0.2054
-------------------------------------------------------------------------------------

THE ABOVE MODEL IS THE BEST  3 VARIABLE MODEL FOUND.

STEP 4     VARIABLE X3 ENTERED                  R SQUARE = 0.98237562
                                                C(P) =      5.00000000

                   DF          SUM OF SQUARES       MEAN SQUARE             F       PROB>F

REGRESSION          4           2667.89943757      666.97485939        111.48       0.0001
ERROR               8             47.86363935        5.98295492
TOTAL              12           2715.76307692

                      B VALUE       STD ERROR       TYPE II SS              F       PROB>F

INTERCEPT     62.40536930
X1             1.55110265      0.74476987       25.95091138          4.34       0.0708
X2             0.51016758      0.72378800        2.97247824          0.50       0.5009
X3             0.10190940      0.75470905        0.10909005          0.02       0.8959
X4            -0.14406103      0.70905206        0.24697472          0.04       0.8441
-------------------------------------------------------------------------------------

THE ABOVE MODEL IS THE BEST  4 VARIABLE MODEL FOUND.
```

Figure 7.11 *Continued*

The BMD–P regression programs (see Dixon [1977]) also have several interesting variations of the basic stepwise procedure. The BMDP2R program can perform the stepwise, forward selection, and backward elimination techniques by specifying the values for F_{IN} and F_{OUT} while operating in the *F-Method* mode. There are three additional operating modes. The *FSWAP* method enters and removes regressors as in the *F*-Method, except that at each step a comparison is made to determine whether any regressor in the model can be swapped with a regressor currently excluded to increase R^2. If such an exchange is possible, it is performed. Note that the procedure is not the same as either of the SAS R^2 improvement procedures, because regressors are entered or removed according to the partial *F*-statistic. In the SAS routines, regressors are entered based on the associated changes in R^2, and are not removed if their partial *F*-values drop below a cutoff value.

The second BMD–P variation of stepwise regression is called the *R–Method*. In this option, regressors are entered according to the value of the partial *F*-statistic. However, if two or more regressors are in the model, the one with the smallest partial F is dropped if its removal yields a larger R^2 than was observed previously for that subset size. The final variable selection option is called *RSWAP*. This procedure is similar to the *R*-Method, except that instead of removing a regressor and reducing the number of terms in the model, any regressor removed is exchanged for another not presently in the model.

None of the procedures discussed in this section will necessarily produce the same final model as all possible regressions. For an interesting comparison of these techniques on one data set, see Younger [1979].

7.3 Some final considerations

This chapter has discussed several procedures for variable selection in linear regression. The methods may be generally classified as stepwise-type methods or all possible regressions (with variations). The primary advantage of the stepwise-type methods is that they are fast, easy to implement on digital computers, and readily available for almost all computer systems. Their disadvantages are that they do not produce subset models that are necessarily "best" with respect to any standard criterion, and furthermore as they are oriented towards producing a single final equation, the unsophisticated user may be led to believe that this model is in some sense optimal. On the other hand all possible regressions will identify the subset models that are best with respect to whatever criterion the analyst imposes. For up to about 20 or 30 candidate regressors, the cost of computing with all possible regressions is approximately the same as the stepwise-type procedures.

However, all possible regression codes are not as widely available as stepwise-type codes (at least at present).

When the number of candidate regressors is too large to initially employ the all possible regressions approach, we recommend a two-stage strategy. Stepwise-type methods can be used to "screen" the candidate regressors, eliminating those that have negligible effects so that a smaller list of candidate regressors results. This reduced set of candidate regressors can then be investigated by all possible regressions. The analyst should always use knowledge of the problem environment and common sense in evaluating candidate regressors. When confronted with a large list of candidate regressors it is usually profitable to invest in some serious thinking before resorting to the computer. Often we find that some variables can be eliminated on the basis of logic or engineering sense.

We have discussed several formal criteria for evaluating subset regression models, such as Mallows' C_p statistic and the residual mean square. Usually however, the choice of a final model is not clear-cut. In addition to the formal evaluation criteria, we would suggest that the analyst ask the following questions:

1. Is the equation reasonable? That is, do the regressors in the model make sense in light of the problem environment?

2. Is the model usable for its intended purpose? For example, a model intended for prediction that contains a regressor that is unobservable at the time the prediction is required is unusable. If the cost of collecting data on a regressor is prohibitive, this would also render the model unusable.

3. Are the regression coefficients reasonable? That is, are the signs and magnitudes of the coefficients realistic and are the standard errors relatively small?

4. Are the usual diagnostic checks for model adequacy satisfactory? For example, do the residual plots indicate unexplained structure or outliers, or are there one or more high-leverage points that may be controlling the fit?

If these four questions are taken seriously and the answers strictly applied, in some (perhaps many) instances there would be *no* final satisfactory regression equation. Clearly, judgment and experience in the model's intended operating environment are required. Finally, although the equation fits the data well and passes the usual diagnostic checks, there is no assurance that it will predict new observations accurately. We recommend that the predictive ability of a model be assessed by observing its performance on new data not used to build the model. If this cannot be done easily then some of the original data may be set side for this purpose. We will discuss this in more detail in Chapter 10.

Problems

7.1 Consider the National Football League data in Appendix Table B.1.

a. Use the forward selection algorithm to select a subset regression model.

b. Use the backward elimination algorithm to select a subset regression model.

c. Use stepwise regression to select a subset regression model.

d. Comment on the final model chosen by these three procedures.

7.2 Consider the National Football League data in Appendix Table B.1. Restricting your attention to regressors x_1 (rushing yards), x_2 (passing yards), x_4 (field goal percentage), x_7 (percent rushing), x_8 (opponents' rushing yards), and x_9 (opponents' passing yards), apply the all possible regressions procedure. Evaluate R_p^2, C_p, and MS_E for each model. Which subset of regressors do you recommend?

7.3 In stepwise regression, we specify that $F_{\text{IN}} \geqslant F_{\text{OUT}}$. Justify this choice of cutoff values.

7.4 Consider the solar thermal energy test data in Appendix Table B.2.

a. Use forward selection to specify a subset regression model.

b. Use backward elimination to specify a subset regression model.

c. Use stepwise regression to specify a subset regression model.

d. Apply all possible regressions to the data. Evaluate R_p^2, C_p, and MS_E for each model. Which subset model do you recommend?

e. Compare and contrast the models produced by the variable selection strategies in Parts (a) through (d).

7.5 Consider the gasoline mileage performance data in Appendix Table B.3.

a. Use the directed search on t method to find the "best" subset regression model.

b. Use stepwise regression to specify a subset regression model. Does this lead to the same model found in Part (a)?

7.6 Consider the property valuation data found in Appendix Table B.4.

a. Use the directed search on t method to find the "best" set of regressors.

b. Use stepwise regression to select a subset regression model. Does this model agree with the one found in Part (a)?

7.7 Use stepwise regression with $F_{IN} = F_{OUT} = 4.0$ to find the "best" set of regressor variables for the Belle Ayr liquefaction data in Appendix Table B.5. Repeat the analysis with $F_{IN} = F_{OUT} = 2.0$. Are there any substantial differences in the models obtained?

7.8 Use the directed search on t method to select a subset regression model for the Belle Ayr liquefaction data in Appendix Table B.5. Evaluate the subset models using the C_p criterion. Justify your choice of final model using the standard checks for model adequacy.

7.9 Analyze the tube-flow reactor data in Appendix Table B.6 using all possible regressions. Evaluate the subset models using the R_p^2, C_p, and MS_E criteria. Justify your choice of final model using the standard checks for model adequacy.

7.10 Analyze the data in Appendix Table B.7 on factors affecting CPU time using all possible regressions. Evaluate the subset models using the R_p^2, C_p, and MS_E criteria. Use residual analysis to investigate the adequacy of any reasonable final model(s).

7.11 Apply forward selection and backward elimination to the data in Appendix Table B.7 on factors affecting CPU time. Justify your choice of cutoff values for F_{IN} and F_{OUT}. Are there any differences in the final models recommended by these two procedures?

7.12 Data concerning the effectiveness of insecticides on cricket eggs is presented in Appendix Table B.8.

a. Analyze these data using stepwise regression. What conclusions can you draw about the adequacy of the final model?

b. Expand the set of candidate regressors to include all possible cross-products and squared effects of the original four regressors. Apply stepwise regression to this expanded set of regressors. How does the final model differ from the one found in Part (a)? Which model do you prefer?

c. Use the directed search on t method to investigate the expanded set of candidate regressors in Part (b). Compare the final model obtained with the one found in Part (b).

7.13 Use stepwise regression, backward elimination, and forward selection to analyze the data presented in Example 7.3. In each procedure, explain your choice of the cutoff values F_{IN} and/or F_{OUT}. Compare the final models produced by each procedure.

7.14 Consider the all possible regressions analysis of Hald's cement data in Example 7.1. If the objective is to develop a model to predict new observations, which equation would you recommend and why?

7.15 Consider the all possible regressions analysis of the National Football League data in Problem 7.2. Identify the subset regression models that are R^2 adequate (.05).

7.16 Suppose that the full model is $y_i = \beta_0 + \beta_1 x_{i1} + \beta_2 x_{i2} + \varepsilon_i$, $i = 1, 2, \ldots, n$, where x_{i1} and x_{i2} have been coded so that $S_{11} = S_{22} = 1$. We will also consider fitting a subset model, say $y_i = \beta_0 + \beta_1 x_{i1} + \varepsilon_i$.

a. Let $\hat{\beta}_1^*$ be the least squares estimate of β_1 from the full model. Show that $V(\hat{\beta}_1^*) = \sigma^2/(1 - r_{12}^2)$, where r_{12} is the correlation between x_1 and x_2.

b. Let $\hat{\beta}_1$ be the least squares estimate of β_1 from the subset model. Show that $V(\hat{\beta}_1) = \sigma^2$. Is β_1 estimated more precisely from the subset model or from the full model?

c. Show that $E(\hat{\beta}_1) = \beta_1 + r_{12}\beta_2$. Under what circumstances is $\hat{\beta}_1$ an unbiased estimator of β_1?

d. Find the mean square error for the subset estimator $\hat{\beta}_1$. Compare $\text{MSE}(\hat{\beta}_1)$ with $V(\hat{\beta}_1^*)$. Under what circumstances is $\hat{\beta}_1$ a preferable estimator, with respect to MSE?

You may find it helpful to reread Section 7.1.2.

8

MULTICOLLINEARITY

8.1 Introduction

The use and interpretation of a multiple regression model often depends explicitly or implicitly on the estimates of the individual regression coefficients. Some examples of inferences that are frequently made include

1. Identifying the relative effects of the regressor variables
2. Prediction and/or estimation
3. Selection of an appropriate set of variables for the model.

If there is no linear relationship between the regressors, they are said to be *orthogonal*. When the regressors are orthogonal, inferences such as those illustrated above can be made relatively easily. Unfortunately in most applications of regression, the regressors are not orthogonal. Sometimes the lack of orthogonality is not serious. However, in some situations the regressors are nearly perfectly linearly related, and in such cases the inferences based on the regression model can be misleading or erroneous. When there are near linear dependencies between the regressors, the problem of *multicollinearity* is said to exist.

This chapter will discuss a variety of problems and techniques related to multicollinearity. Specifically, we will examine the causes of multicollinearity, some of its specific effects on inference, methods of detecting the presence of multicollinearity, and some techniques for dealing with the problem.

8.2 Sources of multicollinearity

We write the multiple regression model as

$$\mathbf{y}=\mathbf{X}\boldsymbol{\beta}+\boldsymbol{\varepsilon} \tag{8.1}$$

where $\mathbf{y}$ is an $(n\times 1)$ vector of responses, $\mathbf{X}$ is an $(n\times p)$ matrix of the regressor variables, $\boldsymbol{\beta}$ is a $(p\times 1)$ vector of unknown constants, and $\boldsymbol{\varepsilon}$ is an $(n\times 1)$ vector of random errors, with $\varepsilon_i \sim \text{NID}(0,\sigma^2)$. It will be convenient to assume that the regressor variables and the response have been centered and scaled to unit length as in Section 4.8. Consequently $\mathbf{X}'\mathbf{X}$ is a $(p\times p)$ matrix of correlations* between the regressors and $\mathbf{X}'\mathbf{y}$ is a $(p\times 1)$ vector of correlations between the regressors and the response.

Let the jth column of the $\mathbf{X}$ matrix be denoted $\mathbf{X}_j$, so that $\mathbf{X}=[\mathbf{X}_1,\mathbf{X}_2,\ldots,\mathbf{X}_p]$. Thus $\mathbf{X}_j$ contains the n levels of the jth regressor variable. We may formally define multicollinearity in terms of the linear dependence of the columns of $\mathbf{X}$. The vectors $\mathbf{X}_1,\mathbf{X}_2,\ldots,\mathbf{X}_p$ are linearly dependent if there is a set of constants $t_1, t_2,\ldots, t_p$, not all zero, such that†

$$\sum_{j=1}^{p} t_j\mathbf{X}_j=\mathbf{0} \tag{8.2}$$

If (8.2) holds exactly for a subset of the columns of $\mathbf{X}$, then the rank of the $\mathbf{X}'\mathbf{X}$ matrix is less than p and $(\mathbf{X}'\mathbf{X})^{-1}$ does not exist. However, suppose that (8.2) is approximately true for some subset of the columns of $\mathbf{X}$. Then there will be a near linear dependency in $\mathbf{X}'\mathbf{X}$ and the problem of multicollinearity is said to exist. Note that multicollinearity is a form of ill-conditioning in the $\mathbf{X}'\mathbf{X}$ matrix. Furthermore the problem is one of degree; that is, every data set will suffer from multicollinearity to some extent unless the columns of $\mathbf{X}$ are orthogonal ($\mathbf{X}'\mathbf{X}$ is a diagonal matrix). Generally this will happen only in a designed experiment. As we shall see, the presence of multicollinearity can make the usual least squares analysis of the regression model dramatically inadequate.

*It is customary to refer to the off-diagonal elements of $\mathbf{X}'\mathbf{X}$ as correlation coefficients, although the regressors are not necessarily random variables.

†If the regressors are not centered, then $\mathbf{0}$ in (8.2) becomes a vector of constants m, not necessarily equal to 0.

There are four primary sources of multicollinearity:

1. The data collection method employed.
2. Constraints on the model or in the population.
3. Model specification.
4. An over-defined model.

It is important to understand the differences among these sources of multicollinearity, because the recommendations for analysis of the data and interpretation of the resulting model depend to some extent on the cause of the problem (see Mason, Gunst and Webster [1975] for further discussion of the source of multicollinearity).

The data collection method can lead to multicollinearity problems when the analyst samples only a subspace of the region of the regressors defined (approximately) by (8.2). For example, consider the soft drink delivery time data discussed in Example 4.1. The space of the regressor variables cases and distance, as well as the subspace of this region that has been sampled, is shown in Figure 4.6. Note that the sample (cases, distance) pairs fall approximately along a straight line. In general if there are more than two regressors, the data will lie approximately along a hyperplane defined by (8.2). In this example observations with a small number of cases generally also have a short distance, while observations with a large number of cases usually also have a long distance. Thus cases and distance are positively correlated, and if this positive correlation is strong enough, a multicollinearity problem will occur. Multicollinearity caused by the sampling technique is not inherent in the model or the population being sampled. For example, in the delivery time problem we could collect data with a small number of cases and a long distance. There is nothing in the physical structure of the problem to prevent this.

Constraints on the model or in the population being sampled can cause multicollinearity. For example, suppose that an electric utility is investigating the effect of family income (x_1) and house size (x_2) on residential electricity consumption. The levels of the two regressor variables obtained in the sample data are shown in Figure 8.1. Note that the data lies approximately along a straight line, indicating a potential multicollinearity problem. In this example a physical constraint in the population has caused this phenomena; namely, families with higher incomes generally have larger homes than families with lower incomes. When physical constraints such as this are present, multicollinearity will exist *regardless* of the sampling

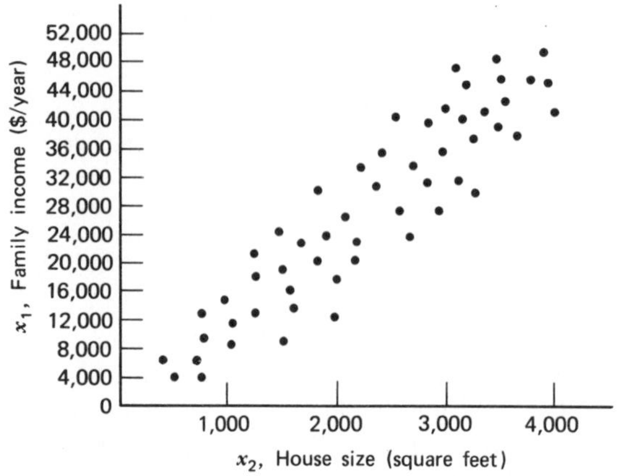

Figure 8.1 Levels of family income and house size for a study on residential electricity consumption.

method employed. Constraints often occur in problems involving production or chemical processes, where the regressors are the components of a product, and these components add to a constant.

Multicollinearity may also be induced by the choice of model. For example, we know from Chapter 5 that adding polynomial terms to a regression model causes ill-conditioning in $\mathbf{X}'\mathbf{X}$. Furthermore if the range of x is small, adding an x^2 term can result in significant multicollinearity. We often encounter situations such as these where two or more regressors are nearly linearly dependent, and retaining all these regressors may contribute to multicollinearity. In these cases some subset of the regressors is usually preferable from the standpoint of multicollinearity.

An over-defined model has more regressor variables than observations. These models are sometimes encountered in medical and behavioral research, where there may be only a small number of subjects (sample units) available, and information is collected for a large number of regressors on each subject. The usual approach to dealing with multicollinearity in this context is to eliminate some of the regressor variables from consideration. Mason, Gunst and Webster [1975] give three specific recommendations: (1) redefine the model in terms of a smaller set of regressors, (2) perform preliminary studies using only subsets of the original regressors, and (3) use principal components type regression methods to decide which regressors to remove from the model. The first two methods ignore the interrelationships between the regressors, and, consequently can lead to unsatisfactory results. Principal components regression will be discussed in Section 8.5.5, although not in the context of over-defined models.

8.3 Effects of multicollinearity

The presence of multicollinearity has a number of potentially serious effects on the least squares estimates of the regression coefficients. Some of these effects may be easily demonstrated. Suppose that there are only two regressor variables, x_1 and x_2. The model, assuming that x_1, x_2, and y are scaled to unit length, is

$$y=\beta_1 x_1+\beta_2 x_2+\varepsilon$$

and the least squares normal equations are

$$(\mathbf{X}'\mathbf{X})\hat{\boldsymbol{\beta}}=\mathbf{X}'\mathbf{y}$$

$$\begin{bmatrix} 1 & r_{12} \\ r_{12} & 1 \end{bmatrix}\begin{bmatrix} \hat{\beta}_1 \\ \hat{\beta}_2 \end{bmatrix}=\begin{bmatrix} r_{1y} \\ r_{2y} \end{bmatrix}$$

where r_{12} is the simple correlation between x_1 and x_2 and r_{jy} is the simple correlation between x_j and y, $j=1,2$. Now the inverse of $(\mathbf{X}'\mathbf{X})$ is

$$\mathbf{C}=(\mathbf{X}'\mathbf{X})^{-1}=\begin{bmatrix} \dfrac{1}{(1-r_{12}^2)} & \dfrac{-r_{12}}{(1-r_{12}^2)} \\ \dfrac{-r_{12}}{(1-r_{12}^2)} & \dfrac{1}{(1-r_{12}^2)} \end{bmatrix} \tag{8.3}$$

and the estimates of the regression coefficients are

$$\hat{\beta}_1=\frac{r_{1y}-r_{12}r_{2y}}{(1-r_{12}^2)}$$

$$\hat{\beta}_2=\frac{r_{2y}-r_{12}r_{1y}}{(1-r_{12}^2)} \tag{8.4}$$

If there is strong multicollinearity between x_1 and x_2, then the correlation coefficient r_{12} will be large. From (8.3) we see that as $|r_{12}|\rightarrow 1$, $V(\hat{\beta}_j)=C_{jj}\sigma^2 \rightarrow \infty$ and $\operatorname{Cov}(\hat{\beta}_1,\hat{\beta}_2)=C_{12}\sigma^2 \rightarrow \pm\infty$ depending on whether $r_{12}\rightarrow +1$ or $r_{12}\rightarrow -1$. Therefore strong multicollinearity between x_1 and x_2 results in large variances and covariances for the least squares estimators of the regression coefficients.* This implies that different samples taken at the same x levels could lead to widely different estimates of the model parameters.

*Multicollinearity is not the only cause of large variances and covariances of regression coefficients.

When there are more than two regressor variables, multicollinearity produces similar effects. It can be shown that the diagonal elements of the $\mathbf{C}=(\mathbf{X}'\mathbf{X})^{-1}$ matrix are

$$C_{jj}=\frac{1}{\left(1-R_j^2\right)}, \qquad j=1,2,\ldots,p \tag{8.5}$$

where R_j^2 is the coefficient of multiple determination from the regression of x_j on the remaining $p-1$ regressor variables. If there is strong multicollinearity between x_j and any subset of the other $p-1$ regressors, then the value of R_j^2 will be close to unity. Since the variance of $\hat{\beta}_j$ is $V(\hat{\beta}_j)=C_{jj}\sigma^2=(1-R_j^2)^{-1}\sigma^2$, strong multicollinearity implies that the variance of the least squares estimate of the regression coefficient β_j is very large. Generally the covariance of $\hat{\beta}_i$ and $\hat{\beta}_j$ will also be large if the regressors x_i and x_j are involved in a multicollinear relationship.

Multicollinearity also tends to produce least squares estimates $\hat{\beta}_j$ that are too large in absolute value. To see this, consider the squared distance from $\hat{\boldsymbol{\beta}}$ to the true parameter vector $\boldsymbol{\beta}$, for example

$$L_1^2=(\hat{\boldsymbol{\beta}}-\boldsymbol{\beta})'(\hat{\boldsymbol{\beta}}-\boldsymbol{\beta}) \tag{8.6}$$

The expected squared distance, $E(L_1^2)$, is

$$\begin{aligned}E\left(L_1^2\right)&=E(\hat{\boldsymbol{\beta}}-\boldsymbol{\beta})'(\hat{\boldsymbol{\beta}}-\boldsymbol{\beta})\\&=\sum_{j=1}^{p} E\left(\hat{\beta}_j-\beta_j\right)^2\\&=\sum_{j=1}^{p} V\left(\hat{\beta}_j\right)\\&=\sigma^2 Tr(\mathbf{X}'\mathbf{X})^{-1}\end{aligned} \tag{8.7}$$

where the trace of a matrix (denoted Tr) is just the sum of the main diagonal elements. When there is multicollinearity present, some of the eigenvalues of $\mathbf{X}'\mathbf{X}$ will be small. Since the trace of a matrix is also equal to the sum of its eigenvalues, (8.7) becomes

$$E\left(L_1^2\right)=\sigma^2\sum_{j=1}^{p}(1/\lambda_j) \tag{8.8}$$

where $\lambda_j>0, j=1,2,\ldots,p$ are the eigenvalues of $\mathbf{X}'\mathbf{X}$. Thus if the $\mathbf{X}'\mathbf{X}$ matrix is ill-conditioned because of multicollinearity, at least one of the λ_j

will be small, and (8.8) implies that the distance from the least squares estimate $\hat{\boldsymbol{\beta}}$ to the true parameters $\boldsymbol{\beta}$ may be large. Equivalently we can show that

$$E(L_1^2)=E(\hat{\boldsymbol{\beta}}-\boldsymbol{\beta})'(\hat{\boldsymbol{\beta}}-\boldsymbol{\beta})$$
$$=E(\hat{\boldsymbol{\beta}}'\hat{\boldsymbol{\beta}}-2\hat{\boldsymbol{\beta}}'\boldsymbol{\beta}+\boldsymbol{\beta}'\boldsymbol{\beta})$$

or

$$E(\hat{\boldsymbol{\beta}}'\hat{\boldsymbol{\beta}})=\boldsymbol{\beta}'\boldsymbol{\beta}+\sigma^2 Tr(\mathbf{X}'\mathbf{X})^{-1} \tag{8.9}$$

That is, the vector $\hat{\boldsymbol{\beta}}$ is generally longer than the vector $\boldsymbol{\beta}$. This implies that the method of least squares produces estimated regression coefficients that are too large in absolute value.

While the method of least squares will generally produce poor estimates of the individual model parameters when strong multicollinearity is present, this does not necessarily imply that the fitted model is a poor predictor. If predictions are confined to regions of the x-space where the multicollinearity holds approximately, the fitted model often produces satisfactory predictions. This can occur because the linear combination $\Sigma_{j=1}^{p}\beta_j x_{ij}$ may be estimated quite well, even though the individual parameters β_j are estimated poorly. That is, if the original data lie approximately along the hyperplane defined by (8.2), then future observations that also lie near this hyperplane can often be precisely predicted despite the inadequate estimates of the individual model parameters.

• **Example 8.1** Table 8.1 presents data concerning the percent conversion of n-heptane to acetylene and three explanatory variables (Himmelblau [1970], Kunugi et al. [1961], and Marquardt and Snee [1975]). This is typical chemical process data for which a full quadratic response surface in all three regressors is often considered to be an appropriate tentative model. A plot of contact time versus reactor temperature is shown in Figure 8.2. Since these two regressors are highly correlated, there are potential multicollinearity problems in these data.

The full quadratic model for the acetylene data is

$$P=\gamma_0+\gamma_1 T+\gamma_2 H+\gamma_3 C+\gamma_{12}TH+\gamma_{13}TC+\gamma_{23}HC$$
$$+\gamma_{11}T^2+\gamma_{22}H^2+\gamma_{33}C^2+\varepsilon$$

where

$$P=\text{Percent Conversion}$$
$$T=\frac{\text{Temperature}-1212.50}{80.623}$$
$$H=\frac{H_2/(n-\text{Heptane})-12.44}{5.662}$$

Table 8.1 Acetylene data for Example 8.1

Observation, i	Conversion of n-Heptane to Acetylene (%)	Reactor Temperature (°C)	Ratio of H_2 to n-Heptane (mole ratio)	Contact Time (sec)
1	49.0	1300	7.5	0.0120
2	50.2	1300	9.0	0.0120
3	50.5	1300	11.0	0.0115
4	48.5	1300	13.5	0.0130
5	47.5	1300	17.0	0.0135
6	44.5	1300	23.0	0.0120
7	28.0	1200	5.3	0.0400
8	31.5	1200	7.5	0.0380
9	34.5	1200	11.0	0.0320
10	35.0	1200	13.5	0.0260
11	38.0	1200	17.0	0.0340
12	38.5	1200	23.0	0.0410
13	15.0	1100	5.3	0.0840
14	17.0	1100	7.5	0.0980
15	20.5	1100	11.0	0.0920
16	29.5	1100	17.0	0.0860

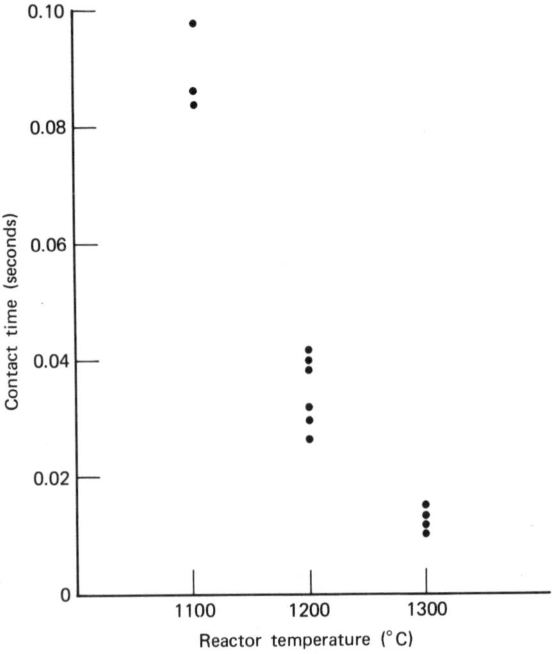

Figure 8.2 Contact time versus reactor temperature, acetylene data. (From Marquardt and Snee [1975], with permission of the publisher.)

and

$$C = \frac{\text{Contact Time} - 0.0403}{0.03164}$$

Each of the original regressors has been scaled using the unit normal scaling of Section 4.8 (subtracting the average [centering] and dividing by the standard deviation). The squared and cross-product terms are generated from the scaled linear terms. As we noted in Chapter 5, centering the linear terms is helpful in removing nonessential ill-conditioning when fitting polynomials. The least squares fit is

$$\hat{P} = 35.897 + 4.019T + 2.781H - 8.031C - 6.457TH - 26.982TC$$

$$-3.768HC - 12.524T^2 - 0.973H^2 - 11.594C^2$$

The summary statistics for this model are displayed in Table 8.2. The regression coefficients are reported in terms of both the original centered regressors and standardized regressors.

The fitted values for the six points (A, B, E, F, I, and J) that define the boundary of the regressor variable hull of contact time and reactor temperature are shown in Figure 8.3, along with the corresponding observed values of percent conversion. The predicted and observed values agree very closely; consequently the model seems adequate for interpolation within the range of the original data. Now consider using the model for extrapolation. Figure 8.3 (Points C, D, G, and H) also shows predictions made at the corners of the region defined by the range of the original data. These points represent

Table 8.2 Summary statistics for the least squares acetylene model

Term	Regression Coefficient	Standard Error	t_0	Standardized Regression Coefficient
Intercept	35.8971	1.0903	32.93	
T	4.0187	4.5012	0.89	0.3377
H	2.7811	0.3074	9.05	0.2337
C	−8.0311	6.0657	−1.32	−0.6749
TH	−6.4568	1.4660	−4.40	−0.4799
TC	−26.9818	21.0224	−1.28	−2.0344
HC	−3.7683	1.6554	−2.28	−0.2657
T^2	−12.5237	12.3239	−1.02	−0.8346
H^2	−0.9721	0.3746	−2.60	−0.0904
C^2	−11.5943	7.7070	−1.50	−1.0015

$MS_E = 0.8126$, $R^2 = 0.998$, $F_0 = 289.72$

When the response is standardized, $MS_E = 0.00038$ for the least squares model

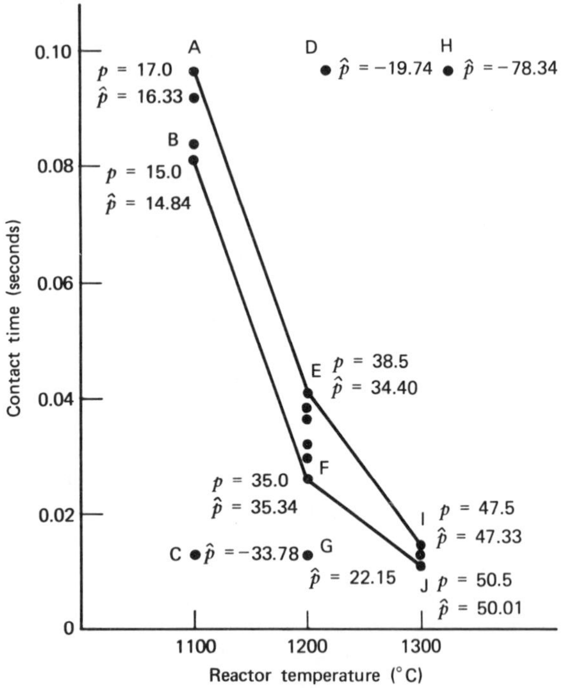

Figure 8.3 Predictions of percent conversion within the range of the data and extrapolation for the least squares acetylene model. (Adapted from Marquardt and Snee [1975], with permission of the publisher.)

relatively mild extrapolation, since the original ranges of the regressors have not been exceeded. The predicted conversions at three of the four extrapolation points are negative, an obvious impossibility. It seems that the least squares model fits the data reasonably well, but extrapolates very poorly. A likely cause of this in view of the strong apparent correlation between contact time and reactor temperature, is multicollinearity. In general if a model is to extrapolate well, good estimates of the individual coefficients are required. When multicollinearity is suspected, the least squares estimates of the regression coefficients may be very poor. This may seriously limit the usefulness of the regression model for inference and prediction.

8.4 Multicollinearity diagnostics

Several techniques have been proposed for detecting multicollinearity. We will now discuss and illustrate some of these diagnostic measures. Desirable characteristics of a diagnostic procedure are that it directly reflect the degree of the multicollinearity problem and provide information helpful in determining which regressors are involved.

8.4.1 Examination of the correlation matrix

A very simple measure of multicollinearity is inspection of the off-diagonal elements r_{ij} in $\mathbf{X'X}$. If regressors x_i and x_j are nearly linearly dependent, then $|r_{ij}|$ will be near unity. To illustrate this procedure, consider the acetylene data from Example 8.1. Table 8.3 shows the nine regressor variables and the response in standardized form; that is, each of the variables has been centered by subtracting the mean for that variable and dividing by the square root of the corrected sum of squares for that variable. The $\mathbf{X'X}$ matrix in correlation form for the acetylene data is

$$\mathbf{X'X}=\begin{bmatrix} 1.000 & 0.224 & -0.958 & -0.132 & 0.443 & 0.205 & -0.271 & 0.031 & -0.577 \\ & 1.000 & -0.240 & 0.039 & 0.192 & -0.023 & -0.148 & 0.498 & -0.224 \\ & & 1.000 & 0.194 & -0.661 & -0.274 & 0.501 & -0.018 & 0.765 \\ & & & 1.000 & -0.265 & -0.975 & 0.246 & 0.398 & 0.274 \\ & & & & 1.000 & 0.323 & -0.972 & 0.126 & -0.972 \\ & & & & & 1.000 & -0.279 & -0.374 & 0.358 \\ & & & & & & 1.000 & -0.124 & 0.874 \\ & & & & & & & 1.000 & -0.158 \\ & & & & & & & & 1.000 \\ \text{Symmetric} & & & & & & & & \end{bmatrix}$$

The $\mathbf{X'X}$ matrix reveals the high correlation between reactor temperature (x_1) and contact time (x_3) suspected earlier from inspection of Figure 8.2, since $r_{13}=-0.958$. Furthermore there are other large correlation coefficients between x_1x_2 and x_2x_3, x_1x_3 and x_1^2, and x_1^2 and x_3^2. This is not surprising as these variables are generated from the linear terms and they involve the highly correlated regressors x_1 and x_3. Thus inspection of the correlation matrix indicates that there are several near linear dependencies in the acetylene data.

Examining the simple correlations r_{ij} between the regressors is helpful in detecting near linear dependency between pairs of regressors only. Unfortunately when more than two regressors are involved in a near linear dependency, there is no assurance that any of the pairwise correlations r_{ij} will be large. As an illustration, consider the data in Table 8.4. These data were artificially generated by Webster, Gunst and Mason [1974]. They required that $\Sigma_{j=1}^{4} x_{ij}=10$ for Observations 2 through 12, while $\Sigma_{j=1}^{4} x_{ij}=11$ for Observation 1. Regressors 5 and 6 were obtained from a table of normal random numbers. The responses y_i were generated by the relationship

$$y_i=10+2.0x_{i1}+1.0x_{i2}+0.2x_{i3}-2.0x_{i4}+3.0x_{i5}+10.0x_{i6}+\varepsilon_i$$

Table 8.3 Standardized acetylene data[a]

Observation i	y	x_1	x_2	x_3	x_1x_2	x_1x_3	x_2x_3	x_1^2	x_2^2	x_3^2
1	.27979	.28022	−.22544	−.23106	−.33766	−.02085	.30952	.07829	−.04116	−.03452
2	.30583	.28022	−.15704	−.23106	−.25371	−.02085	.23659	.07829	−.13270	−.03452
3	.31234	.28022	−.06584	−.23514	−.14179	−.02579	.14058	.07829	−.20378	−.02735
4	.26894	.28022	.04817	−.22290	.00189	−.01098	.01960	.07829	−.21070	−.04847
5	.24724	.28022	.20777	−.21882	.19398	−.00605	−.14065	.07829	−.06745	−.05526
6	.18214	−.04003	.48139	−.23106	.52974	−.02085	−.44415	.07829	.59324	−.03452
7	.17590	−.04003	−.32577	−.00255	−.00413	.25895	.07300	−.29746	.15239	−.23548
8	−.09995	−.04003	−.22544	−.01887	−.02171	.26177	.08884	−.29746	−.04116	−.23418
9	−.03486	−.04003	−.06584	−.06784	−.04970	.27023	.08985	−.29746	−.20378	−.21822
10	−.02401	−.04003	.04817	−.11680	−.06968	.27869	.04328	−.29746	−.21070	−.18419
11	.04109	−.04003	.20777	−.05152	−.09766	.26741	.01996	−.29746	−.06745	−.22554
12	.05194	−.04003	.48139	.00561	−.14563	.25754	.08202	−.29746	.59329	−.23538
13	.45800	−.36029	−.32577	.35653	.45252	−.29615	−.46678	.32879	.15239	.24374
14	.41460	−.36029	−.22544	.47078	.29423	−.47384	−.42042	.32879	−.04116	.60000
15	−.33865	−.36029	−.06584	.42187	.04240	−.39769	−.05859	.32879	−.20378	.43527
16	−.14335	−.36029	.20777	.37285	−.38930	−.32153	−.42738	.32879	−.06745	.28861

[a] The standardized data was constructed from the centered and scaled form of the original data in Table 8.1.

Table 8.4 Unstandardized regressor and response variables from Webster, Gunst, and Mason [1974]

Observation i	y_i	x_{i1}	x_{i2}	x_{i3}	x_{i4}	x_{i5}	x_{i6}
1	10.006	8.000	1.000	1.000	1.000	0.541	−0.099
2	9.737	8.000	1.000	1.000	0.000	0.130	0.070
3	15.087	8.000	1.000	1.000	0.000	2.116	0.115
4	8.422	0.000	0.000	9.000	1.000	−2.397	0.252
5	8.625	0.000	0.000	9.000	1.000	−0.046	0.017
6	16.289	0.000	0.000	9.000	1.000	0.365	1.504
7	5.958	2.000	7.000	0.000	1.000	1.996	−0.865
8	9.313	2.000	7.000	0.000	1.000	0.228	−0.055
9	12.960	2.000	7.000	0.000	1.000	1.380	0.502
10	5.541	0.000	0.000	0.000	10.000	−0.798	−0.399
11	8.756	0.000	0.000	0.000	10.000	0.257	0.101
12	10.937	0.000	0.000	0.000	10.000	0.440	0.432

where $\varepsilon_i \sim N(0,1)$. The $\mathbf{X'X}$ matrix in correlation form for these data is

$$\mathbf{X'X}=\begin{bmatrix} 1.000 & 0.052 & -0.343 & -0.498 & 0.417 & -0.192 \\ & 1.000 & -0.432 & -0.371 & 0.485 & -0.317 \\ & & 1.000 & -0.355 & -0.505 & 0.494 \\ & & & 1.000 & -0.215 & -0.087 \\ & & & & 1.000 & -0.123 \\ & & & & & 1.000 \\ \text{Symmetric} & & & & & \end{bmatrix}$$

None of the pairwise correlations r_{ij} are suspiciously large, and consequently we have no indication of the near linear dependency among the regressors. Generally inspection of the r_{ij} is not sufficient for detecting anything more complex than pairwise multicollinearity.

8.4.2 Variance inflation factors

The diagonal elements of the $\mathbf{C}=(\mathbf{X'X})^{-1}$ matrix are very useful in detecting multicollinearity. Recall from (8.5) that C_{jj}, the jth diagonal element of $\mathbf{C}$, can be written as $C_{jj}=(1-R_j^2)^{-1}$, where R_j^2 is the coefficient of determination obtained when x_j is regressed on the remaining $p-1$ regressors. If x_j is nearly orthogonal to the remaining regressors, R_j^2 is small and C_{jj} is close to unity, while if x_j is nearly linearly dependent on some subset of the remaining regressors, R_j^2 is near unity and C_{jj} is large. Since the variance of the jth regression coefficient is $C_{jj}\sigma^2$, we can view C_{jj} as the factor by which the variance of $\hat{\beta}_j$ is increased due to near linear dependencies among the regressors. Marquardt [1970] has called $C_{jj}=(1-R_j^2)^{-1}$ the "variance inflation factor" (VIF). The VIF for each term in the model measures the combined effect of the dependencies among the regressors on the variance of that term. One or more large VIFs indicate multicollinearity. Practical experience indicates that if any of the VIFs exceeds 5 or 10, it is an indication that the associated regression coefficients are poorly estimated because of multicollinearity.

The VIFs have another interesting interpretation. The length of the normal-theory confidence interval on the jth regression coefficient may be written as

$$L_j=2\left(C_{jj}\hat{\sigma}^2\right)^{1/2}t_{\alpha/2,\,n-p-1}$$

and the length of the corresponding interval based on an *orthogonal reference design* with the same sample size and root-mean-square (rms) values [i.e., rms$=\sum_{i=1}^{n}(x_{ij}-\bar{x}_j)^2/n$ is a measure of the spread of the regressor x_j] as the original design is

$$L^*=2\hat{\sigma}t_{\alpha/2,\,n-p-1}$$

The ratio of these two confidence intervals is $L_j/L^*=C_{jj}^{1/2}$. Thus the square root of the jth VIF indicates how much longer the confidence interval for the jth regression coefficient is because of multicollinearity.

The VIFs for the acetylene data are shown in Panel A of Table 8.5. These VIFs are the main diagonal elements of $(\mathbf{X}'\mathbf{X})^{-1}$, assuming that the linear terms in the model are centered and the second-order terms generated directly from the linear terms. The maximum VIF$=$6565.91, so we conclude that a multicollinearity problem exists. Furthermore the VIFs for several of the other cross-product and squared variables involving x_1 and x_3 are large. Thus the VIFs can help identify which regressors are involved in the multicollinearity. Note that the VIFs in polynomial models are affected by centering the linear terms. Panel B of Table 8.5 shows the VIFs for the acetylene data, assuming that the linear terms are *not* centered. These VIFs are much larger than those for the centered data. Thus centering the linear terms in a polynomial model removes some of the nonessential ill-conditioning caused by the choice of origin for the regressors.

The VIFs for the Webster, Gunst, and Mason data are shown in Panel C of Table 8.5. Since the maximum VIF$=$297.72, multicollinearity is clearly

Table 8.5 VIFs for acetylene data and Webster, Gunst, and Mason data

Acetylene Data		Webster, Gunst, and Mason Data
Centered Data	Uncentered Data	
A	B	C
Term VIF	Term VIF	Term VIF
$x_1=374$	$x_1=2856749$	$x_1=182.05$
$x_2=1.74$	$x_2=10956.1$	$x_2=161.36$
$x_3=679.11$	$x_3=2017163$	$x_3=266.26$
$x_1x_2=31.03$	$x_1x_2=2501945$	$x_4=297.72$
$x_1x_3=6565.91$	$x_1x_3=65.73$	$x_5=1.92$
$x_2x_3=35.60$	$x_2x_3=12667.1$	$x_6=1.46$
$x_1^2=1762.58$	$x_1^2=9802.9$	
$x_2^2=3.17$	$x_3^2=1428092$	
$x_3^2=1158.13$	$x_3^2=240.36$	
Maximum VIF=6565.91	Maximum VIF=2856749	Maximum VIF=297.72

indicated. Once again, note that the VIFs corresponding to the regressors involved in the multicollinearity are much larger than those for x_5 and x_6.

8.4.3 Eigensystem analysis of X′X

The characteristic roots or eigenvalues of **X′X**, say $\lambda_1, \lambda_2, \ldots, \lambda_p$, can be used to measure the extent of multicollinearity in the data.* If there are one or more near linear dependencies in the data, then one or more of the characteristic roots will be small. One or more small eigenvalues imply that there are near linear dependencies among the columns of **X**. Some analysts prefer to examine the condition number of **X′X**, defined as

$$\kappa = \frac{\lambda_{\max}}{\lambda_{\min}} \tag{8.10}$$

This is just a measure of the spread in the eigenvalues spectrum of **X′X**. Generally if the condition number is less than 100, there is no serious

*Recall that the eigenvalues of a $p \times p$ matrix **A** are the p roots of the equation $|\mathbf{A}-\lambda\mathbf{I}|=0$. Eigenvalues are almost always calculated by computer routines. Methods for computing eigenvalues and eigenvectors are discussed in Smith et al. [1974], Stewart [1973], and Wilkinson [1965].

problem with multicollinearity. Condition numbers between 100 and 1000 imply moderate to strong multicollinearity, and if κ exceeds 1000, severe multicollinearity is indicated.

The eigenvalues of $\mathbf{X'X}$ for the acetylene data are $\lambda_1=4.2048$, $\lambda_2=2.1626$, $\lambda_3=1.1384$, $\lambda_4=1.0413$, $\lambda_5=0.3845$, $\lambda_6=0.0495$, $\lambda_7=0.0136$, $\lambda_8=0.0051$, and $\lambda_9=0.0001$. There are four very small eigenvalues, a symptom of seriously ill-conditioned data. The condition number is

$$\kappa=\frac{\lambda_{\max}}{\lambda_{\min}}=\frac{4.2048}{0.0001}=42{,}048$$

which indicates severe multicollinearity. The eigenvalues for the Webster, Gunst, and Mason data are $\lambda_1=2.24879$, $\lambda_2=1.54615$, $\lambda_3=0.92208$, $\lambda_4=0.79399$, $\lambda_5=0.30789$, and $\lambda_6=0.00111$. The small eigenvalue indicates the near linear dependency in the data. The condition number is

$$\kappa=\frac{\lambda_{\max}}{\lambda_{\min}}=\frac{2.24879}{0.00111}=2025.94$$

which also indicates strong multicollinearity.

Eigensystem analysis can also be used to identify the nature of the near linear dependencies in the data. The $\mathbf{X'X}$ matrix may be decomposed as

$$\mathbf{X'X}=\mathbf{T\Lambda T'}$$

where $\mathbf{\Lambda}$ is a $p\times p$ diagonal matrix whose main diagonal elements are the eigenvalues $\lambda_j (j=1,2,\ldots,p)$ of $\mathbf{X'X}$ and $\mathbf{T}$ is a $p\times p$ orthogonal matrix whose columns are the eigenvectors of $\mathbf{X'X}$. Let the columns of $\mathbf{T}$ be denoted by $\mathbf{t}_1,\mathbf{t}_2,\ldots,\mathbf{t}_p$. If the eigenvalue λ_j is close to zero, indicating a near linear dependency in the data, the elements of the associated eigenvector $\mathbf{t}_j$ describe the nature of this linear dependency. Specifically, the elements of the vector $\mathbf{t}_j$ are the coefficients $t_1, t_2,\ldots, t_p$ in (8.2).

Table 8.6 displays the eigenvectors for the Webster, Gunst, and Mason data. The smallest eigenvalue is $\lambda_6=0.00111$, so the elements of the eigenvector $\mathbf{t}_6$ are the coefficients of the regressors in Equation (8.2). This implies that

$$-0.44768x_1-0.42114x_2-0.54169x_3-0.57337x_4$$
$$-0.00605x_5-0.00217x_6=0$$

Assuming that -0.00605 and -0.00217 are approximately zero and re-

Table 8.6 Eigenvectors for the Webster, Gunst and Mason data

$\mathbf{t}_1$	$\mathbf{t}_2$	$\mathbf{t}_3$	$\mathbf{t}_4$	$\mathbf{t}_5$	$\mathbf{t}_6$
−.39072	−.33968	.67980	.07990	−.25104	−.44768
−.45560	−.05392	−.70013	.05769	−.34447	−.42114
.48264	−.45333	−.16078	.19103	.45364	−.54169
.18766	.73547	.13587	−.27645	.01521	−.57337
−.49773	−.09714	−.03185	−.56356	.65128	−.00605
.35195	−.35476	−.04864	−.74818	−.43375	−.00217

arranging terms gives

$$x_1 \simeq -.941x_2 - 1.210x_3 - 1.281x_4$$

That is, the first four regressors add approximately to a constant. Thus the elements of $\mathbf{t}_6$ directly reflect the relationship used to generate x_1, x_2, x_3, and x_4.

Belsley, Kuh and Welsch [1980] propose a similar approach for diagnosing multicollinearity. The $n \times p$ $\mathbf{X}$ matrix may be decomposed as

$$\mathbf{X} = \mathbf{UDT}'$$

where $\mathbf{U}$ is $n \times p$, $\mathbf{T}$ is $p \times p$, $\mathbf{U}'\mathbf{U} = \mathbf{I}$, $\mathbf{T}'\mathbf{T} = \mathbf{I}$, and $\mathbf{D}$ is a $p \times p$ diagonal matrix with nonnegative diagonal elements μ_j, $j = 1, 2, \ldots, p$. The μ_j are called the *singular values* of $\mathbf{X}$ and $\mathbf{X} = \mathbf{UDT}'$ is called the *singular-value decomposition* of $\mathbf{X}$. The singular-value decomposition is closely related to the concepts of eigenvalues and eigenvectors, since $\mathbf{X}'\mathbf{X} = (\mathbf{UDT}')'\mathbf{UDT}' = \mathbf{TD}^2\mathbf{T}' = \mathbf{T\Lambda T}'$, so that the squares of the singular values of $\mathbf{X}$ are the eigenvalues of $\mathbf{X}'\mathbf{X}$. $\mathbf{T}$ is the matrix of eigenvectors of $\mathbf{X}'\mathbf{X}$ defined earlier, and $\mathbf{U}$ is a matrix whose columns are the eigenvectors associated with the p nonzero eigenvalues of $\mathbf{XX}'$.

Ill-conditioning in $\mathbf{X}$ is reflected in the size of the singular values. There will be one small singular value for each near-linear dependency. The extent of ill-conditioning depends on how small the singular value is relative to the maximum singular value $\mu_{\max}$. Belsley, Kuh, and Welsch [1980] define the *condition indices* of the $\mathbf{X}$ matrix as

$$\eta_j = \frac{\mu_{\max}}{\mu_j} \qquad j = 1, 2, \ldots, p$$

The largest value for η_j is the condition number of $\mathbf{X}$. Note that this approach deals directly with the data matrix* **X**, with which we are principally concerned, not the matrix of sums of squares and cross-products **X′X**. A further advantage of this approach is that algorithms for generating the singular value decomposition are more stable numerically than those for eigensystem analysis, although in practice this is not likely to be a severe handicap if one prefers the eigensystem approach.

The covariance matrix of $\hat{\boldsymbol{\beta}}$ is

$$V(\hat{\boldsymbol{\beta}}) = \sigma^2(\mathbf{X}'\mathbf{X})^{-1} = \sigma^2\mathbf{T}\boldsymbol{\Lambda}^{-1}\mathbf{T}'$$

and the variance of the jth regression coefficient is the jth diagonal element of this matrix, or

$$V(\hat{\beta}_j) = \sigma^2 \sum_{i=1}^{p} \frac{t_{ji}^2}{\mu_i^2} = \sigma^2 \sum_{i=1}^{p} \frac{t_{ji}^2}{\lambda_i}$$

Note also that apart from σ^2, the jth diagonal element of $\mathbf{T}\boldsymbol{\Lambda}^{-1}\mathbf{T}'$ is the jth variance inflation factor, so

$$\text{VIF}_j = \sum_{i=1}^{p} \frac{t_{ji}^2}{\mu_i^2} = \sum_{i=1}^{p} \frac{t_{ji}^2}{\lambda_i}$$

Clearly, one or more small singular values (or small eigenvalues) can dramatically inflate the variance of $\hat{\beta}_j$. Belsley, Kuh, and Welsch suggest using variance-decomposition proportions, for example

$$\pi_{ij} = \frac{\left(t_{ji}^2/\mu_i^2\right)}{\text{VIF}_j}, \qquad i, j = 1, 2, \ldots, p$$

as measures of multicollinearity. If we array the π_{ij} in a $p \times p$ matrix $\boldsymbol{\pi}$, then the elements of each column of $\boldsymbol{\pi}$ are just the proportions of the variance of each $\hat{\beta}_j$ (or each variance inflation factor) contributed by the ith singular value (or eigenvalue). If a high proportion of the variance for two or more regression coefficients is associated with one small singular value, multicollinearity is indicated. For example if π_{32} and π_{34} are large, the third singular value is associated with a multicollinearity that is inflating the variances of $\hat{\beta}_2$ and $\hat{\beta}_4$. Condition indices greater than 30 and variance-decomposition proportions greater than 0.5 are recommended guidelines.

*Belsley, Kuh, and Welsch [1980] also suggest that the columns of **X** should be scaled to unit length, but not centered, so that the role of the intercept in near-linear dependencies can be diagnosed.

8.4.4 Other diagnostics

There are several other techniques that are occasionally useful in diagnosing multicollinearity. The determinant of $\mathbf{X'X}$ can be used as an index of multicollinearity. Since the $\mathbf{X'X}$ matrix is in correlation form, the possible range of values of the determinant is $0 \leqslant |\mathbf{X'X}| \leqslant 1$. If $|\mathbf{X'X}| = 1$, the regressors are orthogonal, while if $|\mathbf{X'X}| = 0$, there is an exact linear dependency among the regressors. The degree of multicollinearity becomes more severe as $|\mathbf{X'X}|$ approaches zero. While this measure of multicollinearity is easy to apply, it does not provide any information on the source of the multicollinearity.

Willan and Watts [1978] suggest another interpretation of this diagnostic. The joint $100(1-\alpha)$ percent confidence region for $\boldsymbol{\beta}$ based on the observed data is

$$(\boldsymbol{\beta}-\hat{\boldsymbol{\beta}})'\mathbf{X'X}(\boldsymbol{\beta}-\hat{\boldsymbol{\beta}}) \leqslant p\hat{\sigma}^2 F_{\alpha, p, n-p-1}$$

while the corresponding confidence region for $\hat{\boldsymbol{\beta}}$ based on the orthogonal reference design described earlier is

$$(\boldsymbol{\beta}-\hat{\boldsymbol{\beta}})'(\boldsymbol{\beta}-\hat{\boldsymbol{\beta}}) \leqslant p\hat{\sigma}^2 F_{\alpha, p, n-p-1}$$

The orthogonal reference design produces the smallest joint confidence region for fixed sample size and rms values and a given α. The ratio of the volumes of the two confidence regions is $|\mathbf{X'X}|^{1/2}$, so that $|\mathbf{X'X}|^{1/2}$ measures the loss of estimation power due to multicollinearity. Put another way, $100(|\mathbf{X'X}|^{-1/2}-1)$ reflects the percentage increase in the volume of the joint confidence region for $\boldsymbol{\beta}$ because of the near linear dependencies in $\mathbf{X}$. For example, if $|\mathbf{X'X}| = 0.25$, then the volume of the joint confidence region is $100[(0.25)^{-1/2}-1] = 100$ percent larger than it would be if an orthogonal design had been used.

The F-statistic for significance of regression and the individual t (or partial F)-statistics can sometimes indicate the presence of multicollinearity. Specifically, if the overall F-statistic is significant but the individual t-statistics are all nonsignificant, multicollinearity is present. Unfortunately, many data sets that have significant multicollinearity will not exhibit this behavior, and so the usefulness of this measure of multicollinearity is questionable.

The signs and magnitudes of the regression coefficients will sometimes provide an indication that multicollinearity is present. In particular, if adding or removing a regressor produces large changes in the estimates of the regression coefficients, multicollinearity is indicated. If the deletion of one or more data points results in large changes in the regression coefficients, there may be multicollinearity present. Finally, if the signs or magnitudes of the regression coefficients in the regression model are contrary to prior expectation, we should be alert to possible multicollinearity. For example, the least squares model for the acetylene data has large standardized regression coefficients for the x_1x_3 interaction and for the squared terms x_1^2 and x_3^2. It is somewhat unusual for quadratic models to display large regression coefficients for the higher-order terms, and so this may be an indication of multicollinearity. However, one should be cautious in using the signs and magnitudes of the regression coefficients as indications of multicollinearity, as many seriously ill-conditioned data sets do not exhibit behavior that is obviously unusual in this respect.

We believe that the VIFs and the procedures based on the eigenvalues of $\mathbf{X'X}$ are the best currently available multicollinearity diagnostics. They are easy to compute, straightforward to interpret, and useful in investigating the specific nature of the multicollinearity. For additional information on these and other methods of detecting multicollinearity, see Belsley, Kuh, and Welsch [1980], Farrar and Glauber [1967], and Willan and Watts [1978].

8.5 Methods for dealing with multicollinearity

Several techniques have been proposed for dealing with the problems caused by multicollinearity. The general approaches include collecting additional data, model respecification, and the use of estimation methods other than least squares that are specifically designed to combat the problems induced by multicollinearity.

8.5.1 Collecting additional data

Collecting additional data has been suggested as the best method of combating multicollinearity (e.g., see Farrar and Glauber [1967] and Silvey [1969]). The additional data should be collected in a manner designed to break up the multicollinearity in the existing data. For example, consider the delivery time data first introduced in Example 4.1. A plot of the regressors cases (x_1) versus distance (x_2) is shown in Figure 4.6. We have remarked previously that most of this data lies along a line from low values of cases and distance to high values of cases and distance, and consequently there may be some problem with multicollinearity. This could be avoided by collecting some additional data at points designed to break up any potential multicollinearity; that is, at points where cases is small and distance is large and points where cases is large and distance is small.

Unfortunately, collecting additional data is not always possible because of economic constraints or because the process being studied is no longer available for sampling. Even when additional data is available, it may be inappropriate to use it if the new data extends the range of the regressor variables far beyond the analyst's region of interest. Furthermore if the new data points are unusual or atypical of the process being studied, their presence in the sample could be highly influential on the fitted model. Finally, note that collecting additional data is not a viable solution to the multicollinearity problem when the multicollinearity is due to constraints on the model or in the population. For example, consider the factors family income (x_1) and house size (x_2) plotted in Figure 8.1. Collection of additional data would be of little value here, since the relationship between family income and house size is a structural characteristic of the population. Virtually all the data in the population will exhibit this behavior.

8.5.2 Model respecification

Multicollinearity is often caused by the choice of model, such as when two highly correlated regressors are used in the regression equation. In these situations some respecification of the regression equation may lessen the impact of multicollinearity. One approach to model respecification is to redefine the regressors. For example, if x_1, x_2 and x_3 are nearly linearly dependent it may be possible to find some function such as $x=(x_1+x_2)/x_3$ or $x=x_1x_2x_3$ that preserves the information content in the original regressors but reduces the ill-conditioning.

Another widely used approach to model respecification is variable elimination. That is, if x_1, x_2, and x_3 are nearly linearly dependent, eliminating one regressor (say x_3) may be helpful in combating multicollinearity. Variable elimination is often a highly effective technique. However, it may not provide a satisfactory solution if the regressors dropped from the model have significant explanatory power relative to the response y. That is, eliminating regressors to reduce multicollinearity may damage the predictive power of the model. Care must be exercised in variable selection because many of the selection procedures are seriously distorted by multicollinearity, and there is no assurance that the final model will exhibit any lesser degree of multicollinearity than was present in the original data.

The strong multicollinearity exhibited by the acetylene data in Example 8.1 is likely caused by the choice of model, since x_1 (temperature) and x_3

Table 8.7 Subset regression models for the acetylene data

	Model A		Model B		Model C		Model D		Model E	
Regressor	$\hat{\beta}$	VIF	$\hat{\beta}$	VIF	$\hat{\beta}$	VIF	$\hat{\beta}$	VIF	$\hat{\beta}$	VIF
x_1	0.8595	12.23	0.8930	1.14					0.8355	3.09
x_2	0.1657	1.06	0.0265	1.45	0.1369	1.48	0.2365	1.72	0.2280	1.46
x_3	−0.0506	12.33			0.7225	2.64	−1.1281	5.74		
x_1x_2			0.0451	1.37			−0.4805	31.03	−0.4788	30.37
x_1x_3							−3.3111	1221.64		
x_2x_3					0.0555	1.52	−0.2728	35.43	−0.2529	33.35
x_1^2			0.0005	1.20			−1.4902	353.20	0.2161	12.39
x_2^2			0.0294	1.72	0.0238	1.86	−0.0919	3.16	−0.0844	1.84
x_3^2					0.0420	2.98	−1.5068	320.71	−0.1879	19.70
	$R^2 = 0.9198$		$R^2 = 0.9946$		$R^2 = 0.9807$		$R^2 = 0.9974$		$R^2 = 0.9970$	
	$MS_E = 0.00668$		$MS_E = 0.00054$		$MS_E = 0.00193$		$MS_E = 0.00037$		$MS_E = 0.00037$	
	$C_p = 201.5626$		$C_p = 9.9948$		$C_p = 46.3566$		$C_p = 8.7971$		$C_p = 7.7876$	

(contact time) are highly correlated and the original equation is a full quadratic in all three regressors. Variable selection may be effective in eliminating this ill-conditioning.

Table 8.7 presents summary information concerning five different subset regression models for the acetylene data. Model A is an all linear term model that attempts to reduce multicollinearity by deleting the second-order terms. Model B is a full quadratic in x_1 and x_2 only. This is a logical candidate subset model resulting from deleting x_3 (and all its second-order terms) because of its high correlation with x_1. Application of stepwise regression and forward selection with a selection level for variable entry of $\alpha=.25$ also leads to Model B. Model C, a full quadratic in x_2 and x_3 is an obvious companion to Model B. Model D, which contains eight terms (only x_1 is dropped), is produced by backward elimination with the selection level for variable removal $\alpha=.10$. Model E is the minimum C_p equation from all possible regressions.

The three-term linear Model A is not very satisfactory. It still has moderately strong ill-conditioning, and the R^2 and residual mean square statistics do not compare favorably with the full model fit. The C_p statistic indicates that substantial bias remains in this model. Furthermore the residual plots exhibit a pattern indicating that the equation systematically over-predicts low percent conversions and under-predicts high percent conversions. Scatter plots of the data also indicate that at least some of the second-order terms are necessary. We conclude that Model A is inadequate.

Models B and C, which are complete quadratic equations found by deleting either x_1 or x_3, appear to be more reasonable subset models, although the C_p statistic for Model C is quite large. Neither equation exhibits significant multicollinearity. The only potential difficulty with Model B (which involves only x_1 and x_2) is that the analyst could not use it to predict changes in percent conversion as a function of changes in contact time (x_3). Similarly, Model C is noninformative concerning the effect of temperature (x_1) on the process. However, perhaps *both* of these equations should be reported to the user. The joint use of both equations may be more informative than a single equation. Notice however, that even with two equations it is not obvious how the user would predict the effect of simultaneous changes in both x_1 and x_3.

Models D and E appear less satisfactory than B and C. Both equations have large variance inflation factors, implying that variable elimination has not been as effective in combating multicollinearity as it was in Models B and C.

We may also investigate the predictive performance of these subset models. The predicted percent conversion at the 10 points in Figure 8.3 for all five equations is shown below:

Point	Predicted Percent Conversion				
	Model A	Model B	Model C	Model D	Model E
Interpolation					
A	19.01	16.98	16.66	16.18	16.51
B	18.52	13.48	13.61	14.91	14.84
E	32.04	29.54	29.35	28.48	28.87
F	35.16	35.41	38.94	35.29	35.56
I	49.30	47.99	46.39	46.89	47.96
J	47.25	49.40	49.04	50.31	49.50
Extrapolation					
C	21.87	22.08	48.65	−68.86	23.35
D	37.10	38.21	44.38	−44.71	19.36
G	33.34	31.43	49.76	19.71	26.57
H	45.61	49.40	22.47	−157.56	46.57

As noted before, the three–term linear Model A tends to over-predict low percent conversions and under-predict high percent conversions. The other four subset models appear to perform satisfactorily for interpolation. However, Model D gives negative predictions of percent conversion at 3 of the 4 extrapolation points, a clearly unreasonable result. Models B, C, and E at least give feasible predictions at the extrapolation points, although there is not spectacular consistency across the three equations.

It is difficult to recommend a "final" equation at this point. Possibly Models B and C could be jointly used, or perhaps Model E will be satisfactory. Further analysis is necessary before any final recommendation. It *is* clear, however, that model respecification, if carefully performed, can be helpful in reducing the effects of multicollinearity.

8.5.3 Ridge regression

When the method of least squares is applied to nonorthogonal data, very poor estimates of the regression coefficients are usually obtained. We saw in Section 8.3 that the variance of the least squares estimates of the regression coefficients may be considerably inflated, and the length of the vector of least squares parameter estimates is too long on the average. This implies that the absolute value of the least squares estimates are too large, and that they are very unstable; that is, their magnitudes and signs may change considerably given a different sample.

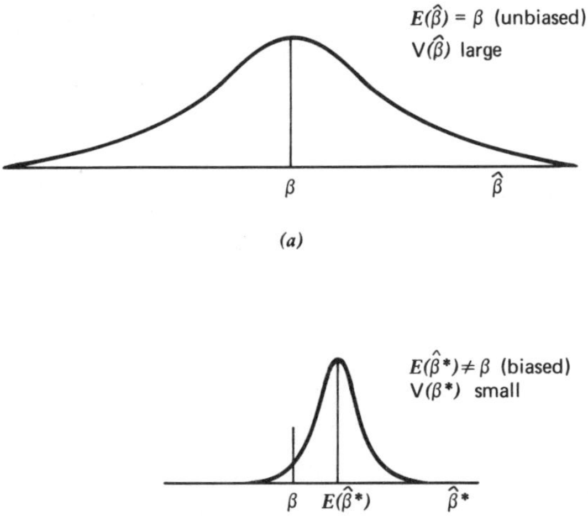

(a)

(b)

Figure 8.4 Sampling distributions of (**a**) unbiased and (**b**) biased estimators of β. (Adapted from Marquardt and Snee [1975], with permission of the publisher.)

The problem with the method of least squares is the requirement that $\hat{\boldsymbol{\beta}}$ be an unbiased estimator of $\boldsymbol{\beta}$. The Gauss–Markoff property referred to in Section 4.2.3 assures us that the least squares estimator has minimum variance in the class of unbiased linear estimators, but there is no guarantee that this variance will be small. The situation is illustrated in Figure 8.4a, where the sampling distribution of $\hat{\beta}$, the unbiased estimator of β, is shown. The variance of $\hat{\beta}$ is large, implying that confidence intervals on β would be wide and the point estimate $\hat{\beta}$ is very unstable.

One way to alleviate this problem is to drop the requirement that the estimator of $\boldsymbol{\beta}$ be unbiased. Suppose that we can find a biased estimator of $\boldsymbol{\beta}$, say $\hat{\boldsymbol{\beta}}^*$, that has a smaller variance than the unbiased estimator $\hat{\boldsymbol{\beta}}$. The mean square error of the estimator $\hat{\boldsymbol{\beta}}^*$ is defined as

$$\text{MSE}(\hat{\boldsymbol{\beta}}^*) = E(\hat{\boldsymbol{\beta}}^* - \boldsymbol{\beta})^2 = V(\hat{\boldsymbol{\beta}}^*) + \left[E(\hat{\boldsymbol{\beta}}^*) - \boldsymbol{\beta}\right]^2 \tag{8.11}$$

or

$$\text{MSE}(\hat{\boldsymbol{\beta}}^*) = \text{Variance}\,(\hat{\boldsymbol{\beta}}^*) + (\text{Bias in } \hat{\boldsymbol{\beta}}^*)^2$$

Note that the MSE is just the expected squared distance from $\hat{\boldsymbol{\beta}}^*$ to $\boldsymbol{\beta}$ [see (8.7)]. By allowing a small amount of bias in $\hat{\boldsymbol{\beta}}^*$ the variance of $\hat{\boldsymbol{\beta}}^*$ can be made small such that the MSE of $\hat{\boldsymbol{\beta}}^*$ is less than the variance of the unbiased estimator $\hat{\boldsymbol{\beta}}$. Figure 8.4b illustrates a situation where the variance of the biased estimator is considerably smaller than the variance of the unbiased estimator (Figure 8.4a). Consequently, confidence intervals on β would be much narrower using the biased estimator. The small variance for the biased estimator also implies that $\hat{\beta}^*$ is a more stable estimator of β than is the unbiased estimator $\hat{\beta}$.

A number of procedures have been developed for obtaining biased estimators of regression coefficients. One of these procedures is ridge regression, originally proposed by Hoerl and Kennard [1970a, b]. The ridge estimator is found by solving a slightly modified version of the normal equations. Specifically, we define the ridge estimator $\hat{\boldsymbol{\beta}}_R$ as the solution to

$$(\mathbf{X}'\mathbf{X}+k\mathbf{I})\hat{\boldsymbol{\beta}}_R=\mathbf{X}'\mathbf{y} \tag{8.12}$$

or

$$\hat{\boldsymbol{\beta}}_R=(\mathbf{X}'\mathbf{X}+k\mathbf{I})^{-1}\mathbf{X}'\mathbf{y} \tag{8.13}$$

where $k \geq 0$ is a constant selected by the analyst. The procedure is called ridge regression because the underlying mathematics are similar to the method of ridge analysis used earlier by Hoerl [1959] for describing the behavior of second-order response surfaces. Note that when $k=0$ the ridge estimator is the least squares estimator.

The ridge estimator is a linear transformation of the least squares estimator since

$$\begin{aligned}\hat{\boldsymbol{\beta}}_R&=(\mathbf{X}'\mathbf{X}+k\mathbf{I})^{-1}\mathbf{X}'\mathbf{y}\\&=(\mathbf{X}'\mathbf{X}+k\mathbf{I})^{-1}(\mathbf{X}'\mathbf{X})\hat{\boldsymbol{\beta}}\\&=\mathbf{Z}_k\hat{\boldsymbol{\beta}}\end{aligned}$$

Therefore since $E(\hat{\boldsymbol{\beta}}_R)=E(\mathbf{Z}_k\hat{\boldsymbol{\beta}})=\mathbf{Z}_k\boldsymbol{\beta}$, $\hat{\boldsymbol{\beta}}_R$ is a biased estimator of $\boldsymbol{\beta}$. We usually refer to the constant k as the *biasing parameter*. The covariance matrix of $\hat{\boldsymbol{\beta}}_R$ is

$$V(\hat{\boldsymbol{\beta}}_R)=\sigma^2(\mathbf{X}'\mathbf{X}+k\mathbf{I})^{-1}\mathbf{X}'\mathbf{X}(\mathbf{X}'\mathbf{X}+k\mathbf{I})^{-1} \tag{8.14}$$

The mean square error of the ridge estimator is

$$\begin{aligned}\text{MSE}(\hat{\boldsymbol{\beta}}_R)&=\text{Variance}(\hat{\boldsymbol{\beta}}_R)+(\text{Bias in }\hat{\boldsymbol{\beta}}_R)^2\\&=\sigma^2 Tr\left[(\mathbf{X}'\mathbf{X}+k\mathbf{I})^{-1}\mathbf{X}'\mathbf{X}(\mathbf{X}'\mathbf{X}+k\mathbf{I})^{-1}\right]\\&\quad+k^2\boldsymbol{\beta}'(\mathbf{X}'\mathbf{X}+k\mathbf{I})^{-2}\boldsymbol{\beta}\\&=\sigma^2\sum_{j=1}^{p}\frac{\lambda_j}{(\lambda_j+k)^2}+k^2\boldsymbol{\beta}'(\mathbf{X}'\mathbf{X}+k\mathbf{I})^{-2}\boldsymbol{\beta}\end{aligned} \tag{8.15}$$

where $\lambda_1, \lambda_2, \ldots, \lambda_p$ are the eigenvalues of $\mathbf{X}'\mathbf{X}$. The first term on the right-hand side of (8.15) is the sum of variances of the parameters in $\hat{\boldsymbol{\beta}}_R$ and the second term is the square of the bias. If $k>0$, note that the bias in $\hat{\boldsymbol{\beta}}_R$ increases with k. However, the variance decreases as k increases.

In using ridge regression we would like to choose a value of k such that the reduction in the variance term is greater than the increase in the squared bias. If this can be done, the mean square error of the ridge estimator $\hat{\boldsymbol{\beta}}_R$ will be less than the variance of the least squares estimator $\hat{\boldsymbol{\beta}}$. Hoerl and Kennard proved that there exists a nonzero value of k for which the MSE of $\hat{\boldsymbol{\beta}}_R$ is less than the variance of the least squares estimator $\hat{\boldsymbol{\beta}}$, provided that $\boldsymbol{\beta}'\boldsymbol{\beta}$ is bounded. The residual sum of squares is

$$
\begin{aligned}
SS_E &= (\mathbf{y}-\mathbf{X}\hat{\boldsymbol{\beta}}_R)'(\mathbf{y}-\mathbf{X}\hat{\boldsymbol{\beta}}_R) \\
&= (\mathbf{y}-\mathbf{X}\hat{\boldsymbol{\beta}})'(\mathbf{y}-\mathbf{X}\hat{\boldsymbol{\beta}}) + (\hat{\boldsymbol{\beta}}_R-\hat{\boldsymbol{\beta}})'\mathbf{X}'\mathbf{X}(\hat{\boldsymbol{\beta}}_R-\hat{\boldsymbol{\beta}})
\end{aligned} \tag{8.16}
$$

Since the first term on the right-hand side of (8.16) is the residual sum of squares for the least squares estimates $\hat{\boldsymbol{\beta}}$, we see that as k increases the residual sum of squares increases. Consequently, because the total sum of squares is fixed, R^2 decreases as k increases. Therefore the ridge estimate will not necessarily provide the best "fit" to the data, but this should not overly concern us, since we are more interested in obtaining a stable set of parameter estimates. The ridge estimates may result in an equation that does a better job of predicting future observations than would least squares (although there is no conclusive *proof* that this will happen).

Hoerl and Kennard have suggested that an appropriate value of k may be determined by inspection of the *ridge trace*. The ridge trace is a plot of the elements of $\hat{\boldsymbol{\beta}}_R$ versus k, for values of k usually in the interval zero to one. Marquardt and Snee [1975] suggest using up to about 25 values of k, spaced approximately logarithmically over the interval [0, 1]. If multicollinearity is severe, the instability in the regression coefficients will be obvious from the ridge trace. As k is increased, some of the ridge estimates will vary dramatically. At some value of k, the ridge estimates $\hat{\boldsymbol{\beta}}_R$ will stabilize. The objective is to select a reasonably small value of k at which the ridge estimates $\hat{\boldsymbol{\beta}}_R$ are stable. Hopefully this will produce a set of estimates with smaller MSE than the least squares estimates.

- **Example 8.2** To obtain the ridge solution for the acetylene data, we must solve the equations $(\mathbf{X}'\mathbf{X}+k\mathbf{I})\hat{\boldsymbol{\beta}}_R=\mathbf{X}'\mathbf{y}$ for several values $0\leqslant k\leqslant 1$, with $\mathbf{X}'\mathbf{X}$ and $\mathbf{X}'\mathbf{y}$ in correlation form. The ridge trace is shown in Figure 8.5, and the ridge coefficients for several values of k are listed in Table 8.8. This table also shows the residual mean square and R^2 for each ridge model. Notice that as k increases, MS_E increases and R^2 decreases. The ridge trace illustrates the

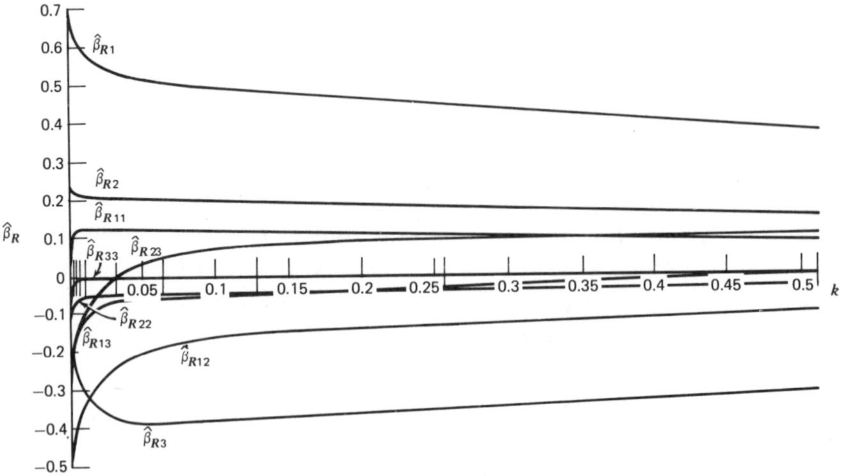

Figure 8.5 Ridge trace for acetylene data using nine regressors.

instability of the least squares solution, as there are large changes in the regression coefficients for small values of k. However, the coefficients stabilize rapidly as k increases.

Judgment is required to interpret the ridge trace and select an appropriate value of k. We want to choose k large enough to provide stable coefficients, but not unnecessarily large as this introduces additional bias and increases the residual mean square. From Figure 8.5, we see that reasonable coefficient stability is achieved in the region $0.008<k<0.064$, without a severe increase in the residual mean square (or loss in R^2). If we choose $k=0.032$, the ridge regression model is

$$\hat{y}=0.5392x_1+0.2117x_2-0.3735x_3-0.2329x_1x_2-0.0675x_1x_3$$
$$+0.0123x_2x_3+0.1249x_1^2-0.0481x_2^2-0.0267x_3^2$$

Note that in this model the estimates of β_{13}, β_{11}, and β_{23} are considerably smaller than the least squares estimates and the original negative estimates of β_{23} and β_{11} are now positive. The ridge model expressed in terms of the original regressors is

$$\hat{P}=0.7598+0.1392T+0.0547H-0.0965C-0.0680TH-0.0194TC$$
$$+0.0039CH+0.0407T^2-0.0112H^2-0.0067C^2$$

Figure 8.6 shows the performance of the ridge model in prediction for both interpolation (Points A, B, E, F, I, and J) and extrapolation (Points C, D, G, and H). Comparing Figures 8.6 and 8.3, we note that the ridge model predicts

Table 8.8 Coefficients at various values of *k*

k	.000	.001	.002	.004	.008	.016	.032	.064	.128	.256	.512
$\beta_{R,1}$	.3377	.6770	.6653	.6362	.6003	.5672	.5392	.5122	.4806	.4379	.3784
$\beta_{R,2}$	.2337	.2242	.2222	.2199	.2173	.2148	.2117	.2066	.1971	.1807	.1554
$\beta_{R,3}$	−.6749	−.2129	−.2284	−.2671	−.3134	−.3515	−.3735	−.3800	−.3724	−.3500	−.3108
$\beta_{R,12}$	−.4799	−.4479	−.4258	−.3913	−.3437	−.2879	−.2329	−.1862	−.1508	−.1249	−.1044
$\beta_{R,13}$	−2.0344	−.2774	−.1887	−.1350	−.1017	−.0809	−.0675	−.0570	−.0454	−.0299	−.0092
$\beta_{R,23}$	−.2675	−.2173	−.1920	−.1535	−.1019	−.0433	.0123	.0562	.0849	.0985	.0991
$\beta_{R,11}$	−.8346	.0643	.1035	.1214	.1262	.1254	.1249	.1258	.1230	.1097	.0827
$\beta_{R,22}$	−.0904	−.0732	−.0682	−.0621	−.0558	−.0509	−.0481	−.0464	−.0444	−.0406	−.0341
$\beta_{R,33}$	−1.0015	−.2451	−.1853	−.1313	−.0825	−.0455	−.0267	−.0251	−.0339	−.0464	−.0586
MS_E	.00038	.00047	.00049	.00054	.00062	.00074	.00094	.00127	.00206	.00425	.01002
R^2	.998	.997	.997	.997	.996	.996	.994	.992	.988	.975	.940

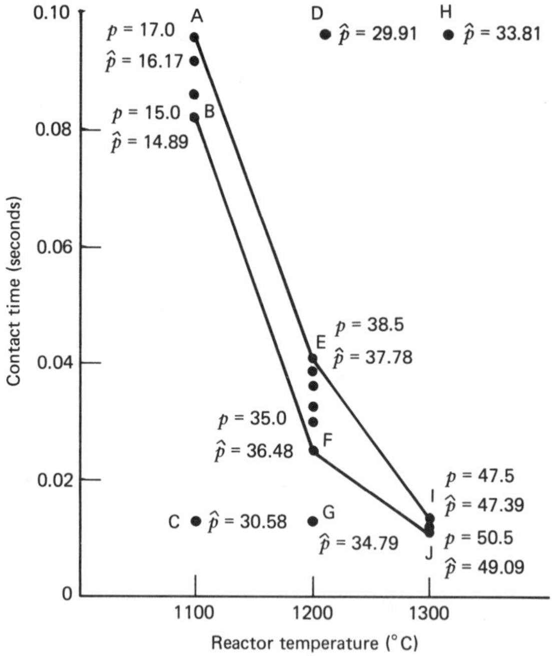

Figure 8.6 Performance of the ridge model with $k=0.032$ in prediction and extrapolation for the acetylene data. (Adapted from Marquardt and Snee [1975], with permission of the publisher.)

as well as the nine–term least squares model at the boundary of the region covered by the data. However, the ridge model gives much more realistic predictions when extrapolating than does least squares. Furthermore the extrapolation performance of the ridge model is roughly comparable to the least squares subset Models B $(x_1, x_2, x_1x_2, x_1^2, x_2^2)$, C $(x_2, x_3, x_2x_3, x_2^2, x_3^2)$, and E $(x_1, x_2, x_1x_2, x_2x_3, x_1^2, x_3^2, x_3^2)$ developed in Section 8.5.2. Thus we conclude that the ridge regression approach in this example has produced a model that is superior to the original nine–term least squares fit, and comparable to least squares models developed by variable selection. •

The ridge regression estimates may be computed by using an ordinary least squares computer program and augmenting the standardized data as follows:

$$\mathbf{X}_A = \begin{bmatrix} \mathbf{X} \\ \sqrt{k}\,\mathbf{I}_p \end{bmatrix} \qquad \mathbf{y}_A = \begin{bmatrix} \mathbf{y} \\ \mathbf{0}_p \end{bmatrix}$$

where $\sqrt{k}\,\mathbf{I}_p$ is a $p \times p$ diagonal matrix with diagonal elements equal to the

Table 8.9 Augmented matrix $\mathbf{X}_A$ and vector $\mathbf{y}_A$ for generating the ridge solution for the acetylene data with $k = 0.032$

$\mathbf{X}_A =$ (first nine columns), $\mathbf{y}_A =$ (last column)

.280224	−.22544	−.23106	−.33766	−.02085	.309525	.078278	−.04116	−.03452	0.27979
.280224	−.15704	−.23106	−.25371	−.02085	.236588	.078278	−0.1327	−.03452	.305829
.280224	−.06584	−.23514	−.14179	−.02579	.140577	.078278	−.20378	−.02735	.312339
.280224	.048167	−0.2229	−.00189	−.01098	0.0196	.078278	−0.2107	−.04847	0.26894
.280224	.207774	−.21882	.193976	−.00605	−.14065	.078278	−.06745	−.05526	0.24724
.280244	.481385	−.23106	.529744	−.02085	−.44415	.078278	.593235	−.03452	.182141
−.04003	−.32577	−.00255	−.00413	.258949	.073001	−.29746	.152387	−.23548	−0.1759
−.04003	−.22544	−.01887	−.02171	.261769	.088842	−.29746	−.04116	−.23418	−.09995
−.04003	−.06584	−.06784	−0.0497	.270231	.089856	−.29746	−.20378	−.21822	−.03486
−.04003	.048167	−0.1168	−.06968	.278693	.043276	−.29746	−0.2107	−.18419	−.02401
−.04003	.207774	−.05152	−.09766	.267411	.019961	−.29746	−.06745	−.22554	.041094
−.04003	.481385	.005609	−.14563	.257539	.082021	−.29746	.593235	−.23538	.051944
−.36029	−.32577	.356528	.452517	−.29615	−.46678	.328768	.152387	.243742	−0.458
−.36029	−.22544	.470781	.294227	−.47384	−.42042	.328768	−.04116	.599999	−0.4146
−.36029	−.06584	.421815	.042401	−.39769	−.05859	.328768	−.20378	.435271	−.33865
−.36029	.207774	0.37285	−0.3893	−.32153	.427375	.328768	−.06745	.288613	−.14335
0.17888	0	0	0	0	0	0	0	0	0
0	0.17888	0	0	0	0	0	0	0	0
0	0	0.17888	0	0	0	0	0	0	0
0	0	0	0.17888	0	0	0	0	0	0
0	0	0	0	0.17888	0	0	0	0	0
0	0	0	0	0	0.17888	0	0	0	0
0	0	0	0	0	0	0.17888	0	0	0
0	0	0	0	0	0	0	0.17888	0	0
0	0	0	0	0	0	0	0	0.17888	0

square root of the biasing parameter and $\mathbf{0}_p$ is a $p \times 1$ vector of zeros. The ridge estimates are then computed from

$$\hat{\boldsymbol{\beta}}_R = (\mathbf{X}_A'\mathbf{X}_A)^{-1}\mathbf{X}_A'\mathbf{y}_A = (\mathbf{X}'\mathbf{X} + k\mathbf{I}_p)^{-1}\mathbf{X}'\mathbf{y}$$

Table 8.9 shows the augmented matrix $\mathbf{X}_A$ and vector $\mathbf{y}_A$ required to produce the ridge solution for the acetylene data with $k=0.032$.

Some other properties of ridge regression. Figure 8.7 illustrates the geometry of ridge regression for a two–regressor problem. The point $\hat{\boldsymbol{\beta}}$ at the center of the ellipses corresponds to the least squares solution, where the residual sum of squares takes on its minimum value. The small ellipse represents the locus of points in the β_1, β_2 plane where the residual sum of squares is constant at some value greater than the minimum. The ridge estimate $\hat{\boldsymbol{\beta}}_R$ is the shortest vector from the origin that produces a residual sum of squares equal to the value represented by the small ellipse. That is, the ridge estimate $\hat{\boldsymbol{\beta}}_R$ produces the vector of regression coefficients with the smallest norm consistent with a specified increase in the residual sum of squares. We note that the ridge estimator shrinks the least squares estimator towards the origin. Consequently ridge estimators (and other biased estimators, generally) are sometimes called *shrinkage* estimators. Hocking [1976] has observed that the ridge estimator shrinks the least squares estimator with respect to the contours of $\mathbf{X}'\mathbf{X}$. That is, $\hat{\boldsymbol{\beta}}_R$ is the solution to

$$\begin{aligned} &\underset{\boldsymbol{\beta}}{\text{minimize}}\,(\boldsymbol{\beta} - \hat{\boldsymbol{\beta}})'\mathbf{X}'\mathbf{X}(\boldsymbol{\beta} - \hat{\boldsymbol{\beta}}) \\ &\text{subject to } \boldsymbol{\beta}'\boldsymbol{\beta} \leqslant d^2 \end{aligned} \tag{8.17}$$

where the radius d depends on k.

Many of the properties of the ridge estimator assume that the value of k is fixed. In practice, since k is estimated from the data by inspection of the ridge trace, k is stochastic. It is of interest to ask if the optimality properties cited by Hoerl and Kennard hold if k is stochastic. Several authors have shown through simulations that ridge regression generally offers improvement in mean square error over least squares when k is estimated from the data. Theobald [1974] has generalized the conditions under which ridge regression leads to smaller MSE than least squares. The expected improvement depends on the orientation of the $\boldsymbol{\beta}$ vector relative to the eigenvectors of $\mathbf{X}'\mathbf{X}$. The expected improvement is greatest when $\boldsymbol{\beta}$ coincides with the eigenvector associated with the largest eigenvalue of $\mathbf{X}'\mathbf{X}$. Other interesting results appear in Mayer and Willke [1973] and Lowerre [1974].

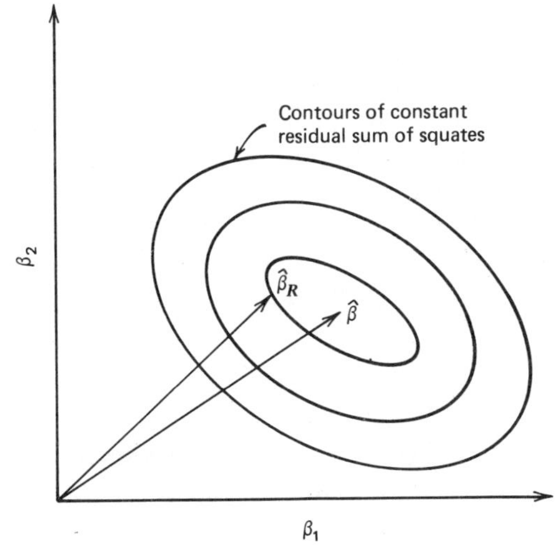

Figure 8.7 A geometric interpretation of ridge regression.

Obenchain [1977] has shown that nonstochastically shrunken ridge estimators yield the same t- and F-statistics for testing hypotheses as does least squares. Thus, although ridge regression leads to biased point estimates, it does not generally require a new distribution theory. However, distributional properties are still unknown for stochastic choice of k.

Relationship to other estimators. Ridge regression is closely related to Bayesian estimation. Generally, if prior information about $\boldsymbol{\beta}$ can be described by a p-variate normal distribution with mean vector $\boldsymbol{\beta}_0$ and covariance matrix $\mathbf{V}_0$, then the Bayes estimator of $\boldsymbol{\beta}$ is

$$\hat{\boldsymbol{\beta}}_B = \left(\frac{1}{\sigma^2}\mathbf{X}'\mathbf{X} + \mathbf{V}_0^{-1} \right)^{-1} \left(\frac{1}{\sigma^2}\mathbf{X}'\mathbf{y} + \mathbf{V}_0^{-1}\boldsymbol{\beta}_0 \right)$$

The use of Bayesian methods in regression is discussed in Leamer [1973, 1978] and Zellner [1971]. Two major drawbacks of this approach are that the data analyst must make an explicit statement about the form of the prior distribution and the statistical theory is not widely understood. However if we choose the prior mean $\boldsymbol{\beta}_0 = \mathbf{0}$ and $\mathbf{V}_0 = \sigma_0^2\mathbf{I}$, then we obtain

$$\hat{\boldsymbol{\beta}}_B = (\mathbf{X}'\mathbf{X} + k\mathbf{I})^{-1}\mathbf{X}'\mathbf{y} \equiv \hat{\boldsymbol{\beta}}_R, \; k = \frac{\sigma^2}{\sigma_0^2}$$

the usual ridge estimator. In effect, the method of least squares can be viewed as a Bayes estimator using an unbounded uniform prior distribution for $\boldsymbol{\beta}$. The ridge estimator results from a prior distribution that places weak boundedness conditions on $\boldsymbol{\beta}$. Also see Lindley and Smith [1972].

Theil and Goldberger [1961] and Theil [1963] have introduced a procedure called mixed estimation. The technique uses prior or additional information to augment the data directly instead of through a prior distribution. Mixed estimation starts with the usual regression model $\mathbf{y}=\mathbf{X}\boldsymbol{\beta}+\boldsymbol{\varepsilon}$ and assumes that the analyst can write a set of $r<p$ prior restrictions on $\boldsymbol{\beta}$ such as

$$\mathbf{a}=\mathbf{D}\boldsymbol{\beta}+\boldsymbol{\delta}$$

where $E(\boldsymbol{\delta})=\mathbf{0}$, $V(\boldsymbol{\delta})=\mathbf{V}$, $\mathbf{D}$ is an $r\times p$ matrix of known constants with rank r, and $\mathbf{a}$ is an $r\times 1$ vector of random variables. If we augment $\mathbf{y}$ and $\mathbf{X}$ to give

$$\begin{bmatrix}\mathbf{y}\\ \mathbf{a}\end{bmatrix}=\begin{bmatrix}\mathbf{X}\\ \mathbf{D}\end{bmatrix}\boldsymbol{\beta}+\begin{bmatrix}\boldsymbol{\varepsilon}\\ \boldsymbol{\delta}\end{bmatrix}$$

and apply least squares, we obtain the unbiased mixed estimator

$$\hat{\boldsymbol{\beta}}_{ME}=\left(\frac{1}{\sigma^2}\mathbf{X}'\mathbf{X}+\mathbf{D}'\mathbf{V}^{-1}\mathbf{D}\right)^{-1}\left(\frac{1}{\sigma^2}\mathbf{X}'\mathbf{y}+\mathbf{D}'\mathbf{V}^{-1}\mathbf{a}\right)$$

Now if we let $\mathbf{D}=\mathbf{A}$ (where $\mathbf{A}'\mathbf{A}=\mathbf{I}$), $\mathbf{a}=\mathbf{0}$, and $\mathbf{V}=\sigma_1^2\mathbf{I}$, then

$$\hat{\boldsymbol{\beta}}_{ME}=(\mathbf{X}'\mathbf{X}+k\mathbf{I})^{-1}\mathbf{X}'\mathbf{y}\equiv\hat{\boldsymbol{\beta}}_R,\ k=\frac{\sigma^2}{\sigma_1^2}$$

While mixed estimation and ridge regression can be numerically equivalent, there is some difference in the viewpoint adopted. In mixed estimation, $\mathbf{a}$ is a random variable, while in ridge regression the elements of $\mathbf{a}$ are specified constants which gives a biased estimator. Mixed estimation is less formal than Bayesian estimation because it allows introduction of prior information without complete specification of a prior distribution for $\boldsymbol{\beta}$. An application of mixed estimation to combat multicollinearity is in Belsley, Kuh, and Welch [1980].

Methods for choosing k. Much of the controversy concerning ridge regression centers around the choice of the biasing parameter k. Choosing k by inspection of the ridge trace is a subjective procedure requiring judgment on the part of the analyst. Several authors have proposed procedures for choosing k that are more analytical. Hoerl, Kennard and Baldwin [1975] have suggested that an appropriate choice for k is

$$k = \frac{p\hat{\sigma}^2}{\hat{\boldsymbol{\beta}}'\hat{\boldsymbol{\beta}}} \tag{8.18}$$

where $\hat{\boldsymbol{\beta}}$ and $\hat{\sigma}^2$ are found from the least squares solution. They showed via simulation that the resulting ridge estimator had significant improvement in MSE over least squares. In a subsequent paper, Hoerl and Kennard [1976] proposed an iterative estimation procedure based on (8.18). Specifically, they suggested the following sequence of estimates of $\boldsymbol{\beta}$ and k;

$$\begin{array}{ll}
\hat{\boldsymbol{\beta}} & k_0 = \dfrac{p\hat{\sigma}^2}{\hat{\boldsymbol{\beta}}'\hat{\boldsymbol{\beta}}} \\[2ex]
\hat{\boldsymbol{\beta}}_R(k_0) & k_1 = \dfrac{p\hat{\sigma}^2}{\hat{\boldsymbol{\beta}}_R'(k_0)\hat{\boldsymbol{\beta}}_R(k_0)} \\[2ex]
\hat{\boldsymbol{\beta}}_R(k_1) & k_2 = \dfrac{p\hat{\sigma}^2}{\hat{\boldsymbol{\beta}}_R'(k_1)\boldsymbol{\beta}_R(k_1)} \\
 & \quad\vdots
\end{array}$$

The relative change in k_j is used to terminate the procedure. If

$$\frac{k_{j+1}-k_j}{k_j} > 20T^{-1.3}$$

the algorithm should continue; otherwise terminate and use $\hat{\boldsymbol{\beta}}_R(k_j)$, where $T = Tr(\mathbf{X}'\mathbf{X})^{-1}/p$. This choice of termination criterion has been selected because T increases with the spread in the eigenvalues of $\mathbf{X}'\mathbf{X}$, allowing further shrinkage as the degree of ill-conditioning in the data increases. The authors cite simulation studies in which this termination rule performed well.

McDonald and Galarneau [1975] suggest choosing k so that

$$\hat{\boldsymbol{\beta}}_R'\hat{\boldsymbol{\beta}}_R = \hat{\boldsymbol{\beta}}'\hat{\boldsymbol{\beta}} - \hat{\sigma}^2 \sum_{j=1}^{p} \left(\frac{1}{\lambda_j}\right) \tag{8.19}$$

For cases where the right-hand side of (8.19) is negative, they investigated letting either $k=0$ (least squares) or $k=\infty(\hat{\boldsymbol{\beta}}_R = \mathbf{0})$. Neither method was better than least squares in all cases. Mallows [1973] modified the C_p

statistic to a C_k statistic that can be used to determine k. He proposed plotting C_k against V_k, where

$$C_k = \frac{SS_E(k)}{\hat{\sigma}^2} - n + 2 + 2\,Tr(\mathbf{XL})$$

$$V_k = 1 + Tr(\mathbf{X'XLL'})$$

$$\mathbf{L} = (\mathbf{X'X} + k\mathbf{I})^{-1}\mathbf{X'}$$

and $SS_E(k)$ is the residual sum of squares as a function of k. The suggestion is to choose k to minimize C_k. Marquardt [1970] has proposed using a value of k such that the maximum VIF is between one and ten, preferably closer to one. Other methods of choosing k have been suggested by Dempster et al. [1977], Goldstein and Smith [1974], Lawless and Wang [1976], Lindley and Smith [1972], and Obenchain [1975].

There is no assurance that any of these procedures will produce similar choices for k. Furthermore there is no guarantee that these methods are superior to straightforward inspection of the ridge trace.

• **Example 8.3** We will illustrate two alternatives to inspection of the ridge trace for selection of k using the acetylene data. The value of k from (8.18) for the acetylene data is

$$k = \frac{p\hat{\sigma}^2}{\hat{\boldsymbol{\beta}}'\hat{\boldsymbol{\beta}}} = \frac{9(0.00038)}{6.77} = 0.0005$$

where $\hat{\boldsymbol{\beta}}$ and $\hat{\sigma}^2$ are taken from the standardized least squares solution in Table 8.2. Note that this value is k is considerable smaller than the value chosen by inspection of the ridge trace. Column 1 of Table 8.10 shows the corresponding ridge regression coefficients.

The iterative estimate of k suggested by Hoerl and Kennard [1976] may be found by starting with $k_0 = 0.0005$, yielding $\hat{\boldsymbol{\beta}}_R(k_0)$ in column (1) of Table 8.10. Note that $\hat{\boldsymbol{\beta}}_R'(k_0)\hat{\boldsymbol{\beta}}_R(k_0) = 1.0999$; that is, the squared length of the vector of ridge regression coefficients with $k_0 = 0.0005$ is much shorter than the squared length of the least squares vector. The new estimate of k is now

$$k_1 = \frac{p\hat{\sigma}^2}{\hat{\boldsymbol{\beta}}_R'(k_0)\hat{\boldsymbol{\beta}}_R(k_0)} = \frac{9(0.00038)}{1.0999} = 0.0031$$

The criteria for termination is to compare the relative change in k_0 to 20 $T^{-1.3}$ where

$$T = \frac{Tr(\mathbf{X'X})^{-1}}{p} = \frac{\sum_{j=1}^{p} (1/\lambda_j)}{p} = \frac{10611.6918}{9} = 1179.08$$

Table 8.10 Ridge models for two choices of k

Term	(1) $k=0.0005$	(2) $k=0.005$
x_1	0.6650	0.6283
x_2	0.2258	0.2193
x_3	−0.2295	−0.2775
x_1x_2	−0.4608	−0.3816
x_1x_3	−0.4234	−0.1261
x_2x_3	−0.2329	−0.1429
x_1^2	−0.0072	0.1233
x_2^2	−0.0769	−0.0606
x_3^2	−0.3197	−0.1197
MS_E	0.00045	0.00055
R^2	0.9973	0.9967

and λ_j are the eigenvalues of $\mathbf{X}'\mathbf{X}$. Therefore, since

$$\frac{k_1-k_0}{k_0}=\frac{0.0031-0.0005}{0.0005}=5.20>20T^{-1.3}=20(1179.08)^{-1.3}=0.0020$$

another iteration should be performed. It can be shown that the final value of k determined by this procedure is $k=0.005$. Column 2 of Table 8.10 presents the corresponding ridge coefficients.

Comparing the ridge regression coefficients for these two values of k with the corresponding coefficients for $k=0.032$ selected from the ridge trace, we note striking similarities. For both $k=0.0005$ and $k=0.005$, the large initial least squares estimates of β_{13}, β_{11} and β_{33} have been significantly reduced. For $k=0.005$ the ridge estimate of β_{11} is still negative, while for $k=0.005$ the estimate is now positive. Note that the larger value of k selected from the ridge trace produces more shrinkage in the estimates, resulting in a further sign change for β_{23}. However, all three models are quite similar, and are apparently more reasonable relationships for the acetylene data than the nine-term ordinary least squares model.

Ridge regression and variable selection. Standard variable selection algorithms often do not perform well when the data is highly multicollinear. However, variable selection usually works quite well when the regressors are orthogonal, or nearly orthogonal. If the regressors have been made more nearly orthogonal by the use of biased estimators, then variable selection may be a good strategy. Hoerl and Kennard [1970b] suggest that the ridge trace can be used as a guide for variable selection. They propose the following rules for removing regressors from the full model:

1. Remove regressors that are stable but have small prediction power; that is, regressors with small standardized coefficients.

2. Remove regressors with unstable coefficients that do not hold their prediction power; that is, unstable coefficients that are driven to zero.

3. Remove one or more of the remaining regressors that have unstable coefficients.

The subset of remaining regressors, say p in number, are used in the "final" model. We may examine these regressors to see if they form a nearly orthogonal subset. This may be done by plotting $\hat{\boldsymbol{\beta}}_R'(k)\hat{\boldsymbol{\beta}}_R(k)$, the squared length of the coefficient vector as a function of k, versus k. If the regressors are orthogonal, the squared length of the vector of ridge estimates should be $\hat{\boldsymbol{\beta}}'\hat{\boldsymbol{\beta}}/(1+k)^2$, where $\hat{\boldsymbol{\beta}}$ is the ordinary least squares estimate of $\boldsymbol{\beta}$. Therefore if the subset model contains nearly orthogonal regressors, the functions $\hat{\boldsymbol{\beta}}_R'(k)\hat{\boldsymbol{\beta}}_R(k)$ and $\hat{\boldsymbol{\beta}}'\hat{\boldsymbol{\beta}}/(1+k)^2$ plotted against k should be very similar.

• **Example 8.4** To illustrate how ridge regression may be helpful in variable selection, consider the ridge model for the aceylene data. Table 8.8 indicates that as k increases, the large coefficients for x_1x_3 and x_3^2 shrink rapidly toward zero. At $k=0.032$, the coefficients of x_2x_3 and x_2^2 are also small. Since these coefficients have been computed using standardized data, their magnitudes reflect the relative contribution of the corresponding regressors. Suppose we decide to eliminate these four terms on the basis of their small

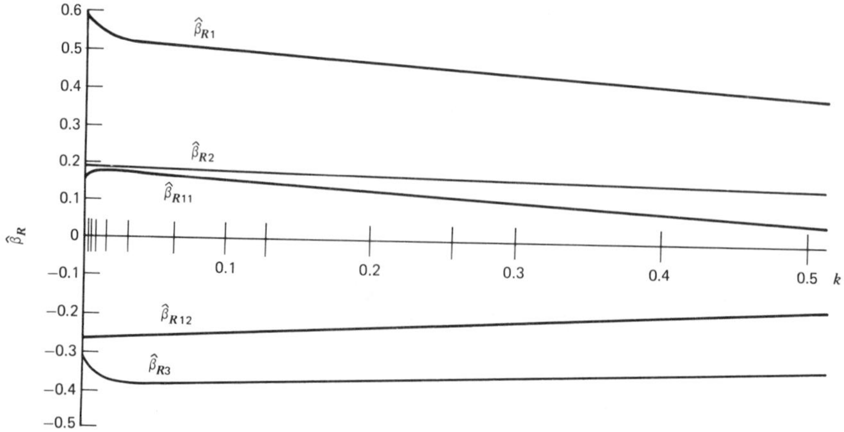

Figure 8.8 Ridge trace for acetylene data using five regressors.

Table 8.11 Coefficients at various values of *k*

k	.000	.001	.002	.004	.008	.0116	.032	.064	.128	.256	.512
$\beta_{R,1}$	.6036	.5946	.5878	.5771	.5622	.5449	.5268	.5070	.4815	.4437	.3882
$\beta_{R,2}$	.1940	.1935	.1934	.1931	.1925	.1913	.1891	.1849	.1773	.1645	.1442
$\beta_{R,3}$	−.3240	−.3317	−.3386	−.3493	−.3634	−.3778	−.3883	−.3916	−.3853	−.3667	−.3318
$\beta_{R,12}$	−.2738	−.2729	−.2725	−.2717	−.2703	−.2677	−.2628	−.2540	−2384	−.2135	−.1785
$\beta_{R,11}$	.1728	.1742	.1755	.1774	.1793	.1795	.1756	.1642	.1416	.1052	.0583
MS_E	.00063	.00063	.00063	.00063	.00064	.00065	.00069	.00082	.00127	.00266	.00632
R^2	.994	.994	.994	.994	.994	.994	.993	.992	.987	.973	.937

regression coefficients in the ridge model. Applying ridge regression to the remaining five regressors ($x_1, x_2, x_3, x_1^2, x_1x_2$) produces the ridge trace shown in Figure 8.8. Table 8.11 presents the regression coefficients, residual mean square, and R^2 for several values of k. The ridge trace is much more stable than when all nine regressors are considered. That is, the introduction of further bias by increasing k does not change the regression coefficients dramatically. Furthermore there is little immediate change in MS_E or R^2. The maximum variance inflation factor for the least squares fit to these five regressors is 53.52 (still quite large, but a considerable improvement over the original model). Elimination of x_1x_3, x_2x_3, x_2^2 and x_3^2 has greatly improved the conditioning of the data. We conclude that this subset least squares equation is a reasonable model for the data. Note that this is a different subset model from those found by the standard variable selection methods in Section 8.5.2. Snee [1973] also suggested this subset model from graphical analysis of the data.

Figure 8.9 shows the prediction performance of this subset least squares model. Comparing this display with Figures 8.3 and 8.6, we observe that this subset model predicts about as well as the ridge model in prediction and extrapolation. Its prediction and performance is also roughly comparable to

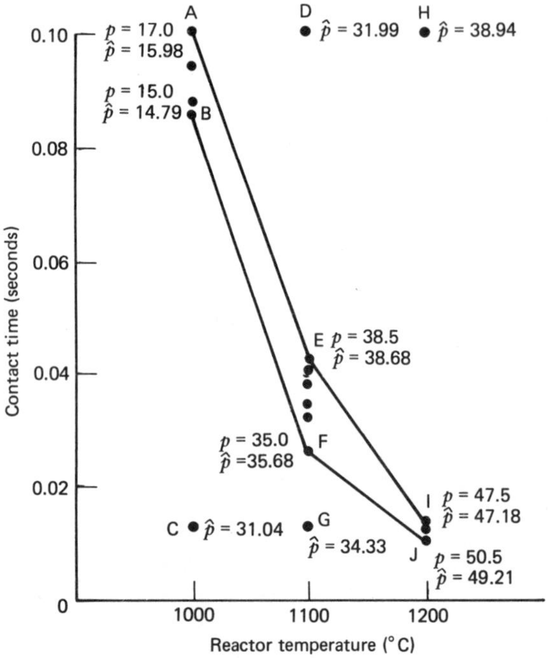

Figure 8.9 Prediction and extrapolation with the five-variable least squares model of the acetylene data. (Adapted from Marquardt and Snee [1975], with permission of the publisher.)

the subset models B, C and E in Section 8.5.2. Thus it seems that ridge regression has greatly improved the results for the original least squares nine-term model, either directly with the nine-term ridge model, or indirectly as an aid to variable elimination.

8.5.4 Generalized ridge regression

Hoerl and Kennard [1970a] proposed an extension of the ordinary ridge regression procedure that allows separate biasing parameters for each regressor. This procedure is known as generalized ridge regression.

The discussion of generalized ridge regression is somewhat simplified if we transform the data to the space of orthogonal regressors. To do this, recall that if $\mathbf{\Lambda}$ is the $p \times p$ diagonal matrix whose main diagonal elements are the eigenvalues $(\lambda_1, \lambda_2, \ldots, \lambda_p)$ of $\mathbf{X'X}$ and if $\mathbf{T}$ is the corresponding orthogonal matrix of eigenvectors, then

$$\mathbf{T'X'XT} = \mathbf{\Lambda} \tag{8.20}$$

Letting

$$\mathbf{Z} = \mathbf{XT} \tag{8.21}$$

and

$$\boldsymbol{\alpha} = \mathbf{T}'\boldsymbol{\beta} \tag{8.22}$$

the linear model becomes

$$\begin{aligned} \mathbf{y} &= \mathbf{X}\boldsymbol{\beta} + \boldsymbol{\varepsilon} \\ &= (\mathbf{ZT}')(\mathbf{T}\boldsymbol{\alpha}) + \boldsymbol{\varepsilon} \\ &= \mathbf{Z}\boldsymbol{\alpha} + \boldsymbol{\varepsilon} \end{aligned} \tag{8.23}$$

The least squares estimator of $\boldsymbol{\alpha}$ is the solution to

$$(\mathbf{Z'Z})\hat{\boldsymbol{\alpha}} = \mathbf{Z'y} \tag{8.24}$$

which is equivalent to

$$\mathbf{\Lambda}\hat{\boldsymbol{\alpha}} = \mathbf{Z'y} \tag{8.25}$$

or

$$\hat{\boldsymbol{\alpha}} = \mathbf{\Lambda}^{-1}\mathbf{Z'y} \tag{8.26}$$

The vector of original parameter estimates is given by using (8.22), that is,

$$\hat{\boldsymbol{\beta}} = \mathbf{T}\hat{\boldsymbol{\alpha}} \tag{8.27}$$

We often refer to (8.23) as the *canonical form* of the model. In terms of the canonical form, the generalized ridge estimator is the solution to

$$(\boldsymbol{\Lambda} + \mathbf{K})\hat{\boldsymbol{\alpha}}_{GR} = \mathbf{Z}'\mathbf{y} \tag{8.28}$$

where $\mathbf{K}$ is a diagonal matrix with elements $(k_1, k_2, \ldots, k_p)$. In terms of the original model, the generalized ridge coefficients are

$$\hat{\boldsymbol{\beta}}_{GR} = \mathbf{T}\hat{\boldsymbol{\alpha}}_{GR} \tag{8.29}$$

Now consider the choice of the biasing parameters in $\mathbf{K}$. The mean square error for generalized ridge regression is

$$\begin{aligned} \text{MSE}(\hat{\boldsymbol{\beta}}_{GR}) &= E\left[(\hat{\boldsymbol{\beta}}_{GR} - \boldsymbol{\beta})'(\hat{\boldsymbol{\beta}}_{GR} - \boldsymbol{\beta})\right] \\ &= E\left[(\hat{\boldsymbol{\alpha}}_{GR} - \boldsymbol{\alpha})'(\hat{\boldsymbol{\alpha}}_{GR} - \boldsymbol{\alpha})\right] \\ &= \sigma^2 \sum_{j=1}^{p} \frac{\lambda_j}{(\lambda_j + k_j)^2} + \sum_{j=1}^{p} \frac{\alpha_j^2 k_j^2}{(\lambda_j + k_j)^2} \end{aligned} \tag{8.30}$$

The first term on the right-hand side of (8.30) is the sum of the variances of the parameter estimates and the second term is the squared bias. The mean square error (8.30) is minimized by choosing

$$k_j = \frac{\sigma^2}{\alpha_j^2}, \qquad j = 1, 2, \ldots, p \tag{8.31}$$

Unfortunately the optimal k_j depend on unknown parameters σ^2 and α_j. Hoerl and Kennard [1970a] suggest an iterative approach to determining the k_j. Beginning with the least squares solution, we obtain an initial estimate of the k_j, for example

$$k_j^0 = \frac{\hat{\sigma}^2}{\hat{\alpha}_j^2}, \qquad j = 1, 2, \ldots, p$$

Use these initial estimates of the k_j to compute initial generalized ridge estimates from

$$\hat{\boldsymbol{\alpha}}_{GR}^0 = (\boldsymbol{\Lambda} + \mathbf{K}^0)^{-1}\mathbf{Z}'\mathbf{y}$$

where $\mathbf{K}^0 = \text{diag}(k_1^0, k_2^0, \ldots, k_p^0)$. Then use the initial estimates $\hat{\boldsymbol{\alpha}}_{GR}^0$ to revise the estimates of the k_j:

$$k_j^1 = \frac{\hat{\sigma}^2}{\left(\hat{\alpha}_{GR,j}^0\right)^2}, \qquad j=1,2,\ldots,p$$

These new values of k_j^1 may be used to revise the estimates of the $\boldsymbol{\alpha}$. The iterative process should continue until stable parameter estimates result. One measure of stability often used is the squared length of the vector $\hat{\boldsymbol{\alpha}}'_{GR}\hat{\boldsymbol{\alpha}}_{GR}$. Specifically, if the squared length of the vector of parameter estimates does not change significantly from iteration $i-1$ to iteration i, then terminate. Otherwise the iterative estimation procedure should continue. Note that there is no helpful graphical display of the coefficients such as the ridge trace in generalized ridge regression.

We may use (8.31) to justify our choice of the biasing parameter k in ordinary ridge regression. The value k in (8.18) is a weighted average of the k_j from (8.31). Clearly if the k_j are combined to produce a single biasing parameter, we should not use an ordinary average because a small α_j would produce a large value of k inducing too much bias in the parameter estimates. However, the harmonic mean of the k_j is

$$k_h = \frac{p}{\sum_{j=1}^{p} (1/k_j)} = \frac{p}{\sum_{j=1}^{p} \left(\alpha_j^2/\sigma^2\right)} = \frac{p\sigma^2}{\sum_{j=1}^{p} \alpha_j^2} = \frac{p\sigma^2}{\boldsymbol{\alpha}'\boldsymbol{\alpha}} = \frac{p\sigma^2}{\boldsymbol{\beta}'\boldsymbol{\beta}} = k$$

as given in (8.18).

Hemmerle [1975] showed that Hoerl and Kennard's iterative procedure for estimating the k_j has an explicit closed form solution so that in general, iteration is unnecessary. Specifically, let

$$\hat{\boldsymbol{\alpha}}_{GR} = \mathbf{B}\hat{\boldsymbol{\alpha}} \tag{8.32}$$

where $\hat{\boldsymbol{\alpha}}$ is the least squares estimator and $\mathbf{B}$ is a diagonal matrix of nonnegative elements $b_1, b_2, \ldots, b_p$. Hocking et al. [1976] show that Hemmerle's results are to choose

$$b_j = 0 \quad \text{if} \quad \tau_j^2 < 4$$

$$b_j = 0.5 + \left[0.25 - \left(1/\tau_j^2\right)\right]^{1/2} \quad \text{if} \quad \tau_j^2 \geq 4 \tag{8.33}$$

where $\tau_j^2 = \hat{\alpha}_j^2 \lambda_j / \hat{\sigma}^2$. Noting that τ_j is the t-statistic associated with the jth regressor, we observe that if the t-statistic is "small" the corresponding generalized ridge coefficient is set equal to zero, while if the t-statistic is "large", the corresponding generalized ridge coefficient is a fraction b_j of the least squares coefficient. In other words, nonsignificant coefficients are shrunk to zero, while significant coefficients are shrunk less severely. We refer to this solution as the fully iterated generalized ridge solution.

Hemmerle noted that the fully iterated generalized ridge solution often results in the introduction of too much bias (or too much shrinkage) in the final parameter estimates. Hemmerle proposed a technique to avoid this based on constraining the residual sum of squares to prevent an undesired significant increase. He recommended that a limit be placed on the total loss in R^2, and that this loss be allocated proportionally to the individual regressors. His procedure results in modified values of b_j, say b_j^*, given by

$$b_j^* = 1 - \sqrt{m}\,(1 - b_j) \tag{8.34}$$

where m is the ratio of the allowable loss in R^2 to the loss in R^2 if b_j from (8.33) is used. Hocking et al. [1976] object to the use of (8.34) because it forces all the b_j^* to be nonzero. By setting some of the b_j to zero, the strong influence of a small eigenvalue on variance inflation was eliminated. Using (8.34) allows the influence of that eigenvalue to return.

In a subsequent paper, Hemmerle and Brantle [1978] suggest selecting the k_j based on minimizing an estimator of the mean square error criterion. An explicit closed form solution is developed for the resulting vector of parameter estimates. A procedure is also given for obtaining constrained generalized ridge estimates, where the constraints are chosen to utilize prior information about the signs of the regression coefficients. However, a Monte Carlo simulation failed to show any obvious superiority for this method.

Unfortunately there is no clear-cut "best" choice of the k_j for generalized ridge regression. We agree with Hemmerle [1975] that fully iterated generalized ridge often results in too much shrinkage, and some type of constrained procedure is appropriate, particularly for data that is severely ill-conditioned. Constraining the maximum increase in the residual sum of squares to between one and twenty percent often works well in practice. However, more work needs to be done to develop better guidelines for choosing the parameters k_j and controlling the amount of shrinkage.

- **Example 8.5** We will illustrate the fully iterated generalized ridge regression solution for the acetylene data. To transform the data to orthogonal regressors, we require the matrix $\mathbf{T}$ of eigenvectors associated with the eigenvalues of $\mathbf{X'X}$. The $\mathbf{T}$ matrix is shown in Table 8.12. The matrix $\mathbf{Z=XT}$ of

Table 8.12 Matrix T of eigenvectors for the acetylene data

.3387	.1057	.6495	.0073	.1428	−.2488	−.2077	−.5436	.1768
.1324	.3391	−.0068	−.7243	−.5843	.0205	−.0102	−.0295	−.0035
−.4137	−.0978	−.4696	−.0718	−.0182	.0160	−.1468	−.7172	.2390
−.2191	.5403	.0897	.3612	−.1661	.3733	−.5885	.0909	.0003
.4493	.0860	−.2863	.1912	−.0943	.0333	.0575	.1543	.7969
.2524	−.5172	−.0570	−.3447	.2007	.3232	−.6209	.1280	.0061
−.4056	−.0742	.4404	−.2230	.1443	.5393	.3233	.0565	.4087
.0258	.5316	−.2240	−.3417	.7342	−.0705	−.0057	.0761	.0050
−.4667	−.0969	.1421	−.1337	−.0350	−.6299	−.3089	.3631	.3309

orthogonalized data is shown in Table 8.13. These calculations may be conveniently performed on almost any modern digital computer (for example, the quantities shown in Table 8.12 and 8.13 can be obtained using the MATRIX procedure in SAS).

The least squares regression coefficients $\hat{\alpha}_j$, computed from $\hat{\boldsymbol{\alpha}}=(\mathbf{Z}'\mathbf{Z})^{-1}\mathbf{Z}'\mathbf{y} = \boldsymbol{\Lambda}^{-1}\mathbf{Z}'\mathbf{y}$, are shown in Table 8.14. This table also contains the eigenvalues of $\mathbf{X}'\mathbf{X}$, the values of τ_j^2, b_j, and the corresponding generalized ridge coefficients. To illustrate the calculations note that

$$\tau_1^2 = \frac{\hat{\alpha}_1^2\lambda_1}{\hat{\sigma}^2} = \frac{(-0.35225)^2 4.20480}{0.0003826} = 1363.71$$

since $\tau_1^2 \geqslant 4$, we compute b_1 from (8.33) as follows:

$$\begin{aligned} b_1 &= 0.5 + \left[0.25 - \left(1/\tau_1^2\right)\right]^{1/2} \\ &= 0.5 + \left[0.25 - (1/1363.71)\right]^{1/2} \\ &= 0.999266 \end{aligned}$$

Therefore the corresponding generalized ridge coefficient is

$$\begin{aligned} \hat{\alpha}_{GR,1} &= b_1\hat{\alpha}_1 \\ &= (0.999266)(-0.35225) \\ &= -0.351991 \end{aligned}$$

Notice that the fully iterated generalized ridge solution shrinks four of the orthogonal coefficients to zero. Table 8.14 also shows the MS_E and R^2 values for the least squares and the generalized ridge solutions. As anticipated, the fully iterated generalized ridge procedure has allowed the residual sum of

Table 8.13 Matrix Z = XT of orthogonalized acetylene data

Observation	Z_1	Z_2	Z_3	$Z_4(=Z_{X_1X_2})$	$Z_5(=Z_{X_1X_3})$	$Z_6(=Z_{X_2X_3})$	$Z_7(=Z_{X_1^2})$	$Z_8(=Z_{X_2^2})$	$Z_9(=Z_{X_3^2})$
1	.5415	−1.0347	1.0487	−.1880	1.7389	−.6593	.6492	.7822	.2402
2	.4846	−.8830	1.1638	−.0468	.8909	−.3874	.5067	.2045	−.1939
3	.4046	−.6129	1.2914	.0676	−.0025	−.1631	.2187	−.0898	−1.6609
4	.3388	−.1513	1.3176	.1315	−.7526	.3579	.1269	−1.2150	.9250
5	.2353	.6905	1.2785	−.0089	−1.0842	.6884	−.4181	−1.2768	1.6754
6	.0310	2.7455	.9535	−.7783	.2235	.2093	−1.1200	1.3128	−1.1453
7	.5940	−.0165	−1.0885	1.1554	1.5790	.1926	−1.3363	−.4626	.5964
8	.6385	−.2399	−.9170	1.0916	.3634	.4238	−1.2453	−.7138	−.3611
9	.7139	−.3558	−.7151	.8354	−.9374	.3207	−.6525	.5144	−.7716
10	.7436	−.2228	−.6170	.5668	−1.4297	−.4038	.5657	2.5203	1.4085
11	.7668	.1034	−.8626	−.0706	−1.3472	−.3706	1.5958	−.8815	−1.3485
12	.8726	1.1054	−1.5272	−1.8442	.8129	−.9285	.8411	−.8981	.7053
13	−1.7109	.8164	−.3702	1.2052	.8885	1.9123	2.0708	.2251	−.1036
14	−2.1618	.1860	−.1026	.5619	−.1290	−2.5588	−.3380	−.1080	.8652
15	−1.6050	−.6784	−.2117	−.3325	−.7456	−.0658	−.8259	−.4662	−1.0012
16	−.8875	−1.4521	−.6417	−2.3461	−.0690	1.4324	−.6387	.5524	.1699

Table 8.14 Fully iterated generalized ridge solution for the acetylene data

Term	Least Squares Coefficient $\hat{\alpha}_j$	λ_j	τ_j^2	b_j	Generalized Ridge Coefficient $\hat{\alpha}_{GR,j}$
z_1	−0.35225	4.20480	1363.71	0.999266	−0.351991
z_2	0.0047813	2.16261	0.13	0.0	0.0
z_3	0.60045	1.13839	1072.39	0.999067	0.599890
z_4	−0.23836	1.04130	154.59	0.993489	−0.236808
z_5	0.0094903	0.38453	0.09	0.0	0.0
z_6	0.21713	0.04951	6.10	0.793407	0.172272
z_7	0.38298	0.01363	5.23	0.742115	0.284215
z_8	0.52070	0.00513	3.63	0.0	0.0
z_9	−2.4010	0.00010	1.46	0.0	0.0
MS_E	0.0003826				0.0007606
R^2	0.998				0.995

squares to increase substantially as MS_E for generalized ridge is approximately twice MS_E for least squares. However, because the residual sum of squares is still very small, there has not been a substantial degradation in R^2.

To express the solution in terms of the standardized predictors, we solve $\hat{\boldsymbol{\beta}}_{\text{GR}} = \mathbf{T}\hat{\boldsymbol{\alpha}}_{\text{GR}}$, which yields the results shown in column 1 of Table 8.15. These results differ considerably from the least squares solution. However, the generalized ridge solution differs only slightly from the solution obtained via

Table 8.15 Fully iterated generalized ridge solution for the acetylene data, in terms of standardized and original centered regressors

Parameter	(1) Standardized Estimate	(2) Estimate for Original Centered Regressors
β_0	0	34.3462
β_1	0.4869	5.7929
β_2	0.2076	2.4701
β_3	−0.4611	−5.4886
β_{12}	−0.3398	−4.5715
β_{13}	−0.0465	−0.6175
β_{23}	−0.0941	−1.3318
β_{11}	0.1718	2.5761
β_{22}	−0.0328	−0.3548
β_{33}	−0.0203	−0.2350

ordinary ridge regression with either one–step or iterative estimation of k (see Example 8.3, Table 8.10) or with k chosen from inspection of the ridge trace (see Example 8.2, Table 8.8). As in the ordinary ridge solution, the large least squares estimates of β_{13}, β_{11}, and β_{23} have been dramatically reduced, and the original negative least squares estimate of β_{11} is now positive. Column 2 of Table 8.15 shows the generalized ridge regression coefficients in terms of the original centered regressors. •

8.5.5 Principal components regression

Biased estimators of regression coefficients can also be obtained by using a procedure known as principal components regression. Consider the canonical form of the model,

$$\mathbf{y}=\mathbf{Z}\boldsymbol{\alpha}+\boldsymbol{\varepsilon} \tag{8.35}$$

where

$$\mathbf{Z}=\mathbf{XT}, \qquad \boldsymbol{\alpha}=\mathbf{T}'\boldsymbol{\beta}, \qquad \text{and} \quad \mathbf{T}'\mathbf{X}'\mathbf{X}\mathbf{T}=\mathbf{Z}'\mathbf{Z}=\boldsymbol{\Lambda}$$

Recall that $\boldsymbol{\Lambda}=\text{diag}(\lambda_1, \lambda_2, \ldots, \lambda_p)$ is a $p\times p$ diagonal matrix of the eigenvalues of $\mathbf{X}'\mathbf{X}$ and $\mathbf{T}$ is a $p\times p$ orthogonal matrix whose columns are the eigenvectors associated with $\lambda_1, \lambda_2, \ldots, \lambda_p$. The columns of $\mathbf{Z}$, which define a new set of orthogonal regressors, such as

$$\mathbf{Z}=\left[\mathbf{Z}_1,\mathbf{Z}_2,\ldots,\mathbf{Z}_p\right]$$

are referred to as *principal components*.

The least squares estimator of $\boldsymbol{\alpha}$ is

$$\hat{\boldsymbol{\alpha}}=(\mathbf{Z}'\mathbf{Z})^{-1}\mathbf{Z}'\mathbf{y}=\boldsymbol{\Lambda}^{-1}\mathbf{Z}'\mathbf{y} \tag{8.36}$$

and the covariance matrix of $\hat{\boldsymbol{\alpha}}$ is

$$V(\hat{\boldsymbol{\alpha}})=\sigma^2(\mathbf{Z}'\mathbf{Z})^{-1}=\sigma^2\boldsymbol{\Lambda}^{-1} \tag{8.37}$$

Thus a small eigenvalue of $\mathbf{X}'\mathbf{X}$ means that the variance of the corresponding orthogonal regression coefficient will be large. Since

$$\mathbf{Z}'\mathbf{Z}=\sum_{i=1}^{p}\sum_{j=1}^{p}\mathbf{Z}_i\mathbf{Z}_j'=\boldsymbol{\Lambda}$$

we often refer to the eigenvalue λ_j as the variance of the jth principal component. If all the λ_j are equal to unity, the *original* regressors are orthogonal, while if an λ_j is exactly equal to zero this implies a perfect linear relationship between the *original* regressors. One or more of the λ_j near zero implies that multicollinearity is present. Note also that the covariance matrix of the standardized regression coefficients $\hat{\boldsymbol{\beta}}$ is

$$V(\hat{\boldsymbol{\beta}}) = V(\mathbf{T}\hat{\boldsymbol{\alpha}}) = \mathbf{T}\boldsymbol{\Lambda}^{-1}\mathbf{T}'\sigma^2$$

This implies that the variance of $\hat{\beta}_j$ is $\sigma^2(\sum_{i=1}^{p} t_{ji}^2/\lambda_i)$. Therefore the variance of $\hat{\beta}_j$ is a linear combination of the reciprocals of the eigenvalues. This demonstrates how one or more small eigenvalues can destroy the precision of the least squares estimate $\hat{\beta}_j$.

We have observed previously how the eigenvalues and eigenvectors of $\mathbf{X}'\mathbf{X}$ provide specific information on the nature of the multicollinearity. Since $\mathbf{Z} = \mathbf{X}\mathbf{T}$, we have

$$\mathbf{Z}_i = \sum_{j=1}^{p} t_{ji}\mathbf{X}_j \tag{8.38}$$

where $\mathbf{X}_j$ is the jth column of the $\mathbf{X}$ matrix and t_{ji} are the elements of the ith column of $\mathbf{T}$ (the ith eigenvector of $\mathbf{X}'\mathbf{X}$). If the variance of the ith principal component (λ_i) is small, this implies that $\mathbf{Z}_i$ is nearly a constant, and (8.38) indicates that there is a linear combination of the *original regressors* that is nearly constant. This is the definition of multicollinearity; that is, the t_{ji} are the constants in (8.2). Therefore (8.38) explains why the elements of the eigenvector associated with a small eigenvalue of $\mathbf{X}'\mathbf{X}$ identifies the regressors involved in the multicollinearity.

The principal components regression approach combats multicollinearity by using less than the full set of principal components in the model. To obtain the principal components estimator, assume that the regressors are arranged in order of decreasing eigenvalues, $\lambda_1 \geqslant \lambda_2 \geqslant \cdots \geqslant \lambda_p > 0$. Suppose that the last s of these eigenvalues are approximately equal to zero. In principal components regression the principal components corresponding to near zero eigenvalues are removed from the analysis and least squares applied to the remaining components. That is,

$$\hat{\boldsymbol{\alpha}}_{PC} = \mathbf{B}\hat{\boldsymbol{\alpha}} \tag{8.39}$$

where $b_1 = b_2 = \cdots = b_{p-s} = 1$ and $b_{p-s+1} = b_{p-s+2} = \cdots = b_p = 0$. Thus the principal components estimator is

$$\hat{\boldsymbol{\alpha}}_{PC} = \begin{bmatrix} \hat{\alpha}_1 \\ \hat{\alpha}_2 \\ \vdots \\ \hat{\alpha}_{p-s} \\ --- \\ 0 \\ 0 \\ \vdots \\ 0 \end{bmatrix} \begin{matrix} \\ p-s \text{ components} \\ \\ \\ \\ s \text{ components} \\ \\ \end{matrix}$$

or in terms of the standardized regressors

$$\begin{aligned} \hat{\boldsymbol{\beta}}_{PC} &= \mathbf{T}\hat{\boldsymbol{\alpha}}_{PC} \\ &= \sum_{j=1}^{p-s} \lambda_j^{-1} \mathbf{t}_j' \mathbf{X}' \mathbf{y} \mathbf{t}_j \end{aligned} \tag{8.40}$$

A simulation study by Gunst and Mason [1977] showed that principal components regression offers considerable improvement over least squares when the data are ill-conditioned. They also point out that another advantage of principal components is that exact distribution theory and variable selection procedures are available (see Mansfield, et al. [1977]).

- **Example 8.6** We will illustrate the use of principal components regression for the acetylene data. We begin with the linear transformation $\mathbf{Z}=\mathbf{XT}$ that transforms the original standardized regressors into an orthogonal set of variables (the principal components). The $\mathbf{T}$ matrix for the acetylene data is shown in Table 8.12. This matrix indicates that the relationship between z_1 (for example) and the standardized regressors is

$$\begin{aligned} z_1 = {} & 0.3387x_1 + 0.1324x_2 - 0.4137x_3 - 0.2191x_1x_2 + 0.4493x_1x_3 \\ & + 0.2524x_2x_3 - 0.4056x_1^2 + 0.0258x_2^2 - 0.4667x_3^2 \end{aligned}$$

The relationships between the remaining principal components $z_2, z_3, \ldots, z_9$ and the standardized regressors are determined similarly. Table 8.13 shows the elements of the $\mathbf{Z}$ matrix (sometimes called the principal component scores). The least squares estimate $\hat{\boldsymbol{\alpha}}=(\mathbf{Z}'\mathbf{Z})^{-1}\mathbf{Z}'\mathbf{y}=\boldsymbol{\Lambda}^{-1}\mathbf{Z}'\mathbf{y}$, along with the eigenvalues associated with the principal components is shown in Table 8.14.

Table 8.16 Principal components regression for the acetylene data

Parameter	Principal Components in Model									
	A z_1		B z_1, z_2		C z_1, z_2, z_3		D z_1, z_2, z_3, z_4		E z_1, z_2, z_3, z_4, z_5	
	Standardized Estimate	Original Estimate	Standardized Estimate	Original Estimate	Standardized Estimate	Original Estimate	Standardized Estimate	Original Estimate	Standardized Estimate	Original Estimates
β_0	.0000	42.1943	.0000	42.2219	.0000	36.6275	.0000	34.6688	.0000	34.7517
β_1	.1193	1.4194	.1188	1.4141	.5087	6.0508	.5070	6.0324	.5056	6.0139
β_2	.0466	.5530	.0450	.5346	.0409	.4885	.2139	2.5438	.2195	2.6129
β_3	−.1457	−1.7327	−.1453	−1.7281	−.4272	−5.0830	−.4100	−4.8803	−.4099	4.8757
β_{12}	−.0772	−1.0369	−.0798	−1.0738	−.0260	−.3502	−.1123	−1.5115	−.1107	−1.4885
β_{13}	.1583	2.0968	.1578	2.0922	−.0143	−.1843	−.0597	−.7926	−.0588	−.7788
β_{23}	.0889	1.2627	.0914	1.2950	.0572	.8111	.1396	1.9816	.1377	1.9493
β_{11}	−.1429	−2.1429	−.1425	−2.1383	.1219	1.8295	.1751	2.6268	.1738	2.6083
β_{22}	.0091	.0968	.0065	.0691	−.1280	−1.3779	−.0460	−.4977	−.0533	−.5760
β_{33}	−.1644	−1.9033	−.1639	−1.8986	−.0786	−.9125	−.0467	−.5392	−.0463	−.5346
R^2	.5217		.5218		.9320		.9914		.9915	
MS_E	.079713		.079705		.011333		.001427		.00142	

The principal components estimator reduces the effects of multicollinearity by using a subset of the principal components in the model. Since there are four small eigenvalues for the acetylene data, this implies that there are four principal components that should be deleted. We will exclude z_6, z_7, z_8, and z_9, and consider regressions involving only the first five principal components.

Suppose we consider a regression model involving only the first principal component, as in

$$y=\alpha_1 z_1+\varepsilon$$

The fitted model is

$$\hat{y}=-0.35225z_1$$

or $\hat{\boldsymbol{\alpha}}'_{PC}=[-0.35225,0,0,0,0,0,0,0,0]$. The coefficients in terms of the standardized regressors are found from $\hat{\boldsymbol{\beta}}_{PC}=\mathbf{T}\hat{\boldsymbol{\alpha}}_{PC}$. Panel A of Table 8.16 shows the resulting standardized regression coefficients, as well as the regression coefficients in terms of the original centered regressors. Note that even though only one principal component is included, the model produces estimates for all nine standardized regression coefficients.

The results of adding the other principal components z_2, z_3, z_4, and z_5 to the model one at a time, are displayed in panels B, C, D, and E respectively, of Table 8.16. We see that using different numbers of principal components in the model produces substantially different estimates of the regression coefficients. Furthermore the principal components estimates differ considerably from the least squares estimates (for example, see Table 8.8). However, the principal components procedure with either four or five components included results in coefficient estimates that do not differ dramatically from those produced by the other biased estimation methods (refer to the ordinary ridge regression estimates in Tables 8.8 and 8.10). Principal components shrinks the large least squares estimates of β_{13} and β_{33} and changes the sign of the original negative least squares estimate of β_{11}. The five-component model does not substantially degrade the fit to the original data as there has been little loss in R^2 from the least squares model. Thus we conclude that the relationship based on the first five principal components provides a more plausible model for the acetylene data than was obtained via ordinary least squares. •

Marquardt [1970] suggested a generalization of principal components regression. He felt that the assumption of an integral rank for the **X** matrix is too restrictive, and proposed a "fractional rank" estimator that allows the rank to be a piecewise continuous function. Specifically, if the rank of the **X** matrix is in the interval $[r, r+1]$, then Marquardt's fractional rank estimator is

$$\hat{\boldsymbol{\alpha}}_{FR}=(1-c)\hat{\boldsymbol{\alpha}}_r+c\hat{\boldsymbol{\alpha}}_{r+1} \tag{8.41}$$

for $0\leqslant c\leqslant 1$, where $\hat{\boldsymbol{\alpha}}_r$ and $\hat{\boldsymbol{\alpha}}_{r+1}$ are the principal components estimators for

$\boldsymbol{\alpha}$ for assumed ranks r and $r+1$. That is, the last $r-1$ elements of $\hat{\boldsymbol{\alpha}}_{FR}$ are zero, the $(p-r+1)$st element is $c\hat{\alpha}_{p-r+1}$ and the first $p-r$ elements are the least squares estimators $\hat{\alpha}_1, \hat{\alpha}_2, \ldots, \hat{\alpha}_{p-r}$. Criteria for choosing r and c are discussed by Hocking et al. [1976].

8.5.6 Latent root regression analysis

The latent root regression procedure was developed by Hawkins [1973] and Webster et al. [1974], following the same philosophy as principal components. The procedure forms estimators from the eigenvalues (or latent roots) of the correlation matrix of regressor and response variables

$$\mathbf{A}'\mathbf{A} = \begin{bmatrix} 1 & \mathbf{y}'\mathbf{X} \\ \mathbf{X}'\mathbf{y} & \mathbf{X}'\mathbf{X} \end{bmatrix}$$

Let $0 \leqslant l_0 \leqslant l_1 \leqslant \cdots \leqslant l_p$ and $\mathbf{v}_0, \mathbf{v}_1, \ldots, \mathbf{v}_p$ be the eigenvalues and eigenvectors of $\mathbf{A}'\mathbf{A}$, and denote the last p elements of $\mathbf{v}_j$ by $\boldsymbol{\delta}_j$ so that $\mathbf{v}_j' = [v_{0j}, \boldsymbol{\delta}_j']$. The latent root estimator is

$$\hat{\boldsymbol{\beta}}_{LR} = \sum_{j=s}^{p} l_j^{-1} \phi_j \boldsymbol{\delta}_j \tag{8.42}$$

$$\phi_j = -S_{yy}^{1/2} v_{0j} \sum_{q=s}^{p} v_{0q}^2 l_q^{-1} \tag{8.43}$$

The s terms corresponding to $j = 0, 1, \ldots, s-1$ deleted from (8.42) correspond to those eigenvectors for which both $|v_{0j}|$ and l_j are nearly zero. Thus like principal components regression, latent root regression attempts to identify and eliminate multicollinearities that do not aid in prediction. Latent root regression reduces to least squares when no terms are deleted $(s=0)$.

Gunst et al. [1976] and Gunst and Mason [1977] indicate that latent root regression may provide considerable improvement in mean square error over least squares. Gunst [1979] points out that latent root regression can produce regression coefficients that are very similar to those found by principal components, particularly when there are only one or two strong multicollinearities in **X**. A number of large-sample properties of latent root regression are in White and Gunst [1979].

8.5.7 Comparison and evaluation of biased estimators

A number of Monte Carlo simulation studies have been conducted to examine the effectiveness of biased estimators and to attempt to determine which procedures perform best. For example, see McDonald and Galarneau [1975], Hoerl and Kennard [1976], Hoerl, Kennard and Baldwin [1975] (who compare least squares and ridge), Gunst et al. [1976] (latent root versus least squares), Lawless [1978], Hemmerle and Brantle [1978] (ridge, generalized ridge, and least squares), Lawless and Wang [1976] (least squares, ridge, and principal components), Wichern and Churchill [1978], Gibbons [1979] (various forms of ridge), Gunst and Mason [1977] (ridge, principal components, latent root, and others), and Dempster et al. [1977]. The Dempster et al. [1977] study compared 57 different estimators for 160 different model configurations. While no single procedure emerges from these studies as best overall, there is considerable evidence indicating the superiority of biased estimation to least squares if multicollinearity is present. Our own preference in practice is for ordinary ridge regression with k selected by inspection of the ridge trace. The procedure is straightforward, easy to implement on a standard least squares computer program, and the analyst can learn to interpret the ridge trace very quickly. It is also occasionally useful to find the "optimum" value of k suggested by Hoerl, Kennard, and Baldwin [1975] and the iteratively estimated "optimum" k of Hoerl and Kennard [1976] and compare the resulting models with the one obtained via the ridge trace.

As we have noted previously, if the mean square error is regarded as a function of $\boldsymbol{\beta}$, then the mean square error is minimized when $\boldsymbol{\beta}$ is aligned with the normalized eigenvector corresponding to the largest eigenvalue of $\mathbf{X}'\mathbf{X}$. Similarly, the mean square error is maximized when $\boldsymbol{\beta}$ is aligned with the normalized eigenvector corresponding to the smallest eigenvalue of $\mathbf{X}'\mathbf{X}$. This implies that potential improvements in mean square error from biased estimation depend on the orientation of the parameter vector. In her simulation study, Gibbons [1979] reports that if $\boldsymbol{\beta}$ is favorably aligned, then ridge-type estimators are always superior to least squares, while if $\boldsymbol{\beta}$ is unfavorably aligned, ridge-type estimators are not always superior to least squares. Thus if the analyst had some prior information about the alignment of $\boldsymbol{\beta}$ with the eigenvectors of $\mathbf{X}'\mathbf{X}$, a decision regarding the potential utility of biased estimation for that particular problem could be made. However, it does not seem to be a simple matter to obtain such information.

The use of biased estimators in regression is not without controversy. Several authors have been critical of ridge regression and other related biased estimation techniques. Conniffe and Stone [1973, 1975] have criticized the use of the ridge trace to select the biasing parameter, since $\hat{\boldsymbol{\beta}}_R$ will change slowly and eventually stablilize as k increases even for orthogonal regressors. They also claim that if the data are not adequate to support a least squares analysis, then it is unlikely that ridge regression will be of any substantive help, since the parameter estimates will be nonsensical.

Marquardt and Snee [1975] and Smith and Goldstein [1975] do not accept these conclusions, and feel that biased estimators are a valuable tool for the data analyst confronted by ill-conditioned data. Several authors have noted that while we can prove that there exists a k such that the mean square error of the ridge estimator is always less than the mean square error of the least squares estimator, there is no assurance that the ridge trace (or any other method that selects the biasing parameter stochastically by analysis of the data) will produce the optimal k.

Draper and Van Nostrand [1977a, b, 1979] are also critical of biased estimators. They find fault with a number of the technical details of the simulation studies used as the basis of claims of improvement in MSE for biased estimation, suggesting that the simulations have been designed to favor the biased estimators. They note that ridge regression is really only appropriate in situations where external information is added to a least squares problem. This may take the form of either the Bayesian formulation and interpretation of the procedure, or a constrained least squares problem in which the constraints on $\boldsymbol{\beta}$ are chosen to reflect the analyst's knowledge of the regression coefficients to "improve the conditioning" of the data.

Smith and Campbell [1980] suggest using explicit Bayesian analysis or mixed estimation to resolve multicollinearity problems. They reject ridge methods as weak and imprecise because they only loosely incorporate prior beliefs and information into the analysis. When explicit prior information is known, then Bayesian or mixed estimation should certainly be used. However, often the prior information is not easily reduced to a specific prior distribution, and ridge regression methods offer a method to incorporate, at least approximately, this knowledge.

There has also been some controversy surrounding whether the regressors and the response should be centered and scaled so that $\mathbf{X}'\mathbf{X}$ and $\mathbf{X}'\mathbf{y}$ are in correlation form. This results in an artifical removal of the intercept from the model. Effectively, the intercept in the ridge model is estimated by $\bar{y}$. Hoerl and Kennard [1970 a, b] use this approach, as do Marquardt and Snee [1975], who note that centering tends to minimize any nonessential ill-conditioning when fitting polynomials. On the other hand, Brown [1977] feels that the variables should not be centered, because centering affects only the intercept estimate and not the slopes. Belsley, Kuh, and Welsch [1980] suggest not centering the regressors so that the role of the intercept in any near linear dependencies may be diagnosed. Centering and scaling allows the analyst to think of the parameter estimates as standardized regression coefficients, which is often intuitively appealing. Furthermore, centering the regressors can remove nonessential ill-conditioning thereby reducing variance inflation in the parameter estimates. Consequently we recommend both centering and scaling the data.

Despite the objections noted, we believe that biased estimation methods are useful techniques that the analyst should consider when dealing with multicollinearity. Biased estimation methods certainly compare very favorably to other methods for handling multicollinearity, such as variable elimination. As Marquardt and Snee [1975] note, it is often better to use some of the information in all of the regressors as ridge regression does, than to use all of the information in some regressors and none of the information in others as variable elimination does. Furthermore variable elimination can be thought of as a form of biased estimation because, as we noted in Chapter 7, subset regression models often produce biased estimates of the regression coefficients. In effect, variable elimination often shrinks the vector of parameter estimates as does ridge regression. We do not recommend the mechanical or automatic use of ridge regression without thoughtful study of the data and careful analysis of the adequacy of the final model. Properly used, biased estimation methods are a valuable tool in the data analyst's kit.

Problems

8.1. Consider the soft drink delivery time data in Example 4.1.

a. Find the simple correlation between cases (x_1) and distance (x_2).

b. Find the variance inflation factors.

c. Find the condition number of **X′X**.

Is there evidence of multicollinearity in these data?

8.2. Consider the Hald cement data from Example 7.1.

a. From the matrix of correlations between the regressors, would you suspect that multicollinearity is present?

b. Calculate the variance inflation factors.

c. Find the eigenvalues of **X′X**.

d. Find the condition number of **X′X**.

8.3 Using the Hald cement data (Example 7.1), find the eigenvector associated with the smallest eigenvalue of **X′X**. Interpret the elements of this vector. What can you say about the source of multicollinearity in these data?

8.4 Use the regressors x_2 (passing yardage), x_7 (percent of rushing plays, and x_8 (opponents' yards rushing) for the National Football League data in Appendix Table B.1.

a. Does the correlation matrix give any indication of multicollinearity?

b. Calculate the variance inflation factors and the condition number of **X′X**. Is there any evidence of multicollinearity?

8.5 Consider the gasoline mileage data in Appendix Table B.3.

a. Does the correlation matrix give any indication of multicollinearity?

b. Calculate the variance inflation factors and the condition number of $\mathbf{X'X}$. Is there any evidence of multicollinearity?

8.6 Using the gasoline mileage data in Appendix Table B.3, find the eigenvectors associated with the smallest eigenvalues of $\mathbf{X'X}$. Interpret the elements of these vectors. What can you say about the source of multicollinearity in these data?

8.7 Analyze the housing price data in Appendix Table B.4 for multicollinearity. Use the variance inflation factors and the condition number of $\mathbf{X'X}$.

8.8 Analyze the chemical process data in Appendix Table B.5 for evidence of multicollinearity. Use the variance inflation factors and the condition number of $\mathbf{X'X}$.

8.9 Apply ridge regression to the Hald cement data in Example 7.1.

a. Use the ridge trace to select an appropriate value of k. Is the final model a good one?

b. How much inflation in the residual sum of squares has resulted from the use of ridge regression?

c. Compare the ridge regression model with the two-regressor model involving x_1 and x_2 developed by all possible regressions in Example 7.1.

8.10 Use ridge regression on the Hald cement data (Example 7.1) using the value of k in (8.18). Compare this value of k with the value selected by the ridge trace in Problem 8.9. Does the final model differ greatly from the one in Problem 8.9?

8.11 Use the iterative estimation procedure described in Section 8.5.3 to estimate k in ridge regression for the Hald cement data (Example 7.1). Compare the value of k obtained by this procedure with the values given by the ridge trace (Problem 8.9) and (8.18) (Problem 8.10). Does the final model differ greatly from those found in Problems 8.9 and 8.10?

8.12 Use fully iterated generalized ridge regression on the Hald cement data (Example 7.1). How much increase in the residual sum of squares has resulted? How does the final model compare with the one found by ordinary ridge regression in Problem 8.9?

8.13 Consider the data on the activity of cholinestrase in cricket eggs, presented in Appendix B.8. Suppose that a full quadratic is originally proposed as the model for those data.

a. Find the variance inflation factors. Is there evidence of multicollinearity in those data?

b. Estimate the model parameters using ridge regression, selecting the value of k by inspection of the ridge trace. Is this an adequate model?

c. Use the ridge trace to select a subset of the original regressors that give an adequate model.

8.14 Estimate the parameters in a model for the gasoline mileage data in Appendix Table B.3 using ridge regression.

a. Use the ridge trace to select an appropriate value of k. Is the resulting model adequate?

b. How much inflation in the residual sum of squares has resulted from the use of ridge regression?

c. How much reduction in R^2 has resulted from the use of ridge regression?

8.15 Estimate the parameters in a model for the gasoline mileage data in Appendix Table B.3 using ridge regression with the value of k determined by (8.18). Does this model differ dramatically from the one developed in Problem 8.14?

8.16 Estimate the parameters in a model for the gasoline mileage data in Appendix Table B.3 using ridge regression with k determined by the iterative procedure described in Section 8.5.3. Compare the model obtained with the one developed in Problem 8.14.

8.17 Use generalized ridge regression to develop a model for the gasoline mileage data in Appendix Table B.3.

a. Estimate the biasing parameters using fully iterated generalized ridge regression. Does the final model seem reasonable?

b. How much shrinkage has been induced on the coefficient vector?

c. How much increase in the residual sum of squares has resulted from the use of generalized ridge regression?

d. Compare the amount of shrinkage in the coefficient vector for generalized ridge and the ordinary ridge models developed in Problems 8.14 and 8.15. Which model do you prefer?

8.18 Estimate the model parameters for the Hald cement data (Example 7.1) using principal components regression.

a. What is the loss in R^2 for this model, compared to least squares?

b. How much shrinkage in the coefficient vector has resulted?

c. Compare the principal components model with the ordinary ridge model developed in 8.9. Comment on any apparent differences in the models.

8.19 Estimate the model parameters for the gasoline mileage data using principal components regression.

a. How much has the residual sum of squares increased, compared to least squares?

b. How much shrinkage in the coefficient vector has resulted?

c. Compare the principal components and ordinary ridge models (Problem 8.14). Which model do you prefer?

8.20 Show that the ridge estimator is the solution to the problem

$$\underset{\boldsymbol{\beta}}{\text{minimize}} \quad (\boldsymbol{\beta}-\hat{\boldsymbol{\beta}})'\mathbf{X}'\mathbf{X}(\boldsymbol{\beta}-\hat{\boldsymbol{\beta}})$$
$$\text{subject to} \quad \boldsymbol{\beta}'\boldsymbol{\beta} \leqslant d^2$$

8.21 Pure Shrinkage Estimators (Stein [1960]). The pure shrinkage estimator is defined as $\hat{\boldsymbol{\beta}}_S = c\hat{\boldsymbol{\beta}}$, where $0 \leqslant c \leqslant 1$ is a constant chosen by the analyst. Describe the kind of shrinkage that this estimator introduces, and compare it with the shrinkage that results from ridge regression. Intuitively, which estimator seems preferable?

8.22 Show that the pure shrinkage estimator [Problem (8.21)] is the solution to

$$\underset{\boldsymbol{\beta}}{\text{minimize}} \quad (\boldsymbol{\beta}-\hat{\boldsymbol{\beta}})'(\boldsymbol{\beta}-\hat{\boldsymbol{\beta}})$$
$$\text{subject to} \quad \boldsymbol{\beta}'\boldsymbol{\beta} \leqslant d^2$$

8.23 The mean square error criterion for ridge regression is

$$E(L_1^2) = \sum_{j=1}^{p} \frac{\lambda_j}{(\lambda_j + k)^2} + \sum_{j=1}^{p} \frac{\alpha_j^2 k^2}{(\lambda_j + k)^2}$$

Try to find the value of k that minimizes $E(L_1^2)$. What difficulties are encountered?

8.24 Consider the mean square error criterion for generalized ridge regression, (8.30). Show that the mean square error is minimized by choosing $k_j = \sigma^2/\alpha_j^2$, $j = 1, 2, \ldots, p$.

8.25 Directed Ridge Regression. Suppose that instead of shrinking all the elements of the parameter vector, we shrink only those coefficients corresponding to small eigenvalues. If the shrinkage is done using generalized ridge, Guilkey and Murphy [1975] call the procedure directed ridge regression. Sclove [1968] has suggested a similar technique, but shrinks the appropriate subset uniformly towards the origin. What advantages would those methods have relative to ordinary ridge regression? Which type of shrinkage strategy would you recommend?

8.26 Show that if $\mathbf{X}'\mathbf{X}$ is in correlation form, $\mathbf{\Lambda}$ is the diagonal matrix of eigenvalues of $\mathbf{X}'\mathbf{X}$, and $\mathbf{T}$ is the corresponding matrix of eigenvectors, then the variance inflation factors are the main diagonal elements of $\mathbf{T\Lambda}^{-1}\mathbf{T}'$.

9

TOPICS IN THE USE OF REGRESSION ANALYSIS

This chapter surveys a variety of topics that arise in the use of regression analysis. In several cases only a brief glimpse of the subject is given , along with references to more complete presentations.

9.1 Autocorrelation

9.1.1 Sources and effects of autocorrelation

The fundamental assumptions in linear regression are that the error terms ε_i have mean zero, constant variance, and are uncorrelated [$E(\varepsilon_i)=0$, $V(\varepsilon_i)=\sigma^2$, and $E(\varepsilon_i\varepsilon_j)=0$]. For purposes of testing hypotheses and constructing confidence intervals we often add the assumption of normality, so that the ε_i are NID $(0, \sigma^2)$. Some applications of regression involve regressor and response variables that have a natural sequential order over time. Such data are called *time series data*. Regression models using time series data occur relatively often in economics, business, and some fields of engineering. The assumption of uncorrelated or independent errors for time series data is often not appropriate. Usually the errors in time series data exhibit serial correlation; that is, $E(\varepsilon_i\varepsilon_{i+j})\neq 0$. Such error terms are said to be autocorrelated.

There are several sources of autocorrelation. Perhaps the primary cause of autocorrelation in regression problems involving time series data is failure to include one or more important regressors in the model. For example, suppose that we wish to regress annual sales of a soft drink concentrate against the annual advertising expenditures for that product. Now the growth in population over the period of time used in the study will also influence the product sales. If population size is not included in the model, this may cause the errors in the model to be positively autocorrelated, because population size is positively correlated with product sales.

The presence of autocorrelation in the errors has several affects on the ordinary least squares regression procedure. These are summarized as follows:

1. Ordinary least squares regression coefficients are still unbiased, but they are no longer minimum variance estimates. We say that these estimates are inefficient.

2. When the errors are positively autocorrelated, the residual mean square MS_E may seriously underestimate σ^2. Consequently, the standard errors of the regression coefficients may be too small. Thus confidence intervals are shorter than they really should be, and tests of hypotheses on individual regression coefficients may indicate that one or more regressors contribute significantly to the model when they really do not. Generally, underestimating σ^2 gives the analyst a false impression of accuracy.

3. The confidence intervals and tests of hypotheses based on the t and F distributions are, strictly speaking, no longer appropriate.

There are two general approaches for dealing with the problem of autocorrelation. If autocorrelation is present because of an omitted regressor, and if that regressor can be identified and included in the model, the apparent autocorrelation should disappear. If the autocorrelation problem cannot be resolved by including previously omitted factors, then the analyst must turn to a model that specifically incorporates the autocorrelation structure. Such models usually require special parameter estimation techniques.

Because time series data occurs frequently in business and economics, much of the basic methodology appears in the economics literature. Good references on *econometrics* (mathematical and statistical methods in economics) are Johnston [1972] and Wonnacott and Wonnacott [1970]. Other methods for modeling and analyzing time series data are in Box and Jenkins [1976] and Fuller [1976].

9.1.2 Detecting the presence of autocorrelation

Residual plots can be useful for the detection of autocorrelation. The most meaningful display is the plot of residuals versus time, such as shown in Figure 3.5. Note that if there is positive autocorrelation, residuals of identical sign occur in clusters. That is, there are not enough changes of sign in the pattern of residuals. On the other hand if there is negative autocorrelation, the residuals will alternate signs too rapidly.

Various statistical tests can be used to detect the presence of autocorrelation. The test developed by Durbin and Watson [1950, 1951, 1971] is widely used. This test is based on the assumption that the errors in the regression model are generated by a first-order autoregressive process observed at equally spaced time periods; that is,

$$\varepsilon_t = \rho\varepsilon_{t-1} + a_t \tag{9.1}$$

where ε_t is the error term in the model at time period t, a_t is a NID $(0, \sigma_a^2)$ random variable, and $\rho(|\rho|<1)$ is the autocorrelation parameter. Thus a simple linear regression model with first-order autoregressive errors would be

$$\begin{aligned} y_t &= \beta_0 + \beta_1 x_t + \varepsilon_t \\ \varepsilon_t &= \rho\varepsilon_{t-1} + a_t \end{aligned} \tag{9.2}$$

where y_t and x_t are the observations on the response and regressor variables at time period t.

When the regression model errors are generated by the first-order autoregressive process (9.1), several interesting properties of these errors can be developed. By successively substituting for $\varepsilon_{t-1}, \varepsilon_{t-2}, \ldots,$ on the right-hand side of (9.1), we obtain

$$\varepsilon_t = \sum_{u=0}^{\infty} \rho^u a_{t-u}$$

Thus the error term for period t is just a linear combination of all current and previous realizations of the NID $(0, \sigma_a^2)$ random variables a_t. Furthermore we can show that

$$E(\varepsilon_t) = 0 \tag{9.3a}$$

$$V(\varepsilon_t) = \sigma_a^2\left(\frac{1}{1-\rho^2}\right) \tag{9.3b}$$

$$\mathrm{Cov}(\varepsilon_t, \varepsilon_{t+u}) = \rho^{|u|}\sigma^2\left(\frac{1}{1-\rho^2}\right) \tag{9.3c}$$

That is, the errors have zero mean and constant variance, but are autocorrelated unless $\rho=0$.

Because most regression problems involving time series data exhibit positive autocorrelation, the hypotheses usually considered in the Durbin–Watson test are

$$H_0: \rho=0$$
$$H_1: \rho>0 \tag{9.4}$$

The test statistic is

$$d = \frac{\sum_{t=2}^{n}(e_t - e_{t-1})^2}{\sum_{t=1}^{n} e_t^2} \tag{9.5}$$

where $e_t, t=1,2,\ldots,n$ are the residuals from an ordinary least squares analysis applied to the (y_t, x_t) data. Unfortunately, the distribution of d depends on the **X** matrix. However, Durbin and Watson [1951] show that d lies between two bounds, say d_L and d_U, such that if d is outside these limits a conclusion regarding the hypotheses in (9.4) can be reached. The decision procedure is if

$$d<d_L,\ \text{reject } H_0: \rho=0$$

$$d>d_U,\ \text{do not reject } H_0: \rho=0$$

$$d_L \leqslant d \leqslant d_U,\ \text{the test is inconclusive}$$

Clearly small values of d imply that H_0: $\rho=0$ should be rejected because positive autocorrelation indicates that successive error terms are of similar magnitude, and the differences in the residuals $e_t - e_{t-1}$ will be small. Durbin and Watson suggest several procedures for resolving inconclusive results. In these cases, a reasonable approach is to analyze the data using the methods in Section 9.1.3 to see if any major changes in the results occur. Appendix Table A.6 gives the bounds d_L and d_u for a range of sample sizes, various numbers of regressors, and two type I error rates ($\alpha=0.5$ and $\alpha=.01$).

Situations where negative autocorrelation occurs are not often encountered. However if a test for negative autocorrelation is desired, one can use the statistic $4-d$, where d is defined in (9.5). Then the decision rules for H_0: $\rho=0$ versus H_1: $\rho<0$ are the same as those used in testing for positive autocorrelation. It is also possible to conduct a two–sided test (H_0: $\rho=0$ versus H_1: $\rho\neq0$) by using both one–sided tests simultaneously. If this is done, the two–sided procedure has type I error 2α, where α is the type I error used for each one–sided test.

- **Example 9.1** A soft drink beverage company wishes to predict annual regional concentrate sales for a particular product as a function of the annual regional advertising expenditures for that product. Twenty years of data are shown in columns 1 and 2 of Table 9.1. Assuming that a linear relationship is appropriate, a straight line regression model was fitted using ordinary least squares. The residuals from this straight-line model are shown in column 3 of Table 9.1, and other model summary statistics are shown in Table 9.2. Since the regressor and response variables are time series, we suspect that autocorrelation may be present. The plot of residuals versus time in Figure 9.1 is disturbing; there is a definite upward and then downward drift in the residuals. Autocorrelation could be responsible for such a pattern.

We will also use the Durbin–Watson test for

$$H_0: \rho=0$$

$$H_1: \rho>0$$

Columns 4 and 5 of Table 9.1 contain the necessary calculations for obtaining the test statistic

$$d=\frac{\sum_{t=2}^{20}(e_t-e_{t-1})^2}{\sum_{t=1}^{20}e_t^2}=\frac{8195.2065}{7587.9154}=1.08$$

If we choose $\alpha=.05$, then Appendix Table A.6 gives the critical values corresponding to $n=20$ and one regressor as $d_L=1.20$ and $d_U=1.41$. Since the observed value of $d=1.08$ is less than $d_L=1.20$, we reject H_0 and conclude that the errors are positively autocorrelated. •

Although the Durbin–Watson test is extremely useful, it does have limitations. The procedure assumes that the errors are generated by a first–order autoregressive process. It will not necessarily detect autocorrelation when the autocorrelative structure in the errors is other than first–order autoregressive.

Table 9.1 Data for soft drink concentrate sales example

	t	(1) Annual Regional Concentrate Sales y_t (units)	(2) Annual Advertising Expenditures x_t (\$×1000)	(3) Least Squares Residuals e_t	(4) e_t^2	(5) $(e_t - e_{t-1})^2$	(6) Annual Regional Population z_t
1960	1	3083	75	−32.330	1045.2289		825000
1961	2	3149	78	−26.603	707.7196	32.7985	830445
1962	3	3218	80	2.215	4.9062	830.4771	838750
1963	4	3239	82	−16.967	287.8791	367.9491	842940
1964	5	3295	84	−1.148	1.3179	250.2408	846315
1965	6	3374	88	−2.512	6.3101	1.8605	852240
1966	7	3475	93	−1.967	3.8691	0.2970	860760
1967	8	3569	97	11.669	136.1656	185.9405	865925
1968	9	3597	99	−0.513	0.2632	148.4011	871640
1969	10	3725	104	27.032	730.7290	758.7270	877745
1970	11	3794	109	−4.422	19.5541	989.3541	886520
1971	12	3959	115	40.032	1602.5610	1976.1581	894500
1972	13	4043	120	23.577	555.8749	270.7670	900400
1973	14	4194	127	33.940	1151.9236	107.3918	904005
1974	15	4318	135	−2.787	7.7674	1348.8725	908525
1975	16	4493	144	−8.606	74.0632	33.8608	912160
1976	17	4683	153	0.575	0.3306	84.2908	917630
1977	18	4850	161	6.848	46.8951	39.3505	922220
1978	19	5005	170	−18.971	359.8988	666.6208	925910
1979	20	5236	182	−29.063	844.6580	101.8485	929610
			$\sum_{t=1}^{20} e_t^2 = 7587.9154$	$\sum_{t=2}^{20} (e_t - e_{t-1})^2 = 8195.2065$			

Table 9.2 Summary statistics for the least squares model for Example 9.1

Parameter	Estimate	Standard Error	t-Statistic
β_0	1608.508	17.0223	94.49
β_1	20.091	.1428	140.71
$n=20$	$R^2=.9991$		$MS_E=421.5485$

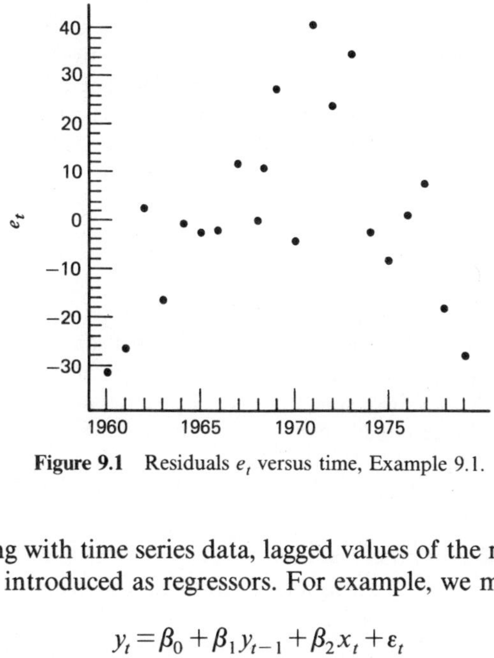

Figure 9.1 Residuals e_t versus time, Example 9.1.

When dealing with time series data, lagged values of the response variable are sometimes introduced as regressors. For example, we might use

$$y_t=\beta_0+\beta_1 y_{t-1}+\beta_2 x_t+\varepsilon_t$$

as a model. If ordinary least squares is used on models of this type, the resulting estimator $\hat{\boldsymbol{\beta}}$ is biased. However, if the errors are uncorrelated, $\hat{\boldsymbol{\beta}}$ is a consistent estimator of $\boldsymbol{\beta}$. If the error terms are correlated, then this is not necessarily true. For further discussion of lagged variable models, see Goldberger [1964]. Furthermore if the model contains lagged variables, the Durbin–Watson test is no longer appropriate. A large-sample test for autocorrelation in lagged variable models is given by Durbin [1970].

9.1.3 Parameter estimation methods

A significant value of the Durbin–Watson statistic or a suspicious residual plot indicates a model specification error. This model misspecification could be either an actual time dependency in the errors, or it could be an "artificial" time dependency caused by the omission of an important regressor. If the apparent autocorrelation results from missing regressors, and if these missing regressors can be identified and incorporated in the model, the apparent autocorrelation problem may be eliminated. This is illustrated in the following example.

- **Example 9.2** Consider the soft drink concentrate sales data discussed in Example 9.1. The Durbin–Watson test has indicated that the errors in the straight-line regression model relating concentrate sales to advertising expenditures exhibit positive autocorrelation. In this problem it is relatively easy to think of other candidate regressors that may be positively correlated with sales. For example, it is very likely that the regional population affects concentrate sales. Data concerning the regional population for the years 1960–1979 is shown in Column 6 of Table 9.1. If we add this regressor to the model, the tentative equation is

$$y_t = \beta_0 + \beta_1 x_t + \beta_2 z_t + \varepsilon_t$$

Table 9.3 presents the summary statistics from the least squares analysis of this data. From Table 9.3 we note that the Durbin–Watson statistic is $d = 3.06$ which, since the five percent critical values are $d_L = 1.10$ and $d_U = 1.54$, would lead us to conclude that there is no evidence of positive autocorrelation in the errors. The plot of residuals versus time, shown in Figure 9.2, is much improved when compared to Figure 9.1. Therefore adding population size to the model has eliminated an apparent autocorrelation problem. •

When the apparent autocorrelation in the errors cannot be removed by adding one or more new regressors to the model, it is necessary to explicitly recognize the autocorrelative structure in the model and devise an appropriate parameter estimation method. There are a number of estimation procedures that can be used (e.g., see Johnston [1972] and Pesaran and

Table 9.3 Summary statistics for the least squares model of Example 9.2

Parameter	Estimate	Standard Error	t-Statistic
β_0	320.340	217.3278	1.47
β_1	18.434	0.2915	63.23
β_2	0.002	0.0003	5.93
$n = 20$	$R^2 = .9997$	$d = 3.06$	$MS_E = 145.3408$

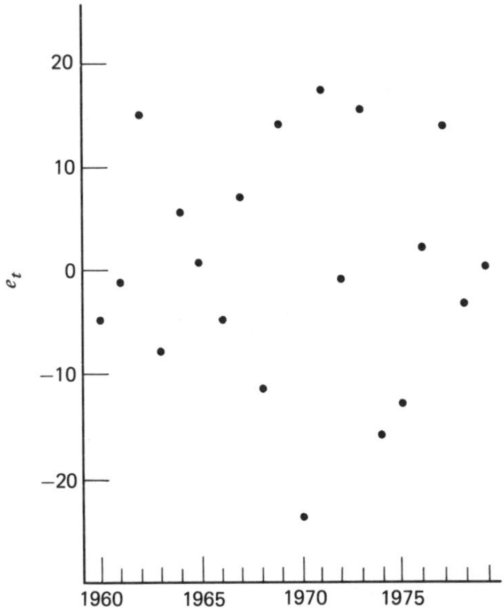

Figure 9.2 Residuals e_t versus time, Example 9.2.

Slater [1980]). We will present the method described by Cochrane and Orcutt [1949].

Consider the simple linear regression model with first-order autocorrelated errors, (9.2). Suppose we transform the response variable so that $y_t' = y_t - \rho y_{t-1}$. Substituting for y_t and y_{t-1}, the model becomes

$$
\begin{aligned}
y_t' &= y_t - \rho y_{t-1} \\
&= \beta_0 + \beta_1 x_t + \varepsilon_t - \rho(\beta_0 + \beta_1 x_{t-1} + \varepsilon_{t-1}) \\
&= \beta_0(1-\rho) + \beta_1(x_t - \rho x_{t-1}) + \varepsilon_t - \rho\varepsilon_{t-1} \\
&= \beta_0' + \beta_1' x_t' + a_t
\end{aligned}
\tag{9.6}
$$

where $\beta_0' = \beta_0(1-\rho)$, $\beta_1' = \beta_1$, $x_t' = x_t - \rho x_{t-1}$, and $a_t = \varepsilon_t - \rho\varepsilon_{t-1}$. Note that the error terms a_t in the reparameterized model are independent random variables [see (9.1)]. Therefore by transforming the regressor and response variables we produce a model that satisfies the usual regression assumptions, and ordinary least squares can be used.

Unfortunately the reparameterized model (9.6) cannot be used directly because the new regressor and response variables x_t' and y_t' are functions of the unknown parameter ρ. However, the first order autoregressive process

$\varepsilon_t = \rho\varepsilon_{t-1} + a_t$ can be viewed as a regression through the origin. Thus ρ can be estimated by obtaining the residuals e_t from an ordinary least squares regression of y_t on x_t, and then regressing e_t on e_{t-1}. The least squares estimate of ρ that results is

$$\hat{\rho} = \frac{\sum_{t=2}^{n} e_t e_{t-1}}{\sum_{t=1}^{n} e_t^2} \tag{9.7}$$

Using this estimate of ρ, obtain the transformed regressor and response variables

$$x_t' = x_t - \hat{\rho} x_{t-1}$$

$$y_t' = y_t - \hat{\rho} y_{t-1}$$

and apply ordinary least squares to the transformed data. The Durbin–Watson test should be applied to the residuals from the reparameterized model. If this procedure indicates that the residuals are uncorrelated, then no additional analysis is required. However, if positive autocorrelation is still indicated, then another iteration is necessary. In the second iteration ρ is estimated with new residuals obtained by using the regression coefficients from the reparameterized model with the original regressor and response variables. This iterative procedure may be continued as necessary until the error terms in the reparameterized model are uncorrelated.

This iterative procedure is not always successful. One reason that the procedure may not eliminate autocorrelaton in the errors is that the estimate $\hat{\rho}$ of ρ is biased downward. We suggest that if one or two iterations does not produce uncorrelated errors, the analyst should consider other estimation techniques. One possibility is to estimate β_0, β_1, and ρ simultaneously by minimizing

$$S(\beta_0, \beta_1, \rho) = \sum_{t=2}^{n} \left[y_t - \rho y_{t-1} - \beta_0(1-\rho) - \beta_1(x_t - \rho x_{t-1}) \right]^2 \tag{9.8}$$

This is a nonlinear least squares problem. However, direct search procedures can be used to minimize $S(\beta_0, \beta_1, \rho)$. For example, any one–dimensional search technique (see Wilde and Beightler [1967, Ch. 6]) could be used to select values for ρ and then linear least squares applied to estimate β_0 and β_1. This process would be repeated using the values of ρ specified by the one-dimensional search until a minimum value of $S(\beta_0, \beta_1, \rho)$ is found. An example of this approach is given by Chatterjee and Price [1977]. Another possibility is to assume that $\rho=1$, yielding the transformed variables $y_t'=y_t-y_{t-1}$ and $x_t'=x_t-x_{t-1}$, and then regressing y_t' on x_t' through the origin. This method is often called the first differences approach. An example is given in Neter and Wasserman [1974].

- **Example 9.3** The data in columns 1 and 2 of Table 9.4 are the share of market for a particular brand of toothpaste y_t and the selling price per pound x_t for 20 consecutive months. We wish to build a regression model relating share of market in period t to the selling price in the same period. A straight-line model is tentatively assumed. Using ordinary least squares, the fitted model is

$$\hat{y}_t=26.90989-24.28977x_t$$

Summary statistics for this model are given in Table 9.5.

Table 9.4 Data for Example 9.3

Period t	(1) Market Share (%) y_t	(2) Price per Pound ($) x_t	(3) e_t	(4) $e_t e_{t-1}$	(5) e_t^2
1	3.63	.97	.2812		.0791
2	4.20	.95	.3654	.1028	.1335
3	3.33	.99	.4670	.1706	.2181
4	4.54	.91	−.2662	−.1243	.0709
5	2.89	.98	−.2159	.0575	.0466
6	4.87	.90	−.1791	.0387	.0321
7	4.90	.89	−.3920	.0702	.1537
8	5.29	.86	−.7307	.2864	.5339
9	6.18	.85	−.0836	.0611	.0070
10	7.20	.82	.2077	−.0174	.0431
11	7.25	.79	−.4710	−.0978	.2218
12	6.09	.83	−.6594	.3106	.4348
13	6.80	.81	−.4352	.2870	.1894
14	8.65	.77	.4432	−.1929	.1964
15	8.43	.76	−.0197	−.0087	.0004
16	8.29	.80	.8119	−.0160	.6592
17	7.18	.83	.4306	.3496	.1854
18	7.90	.79	.1790	.0771	.0320
19	8.45	.76	.0003	.0001	.0000
20	8.23	.78	.2661	.0001	.0708

$$\sum_{t=2}^{20} e_t e_{t-1}=1.3547 \qquad \sum_{t=1}^{20} e_t^2=3.3082$$

Table 9.5 Summary statistics for the ordinary least squares model of Example 9.3

Coefficient	Estimate	Standard Error	t_0
β_0	26.90989	1.1099	24.25
β_1	−24.28977	1.2978	−18.72

$MS_E = 0.1838 \quad R^2 = .95 \quad F_0 = 350.29$

Column 3 of Table 9.4 shows the residuals for this model, and a plot of the residuals in time sequence is given in Figure 9.3. The residual plot is mildly suggestive of positive autocorrelation. This is confirmed by calculation of the Durbin–Watson statistic $d = 1.14$, which when compared with the five percent critical values for $n = 20$ $d_L = 1.20$ and $d_U = 1.41$, indicates that the residuals are positively autocorrelated.

We will use the Cochrane–Orcutt procedure to estimate the model parameters. Columns 4 and 5 of Table 9.4 show the details necessary for estimating the autocorrelation parameter ρ from (9.7):

$$\hat{\rho} = \frac{\sum_{t=2}^{20} e_t e_{t-1}}{\sum_{t=1}^{20} e_t^2} = \frac{1.3547}{3.3082} = 0.409$$

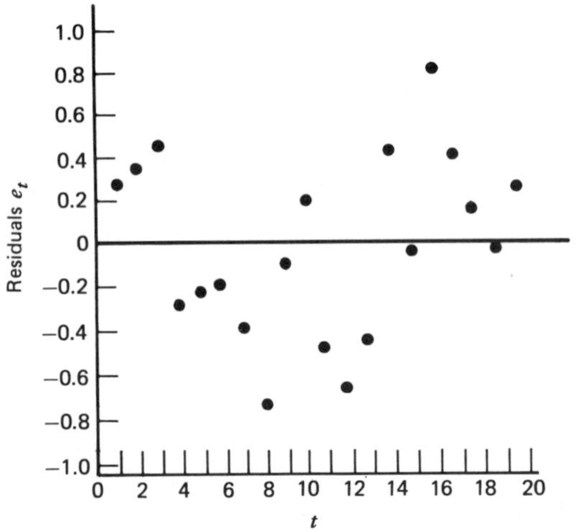

Figure 9.3 Plot of residuals e_t versus time, Example 9.3.

Table 9.6 Transformed regressor and response variables, and residuals

t	(1) x'_t	(2) y'_t	(3) e'_t
2	.553	2.715	.2504
3	.601	1.612	.3176
4	.505	3.178	−.4572
5	.608	1.033	−.1070
6	.499	3.688	−.0908
7	.522	2.908	−.3187
8	.496	3.286	−.5704
9	.498	4.016	.2153
10	.472	4.672	.2419
11	.455	4.305	−.5559
12	.507	3.125	−.4668
13	.471	4.309	−.1655
14	.439	5.869	.6212
15	.445	4.892	−.2010
16	.489	4.842	.8200
17	.503	3.789	.0985
18	.451	4.963	.0029
19	.437	5.219	−.0729
20	.469	4.774	.2660

The transformed variables are then computed from

$$x'_t = x_t - 0.409x_{t-1}$$

$$y'_t = y_t - 0.409y_{t-1}$$

for $t = 2, 3, \ldots, 20$. These transformed variables are shown in Columns 1 and 2 of Table 9.6. An ordinary least squares fit to the transformed variables yields

$$y'_t = 15.85043 - 24.19991x'_t$$

The residuals from this model are shown in Column 3 of Table 9.6. Other model summary statistics are shown in Table 9.7.

The Durbin–Watson statistic for the transformed model is $d = 1.94$. Comparing this with the 5 percent critical values for $n = 19$ $d_L = 1.18$ and $d_U = 1.40$, we conclude that the errors in the transformed model are uncorrelated. Therefore the Cochrane–Orcutt procedure has eliminated the original autocorrelation problem.

Note that β'_1 in the transformed model equals β_1 in terms of the original variables. Therefore comparing Tables 9.5 and 9.7 we see that the Cochrane–

Table 9.7 Summary statistics for the model fit to the transformed variables in Example 9.3

Coefficient	Estimate	Standard Error	t_0
β_0'	15.85043	0.9471	16.74
β_1'	−24.19991	1.9015	−12.73

$MS_E = 0.1547$ $\quad R^2 = .91 \quad F_0 = 161.96$

Orcutt procedure has produced an estimate of the slope that differs only slightly from that found by ordinary least squares. However, when comparing the standard errors, we find that the estimate of the slope from the Cochrane–Orcutt iterative procedure has a larger standard error than the ordinary least squares estimate. This underscores our previous comment that if ordinary least squares is used when the errors are positively autocorrelated, the standard errors of the regression coefficients are likely to be underestimated. In terms of the original variables, the intercept and its standard error are

$$\hat{\beta}_0 = \frac{\hat{\beta}_0}{1-\hat{\rho}} = \frac{15.85043}{1-0.409} = 26.81968$$

and

$$\text{se}(\hat{\beta}_0) = \frac{\text{se}(\hat{\beta}_0')}{1-\hat{\rho}} = \frac{0.9471}{1-0.409} = 1.6025$$

A comparison of these values with Table 9.5 reveals that the iterative estimate of the intercept does not differ dramatically from the ordinary least squares estimate, but the iterative estimate has a larger standard error. •

9.2 Generalized and weighted least squares

The assumptions usually made concerning the linear regression model $\mathbf{y} = \mathbf{X}\boldsymbol{\beta} + \boldsymbol{\varepsilon}$ are that $E(\boldsymbol{\varepsilon}) = \mathbf{0}$ and that $V(\boldsymbol{\varepsilon}) = \sigma^2\mathbf{I}$. Sometimes these assumptions are unreasonable, and in this section we will consider what modifications to the ordinary least squares procedure are necessary when $V(\boldsymbol{\varepsilon}) = \sigma^2\mathbf{V}$, where $\mathbf{V}$ is a known $n \times n$ matrix. This situation has an easy interpretation; if $\mathbf{V}$ is diagonal but with unequal diagonal elements then the observations $\mathbf{y}$ are uncorrelated but have unequal variances, while if some of the off-diagonal elements of $\mathbf{V}$ are nonzero then the observations are correlated.

When the model is

$$\begin{aligned} \mathbf{y} &= \mathbf{X}\boldsymbol{\beta}+\boldsymbol{\varepsilon} \\ E(\boldsymbol{\varepsilon}) &= \mathbf{0},\ V(\boldsymbol{\varepsilon})=\sigma^2\mathbf{V} \end{aligned} \tag{9.9}$$

the ordinary least squares estimator $\hat{\boldsymbol{\beta}}=(\mathbf{X}'\mathbf{X})^{-1}\mathbf{X}'\mathbf{y}$ is no longer appropriate. We will approach this problem by transforming the model to a new set of observations that satisfy the standard least squares assumptions. Then we will use ordinary least squares on the transformed data. Since $\sigma^2\mathbf{V}$ is the covariance matrix of the errors, $\mathbf{V}$ must be nonsingular and positive definite, so there exists an $n \times n$ nonsingular symmetric matrix $\mathbf{K}$, where $\mathbf{K}'\mathbf{K}=\mathbf{K}\mathbf{K}=\mathbf{V}$. The matrix $\mathbf{K}$ is often called the *square root* of $\mathbf{V}$.

Define the new variables

$$\mathbf{z}=\mathbf{K}^{-1}\mathbf{y},\ \mathbf{B}=\mathbf{K}^{-1}\mathbf{X},\ \text{and}\ \mathbf{g}=\mathbf{K}^{-1}\boldsymbol{\varepsilon} \tag{9.10}$$

so that the regression model $\mathbf{y}=\mathbf{X}\boldsymbol{\beta}+\boldsymbol{\varepsilon}$ becomes $\mathbf{K}^{-1}\mathbf{y}=\mathbf{K}^{-1}\mathbf{X}\boldsymbol{\beta}+\mathbf{K}^{-1}\boldsymbol{\varepsilon}$, or

$$\mathbf{z}=\mathbf{B}\boldsymbol{\beta}+\mathbf{g} \tag{9.11}$$

The errors in this transformed model have zero expectation; that is, $E(\mathbf{g})=\mathbf{K}^{-1}E(\boldsymbol{\varepsilon})=\mathbf{0}$. Furthermore the covariance matrix of $\mathbf{g}$ is

$$\begin{aligned} V(\mathbf{g}) &= E\{[\mathbf{g}-E(\mathbf{g})][\mathbf{g}-E(\mathbf{g})]'\} \\ &= E(\mathbf{g}\mathbf{g}') \\ &= E(\mathbf{K}^{-1}\boldsymbol{\varepsilon}\boldsymbol{\varepsilon}'\mathbf{K}^{-1}) \\ &= \mathbf{K}^{-1}E(\boldsymbol{\varepsilon}\boldsymbol{\varepsilon}')\mathbf{K}^{-1} \\ &= \sigma^2\mathbf{K}^{-1}\mathbf{V}\mathbf{K}^{-1} \\ &= \sigma^2\mathbf{K}^{-1}\mathbf{K}\mathbf{K}\mathbf{K}^{-1} \\ &= \sigma^2\mathbf{I} \end{aligned} \tag{9.12}$$

Thus the elements of $\mathbf{g}$ have mean zero, constant variance, and are uncorrelated. Since the errors $\mathbf{g}$ in the model (9.11) satisfy the usual assumptions, we may apply ordinary least squares. The least squares function is

$$\begin{aligned} S(\boldsymbol{\beta}) &= \mathbf{g}'\mathbf{g}=\boldsymbol{\varepsilon}'\mathbf{V}^{-1}\boldsymbol{\varepsilon} \\ &= (\mathbf{y}-\mathbf{X}\boldsymbol{\beta})'\mathbf{V}^{-1}(\mathbf{y}-\mathbf{X}\boldsymbol{\beta}) \end{aligned} \tag{9.13}$$

The least squares normal equations are

$$(\mathbf{X}'\mathbf{V}^{-1}\mathbf{X})\hat{\boldsymbol{\beta}}=\mathbf{X}'\mathbf{V}^{-1}\mathbf{y} \tag{9.14}$$

and the solution to these equations is

$$\hat{\boldsymbol{\beta}}=(\mathbf{X}'\mathbf{V}^{-1}\mathbf{X})^{-1}\mathbf{X}'\mathbf{V}^{-1}\mathbf{y} \tag{9.15}$$

$\hat{\boldsymbol{\beta}}$ is called the generalized least squares estimator of $\boldsymbol{\beta}$.

It is not difficult to show that $\hat{\boldsymbol{\beta}}$ is an unbiased estimator of $\boldsymbol{\beta}$. The covariance matrix of $\hat{\boldsymbol{\beta}}$ is

$$V(\hat{\boldsymbol{\beta}})=\sigma^2(\mathbf{B}'\mathbf{B})^{-1}=\sigma^2(\mathbf{X}'\mathbf{V}^{-1}\mathbf{X})^{-1} \tag{9.16}$$

Furthermore under the assumptions made in (9.9), $\hat{\boldsymbol{\beta}}$ is the best linear unbiased estimator of $\boldsymbol{\beta}$. The analysis of variance in terms of generalized least squares is summarized in Table 9.8.

When the errors $\boldsymbol{\varepsilon}$ are uncorrelated but have unequal variances so that the covariance matrix of $\boldsymbol{\varepsilon}$ is

$$\sigma^2\mathbf{V}=\sigma^2\begin{bmatrix} \frac{1}{w_1} & & & 0 \\ & \frac{1}{w_2} & & \\ & & \ddots & \\ 0 & & & \frac{1}{w_n} \end{bmatrix}$$

Table 9.8 Analysis of variance for generalized least squares

Source	Sum of Squares	Degrees of Freedom	Mean Square	F_0
Regression	$SS_R=\hat{\boldsymbol{\beta}}'\mathbf{B}'\mathbf{z}$			
	$=\mathbf{y}'\mathbf{V}^{-1}\mathbf{X}(\mathbf{X}'\mathbf{V}^{-1}\mathbf{X})^{-1}\mathbf{X}'\mathbf{V}^{-1}\mathbf{y}$	p	SS_R/p	MS_R/MS_E
Error	$SS_E=\mathbf{z}'\mathbf{z}-\hat{\boldsymbol{\beta}}'\mathbf{B}'\mathbf{z}$			
	$=\mathbf{y}'\mathbf{V}^{-1}\mathbf{y}-\mathbf{y}'\mathbf{V}^{-1}\mathbf{X}(\mathbf{X}'\mathbf{V}^{-1}\mathbf{X})^{-1}\mathbf{X}'\mathbf{V}^{-1}\mathbf{y}$	$n-p$	$SS_E/(n-p)$	
Total	$\mathbf{z}'\mathbf{z}=\mathbf{y}'\mathbf{V}^{-1}\mathbf{y}$	n		

for example, the estimation procedure is usually called *weighted least squares*. Let $\mathbf{W}=\mathbf{V}^{-1}$. Since $\mathbf{V}$ is a diagonal matrix, $\mathbf{W}$ is also diagonal with diagonal elements $w_1, w_2, \ldots, w_n$. From (9.14), the weighted least squares normal equations are

$$(\mathbf{X}'\mathbf{W}\mathbf{X})\hat{\boldsymbol{\beta}}=\mathbf{X}'\mathbf{W}\mathbf{y}$$

and

$$\hat{\boldsymbol{\beta}}=(\mathbf{X}'\mathbf{W}\mathbf{X})^{-1}\mathbf{X}'\mathbf{W}\mathbf{y}$$

is the weighted least squares estimator. The w_i are often called weights. Observations with small w_i have larger variance than observations with large w_i.

Weighted least squares estimates may be obtained easily from an ordinary least squares computer program. If we multiply each of the observed values for the ith observation (including the 1 for the intercept) by the square root of the weight for that observation, then we obtain a transformed set of data, such as

$$\mathbf{B}=\begin{bmatrix} 1\sqrt{w_1} & x_{11}\sqrt{w_1} & \cdots & x_{1k}\sqrt{w_1} \\ 1\sqrt{w_2} & x_{21}\sqrt{w_2} & \cdots & x_{2k}\sqrt{w_2} \\ \vdots & \vdots & & \vdots \\ 1\sqrt{w_n} & x_{n1}\sqrt{w_n} & \cdots & x_{nk}\sqrt{w_n} \end{bmatrix} \qquad \mathbf{z}=\begin{bmatrix} y_1\sqrt{w_1} \\ y_2\sqrt{w_2} \\ \vdots \\ y_n\sqrt{w_n} \end{bmatrix}$$

Now if we apply ordinary least squares to this transformed data we obtain

$$\hat{\boldsymbol{\beta}}=(\mathbf{B}'\mathbf{B})^{-1}\mathbf{B}'\mathbf{z}=(\mathbf{X}'\mathbf{W}\mathbf{X})^{-1}\mathbf{X}'\mathbf{W}\mathbf{y}$$

the weighted least squares estimate of $\boldsymbol{\beta}$.

To use weighted least squares, the weights w_i must be known. As noted in Chapter 3, sometimes prior knowledge or experience, or information from a theoretical model can be used to determine the weights (for an example of this approach, see Weisberg [1980]). Alternatively, residual analysis may indicate that the variance of the errors may be a function of one of the regressors, say $V(\varepsilon_i)=\sigma^2 x_{ij}$, so that $w_i=1/x_{ij}$. In some cases, we may have to guess at the weights, perform the analysis, and then re-estimate the weights based on the results. Several iterations may be necessary. We will give an important application of weighted least squares in the next section.

Weighted or generalized least squares methods may also be used in cases where the errors in the regression model are serially correlated. For example, if the observations are arranged in time order, then the ijth element of the matrix $\mathbf{V}$ is ρ_u, where $u=|i-j|$ and $\rho_0=1$. We could estimate ρ_u by finding the simple correlation between observations that are u steps apart. ρ_u is sometimes called the lag u autocorrelation coefficient. The estimates of ρ_u provide an estimate $\hat{\mathbf{V}}$ of $\mathbf{V}$, and the inverse matrix $\hat{\mathbf{V}}^{-1}$ would be used in the estimation procedure.

Since generalized or weighted least squares requires making additional assumptions regarding the errors, it is of interest to ask what happens when we fail to do this and use ordinary least squares in a situation where $V(\boldsymbol{\varepsilon})=\sigma^2\mathbf{V}$ with $\mathbf{V}\neq\mathbf{I}$. If ordinary least squares is used in this case, the resulting estimator $\hat{\boldsymbol{\beta}}=(\mathbf{X}'\mathbf{X})^{-1}\mathbf{X}'\mathbf{y}$ is still unbiased. However, the ordinary least squares estimator is no longer a minimum variance estimator. That is, the covariance matrix of the ordinary least squares estimator is

$$V(\hat{\boldsymbol{\beta}})=\sigma^2(\mathbf{X}'\mathbf{X})^{-1}\mathbf{X}'\mathbf{V}\mathbf{X}(\mathbf{X}'\mathbf{X})^{-1} \tag{9.17}$$

and the covariance matrix of the generalized least squares estimator (9.16) gives smaller variances for the regression coefficients. Thus generalized or weighted least squares is preferable to ordinary least squares whenever $\mathbf{V}\neq\mathbf{I}$.

9.3 Robust regression

9.3.1 The need for robust estimation

When the observations $\mathbf{y}$ in the linear regression model $\mathbf{y}=\mathbf{X}\boldsymbol{\beta}+\boldsymbol{\varepsilon}$ are normally distributed, the method of least squares works well in the sense that it produces an estimate of $\boldsymbol{\beta}$ that has good statistical properties. However, when the observations follow some nonnormal distribution, particularly one that has longer or heavier tails than the normal, the method of least squares may not be appropriate. Heavy-tailed distributions usually generate outliers, and these outliers may have a strong influence on the least squares estimate. In effect, outliers "pull" the least squares fit too much in their direction, and consequently the identification of these outliers is difficult because their residuals have been made artificially small. Skillfull residual analysis coupled with the use of techniques for identifying influential observations such as those in Section 4.7.4 can help the analyst discover these problems. However, the successful use of these diagnostic procedures often requires abilities beyond those of the average analyst.

A number of authors have proposed *robust* regression procedures designed to dampen the effect of observations that would be highly influential if least squares were used. That is, a robust procedure tends to leave the residuals associated with outliers large, thereby making the identification of influential points much easier. In addition to insensitivity to outliers, a robust estimation procedure should be 90–95 percent as efficient as least squares when the underlying distribution is normal. Basic references in robust estimation include Andrews et al. [1972], Andrews [1974], Hill and Holland [1977], Hogg [1974, 1979a, b], and Huber [1972, 1973, 1981].

To motivate the discussion, and to demonstrate why it may be desirable to use an alternative to least squares when the observations are nonnormal, consider the simple linear regression model

$$y_i = \beta_0 + \beta_1 x_i + \varepsilon_i, \qquad i = 1, 2, \ldots, n \tag{9.18}$$

where the errors are independent random variables that follow the double exponential distribution

$$f(\varepsilon_i) = \frac{1}{2\sigma} e^{-|\varepsilon_i|/\sigma}, \qquad -\infty < \varepsilon_i < \infty \tag{9.19}$$

The double exponential distribution is shown in Figure 9.4. The distribution is more pointed in the middle than the normal and tails off to zero as $|\varepsilon_i|$ goes to infinity. However, since the density function goes to zero as $e^{-|\varepsilon_i|}$ goes to zero, and the normal density function goes to zero as $e^{-\varepsilon_i^2}$ goes to zero, we see that the double exponential distribution has heavier tails than the normal.

We will use the method of maximum likelihood to estimate β_0 and β_1. The likelihood function is

$$L(\beta_0, \beta_1) = \prod_{i=1}^{n} \frac{1}{2\sigma} e^{-|\varepsilon_i|/\sigma} = \frac{1}{(2\sigma)^n} e^{-\sum_{i=1}^{n} |\varepsilon_i|/\sigma} \tag{9.20}$$

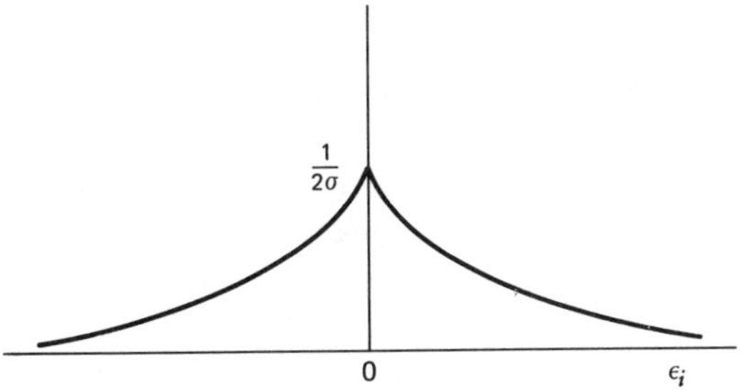

Figure 9.4 The double exponential distribution.

Therefore maximizing the likelihood function would involve minimizing $\sum_{i=1}^{n}|\varepsilon_i|$, the sum of the absolute errors. Recall from Section 2.9 that the method of maximum likelihood applied to the regression model with normal errors leads to the least squares criterion. Thus the assumption of an error distribution with heavier tails than the normal implies that the method of least squares is no longer an optimal estimation technique. Note that the absolute error criterion would weight outliers far less severely than would least squares. Minimizing the sum of the absolute errors is often called the L_1-norm regression problem. Least squares is the L_2-norm regression problem.

The L_1-norm regression problem can be formulated as a linear programming (LP) problem. Let c_i and d_i $(i=1,2,\ldots,n)$ be the positive and negative deviations about the fitted line. Then the regression coefficients $\hat{\beta}_{L_1,0}$ and $\hat{\beta}_{L_1,1}$ that minimize the sum of the absolute errors are the solution to the LP problem

$$\begin{aligned} &\text{minimize} && Z=\sum_{i=1}^{n}(c_i+d_i) \\ &\text{subject to:} && y_i-\hat{\beta}_{L_1,0}-\hat{\beta}_{L_1,1}\ x_i+c_i-d_i=0, \qquad i=1,2,\ldots,n \\ & && \hat{\beta}_{L_1,0},\hat{\beta}_{L_1,1} \text{ unrestricted in sign.} \end{aligned} \tag{9.21}$$

The extension to multiple regression is straightforward. In general, the LP problem has n constraints (one for each observation) and $p+2n$ variables (one variable for each model parameter and $2n$ variables representing the positive and negative deviations). Unfortunately, standard LP algorithms do not ensure that unbiased estimates of $\boldsymbol{\beta}$ are obtained. Hartley and Sielken [1973] have developed an efficient LP algorithm that produces an unbiased solution. For other references on L_1-norm regression, see Barrodale [1968], Barrodale and Roberts [1973], Book, Booker, Hartley and Sielken [1980], Gentle, Kennedy and Sposito [1977], and Wagner [1959].

The L_1-norm regression problem is a special case of L_p-norm regression, in which the model parameters are chosen to minimize $\sum_{i=1}^{n}|\varepsilon_i|^p$ $(1\leqslant p\leqslant 2)$. For $1<p<2$ this reduces to a nonlinear programming problem. Forsythe [1972] has studied this procedure extensively for the straight-line regression model via Monte Carlo simulation, using several nonnormal error distributions. He notes that $p=1.5$ is a good compromise choice leading to substantially better estimates than least squares when the errors are nonnormal. When the error distribution is normal, using $p=1.5$ results in estimates that are at worst 90 percent as efficient as least squares.

9.3.2 *M* estimators

We have noted that the L_1-norm regression problem arises naturally from the maximum likelihood approach with double exponential errors. In general we may define a *class* of robust estimators that minimize a function ρ of the residuals, for example

$$\min_{\boldsymbol{\beta}} \sum_{i=1}^{n} \rho(e_i) = \min_{\boldsymbol{\beta}} \sum_{i=1}^{n} \rho(y_i - \mathbf{x}_i'\boldsymbol{\beta}) \tag{9.22}$$

where $\mathbf{x}_i'$ denotes the ith row of $\mathbf{X}$. An estimator of this type is called an M-estimator, where M stands for maximum likelihood. That is, the function ρ is related to the likelihood function for an appropriate choice of the error distribution. For example, if the method of least squares is used (implying that the error distribution is normal), then $\rho(z)=\frac{1}{2}z^2, -\infty<z<\infty$. The M-estimator is not necessarily scale-invariant (that is, if the residuals $y_i - \mathbf{x}_i'\boldsymbol{\beta}$ were multiplied by a constant, the new solution to (9.22) might not be the same as the old one). To obtain a scale-invariant version of this estimator, we usually solve

$$\min_{\boldsymbol{\beta}} \sum_{i=1}^{n} \rho(e_i/s) = \min_{\boldsymbol{\beta}} \sum_{i=1}^{n} \rho[(y_i - \mathbf{x}_i'\boldsymbol{\beta})/s] \tag{9.23}$$

where s is a robust estimate of scale. A popular choice for s is

$$s = \text{median}|e_i - \text{median}\,(e_i)|/0.6745$$

The constant 0.6745 makes s an approximately unbiased estimator of σ if n is large and the error distribution is normal.

To minimize (9.23), equate the first partial derivatives of ρ with respect to β_j $(j=0,1,\ldots,k)$ equal to zero, yielding a necessary condition for a minimum. This gives the system of $p=k+1$ equations

$$\sum_{i=1}^{n} x_{ij}\psi[(y_i - \mathbf{x}_i'\boldsymbol{\beta})/s] = 0, \; j=0,1,\ldots,k \tag{9.24}$$

where $\psi=\rho'$ and x_{ij} is the ith observation on the jth regressor and $x_{i0}=1$. In general the ψ function is nonlinear and (9.24) must be solved by iterative methods. While several nonlinear optimization techniques could be employed, iteratively reweighted least squares is most widely used. This approach is usually attributed to Beaton and Tukey [1974]. To use iteratively reweighted least squares, suppose that an initial estimate $\hat{\boldsymbol{\beta}}_0$ is available and that s is an estimate of scale. Then write the $p=k+1$ equations in (9.24)

$$\sum_{i=1}^{n} x_{ij}\psi[(y_i-\mathbf{x}_i'\boldsymbol{\beta})/s]=\sum_{i=1}^{n} x_{ij}\frac{\psi[(y_i-\mathbf{x}_i'\boldsymbol{\beta})/s]}{(y_i-\mathbf{x}_i'\boldsymbol{\beta})/s}(y_i-\mathbf{x}_i'\boldsymbol{\beta})/s=0,$$

$$j=0,1,\ldots,k \qquad (9.25)$$

as

$$\sum_{i=1}^{n} x_{ij}w_{i0}(y_i-\mathbf{x}_i'\boldsymbol{\beta})=0, \qquad j=0,1,\ldots,k \qquad (9.26)$$

where

$$w_{i0}=\begin{cases}\dfrac{\psi[(y_i-\mathbf{x}_i'\hat{\boldsymbol{\beta}}_0)/s]}{(y_i-\mathbf{x}_i'\hat{\boldsymbol{\beta}}_0)/s} & \text{if } y_i\neq\mathbf{x}_i'\hat{\boldsymbol{\beta}}_0\\ 1 & \text{if } y_i=\mathbf{x}_i'\hat{\boldsymbol{\beta}}_0\end{cases} \qquad (9.27)$$

In matrix notation, (9.26) becomes

$$\mathbf{X}'\mathbf{W}_0\mathbf{X}\boldsymbol{\beta}=\mathbf{X}'\mathbf{W}_0\mathbf{y} \qquad (9.28)$$

where $\mathbf{W}_0$ is an $n\times n$ diagonal matrix of "weights" with diagonal elements $w_{10}, w_{20},\ldots, w_{n0}$ given by (9.27). We recognize (9.28) as the usual weighted least squares normal equations. Consequently the one-step estimator is

$$\hat{\boldsymbol{\beta}}_1=(\mathbf{X}'\mathbf{W}_0\mathbf{X})^{-1}\mathbf{X}'\mathbf{W}_0\mathbf{y} \qquad (9.29)$$

At the next step, we recompute the weights from (9.27) but using $\hat{\boldsymbol{\beta}}_1$ instead of $\hat{\boldsymbol{\beta}}_0$. Usually only a few iterations are required to achieve convergence. As Holland and Welsch [1977] note, the iteratively reweighted least squares procedure requires only a standard weighted least squares computer program.

A number of popular robust criterion functions are shown in Table 9.9. The behavior of these ρ-functions and their corresponding ψ-functions are illustrated in Figures 9.5 and 9.6, respectively. Robust regression procedures can be classified by the behavior of their ψ-function. The ψ-function controls the weight given to each residual, and (apart from a constant of

Table 9.9 Robust criterion functions

Criterion	$\rho(z)$	$\psi(z)$	$w(z)$	Range
Least squares	$\frac{1}{2}z^2$	z	1.0	$\lvert z\rvert<\infty$
Huber's t function	$\frac{1}{2}z^2$	z	1.0	$\lvert z\rvert\leqslant t$
$t=2$	$\lvert z\rvert t-\frac{1}{2}t^2$	$t\,\text{sign}(z)$	$\frac{t}{\lvert z\rvert}$	$\lvert z\rvert>t$
Ramsay's E_a function $a=0.3$	$a^{-2}[1-\exp(-a\lvert z\rvert)\cdot(1+a\lvert z\rvert)]$	$z\exp(-a\lvert z\rvert)$	$\exp(-a\lvert z\rvert)$	$\lvert z\rvert<\infty$
Andrew's wave function	$a[1-\cos(z/a)]$	$\sin(z/a)$	$\frac{\sin(z/a)}{z/a}$	$\lvert z\rvert\leqslant a\pi$
$a=1.339$	$2a$	0	0	$\lvert z\rvert>a\pi$
Hampel's 17A function	$\frac{1}{2}z^2$	z	1.0	$\lvert z\rvert\leqslant a$
$a=1.7$	$a\lvert z\rvert-\frac{1}{2}a^2$	$a\,\text{sign}(z)$	$a/\lvert z\rvert$	$a<\lvert z\rvert\leqslant b$
$b=3.4$				
$c=8.5$				
	$\frac{a\left(c\lvert z\rvert-\frac{1}{2}z^2\right)}{c-b}-(7/6)a^2$	$\frac{a\,\text{sign}(z)(c-\lvert z\rvert)}{c-b}$	$\frac{a(c-\lvert z\rvert)}{\lvert z\rvert(c-b)}$	$b<\lvert z\rvert\leqslant c$
	$a(b+c-a)$	0	0	$\lvert z\rvert>c$

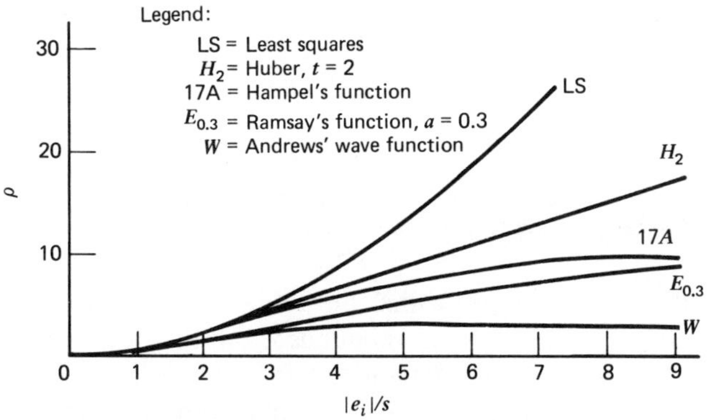

Figure 9.5 Robust criterion functions.

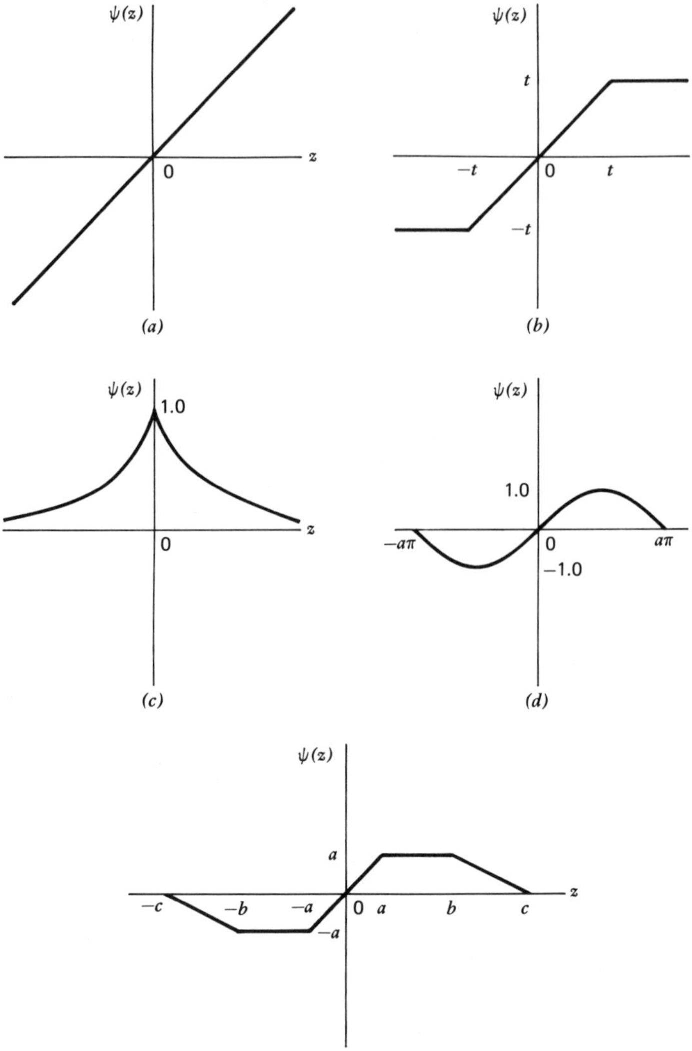

Figure 9.6 Robust influence functions. (**a**) Least squares. (**b**) Huber's *t* function. (**c**) Ramsay's E_a function. (**d**) Andrews' wave function. (**e**) Hampel's 17A function.

proportionality) is sometimes called the influence function. For example, the ψ-function for least squares is unbounded, and thus least squares tend to be nonrobust when used with data arising from a heavy-tailed distribution. The Huber *t* function (Huber [1964]) has a monotone ψ–function, and does not weight large residuals as heavily as least squares. The last three influence functions actually redescend as the residual becomes larger.

Ramsay's E_a function (see Ramsay [1977]) is a soft redescender; that is, the ψ-function is asymptotic to zero for large $|z|$. Andrew's wave function and Hampel's 17A function (see Andrews et al. [1972] and Andrews [1974]) are hard redescenders; that is, the ψ-function equals zero for sufficiently large $|z|$. We should note that the ρ-functions associated with the redescending ψ–functions are nonconvex, and this in theory can cause convergence problems in the iterative estimation procedure. However, this is not a common occurrence. Furthermore each of the robust criterion functions requires the analyst to specify certain "tuning constants" for the ψ–functions. We have shown typical values of these tuning constants in Table 9.9.

Several authors (Andrews [1974], Hogg [1979a], Hocking [1978]) have noted that the starting value $\hat{\boldsymbol{\beta}}_0$ used in robust estimation must be chosen carefully. Using the least squares solution can disguise the high-leverage points. The L_1-norm estimates would be a good choice of starting values. Andrews [1974] and Dutter [1977] also suggest procedures for choosing the starting values.

At present it is difficult to give strong recommendations concerning the error structure of the final robust regression estimates $\hat{\boldsymbol{\beta}}$. Determining the covariance matrix of $\hat{\boldsymbol{\beta}}$ is important if we are to construct confidence intervals or make other model inferences. Huber [1973] has shown that asymptotically $\hat{\boldsymbol{\beta}}$ has an approximate normal distribution with covariance matrix

$$\sigma^2 \frac{E[\psi^2(\varepsilon/\sigma)]}{\{E[\psi'(\varepsilon/\sigma)]\}^2}(\mathbf{X}'\mathbf{X})^{-1}$$

Therefore a resonable approximation for the covariance matrix of $\hat{\boldsymbol{\beta}}$ is

$$\frac{(ns)^2}{n-p} \frac{\sum_{i=1}^{n} \psi^2[(y_i - \mathbf{x}_i'\hat{\boldsymbol{\beta}})/s]}{\left\{\sum_{i=1}^{n} \psi'[(y_i - \mathbf{x}_i'\hat{\boldsymbol{\beta}})/s]\right\}^2}(\mathbf{X}'\mathbf{X})^{-1}$$

The weighted least squares computer program also produces an estimate of the covariance matrix

$$\frac{\sum_{i=1}^{n} w_i (y_i - \mathbf{x}_i'\hat{\boldsymbol{\beta}})^2}{n-p}(\mathbf{X}'\mathbf{W}\mathbf{X})^{-1}$$

Other suggestions are in Welsch [1975] and Hill [1979]. There is no general agreement about which approximation to the covariance matrix of $\hat{\boldsymbol{\beta}}$ is best.

Both Welsch and Hill note that these covariance matrix estimates perform poorly for **X** matrices that have outliers. Ill conditioning (multicollinearity) also distorts robust regression estimates. However, there are indications that in many cases we can make approximate inferences about $\hat{\boldsymbol{\beta}}$ using procedures similar to the usual normal theory.

Robust regression methods have much to offer the data analyst. They can be extremely helpful in locating outliers and highly influential observations. Whenever a least squares analysis is performed it would be useful to perform a robust fit also. If the results of the two procedures are in substantial agreement, then use the least squares results, because inferences based on least squares are at present better understood. However, if the results of the two analyses differ, then reasons for these differences should be identified. Observations that are downweighted in the robust fit should be carefully examined.

• **Example 9.4** Consider Example 4.1 in which a least squares model was developed relating delivery time to two regressors, cases (x_1) and distance (x_2). The model summary statistics do not reveal anything unusual about this fit. However, a normal probability plot of the residuals (Figure 4.8) indicates that the normality assumption is questionable, and that the errors possibly come from a heavily-tailed distribution. Further analysis will determine that there are two relatively influential observations, Points 9 and 22 (see Table 4.10).

To illustrate robust regression methods, robust fits were obtained for these data using Huber's t function, Ramsay's exponential function, Andrews' wave function, and Hampel's 17A function. These fits along with the least squares analysis, are summarized in Tables 9.10–9.14. All the robust fits were obtained using iteratively reweighted least squares, starting with the ordinary least squares solution. Since the median of the least squares residuals is 0.436360, the robust estimate of scale used is

$$s=\text{median}|e_i-0.436360|/0.6745=1.627$$

This parameter was held constant in each iteration. The convergence criteria for the robust estimates was to stop the iterative process when the maximum change in any coefficient was less than 0.1 percent.

Tables 9.10–9.14 display the actual and fitted values, the residuals, the weights given to these residuals, and the parameter estimates for each estimation method. Notice that least squares weights all residuals equally (weight= 1.0), even the extreme ones. The Huber t procedures, summarized in Table 9.11, downweights six of the residuals (Points 1, 4, 9, 20, 23 and 24). The weight given to observation 9 is only 0.327. The effect of this is to reduce the estimates of β_1 and β_2 and increase the estimate of β_0, compared to least squares. Ramsay's E_a function (Table 9.12), which is a soft redescender,

Table 9.10 Least squares fit to the delivery time data

Observation i	y_i	$\hat{y}_i$	e_i	Weight
1	.166800E+02	.217081E+02	−.502808E+01	.100000E+01
2	.115000E+02	.103536E+02	.114639E+01	.100000E+01
3	.120300E+02	.120798E+02	−.497937E−01	.100000E+01
4	.148800E+02	.995565E+01	.492435E+01	.100000E+01
5	.137500E+02	.141944E+02	−.444398E+00	.100000E+01
6	.181100E+02	.183996E+02	−.289574E+00	.100000E+01
7	.800000E+01	.715538E+01	.844624E+00	.100000E+01
8	.178300E+02	.166734E+02	.115660E+01	.100000E+01
9	.792400E+02	.718203E+02	.741971E+01	.100000E+01
10	.215000E+02	.191236E+02	.237641E+01	.100000E+01
11	.403300E+02	.380925E+02	.223749E+01	.100000E+01
12	.210000E+02	.215930E+02	−.593041E+00	.100000E+01
13	.135000E+02	.124730E+02	.102701E+01	.100000E+01
14	.197500E+02	.186825E+02	.106754E+01	.100000E+01
15	.240000E+02	.233288E+02	.671202E+00	.100000E+01
16	.290000E+02	.296629E+02	−.662928E+00	.100000E+01
17	.153500E+02	.149136E+02	.436360E+00	.100000E+01
18	.190000E+02	.155514E+02	.344862E+01	.100000E+01
19	.950000E+01	.770681E+01	.179319E+01	.100000E+01
20	.351000E+02	.408880E+02	−.578797E+01	.100000E+01
21	.179000E+02	.205142E+02	−.261418E+01	.100000E+01
22	.523200E+02	.560065E+02	−.368653E+01	.100000E+01
23	.187500E+02	.233576E+02	−.460757E+01	.100000E+01
24	.198300E+02	.244029E+02	−.457285E+01	.100000E+01
25	.107500E+02	.109626E+02	−.212584E+00	.100000E+01

$\hat{\beta}_0 = 2.3412$
$\hat{\beta}_1 = 1.6159$
$\hat{\beta}_2 = 0.014385$

downweights the residuals more severely. The weight given to Observation 9 is now 0.132, and the five other residuals for points 1, 4, 20, 23, and 24 have weights less than 0.5. Consequently the estimates of β_1 and β_2 are further reduced and the estimate of β_0 gets larger still. The hard redescenders, Andrews' wave function and Hampel's 17A function (Tables 9.13 and 9.14) treat Observation 9 even more severely. The wave function removes Point 9 from the data set (weight=0) while 17A almost does so (weight=0.019). In both instances, the estimates of β_1 and β_2 get even smaller while the estimate of β_0 continues to increase.

Table 9.11 Robust fit (Huber $t=2$) to the delivery time data

Observation i	y_i	$\hat{y}_i$	e_i	Weight
1	$.166800E+02$	$.217651E+02$	$-.508511E+01$	$.639744E+00$
2	$.115000E+02$	$.109809E+02$	$.519115E+00$	$.100000E+01$
3	$.120300E+02$	$.126296E+02$	$-.599594E+00$	$.100000E+01$
4	$.148800E+02$	$.105856E+02$	$.429439E+01$	$.757165E+00$
5	$.137500E+02$	$.146038E+02$	$-.853800E+00$	$.100000E+01$
6	$.181100E+02$	$.186051E+02$	$-.495085E+00$	$.100000E+01$
7	$.800000E+01$	$.794135E+01$	$.586521E-01$	$.100000E+01$
8	$.178300E+02$	$.169564E+02$	$.873625E+00$	$.100000E+01$
9	$.792400E+02$	$.692795E+02$	$.996050E+01$	$.327017E+00$
10	$.215000E+02$	$.193269E+02$	$.217307E+01$	$.100000E+01$
11	$.403300E+02$	$.372777E+02$	$.305228E+01$	$.100000E+01$
12	$.210000E+02$	$.216097E+02$	$-.609734E+00$	$.100000E+01$
13	$.135000E+02$	$.129900E+02$	$.510021E+00$	$.100000E+01$
14	$.197500E+02$	$.188904E+02$	$.859556E+00$	$.100000E+01$
15	$.240000E+02$	$.232828E+02$	$.717244E+00$	$.100000E+01$
16	$.290000E+02$	$.293174E+02$	$-.317449E+00$	$.100000E+01$
17	$.153500E+02$	$.152908E+02$	$.592377E-01$	$.100000E+01$
18	$.190000E+02$	$.158847E+02$	$.311529E+01$	$.100000E+01$
19	$.950000E+01$	$.845286E+01$	$.104714E+01$	$.100000E+01$
20	$.351000E+02$	$.399326E+02$	$-.483256E+01$	$.672828E+00$
21	$.179000E+02$	$.205793E+02$	$-.267929E+01$	$.100000E+01$
22	$.523200E+02$	$.542361E+02$	$-.191611E+01$	$.100000E+01$
23	$.187500E+02$	$.233102E+02$	$-.456023E+01$	$.713481E+00$
24	$.198300E+02$	$.243238E+02$	$-.449377E+01$	$.723794E+00$
25	$.107500E+02$	$.115474E+02$	$-.797359E+00$	$.100000E+01$

$\hat{\beta}_0=3.3736$
$\hat{\beta}_1=1.5282$
$\hat{\beta}_2=0.013739$

It is interesting to compare the robust parameter estimates with the least squares estimates when Point 9 is deleted. In Chapter 4, we discovered that deleting Observation 9 gave least squares estimates of $\hat{\beta}_0=4.477$, $\hat{\beta}_1=1.498$, and $\hat{\beta}_2=0.010$. The robust procedures have approximately the same effect on the parameter estimates, although not generally as severe with respect to β_2. That is, deleting Point 9 and using ordinary least squares or using a robust regression procedure will increase the estimate of the intercept and decrease

Table 9.12 Robust fit (Ramsay's $E_{0.3}$ function) to the delivery time data

Observation i	y_i	$\hat{y}_i$	e_i	Weight
1	$.166800E+02$	$.218009E+02$	$-.512091E+01$	$.388862E+00$
2	$.115000E+02$	$.112454E+02$	$.254584E+00$	$.953571E+00$
3	$.120300E+02$	$.128682E+02$	$-.838215E+00$	$.857082E+00$
4	$.148800E+02$	$.108415E+02$	$.403846E+01$	$.474482E+00$
5	$.137500E+02$	$.147670E+02$	$-.101696E+01$	$.829548E+00$
6	$.181100E+02$	$.186905E+02$	$-.580544E+00$	$.898687E+00$
7	$.800000E+01$	$.826846E+01$	$-.268460E+00$	$.952527E+00$
8	$.178300E+02$	$.170677E+02$	$.762255E+00$	$.868454E+00$
9	$.792400E+02$	$.682279E+02$	$.110121E+02$	$.131649E+00$
10	$.215000E+02$	$.194307E+02$	$.206932E+01$	$.682987E+00$
11	$.403300E+02$	$.369364E+02$	$.339359E+01$	$.535303E+00$
12	$.210000E+02$	$.216035E+02$	$-.603534E+00$	$.895015E+00$
13	$.135000E+02$	$.132081E+02$	$.291877E+00$	$.947139E+00$
14	$.197500E+02$	$.189862E+02$	$.763769E+00$	$.868651E+00$
15	$.240000E+02$	$.232651E+02$	$.734923E+00$	$.873345E+00$
16	$.290000E+02$	$.291901E+02$	$-.190116E+00$	$.964755E+00$
17	$.153500E+02$	$.154431E+02$	$-.931215E-01$	$.983506E+00$
18	$.190000E+02$	$.160129E+02$	$.298707E+01$	$.576113E+00$
19	$.950000E+01$	$.875713E+01$	$.742875E+00$	$.871123E+00$
20	$.351000E+02$	$.395347E+02$	$-.443471E+01$	$.440968E+00$
21	$.179000E+02$	$.205893E+02$	$-.268928E+01$	$.609354E+00$
22	$.523200E+02$	$.534802E+02$	$-.116016E+01$	$.806251E+00$
23	$.187500E+02$	$.232921E+02$	$-.454212E+01$	$.432732E+00$
24	$.198300E+02$	$.243045E+02$	$-.447455E+01$	$.437999E+00$
25	$.107500E+02$	$.117882E+02$	$-.103817E+01$	$.826364E+00$

$\hat{\beta}_0 = 3.8021$
$\hat{\beta}_1 = 1.4894$
$\hat{\beta}_2 = 0.013523$

the estimates of β_1 and β_2. Thus in this example the use of robust regression procedures is a compromise between keeping Point 9 or deleting it from the data and using ordinary least squares. Note also that the robust procedures easily and automatically identify the influential observations in this data.

Which set of robust estimates should be used for the "final" model? In this particular example, it doesn't make a great deal of difference, as all the robust procedures produce coefficient estimates that are quite similar. Some further

Table 9.13 Robust fit (Andrews' wave with $a = 1.48$) to the delivery time data

Observation i	y_i	$\hat{y}_i$	e_i	Weight
1	$.166800E+02$	$.216430E+02$	$-.496300E+01$	$.427594E+00$
2	$.115000E+02$	$.116923E+02$	$-.192338E+00$	$.998944E+00$
3	$.120300E+02$	$.131457E+02$	$-.111570E+01$	$.964551E+00$
4	$.148800E+02$	$.114549E+02$	$.342506E+01$	$.694894E+00$
5	$.137500E+02$	$.152191E+02$	$-.146914E+01$	$.939284E+00$
6	$.181100E+02$	$.188574E+02$	$-.747381E+00$	$.984039E+00$
7	$.800000E+01$	$.890189E+01$	$-.901888E+00$	$.976864E+00$
8	$.178300E+02$	$.174040E+02$	$.425984E+00$	$.994747E+00$
9	$.792400E+02$	$.660818E+02$	$.131582E+02$	0.
10	$.215000E+02$	$.192716E+02$	$.222839E+01$	$.863633E+00$
11	$.403300E+02$	$.363170E+02$	$.401296E+01$	$.597491E+00$
12	$.210000E+02$	$.218392E+02$	$-.839167E+00$	$.980003E+00$
13	$.135000E+02$	$.135744E+02$	$-.744338E-01$	$.999843E+00$
14	$.197500E+02$	$.189979E+02$	$.752115E+00$	$.983877E+00$
15	$.240000E+02$	$.232029E+02$	$.797080E+00$	$.981854E+00$
16	$.290000E+02$	$.286336E+02$	$.366350E+00$	$.996228E+00$
17	$.153500E+02$	$.158247E+02$	$-.474704E+00$	$.993580E+00$
18	$.190000E+02$	$.164593E+02$	$.254067E+01$	$.824146E+00$
19	$.950000E+01$	$.946384E+01$	$.361558E-01$	$.999956E+00$
20	$.351000E+02$	$.387684E+02$	$-.366837E+01$	$.655336E+00$
21	$.179000E+02$	$.209308E+02$	$-.303081E+01$	$.756603E+00$
22	$.523200E+02$	$.523766E+02$	$-.566063E-01$	$.999908E+00$
23	$.187500E+02$	$.232271E+02$	$-.447714E+01$	$.515506E+00$
24	$.198300E+02$	$.240095E+02$	$-.417955E+01$	$.567792E+00$
25	$.107500E+02$	$.123027E+02$	$-.155274E+01$	$.932266E+00$

$\hat{\beta}_0 = 4.6532$
$\hat{\beta}_1 = 1.4582$
$\hat{\beta}_2 = 0.012111$

analysis, such as a study of the residuals from the various models, or an investigation of their relative effectiveness as prediction equations, may provide a basis for final model selection. Suppose we choose the robust fit given by Ramsay's E_a function as the final model. This is often a good compromise choice between the monotone influence function of the Huber t method and the hard redescenders, Andrews' wave function, and Hampel's 17A function.

Table 9.14 Robust fit (Hampel's 17A function) to the delivery time data

Observation i	y_i	$\hat{y}_i$	e_i	Weight
1	$.166800E+02$	$.216086E+02$	$-.492859E+01$	$.561047E+00$
2	$.115000E+02$	$.116605E+02$	$-.160516E+00$	$.100000E+01$
3	$.120300E+02$	$.130997E+02$	$-.106968E+01$	$.100000E+01$
4	$.148800E+02$	$.114491E+02$	$.343090E+01$	$.805460E+00$
5	$.137500E+02$	$.152238E+02$	$-.147384E+01$	$.100000E+01$
6	$.181100E+02$	$.188502E+02$	$-.740196E+00$	$.100000E+01$
7	$.800000E+01$	$.887367E+01$	$-.873671E+00$	$.100000E+01$
8	$.178300E+02$	$.174110E+02$	$.418966E+00$	$.100000E+01$
9	$.792400E+02$	$.661574E+02$	$.130826E+02$	$.192790E-01$
10	$.215000E+02$	$.192131E+02$	$.228695E+01$	$.100000E+01$
11	$.403300E+02$	$.363522E+02$	$.397779E+01$	$.695775E+00$
12	$.210000E+02$	$.218738E+02$	$-.873835E+00$	$.100000E+01$
13	$.135000E+02$	$.135479E+02$	$-.478834E-01$	$.100000E+01$
14	$.197500E+02$	$.189657E+02$	$.784337E+00$	$.100000E+01$
15	$.240000E+02$	$.232006E+02$	$.799403E+00$	$.100000E+01$
16	$.290000E+02$	$.286019E+02$	$.398080E+00$	$.100000E+01$
17	$.153500E+02$	$.158235E+02$	$-.473491E+00$	$.100000E+01$
18	$.190000E+02$	$.164756E+02$	$.252442E+01$	$.100000E+01$
19	$.950000E+01$	$.945380E+01$	$.462003E-01$	$.100000E+01$
20	$.351000E+02$	$.388032E+02$	$-.370325E+01$	$.746183E+00$
21	$.179000E+02$	$.209744E+02$	$-.307436E+01$	$.900263E+00$
22	$.523200E+02$	$.524915E+02$	$-.171475E+00$	$.100000E+01$
23	$.187500E+02$	$.232246E+02$	$-.447458E+01$	$.618068E+00$
24	$.198300E+02$	$.239757E+02$	$-.414568E+01$	$.666865E+00$
25	$.107500E+02$	$.122886E+02$	$-.153862E+01$	$.100000E+01$

$\hat{\beta}_0 = 4.6192$
$\hat{\beta}_1 = 1.4676$
$\hat{\beta}_1 = 0.011993$

A normal probability plot of the residuals from this robust fit is shown in Figure 9.7. If we compare this display to the normal probability plot of the least squares residuals (Figure 4.8), considerable improvement is noted. The residual for Point 9 is clearly identified, while the remaining 24 residuals fall more nearly along a straight line. The robust fit is not greatly influenced by Point 9.

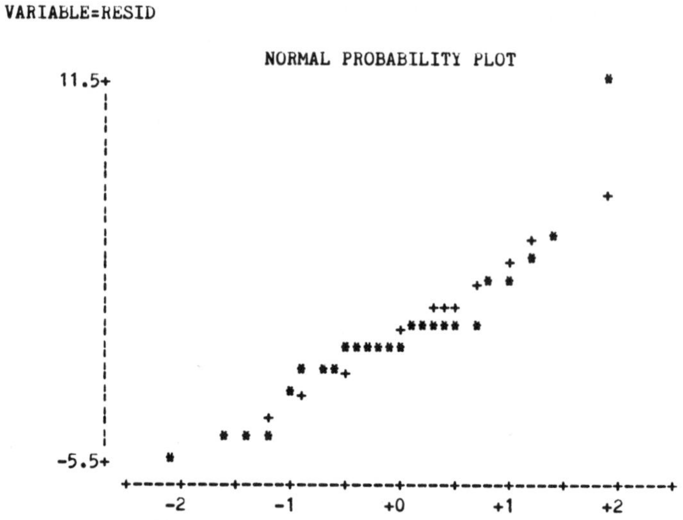

Figure 9.7 Normal probability plot (SAS) for robust fit using Ramsay's E_a function for the delivery time data.

- **Example 9.5** Andrews [1974] uses the stack loss data analyzed by Daniel and Wood [1980] to illustrate robust regression. The data, which are taken from a plant oxidizing ammonia to nitric acid, are shown in Table 9.15. A standard least squares fit to these data gives

$$\hat{y} = -36.9 + 0.72x_1 + 1.30x_2 - 0.15x_3$$

The residuals from this model are shown in column 1 of Table 9.16, and a normal probability plot is shown in Figure 9.8a. Daniel and Wood note that the residual for Point 21 is unusually large, and has considerable influence on the regression coefficients. After an insightful analysis, they delete points 1, 3, 4, and 21 from the data. A least squares fit* to the remaining data yields

$$\hat{y} = -37.6 + 0.80x_1 + 0.58x_2 - 0.07x_3$$

The residuals from this model are shown in column 2 of Table 9.16, and the corresponding normal probability plot is in Figure 9.8b. This plot does not indicate any unusual behavior in the residuals.

Andrews [1974] observes that most users of regression lack the skills of Daniel and Wood, and employs robust regression methods to produce equiva-

*Daniel and Wood [1980] actually fit a model involving x_1, x_2, and x_1^2. Andrews [1974] elected to work with all three original regressors. He notes that if x_3 is deleted and x_1^2 added, smaller residuals result but the general findings are the same.

Table 9.15 Stack loss data from Daniel and Wood [1980]

Observation Number	Stack Loss y	Air Flow x_1	Cooling Water Inlet Temperature x_2	Acid Concentration x_3
1	42	80	27	89
2	37	80	27	88
3	37	75	25	90
4	28	62	24	87
5	18	62	22	87
6	18	62	23	87
7	19	62	24	93
8	20	62	24	93
9	15	58	23	87
10	14	58	18	80
11	14	58	18	89
12	13	58	17	88
13	11	58	18	82
14	12	58	19	93
15	8	50	18	89
16	7	50	18	86
17	8	50	19	72
18	8	50	19	79
19	9	50	20	80
20	15	56	20	82
21	15	70	20	91

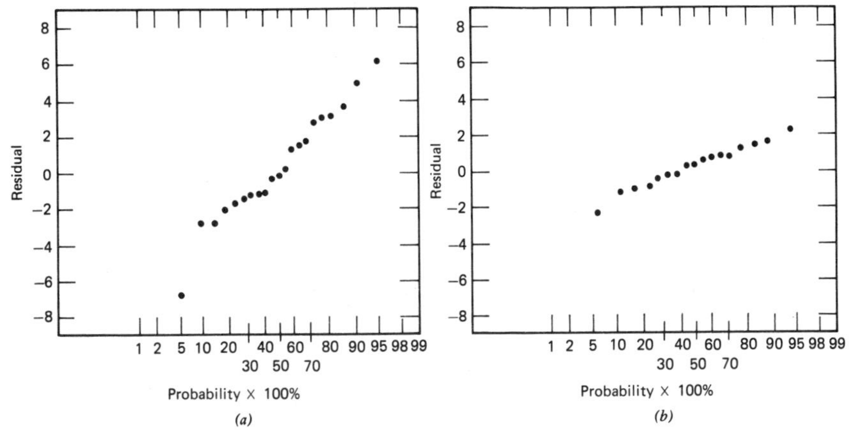

Figure 9.8 Normal probability plots from least squares fits. **(a)** Least squares with all 21 Points. **(b)** Least squares with 1,3,4,21 deleted. (From Andrews [1974] with permission of the publisher.)

lent results. A robust fit to the stack loss data using the wave function with a=1.5 yields

$$\hat{y} = -37.2 + 0.82x_1 + 0.52x_2 - 0.07x_3$$

This is virtually the same equation found by Daniel and Wood after much careful analysis. The residuals from this model are shown in column 3 of Table 9.16, and the normal probability plot is in Figure 9.9a. The four suspicious points are clearly identified in this plot. Finally, Andrews obtains a robust fit to the data with Points 1, 3, 4, and 21 removed. The resulting

Table 9.16 Residuals for various fits to the stack loss data[a]

	Residuals			
	Least Squares		Robust Fit	
Observation	(1) All 21 Points	(2) 1,3,4,21 Out	(3) All 21 Points	(4) 1,3,4,21 Out
1	3.24	6.08[b]	6.11	6.11
2	−1.92	1.15	1.04	1.04
3	4.56	6.44	6.31	6.31
4	5.70	8.18	8.24	8.24
5	−1.71	−0.67	−1.24	−1.24
6	−3.01	−1.25	−0.71	−0.71
7	−2.39	−0.42	−0.33	−0.33
8	−1.39	0.58	0.67	0.67
9	−3.14	−1.06	−0.97	−0.97
10	1.27	0.35	0.14	0.14
11	2.64	0.96	0.79	0.79
12	2.78	0.47	0.24	0.24
13	−1.43	−2.51	−2.71	−2.71
14	−0.05	−1.34	−1.44	−1.44
15	2.36	1.34	1.33	1.33
16	0.91	0.14	0.11	0.11
17	−1.52	−0.37	−0.42	−0.42
18	−0.46	0.10	0.08	0.08
19	−0.60	0.59	0.63	0.63
20	1.41	1.93	1.87	1.87
21	−7.24	− 8.63	−8.91	− 8.91

[a]Adapted from Table 5 in Andrews [1974], with permission of the publisher.
[b]Underlined residuals correspond to points not included in the fit.

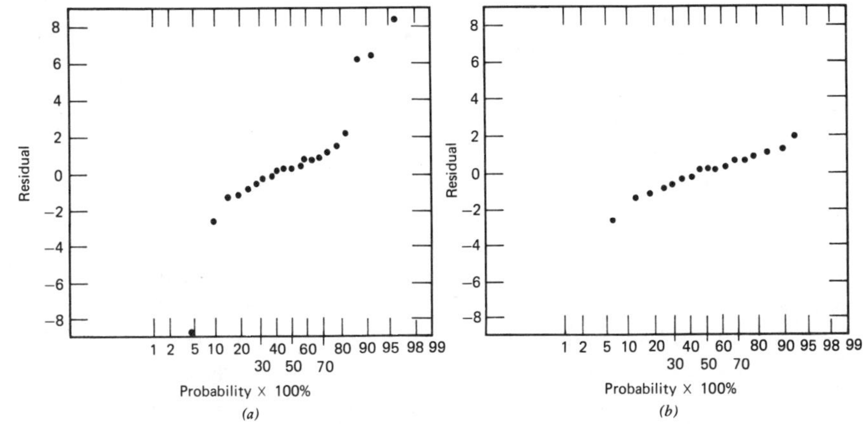

Figure 9.9 Normal probability plots from robust fits. **(a)** Robust fit with all 21 Points. **(b)** Robust fit with 1,3,4,21 deleted. (From Andrews [1974] with permission of the publisher.)

equation is identical to the one found using all 21 data Points. The residuals from this fit and the corresponding normal probability plot are shown in Column 4 of Table 9.16 and Figure 9.9b, respectively. This normal probability plot is virtually identical to the one obtained from the least squares analysis with Points 1, 3, 4, and 21 deleted [Figure 9.8b].

Once again we find that the routine application of robust regression has led to the automatic identification of the suspicious points. It has also produced a fit that does not depend on these points in any important way. Thus robust regression methods can be viewed as procedures for isolating unusually influential points, so that these points may be given further study.

9.3.3 *R* and *L* estimation

In addition to the *M* estimators discussed in the previous section, there are other approaches to robust regression. *R* estimation is a procedure based on ranks. To illustrate the general procedure, consider replacing one factor in the least squares objective function $S(\boldsymbol{\beta})=\Sigma_{i=1}^{n}(y_i-\mathbf{x}_i'\boldsymbol{\beta})^2$ by its rank. Thus if R_i is the rank of $y_i-\mathbf{x}_i'\boldsymbol{\beta}$, then we wish to minimize $\Sigma_{i=1}^{n}(y_i-\mathbf{x}_i'\boldsymbol{\beta})R_i$. More generally, we could replace the ranks (which are the integers $1,2,\ldots,n$) by the score function $a(i)$, $i=1,2,\ldots,n$, so that the objective function becomes

$$\min_{\boldsymbol{\beta}} \sum_{i=1}^{n} (y_i-\mathbf{x}_i'\boldsymbol{\beta})a(R_i)$$

If we set the score function equal to the ranks, that is, $a(i)=i$, the results are called *Wilcoxon* scores. Another possibility is to use *median* scores, that is, $a(i)=-1$ if $i<(n+1)/2$ and $a_i=1$ if $i>(n+1)/2$. Important references on R estimation in regression include Adichie [1967], Hogg and Randles [1975], Jaeckel [1972], and Jurečková [1977].

L estimators are based on *order statistics*. For example, suppose that we wish to estimate the location parameter of a distribution from a random sample $x_1, x_2, \ldots, x_n$. The order statistics of this sample are $x_{[1]} \leqslant x_{[2]} \leqslant \cdots \leqslant x_{[n-1]} \leqslant x_{[n]}$. The sample median would be an L-estimator, since it is a measure of location based on these order statistics. A number of other L-estimators for the location problem are described in Andrews et al. [1972]. The use of L estimation in the regression context is not as simple as M and R estimation. Denby and Larsen [1977] describe slice regression, which for one regressor divides the data into groups and fits the straight line using the centroids of the groups. A stepwise-type extension of this technique could be used for multiple regression. Moussa–Hamouda and Leone [1974, 1977a, b] propose procedures for simple linear regression with repeat observations on y at each x that involve trimming or discarding remote values of y.

While research continues on R and L estimation, we believe that currently the M estimators are more reasonable for the regression problem. They can be easily obtained from a weighted least squares computer program, and the weights in the final iteration identify influential points. Both L- and R-estimators are more difficult to obtain computationally than M-estimators. L-estimators do not always generalize clearly to multiple regression. Furthermore under certain conditions (see Jurečková [1977]) R-estimators are asymptotically equivalent to M-estimators.

9.3.4 Robust ridge regression

Two of the more frequent problems that the regression analyst will encounter are nonnormality of the observations and multicollinearity. Although we usually think of these two problems separately, in a significant number of practical situations nonnormality and multicollinearity occur simultaneously. Several authors have suggested that either robust or biased estimation methods alone may be sufficient for dealing with the combined problem. However, since robust regression estimates are frequently unstable when the $\mathbf{X}$ matrix is ill-conditioned, it would be desirable to have a technique for dealing directly with both problems. Hogg [1979b] has suggested a robust form of ridge regression. Recall that the ridge estimator $\hat{\boldsymbol{\beta}}_R=(\mathbf{X}'\mathbf{X}+k\mathbf{I})^{-1}\mathbf{X}'\mathbf{y}$ can be computed by augmenting the $\mathbf{X}$, $\mathbf{y}$ data with p

pseudo-observations

$$\mathbf{X}_A = \begin{bmatrix} \mathbf{X} \\ \hline \sqrt{k}\, I \end{bmatrix}, \qquad \mathbf{y}_A = \begin{bmatrix} \mathbf{y} \\ \hline \mathbf{0} \end{bmatrix}$$

and then applying ordinary least squares to $\mathbf{X}_A$, $\mathbf{y}_A$, yielding

$$\hat{\boldsymbol{\beta}}_R = (\mathbf{X}_A'\mathbf{X}_A)^{-1}\mathbf{X}_A'\mathbf{y}_A = (\mathbf{X}'\mathbf{X} + k\mathbf{I})^{-1}\mathbf{X}'\mathbf{y}$$

A robust version of ridge regression would replace least squares on $\mathbf{X}_A$, $\mathbf{y}_A$ by an appropriate robust objective function.

Askin and Montgomery [1980] explore this approach and note that generalized ridge, principal components, and fractional rank estimators can be fit using least squares on appropriately augmented data, so robust versions of these biased estimators can be easily computed. They report that the combined estimation procedure gives stable coefficient estimates while simultaneously locating and identifying outliers. The computational properties of the procedure are also good, as the iteratively reweighted least squares algorithm usually converges to the final estimates in fewer iterations that would be required if only the robust criterion were used.

9.4 Why do regression coefficients have the "wrong" sign?

When using multiple regression, occasionally the analyst experiences an apparent contradiction of intuition or theory when one or more of the regression coefficients seems to have the "wrong" sign. For example, the problem situation may imply that a particular regression coefficient should be positive, while the actual estimate of the parameter is negative. This "wrong" sign problem can be disconcerting, as it is usually difficult to explain a negative estimate (say) of a parameter to the model user when that user believes that the coefficient should be positive. Mullet [1976] points out that regression coefficients may have the wrong sign when

1. The range of some of the regressors is too small.
2. Important regressors have not been included in the model.

3. Multicollinearity is present.

4. Computational errors have been made.

It is easy to see how the range of the x's can affect the sign of the regression coefficients. Consider the simple linear regression model. The variance of the regression coefficient $\hat{\beta}_1$ is $V(\hat{\beta}_1)=\sigma^2/S_{xx}=\sigma^2/\sum_{i=1}^{n}(x_i-\bar{x})^2$. Note that the variance of $\hat{\beta}_1$ is inversely proportional to the "spread" of the regressor. Therefore if the levels of x are all close together, the variance of $\hat{\beta}_1$ will be relatively large. In some cases the variance of $\hat{\beta}_1$ could be so large that a negative estimate (for example) of a regression coefficient that is really positive results. The situation is illustrated in Figure 9.10, which plots the sampling distribution of $\hat{\beta}_1$. Examining this figure, we see that the probability of obtaining a negative estimate of $\hat{\beta}_1$ depends on how close the true regression coefficient is to zero and the variance of $\hat{\beta}_1$, which is greatly influenced by the spread of the x's.

In some situations the analyst can control the levels of the regressors. Although it is possible in these cases to decrease the variance of the regression coefficients by increasing the range of the x's, it may not be desirable to spread the levels of the regressors out too far. If the x's cover too large a range and the true response function is nonlinear, the analyst may have to develop a much more complex equation to adequately model the curvature in the system. Furthermore, many problems involve a region of x-space of specific interest to the experimenter, and spreading the regressors out beyond this region of interest may be impractical or impossible. In general we must trade off the precision of estimation, the likely complexity of the model, and the values of the regressors of practical interest when deciding how far to spread out the x's.

Wrong signs can also occur when important regressors have been left out of the model. In these cases, the sign is not really wrong. The partial nature of the regression coefficients cause the sign reversal. To illustrate, consider

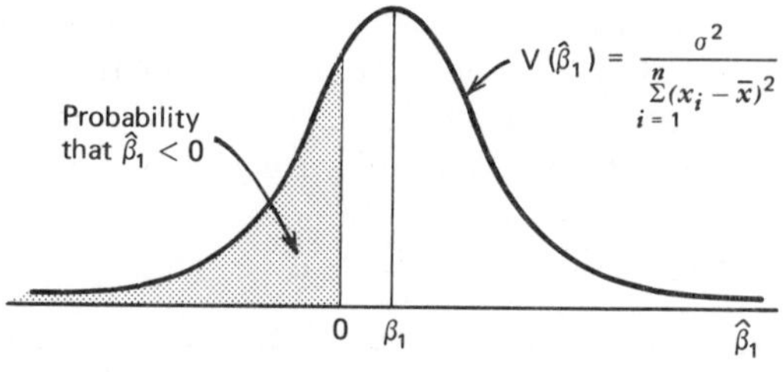

Figure 9.10 The sampling distribution of $\hat{\beta}_1$.

the following data:

x_1	x_2	y
2	1	1
4	2	5
5	2	3
6	4	8
8	4	5
10	4	3
11	6	10
13	6	7

Suppose we fit a model involving only y and x_1. The equation is

$$\hat{y}=1.835+0.463x_1$$

where $\hat{\beta}_1=0.463$ is a "total" regression coefficient. That is, it measures the total effect of x_1 ignoring the information content in x_2. The model involving both x_1 and x_2 is

$$\hat{y}=1.036-1.222x_1+3.649x_2$$

Note that now $\hat{\beta}_1=-1.222$, and a sign reversal has occurred. The reason is that $\hat{\beta}_1=-1.222$ in the multiple regression model is a "partial" regression coefficient; it measures the effect of x_1 *given* that x_2 is also in the model.

The data from this example are plotted in Figure 9.11. The reason for the difference in sign between the partial and total regression coefficients is

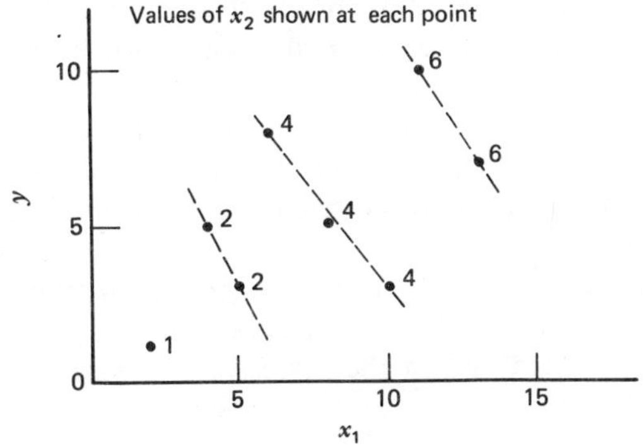

Figure 9.11 Plot of y versus x_1.

obvious from inspection of this figure. If we ignore the x_2 values, the apparent relationship between y and x_1 has a positive slope. However, if we consider the relationship between y and x_1 for constant values of x_2, we note that this relationship really has a negative slope. Thus a "wrong" sign in a regression model may indicate that important regressors are missing. If the analyst can identify these regressors and include them in the model, then the "wrong" signs may disappear.

As noted in Chapter 8, multicollinearity can cause "wrong" signs for regression coefficients. In effect, severe multicollinearity inflates the variances of the regression coefficients, and this increases the probability that one or more regression coefficients will have the wrong sign. Methods for diagnosing and dealing with multicollinearity are summarized in Chapter 8.

Computational error is also a source of "wrong" signs in regression models. Different computer programs handle round-off or truncation problems in different ways, and some programs are more effective than others in this regard. Severe multicollinearity causes the $\mathbf{X'X}$ matrix to be ill-conditioned, and this is also a source of computational error. Longley [1967] has demonstrated that computational error can cause not only sign reversals but regression coefficients to differ by several orders of magnitude. The accuracy of the computer code should be investigated when wrong sign problems are suspected. Methods for evaluating the numerical accuracy of regression computer programs are given by Longley [1967] and Mullet and Murray [1971].

9.5 Effect of measurement errors in the X's

In almost all regression problems, measurement errors affect the regressor and response variables. Often the measurement error is small, and in these cases its effect is usually ignored. However, in some situations the measurement error is not negligible, and its potential influence on the analysis must be taken into account. If the measurement errors effect only the response variable y, no difficulty is encountered as long as the measurement errors are uncorrelated random variables with zero mean and constant variance. That is, these measurement errors are incorporated into the model error term ε.

When the measurement errors affect the regressor variable x, the situation is not so simple. To illustrate the difficulties encountered, suppose that in a simple linear regression problem the regressor is measured with error, so that the observed regressor is

$$X_i = x_i + a_i, \qquad i = 1, 2, \ldots, n$$

where x_i is the true value of the regressor, X_i is the observed value, and a_i is the measurement error with $E(a_i)=0$ and $V(a_i)=\sigma_a^2$. The response variable y_i is also perturbed by an error ε_i, $i=1,2,\ldots,n$ so that the regression model is

$$y_i=\beta_0+\beta_i x_i+\varepsilon_i$$

We assume that the errors ε_i and a_i are uncorrelated; that is, $E(\varepsilon_i a_i)=0$. This is sometimes called the *errors in both variables model.* Since X_i is the observed value of the regressor, we may write

$$\begin{aligned} y_i &= \beta_0+\beta_1(X_i-a_i)+\varepsilon_i \\ &= \beta_0+\beta_1 X_i+(\varepsilon_i-\beta_1 a_i) \end{aligned} \tag{9.30}$$

Initially (9.30) may look like an ordinary linear regression model with error term $\gamma_i=\varepsilon_i-\beta_1 a_i$. However, the regressor variable X_i is a random variable and is correlated with the error term $\gamma_i=\varepsilon_i-\beta_1 a_i$. The correlation between X_i and γ_i is easily seen, since

$$\begin{aligned} \mathrm{Cov}(X_i,\gamma_i) &= E\{[X_i-E(X_i)][\gamma_i-E(\gamma_i)]\} \\ &= E[(X_i-x_i)\gamma_i] \\ &= E[(X_i-x_i)(\varepsilon_i-\beta_1 a_i)] \\ &= E(a_i\varepsilon_i-\beta_1 a_i^2) \\ &= -\beta_1\sigma_a^2 \end{aligned}$$

Thus if $\beta_1\neq 0$, the observed regressor X_i and the error term γ_i are correlated.

The usual assumption when the regressor is a random variable is that the regressor variable and the error component are independent. Violation of this assumption introduces several complexities into the problem. For example, if we apply standard least squares methods to the data (that is, ignoring the measurement error), the estimators of the model parameters are no longer unbiased. In fact, we can show that if $\mathrm{Cov}(x_i, a_i)=0$, then

$$E(\hat{\beta}_1)=\beta_1/(1+\theta)$$

where

$$\theta=\sigma_a^2/\sigma_x^2$$

and

$$\sigma_x^2 = \sum_{i=1}^{n} (x_i - \bar{x})^2/n$$

That is, $\hat{\beta}_1$ is always a biased estimator of β_1 unless $\sigma_a^2 = 0$, which occurs only when there are no measurement errors in the x_i.

Since measurement error is present to some extent in almost all practical regression situations, some advice for dealing with this problem would be helpful. Note that if σ_a^2 is small relative to σ_x^2, the bias in $\hat{\beta}_1$ will be small. This implies that if the variability in the measurement errors is small relative to the variability of the x's, then the measurement errors can be ignored and standard least squares methods applied.

Several alternative estimation methods have been proposed to deal with the problem of measurement errors in the variables. Sometimes these techniques are discussed under the topics *structural* or *functional relationships* in regression. Economists have used a technique called *two-stage least squares* in these cases. Often these methods require more extensive assumptions or information about the parameters of the distribution of measurement errors. Presentations of these methods are in Graybill [1961], Johnston [1972], Sprent [1969], and Wonnacott and Wonnacott [1970]. Other useful references include Davies and Hutton [1975], Dolby [1976], Halperin [1961], Hodges and Moore [1972], Lindley [1974], and Mandansky [1959]. A good review of the subject is in Seber [1977].

Berkson [1950] has investigated a case involving measurement errors in x_i where the method of least squares can be directly applied. His approach consists of setting the observed value of the regressor X_i to a target value. This forces X_i to be treated as fixed, while the true value of the regressor $x_i = X_i - a_i$ becomes a random variable. As an example of a situation where this approach could be used, suppose that the current flowing in an electrical circuit is used as a regressor variable. Current flow is measured with an ammeter, which is not completely accurate, so measurement error is experienced. However by setting the observed current flow to target levels of 100 amps, 125 amps, 150 amps, and 175 amps (for example), the observed current flow can be considered as fixed, and actual current flow becomes a random variable. This type of problem is frequently encountered in engineering and physical science. The regressor is a variable such as temperature, pressure, or flow rate and there is error present in the measuring instrument used to observe the variable. This approach is also sometimes called the controlled-independent-variable model.

If X_i is regarded as fixed at a preassigned target value, then (9.30), found by using the relationship $X_i = x_i + a_i$, is still appropriate. However the error term in this model, $\gamma_i = \varepsilon_i - \beta_1 a_i$, is now independent of X_i because X_i is considered to be a fixed or nonstochastic variable. Thus the errors are uncorrelated with the regressor, and the usual least squares assumptions are satisfied. Consequently a standard least squares analysis is appropriate in this case.

9.6 Simultaneous inference in regression

In Chapters 2 and 4 we discussed procedures for constructing several types of confidence and prediction intervals for the linear regression model. We have noted that these are one-at-a-time intervals; that is, they are the usual type of confidence or prediction interval where the confidence coefficient $1-\alpha$ indicates the proportion of correct statements that results when repeated random samples are selected and the appropriate interval estimate constructed for each sample. Many problems require that several confidence or prediction intervals be constructed using the same sample data. In these cases, the analyst is usually interested in specifying a confidence coefficient that applies simultaneously to the entire set of interval estimates. A set of confidence or prediction intervals that are all true simultaneously with probability $1-\alpha$ are called *simultaneous* or *joint* confidence or prediction intervals. An excellent general reference on simultaneous statistical inference is Miller [1966].

As an example, consider the delivery time data in Example 4.1. Suppose that the analyst wants to draw inferences about the intercept and both regression coefficients β_1 and β_2. One possibility would be to construct 95 percent (for example) confidence intervals about all three parameters, using (4.20). However if these interval estimates are independent, the probability that all statements are correct is $(.95)^3=.7584$. Thus we do not have a confidence level of 95 percent associated with the set of three statements. Furthermore since the intervals are constructed using the same set of sample data, they are not independent. This introduces a further complication into determining the confidence level for the set of statements. In this section, we will present several methods useful for making simultaneous inferences in regression.

9.6.1 Simultaneous inference on model parameters

Consider the estimation of $\beta_0, \beta_1, \ldots, \beta_k$ with a joint confidence region such that we are $100(1-\alpha)$ percent confident that all $p=k+1$ estimates are correct. One possible solution to this problem is to use the region defined by

$$\frac{(\boldsymbol{\beta}-\hat{\boldsymbol{\beta}})'\mathbf{X}'\mathbf{X}(\boldsymbol{\beta}-\hat{\boldsymbol{\beta}})}{p(MS_E)} \leqslant F_{\alpha,p,n-p} \tag{9.31}$$

This inequality, first given as (4.22), describes an elliptically shaped region that is a joint $100(1-\alpha)$ percent confidence region for all p parameters. However, this region is difficult to construct and interpret unless p is small, for example, 2 or 3.

To illustrate the development and construction of this joint confidence region, consider the case of simple linear regression ($p=2$). The model is

$$y=\beta_0+\beta_1 x+\varepsilon$$

$$=\beta_0'+\beta_1(x-\bar{x})+\varepsilon$$

and the least squares estimates of β_0' and β_1 are $\hat{\beta}_0'=\bar{y}$ and $\hat{\beta}_1=S_{xy}/S_{xx}$. These estimates are unbiased, and their variances are $V(\hat{\beta}_0')=\sigma^2/n$ and $V(\hat{\beta}_1)=\sigma^2/S_{xx}$. Under the usual normality assumptions, the random variables

$$\frac{\hat{\beta}_0'-\beta_0'}{(\sigma^2/n)^{1/2}} \quad \text{and} \quad \frac{\hat{\beta}_1-\beta_1}{(\sigma^2/S_{xx})^{1/2}}$$

are independent standard normal random variables. Since the square of a standard normal random variable is a chi-square random variable with one degree of freedom, we have

$$\left[\frac{\hat{\beta}_0'-\beta_0'}{(\sigma^2/n)^{1/2}}\right]^2=\frac{n(\hat{\beta}_0'-\beta_0')^2}{\sigma^2}\sim\chi_1^2$$

and

$$\left[\frac{\hat{\beta}_1-\beta_1}{(\sigma^2/S_{xx})^{1/2}}\right]^2=\frac{S_{xx}(\hat{\beta}_1-\beta_1)^2}{\sigma^2}\sim\chi_1^2$$

Because $\hat{\beta}_0'$ and $\hat{\beta}_1$ are independent these two chi-square random variables are independent. Thus from the additivity property of chi-square, we find that

$$\frac{n(\hat{\beta}_0'-\beta_0')^2}{\sigma^2}+\frac{S_{xx}(\hat{\beta}_1-\beta_1)^2}{\sigma^2}\sim\chi_2^2$$

Now the distribution of $(n-2)MS_E/\sigma^2$ is χ_{n-2}^2, and it can be shown that

MS_E is independent of $\hat{\beta}_0'$ and $\hat{\beta}_1$. Therefore the ratio

$$\frac{\left[\dfrac{n(\hat{\beta}_0'-\beta_0')^2}{\sigma^2}+\dfrac{S_{xx}(\hat{\beta}_1-\beta_1)^2}{\sigma^2}\right]\Big/2}{\left[(n-2)MS_E/\sigma^2\right]/(n-2)}=\frac{n(\hat{\beta}_0'-\beta_0')^2+S_{xx}(\hat{\beta}_1-\beta_1)^2}{2MS_E} \tag{9.32}$$

is distributed as $F_{2,n-2}$. If we now substitute $\hat{\beta}_0'=\hat{\beta}_0+\hat{\beta}_1\bar{x}$ and $\beta_0'=\beta_0+\beta_1\bar{x}$ into (9.32) and simplify we obtain

$$\frac{n(\hat{\beta}_0-\beta_0)^2+2\sum_{i=1}^{n}x_i(\hat{\beta}_0-\beta_0)(\hat{\beta}_1-\beta_1)+\sum_{i=1}^{n}x_i^2(\hat{\beta}_1-\beta_1)^2}{2MS_E}$$

Since the probability statement

$$P\left\{\frac{n(\hat{\beta}_0-\beta_0)^2+2\sum_{i=1}^{n}x_i(\hat{\beta}_0-\beta_0)(\hat{\beta}_1-\beta_1)+\sum_{i=1}^{n}x_i^2(\hat{\beta}_1-\beta_1)^2}{2MS_E}\leqslant F_{\alpha,2,n-2}\right\}=1-\alpha$$

holds for all values of β_0 and β_1, the $100(1-\alpha)$ percent joint confidence region for β_0 and β_1 is

$$\frac{n(\hat{\beta}_0-\beta_0)^2+2\sum_{i=1}^{n}x_i(\hat{\beta}_0-\beta_0)(\hat{\beta}_1-\beta_1)+\sum_{i=1}^{n}x_i^2(\hat{\beta}_1-\beta_1)^2}{2MS_E}\leqslant F_{\alpha,2,n-2} \tag{9.33}$$

Equation (9.33) defines an ellipse that with repeated sampling will contain β_0 and β_1 simultaneously $100(1-\alpha)$ percent of the time.

- **Example 9.6** To illustrate the construction of this confidence region consider the rocket propellant data in Example 2.1. We will find a 95 percent confidence region for β_0 and β_1. Since $\hat{\beta}_0=2{,}627.82$, $\hat{\beta}_1=-37.15$, $\Sigma_{i=1}^{20}x_i=267.25$, $\Sigma_{i=1}^{20}x_i^2=4{,}677.69$, $MS_E=9{,}244.59$, and $F_{.05,2,18}=3.55$, we may sub-

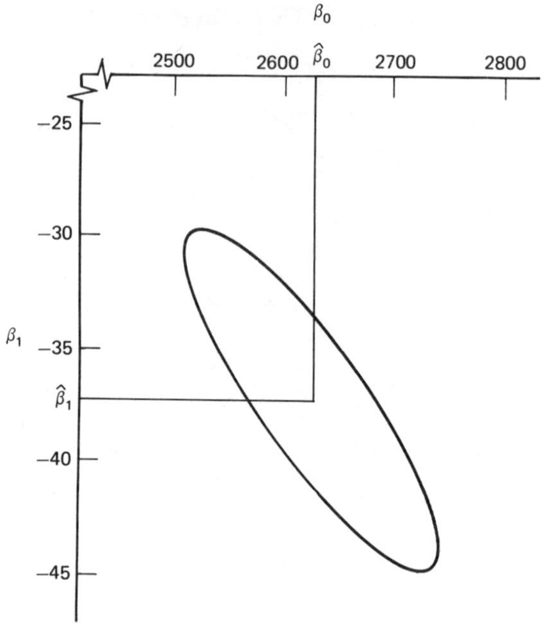

Figure 9.12 Joint 95 percent confidence region for β_0 and β_1 for the rocket propellant data.

stitute into (9.33), yielding

$$\left[20(2{,}627.82-\beta_0)^2+2(267.25)(2{,}627.82-\beta_0)(-37.15-\beta_1)\right.$$
$$\left.+(4{,}677.69)(-37.15-\beta_1)^2\right]/[2(9{,}244.59)]=3.55$$

as the boundary of the ellipse.

The joint confidence region is shown in Figure 9.12. Note that this ellipse is not parallel to the β_1 axis. The tilt of the ellipse is a function of the covariance between $\hat{\beta}_0$ and $\hat{\beta}_1$, which is $-\bar{x}\sigma^2/S_{xx}$. A positive covariance implies that errors in the point estimates of β_0 and β_1 are likely to be in the same direction, while a negative covariance indicates that these errors are likely to be in opposite directions. In our example $\bar{x}$ is positive so $\mathrm{Cov}(\beta_0, \beta_1)$ is negative. Thus if the estimate of the slope is too steep (β_1 is overestimated) the estimate of the intercept is likely to be too small (β_0 is underestimated). The elongation of the region depends on the relative sizes of the variances of β_0 and β_1. Generally if the ellipse is elongated in the β_0 direction (for example), this implies that β_0 is not estimated as precisely as β_1. This is the case in our example. •

There is another general approach for obtaining simultaneous interval estimates of the parameters in a linear regression model. Suppose that the model contains p parameters $\beta_0, \beta_1, \ldots, \beta_k$ (note that $p=k+1$), and we wish to construct simultaneous confidence intervals for r of them, such as $\beta_1, \beta_2, \ldots, \beta_r$, where $r \leqslant p$. These confidence intervals may be constructed by using

$$\hat{\beta}_j \pm \Delta \operatorname{se}(\hat{\beta}_j), \qquad j=1,2,\ldots,r \tag{9.34}$$

where the constant Δ is chosen so that a specified probability that all r intervals are correct is obtained.

Several methods may be used to choose Δ in (9.34). One procedure is the Bonferroni method. In this approach, we set $\Delta = t_{\alpha/(2r), n-p}$ so that (9.34) becomes

$$\hat{\beta}_j \pm t_{\alpha/(2r), n-p} \operatorname{se}(\hat{\beta}_j), \qquad j=1,2,\ldots,r \tag{9.35}$$

The probability is at least $1-\alpha$ that all r intervals are correct. Notice that the Bonferroni confidence intervals look somewhat like the ordinary one-at-a-time confidence intervals based on the t-distribution, except that each individual Bonferroni interval has a confidence coefficient $1-\alpha/r$ instead of $1-\alpha$.

To verify that this approach leads to correct statements, consider the simple linear regression model, and assume that we wish to construct simultaneous confidence intervals on β_0 and β_1. Since $r=p=2$, (9.35) becomes

$$\hat{\beta}_0 \pm t_{\alpha/4, n-2} \operatorname{se}(\hat{\beta}_0)$$

$$\hat{\beta}_1 \pm t_{\alpha/4, n-2} \operatorname{se}(\hat{\beta}_1)$$

That is, each confidence interval has confidence coefficient $1-\alpha/2$. Let E_0 be the event that the confidence interval for β_0 is incorrect and E_1 be the event that the confidence interval for β_1 is incorrect, so that $P(E_0)=P(E_1)$ $=\alpha/2$. Now the probability that either or both statements are incorrect is

$$P(E_0 \cup E_1) = P(E_0) + P(E_1) - P(E_0 \cap E_1)$$

and

$$1 - P(E_0 \cup E_1) = 1 - P(E_0) - P(E_1) + P(E_0 \cap E_1) \tag{9.36}$$

Since $1-P(E_0 \cup E_1)=P(\overline{E_0 \cup E_1})=P(\bar{E}_0 \cap \bar{E}_1)$, the left-hand side of (9.36) is the probability that both confidence intervals are correct. Furthermore since $P(E_0 \cap E_1) \geqslant 0$, we may write (9.36) as

$$\begin{aligned} P(\bar{E}_0 \cap \bar{E}_1) &= P \text{ (both intervals are correct)} \\ &\geqslant 1-P(E_0)-P(E_1) \\ &\geqslant 1-\alpha/2-\alpha/2 \\ &\geqslant 1-\alpha \end{aligned}$$

This expression is called the Bonferroni inequality. Thus if we wish to estimate β_0 and β_1 with confidence intervals such that the joint confidence coefficient is at least $1-\alpha$, then we must construct $100(1-\alpha/2)$ percent confidence intervals for both β_0 and β_1.

• **Example 9.7** We may find 90 percent joint confidence intervals for β_0 and β_1 for the rocket propellant data in Example 2.1 by constructing a 95 percent confidence interval for each parameter. Since

$$\hat{\beta}_0 = 2627.822, \qquad \text{se}(\hat{\beta}_0) = 44.184$$

$$\hat{\beta}_1 = -37.154, \qquad \text{se}(\hat{\beta}_1) = 2.889$$

and $t_{.05/2,18} = t_{.025,18} = 2.101$, the joint confidence intervals are

$$\hat{\beta}_0 - t_{.025,18}\,\text{se}(\hat{\beta}_0) \leqslant \beta_0 \leqslant \hat{\beta}_0 + t_{0.25,18}\,\text{se}(\hat{\beta}_0)$$

$$2627.822-(2.101)(44.184) \leqslant \beta_0 \leqslant 2627.822+(2.101)(44.184)$$

$$2534.911 \leqslant \beta_0 \leqslant 2720.653$$

and

$$\hat{\beta}_1 - t_{.025,18}\,\text{se}(\hat{\beta}_1) \leqslant \beta_1 \leqslant \hat{\beta}_1 + t_{.025,18}\,\text{se}(\hat{\beta}_1)$$

$$-37.154-(2.101)(2.889) \leqslant \beta_1 \leqslant -37.154+(2.101)(2.889)$$

$$-43.224 \leqslant \beta_1 \leqslant -31.084$$

We conclude with 90 percent confidence that this procedure leads to correct interval estimates for both parameters. •

Note that the greater the number of parameters for which Bonferroni–type confidence intervals are constructed, the wider the individual confidence intervals become. In some cases where the number of intervals is relatively large, the individual intervals may be so wide that they are practically useless. Furthermore the confidence ellipse approach (9.31) is a more efficient procedure, as the volume of the ellipse is always less than the volume of the $\beta_0, \beta_1, \ldots, \beta_k$ space covered by the Bonferroni intervals. Thus when at most two or three parameters are involved, the confidence ellipse approach is preferable. However when a statement about more parameters is required, the Bonferroni intervals will be much easier to construct and interpret.

Constructing Bonferroni confidence intervals often requires significance levels not listed in the usual t-tables. Some modern calculators have values of $t_{\alpha,\nu}$ on call as a library function. Appendix Table A.7 gives a brief tabulation of $t_{\alpha/(2p),\nu}$ for $\alpha = .10$, .05, .01; $p = 2(1)10$; and $\nu = 5(1)25(5)50(10)100$. For another table, see Dunn [1961].

The Bonferroni method is not the only approach to choosing Δ in (9.34). Other approaches include the Scheffé S-method (see Scheffé [1953, 1959], for which

$$\Delta = \left(rF_{\alpha,r,n-p}\right)^{1/2}$$

and the maximum modulus t procedure (see Hahn [1972]), for which

$$\Delta = u_{\alpha,r,n-p}$$

where $u_{\alpha,r,n-p}$ is the upper α tail point of the distribution of the maximum absolute value of r independent student-t random variables each based on $n-p$ degrees of freedom. Like the Bonferroni method, these alternative procedures are conservative, in that the overall probability that all r statements are correct is at least $1-\alpha$. If the maximum modulus t procedure is used and the coefficients $\hat{\beta}_j$ $(j = 1, 2, \ldots, r)$ are independent, the joint confidence level associated with the set of confidence intervals is exactly $1-\alpha$. However in most cases the $\hat{\beta}_j$ are not independent. A table of $u_{\alpha,r,n-p}$ taken from Hahn and Hendrickson [1971] is given in Appendix Table A.8. An obvious way to compare these three techniques is in terms of the lengths of the confidence intervals they generate. Generally, the Bonferroni intervals are shorter than the Scheffé intervals and the maximum modulus t intervals are shorter than the Bonferroni intervals.

To illustrate the maximum modulus t procedure for the rocket propellant data, we find $u_{.10,2,18} = 2.082$ from Appendix Table A.8. Therefore the joint confidence intervals with confidence level at least .90 are

$$\hat{\beta}_0 - u_{.10,2,18}\,\text{se}(\hat{\beta}_0) \leqslant \beta_0 \leqslant \hat{\beta}_0 + u_{.10,2,18}\,\text{se}(\hat{\beta}_0)$$

$$2627.822 - (2.082)(44.184) \leqslant \beta_0 \leqslant 2627.822 + (2.082)(44.184)$$

$$2535.831 \leqslant \beta_0 \leqslant 2719.813$$

and

$$\hat{\beta}_1 - u_{.10,2,18}\,\text{se}(\hat{\beta}_1) \leqslant \beta_1 \leqslant \hat{\beta}_1 + u_{.10,2,18}\,\text{se}(\hat{\beta}_1)$$

$$-37.154 - (2.082)(2.889) \leqslant \beta_1 \leqslant -37.154 + (2.082)(2.889)$$

$$-43.169 \leqslant \beta_1 \leqslant -31.139$$

Notice that these confidence intervals are shorter than those produced by the Bonferroni method.

9.6.2 Simultaneous estimation of mean response

Once we have fit the regression model we may be interested in obtaining an interval estimate of the mean response at a particular point. The one-at-a-time confidence interval formula was given previously as (4.25). Now suppose that there are m specific values of $\mathbf{x}$, say $\mathbf{x}_1, \mathbf{x}_2, \ldots, \mathbf{x}_m$ of interest, and we wish to construct a set of m interval estimates on the mean response at these points with a joint confidence coefficient of at least $1-\alpha$. This can be done using any of the three methods discussed in Section 9.6.1. That is, the confidence intervals are

$$\hat{y}_{\mathbf{x}_i} - \Delta\sqrt{\hat{\sigma}^2 \mathbf{x}_i'(\mathbf{X}'\mathbf{X})^{-1}\mathbf{x}_i} \leqslant y_{\mathbf{x}_i} \leqslant \hat{y}_{\mathbf{x}_i} + \Delta\sqrt{\hat{\sigma}^2 \mathbf{x}_i'(\mathbf{X}'\mathbf{X})^{-1}\mathbf{x}_i}, \qquad i = 1, 2, \ldots, m \tag{9.37}$$

If the Bonferroni, maximum modulus t, or Scheffé methods are used, then Δ in Equation (9.37) becomes

$$t_{\alpha/(2m),\,n-p}, \qquad u_{\alpha,\,m,\,n-p}, \qquad \text{or} \quad \left(pF_{\alpha,\,p,\,n-p}\right)^{1/2} \tag{9.38}$$

respectively.

Note that the width of the individual Bonferroni or maximum modulus t intervals increases with m while the width of the Scheffé intervals is constant. In some cases, particularly when m is large, the Scheffé–type confidence intervals will be tighter than either the Bonferroni or the maximum modulus t intervals. On the other hand, when m is small, the Bonferroni or maximum modulus t intervals may be tighter. A comparison of $(pF_{\alpha,p,n-p})^{1/2}$, $u_{\alpha,m,n-p}$, and $t_{\alpha/(2m),n-p}$ will reveal the method that is preferable in the particular application.

It is also possible to obtain a $100(1-\alpha)$ percent joint confidence region for the entire regression surface. Working and Hotelling [1929] have shown that the boundary of the joint confidence region is

$$\hat{y}_{\mathbf{x}} - \sqrt{(F_{\alpha,p,n-p})p\hat{\sigma}^2\mathbf{x}'(\mathbf{X}'\mathbf{X})^{-1}\mathbf{x}} \leqslant y_{\mathbf{x}} \leqslant \hat{y}_{\mathbf{x}} + \sqrt{(F_{\alpha,p,n-p})p\hat{\sigma}^2\mathbf{x}'(\mathbf{X}'\mathbf{X})^{-1}\mathbf{x}} \tag{9.39}$$

This confidence region applies to the entire regression surface for all real values of $\mathbf{x}$. The region that covers any portion of the regression surface that is either not of interest or physically meaningless is ignored. Thus if a portion of the x–space is not of interest (such as values of $x<0$), the probability associated with the confidence region is greater than $1-\alpha$. Note that the Scheffê S-method of obtaining confidence intervals on y at m specific values of $\mathbf{x}$ (9.38) consists of applying (9.39) to the m $\mathbf{x}$ vectors of interest. Since (9.39) has confidence coefficient $1-\alpha$ for all $\mathbf{x}$, (9.39) will cover the mean responses for m specific $\mathbf{x}$ vectors with joint confidence coefficient at least $1-\alpha$. There are a number of other approaches to constructing confidence regions for the regression surface. Refer to Seber [1977, Chapters 5 and 7] for a review.

- **Example 9.8** Consider the rocket propellant data from Example 2.1. Suppose we wish to find 90 percent joint confidence intervals on the mean shear strength of motors made with propellant that is 10 weeks and 18 weeks old. The point estimates of $E(y|x)$ are:

i	x_i	$\widehat{E(y\|x_i)}=\hat{y}_{x_i}=2627.82-37.15x_i$
1	10	2256.282
2	18	1959.050

For a simple linear regression model and $m=2$, (9.37) becomes

$$\hat{y}_{x_i} - \Delta\sqrt{MS_E\left[\frac{1}{n}+\frac{(x_i-\bar{x})}{S_{xx}}\right]} \leqslant E(y|x_i) \leqslant \hat{y}_{x_i}$$

$$+\Delta\sqrt{MS_E\left[\frac{1}{n}+\frac{(x_i-\bar{x})^2}{S_{xx}}\right]w^2} \qquad i=1,2$$

and using the data from Example 2.1, we find that

$$2256.282-\Delta 68.633 \leqslant E(y|x_1=10) \leqslant 2256.282+\Delta 68.633$$

and

$$1959.050-\Delta 69.236 \leqslant E(y|x_2=18) \leqslant 1959.050+\Delta 69.236.$$

For the Bonferroni, maximum modulus t, and Scheffé methods the values of Δ are

$$t_{\alpha/(2m),\, n-p} = t_{.025,18} = 2.101$$

$$u_{\alpha, m, n-p} = u_{.10,2,18} = 2.082$$

$$(pF_{\alpha,p,n-p})^{1/2} = (2F_{.10,2,18})^{1/2} = (6.02)^{1/2} = 2.454$$

respectively. Notice that the maximum modulus t intervals are the narrowest, closely followed by the Bonferroni intervals. The Scheffé intervals are about 20 percent wider. However if $m>2$ confidence statements are required, the maximum modulus t and Bonferroni intervals will get wider, while the width of the Scheffé intervals does not depend on m. Choosing the maximum modulus t method, the 90 percent joint confidence intervals on the mean response are

$$2256.282-(2.082)68.633 \leqslant E(y|x_1=10) \leqslant 2256.282+(2.082)68.633$$

or

$$2087.857 \leqslant E(y|x_1=10) \leqslant 2424.707$$

and

$$1959.050-(2.082)69.236 \leqslant E(y|x_2=18) \leqslant 1959.050+(2.082)69.236$$

or

$$1789.145 \leqslant E(y|x_2=18) \leqslant 2129.955$$

9.6.3 Prediction of m new observations

We now consider the simultaneous prediction of m new observations at m levels of $\mathbf{x}$, such as $\mathbf{x}_1, \mathbf{x}_2, \ldots, \mathbf{x}_m$. The one-at-a-time prediction interval for a single new observation at $\mathbf{x}_0$ is given in (4.36). Any of the three approaches in Section 9.6.1 can be used to obtain a set of m prediction intervals with overall confidence coefficient at least $1-\alpha$. The prediction intervals are

$$\hat{y}_{\mathbf{x}_i} - \Delta\sqrt{\hat{\sigma}^2\left(1+\mathbf{x}_i'(\mathbf{X}'\mathbf{X})^{-1}\mathbf{x}_i\right)} \leqslant y_{\mathbf{x}_i} \leqslant \hat{y}_{\mathbf{x}_i} + \Delta\sqrt{\hat{\sigma}^2\left(1+\mathbf{x}_i'(\mathbf{X}'\mathbf{X})^{-1}\mathbf{x}_i\right)} \qquad i=1,2,\ldots,m \tag{9.40}$$

If the Bonferroni, maximum modulus t, or Scheffé methods are used, then Δ in (9.40) becomes

$$t_{\alpha/(2m),n-p},\ u_{\alpha,m,n-p}, \quad \text{or } \left(mF_{\alpha,m,n-p}\right)^{1/2} \tag{9.41}$$

respectively. A comparison of these values will indicate which procedure produces the shortest prediction intervals in a given situation.

- **Example 9.9** Consider the rocket propellant data from Example 2.1. Suppose that we wish to find 90 percent simultaneous prediction intervals for two future observations on shear strength for motors made from propellant that is 10 weeks and 18 weeks old. Letting $x_1=10$ and $x_2=18$, the point estimates of these future observations are $\hat{y}_{x_1}=2256.282$ psi and $\hat{y}_{x_2}=1959.050$ psi, respectively. For a simple linear regression model and $m=2$, (9.40) becomes

$$\hat{y}_{x_i} - \Delta\sqrt{MS_E\left[1+\frac{1}{n}+\frac{(x_i-\bar{x})^2}{S_{xx}}\right]} \leqslant y_{x_i} \leqslant \hat{y}_{x_i} + \Delta\sqrt{MS_E\left[1+\frac{1}{n}+\frac{(x_i-\bar{x})^2}{S_{xx}}\right]} \qquad i=1,2$$

Using the rocket propellant data, with $x_1=10$ and $x_2=18$, this last equation becomes

$$2256.282-\Delta 118.097 \leqslant y_{x_1} \leqslant 2256.282+\Delta 118.097$$

and

$$1959.050-\Delta 118.449 \leqslant y_{x_2} \leqslant 1959.050+\Delta 118.449$$

The appropriate values of Δ for the Bonferroni, maximum modulus t, and Scheffé methods are

$$t_{\alpha/(2m),n-p} = t_{.025,18} = 2.101$$

$$u_{\alpha,m,n-p} = u_{.10,2,18} = 2.082$$

$$(mF_{\alpha,m,n-p})^{1/2} = (2F_{.10,2,18})^{1/2} = (6.02)^{1/2} = 2.454$$

respectively. The narrowest set of simultaneous prediction intervals are given by the maximum modulus t intervals, which are

$$2256.282 - (2.082)118.097 \leqslant y_{x_1} \leqslant 2256.282 + (2.082)118.097$$

or

$$1966.472 \leqslant y_{x_1} \leqslant 2546.092$$

and

$$1959.050 - (2.082)118.449 \leqslant y_{x_2} \leqslant 1959.050 + (2.082)118.449$$

or

$$1668.376 \leqslant y_{x_2} \leqslant 2249.724$$

9.7 Inverse estimation (calibration or discrimination)

Most regression problems involving prediction or estimation require determining the value of y corresponding to a given x, such as x_0. In this section we consider the inverse problem; that is, given that we have observed a value of y, such as y_0, determine the x–value corresponding to it. For example, suppose we wish to calibrate a thermocouple, and we know that the temperature reading given by the thermocouple is a linear function of the actual temperature, as in

$$\text{observed temperature} = \beta_0 + \beta_1 \text{ (actual temperature)} + \varepsilon$$

or

$$y = \beta_0 + \beta_1 x + \varepsilon \tag{9.42}$$

Now suppose we measure an unknown temperature with the thermocouple and obtain a reading y_0. We would like to estimate the actual temperature;

that is, the temperature x_0 corresponding to the observed temperature reading y_0. This situation arises often in engineering and physical sciences, and is sometimes called the *calibration* problem. It also occurs in bioassay where a standard curve is constructed against which all future assays or *discriminations* are to be run.

Suppose that the thermocouple has been subjected to n controlled and known temperatures, $x_1, x_2, \ldots, x_n$, and a set of corresponding temperature readings $y_1, y_2, \ldots, y_n$ obtained. One method for estimating x given y would be to fit the model (9.42), giving

$$\hat{y} = \hat{\beta}_0 + \hat{\beta}_1 x \tag{9.43}$$

Now let y_0 be the observed value of y. A natural point estimate of the corresponding value of x is

$$\hat{x}_0 = \frac{y_0 - \hat{\beta}_0}{\hat{\beta}_1} \tag{9.44}$$

assuming that $\hat{\beta}_1 \neq 0$. This approach is often called the *classical* estimator. It can be shown (see Graybill [1976], Seber [1977]) that a $100(1-\alpha)$ percent confidence region for x_0 is the set of all x-values satisfying the inequality

$$\frac{\left(y_0 - \hat{\beta}_0 - \hat{\beta}_1 x\right)^2}{\hat{\sigma}^2 A^2} \leqslant t^2_{\alpha/2, n-2} \tag{9.45}$$

where

$$A^2 = 1 + \frac{1}{n} + \frac{(x - \bar{x})^2}{\sum_{i=1}^{n} (x_i - \bar{x})^2}$$

Unfortunately the set of x's satisfying (9.45) can give a finite interval, two semi-infinite intervals, or the entire real line. Only in the first case does a useful confidence interval result, and this will occur only when the F-test (or equivalently the t-test) for the hypothesis H_0: $\beta_1 = 0$ implies rejection. That is, when $\hat{\beta}_1$ is too near zero (the regression line is too flat) the resulting confidence interval is nonsensical. Therefore we recommend that the analyst faced with an inverse estimation or calibration problem fit the model (9.42) and test H_0: $\beta_1 = 0$. If this hypothesis is not rejected, then y and x are not linearly related and it does not make sense to construct either a point estimate or a confidence interval on x. However if H_0: $\beta_1 = 0$ is rejected using the F-test of size α, then the point estimate of x_0 is given by (9.44) and the $100(1-\alpha)$ percent confidence interval for x_0 is

$$\bar{x}+d_1 \leqslant x_0 \leqslant \bar{x}+d_2 \tag{9.46}$$

where d_1 and d_2 are the roots of

$$d^2\left[\hat{\beta}_1^2-\frac{t_{\alpha/2,n-2}^2\hat{\sigma}^2}{\sum_{i=1}^{n}(x_i-\bar{x})^2}\right]-2d\hat{\beta}_1(y_0-\bar{y})+\left[(y_0-\bar{y})^2-t_{\alpha/2,n-2}^2\hat{\sigma}^2\left(1+\frac{1}{n}\right)\right]=0 \tag{9.47}$$

Other confidence interval methods are given by Lieberman, Miller and Hamilton [1967] and Miller [1966].

The classical procedure is not entirely satisfactory. For example, $\hat{x}_0$ is a biased estimator of x_0 and it has infinite mean square error (although Williams [1969] shows that no unbiased estimator of x_0 has a finite variance). Hoadley [1970] notes that the width of the confidence interval on x_0 depends on the magnitude of the F-statistic; a large F gives a short interval and a small F gives a wide one.

An alternate approach to estimating x would be to fit the model

$$x=\alpha_0+\alpha_1 y+\varepsilon \tag{9.48}$$

That is, we could regress x on y to directly obtain a calibration or discrimination equation. This approach is often called the *inverse* estimator. Note however that this assumes that x is a random variable. In most situations the variable x is controlled and is *not* a random variable. Therefore the use of (9.48) seems inappropriate, because the assumptions underlying the linear regression model are violated. If x is an observable random variable however, this approach would be quite reasonable. Unfortunately in many practical problems this assumption is unrealistic.

Despite these cautions, Krutchkoff [1967, 1969] has advocated the use of the inverse estimator (9.48). He has analyzed the performance of this estimator via Monte Carlo simulation, and concludes that it is superior to the classical method. However, Berkson [1969], Halperin [1970], and Williams [1969] have criticized the inverse procedure on several grounds. They note that comparing an estimator with infinite mean square error (the classical estimator) to one with a finite mean square error (the inverse estimator) is not reasonable. Furthermore the inverse method is only superior to the classical procedure when x_0 lies in a small interval around $\bar{x}$. This interval is often very small. Shukla [1972] shows that if the number of observations used to construct the calibration line is small, the inverse procedure will yield smaller mean square error for the estimator of the unknown x than will the classical approach, particularly in cases where extrapolation is required (we point out that extrapolation may be extremely hazardous and is not generally recommended). However, when large samples are used to construct the calibration line, the classical method is preferred. In practice there may be little difference between the classical and inverse estimator when the data are very close to a straight line.

A number of other estimators have been proposed. Graybill [1961, 1976] considers the case where we have repeated observations on y at the unknown value of x. He develops point and interval estimates for x using the classical approach. The probability of obtaining a finite confidence interval for the unknown x is greater when these are repeat observations on y. Hoadley [1970] gives a Bayesian treatment of the problem and derives an estimator that is a compromise between the classical and inverse approaches. He notes that the inverse estimator is the Bayes estimator for a particular choice of prior distribution. Other estimators have been proposed by Kalotay [1971], Naszódi [1978], Perng and Tong [1974], and Tucker [1980]. The paper by Scheffé [1973] is also of interest.

In most calibration studies the analyst can design the data collection experiment. That is, he can specify what x-values are to be observed. Ott and Myers [1968] have considered the choice of an appropriate design for the inverse estimation problem assuming that the unknown x is estimated by the classical approach. They develop designs that are optimal in the sense of minimizing the integrated mean square error. Figures are provided to assist the analyst in design selection.

• **Example 9.10** A mechanical engineer is calibrating a thermocouple. He has chosen 16 levels of temperature evenly spaced over the interval 100°C to 400°C. The actual temperature x (measured by a thermometer of known accuracy) and the observed reading on the thermocouple y are shown in Table 9.17, and a scatter diagram is plotted in Figure 9.13. Inspection of the scatter diagram indicates that the observed temperature on the thermocouple is linearly related to the actual temperature. The straight-line model is

$$\hat{y} = -6.5860 + 0.9525x$$

with $\hat{\sigma}^2 = MS_E = 6.0614$. The F-statistic for this model is $F_0 = 20{,}350.11$, so we reject H_0: $\beta_1 = 0$ and conclude that the slope of the calibration line is not zero.

Table 9.17 Actual and observed temperature

Observation i	Actual Temperature x_i (°C)	Observed Temperature y_i (°C)
1	100	88.8
2	120	108.7
3	140	129.8
4	160	146.2
5	180	161.6
6	200	179.9
7	220	202.4
8	240	224.5
9	260	245.1
10	280	257.7
11	300	277.0
12	320	298.1
13	340	318.8
14	360	334.6
15	380	355.2
16	400	377.0

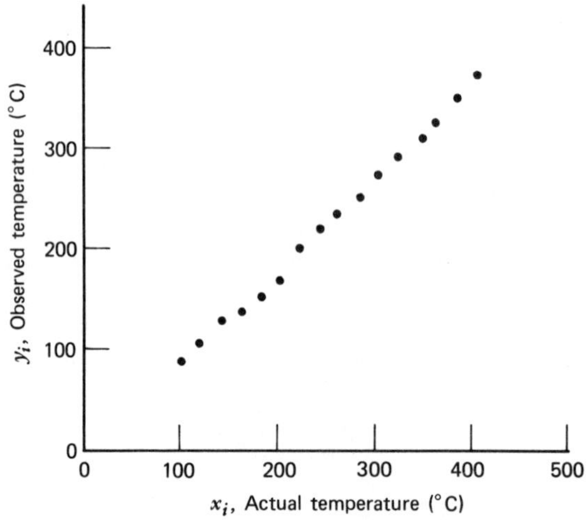

Figure 9.13 Scatter plot of observed and actual temperatures, Example 9.10.

Residual analysis does not reveal any unusual behavior so this model can be used to obtain point and interval estimates of actual temperature from temperature readings on the thermocouple.

Suppose that a new observation on temperature of $y_0 = 200°C$ is obtained using the thermocouple. A point estimate of the actual temperature, from the calibration line, is

$$\hat{x}_0 = \frac{y_0 - \hat{\beta}_0}{\hat{\beta}_1} = \frac{200 - (-6.5860)}{0.9525} = 216.89°C$$

A 95 percent confidence interval on the actual temperature x_0 corresponding to $y_0 = 200°C$ is found by solving (9.47):

$$d^2\left[\hat{\beta}_1^2 - \frac{t^2_{\alpha/2,n-2}\hat{\sigma}^2}{\sum_{i=1}^{n}(x_i - \bar{x})^2}\right] - 2d\hat{\beta}_1(y_0 - \bar{y}) + \left[(y_0 - \bar{y})^2 - t_{\alpha/2,n-2}\hat{\sigma}^2\left(1 + \frac{1}{n}\right)\right] = 0$$

$$d^2\left[(0.9525)^2 - \frac{(2.145)^2(6.0614)}{136{,}000}\right] - 2d(0.9525)(200 - 231.55)$$

$$+\left[(200 - 231.55)^2 - (2.145)^2(6.0614)\left(1 + \frac{1}{16}\right)\right] = 0$$

which reduces to

$$0.907d^2 + 60.103d + 965.771 = 0$$

The roots of this quadratic equation are $d_1 = -38.876$ and $d_2 = -27.389$. Therefore from (9.46) the 95 percent confidence interval on x_0 is

$$\bar{x} + d_1 \leqslant x_0 \leqslant \bar{x} + d_2$$

$$250 - 38.876 \leqslant x_0 \leqslant 250 - 27.389$$

or

$$211.124 \leqslant x_0 \leqslant 222.611$$

•

9.8 Designed experiments for regression

Since so many properties of the fitted regression model depend on the levels of the regressor variables, in situations where these levels can be chosen it is natural to consider the question of experimental design. That is, if we could choose the levels of the regressor variables, what values should we use? The experimental design problem in regression consists of selecting the elements of the **X** matrix. The general approach is to define some reasonable criterion based on the model that is being entertained, and then use it to generate an optimal design.

For example, suppose that we are fitting a simple linear regression model $y=\beta_0+\beta_1 x+\varepsilon$. The variance of the least squares estimate of the slope is

$$V(\hat{\beta}_1)=\frac{\sigma^2}{\sum_{i=1}^{n}(x_i-\bar{x})^2}$$

Assume that the data collection experiment is being designed so that we must choose the levels of the regressor x to run. We wish to find the "best" set of levels or the best design for this situation. Suppose that we are interested in a particular range of values for x. Let the upper and lower limits of this range be $+1$ and -1, respectively. So now the design problem is to choose n levels for x, such as $x_1, x_2, \ldots, x_n$, over the interval $-1 \leqslant x \leqslant 1$ such that a suitable design criterion is optimized.

Suppose that the design criterion specified is to minimize the variance of the regression coefficient $\hat{\beta}_1$. Since σ^2 is constant, $V(\hat{\beta}_1)$ is minimized when $\sum_{i=1}^{n}(x_i-\bar{x})^2$ is maximized. It is easy to see that the optimal design is:

design 1 (n even): $\frac{n}{2}$ runs at $x=-1$, $\frac{n}{2}$ runs at $x=1$

design 2 (n odd): $\frac{(n-1)}{2}$ runs at $x=-1$, 1 run at $x=0$, $\frac{(n-1)}{2}$ runs at $x=1$.

Thus in the optimal design if an even number of runs are to be made, half of them are made at each end of the region of interest, while if an odd number of runs are to be made, one run is made at the center of the region and the remaining $n-1$ runs are allocated equally at the extremes. Note that this strategy spreads the x's out as far as possible so that $\sum_{i=1}^{n}(x_i-\bar{x})^2$ is maximized.

Although this design may be optimal with respect to the chosen criterion, we do not recommend that it be used in practice. Generally speaking, other considerations besides the $V(\hat{\beta}_1)$ must enter into the design decision. We would like for the $V(\hat{\beta}_1)$ to be small, if the straight-line model that we have tentatively entertained is correct. However, we would also like to be able to test the adequacy of the straight-line model, and this cannot be done very effectively unless more than two or three levels of x are used. Furthermore, we may be uncertain about the region of interest for the regressor variable, and choosing the boundaries of this region may have required considerable guesswork. Yet the optimal design places all or nearly all the runs at the boundaries of this region. Thus the design that is "optimal" in the context of minimizing the $V(\hat{\beta}_1)$ is one that requires our assumptions of model structure and region of interest to be absolutely correct. From a practical viewpoint, we must take observations at several other places to provide "insurance" in the event that some of our assumptions are wrong.

Considerable attention has been given to the optimal design problem. Several optimality criteria have been proposed, one of the most popular being to maximize the *determinant* of $\mathbf{X}'\mathbf{X}$. This is a parameter estimation criterion, because if $|\mathbf{X}'\mathbf{X}|$ is a maximum, then $|V(\hat{\boldsymbol{\beta}})|$ is a minimum. A design for which $|\mathbf{X}'\mathbf{X}|$ is maximized is called a D-optimal design. It can be shown that a design that maximizes $|\mathbf{X}'\mathbf{X}|$ also minimizes the maximum variance of $\hat{y}=\mathbf{x}'\hat{\boldsymbol{\beta}}$. A design with the latter property is called G-optimal. Clearly G-optimality is a response estimation criterion. St. John and Draper [1975] review the major results on optimal design and discuss algorithms for generating D-optimal designs.

Unfortunately many of the limitations that we saw in the simple linear regression example above also emerge in more general optimal design problems. For example, D-optimal designs often consist of runs at the same number of points as there are parameters, so that model adequacy checking is impossible. Consequently there are not many practical applications of optimal design theory. One area in which optimal design theory has proved helpful is in the augmentation of existing data for regression. For example, suppose that we are fitting a regression model and a set of n_1 observations are currently available. If n_2 additional observations can be collected, where should these new runs be made? Several authors (Dykstra [1971] and Gaylor and Merrill [1968]) have discussed methods for choosing the n_2 additional runs so that the $|\mathbf{X}'\mathbf{X}|$ is maximized.

Another problem for which useful optimal designs have been developed is the fitting of *response surfaces* (see Montgomery [1976] and Myers [1971]); usually second-order polynomials. An important consideration in the response surface context is the variance of the fitted value $\hat{y}$. A response surface design is *rotatable* if the $V(\hat{y})$ depends on the distance of the point of interest from the design center and not on the direction. This implies that the contours of $V(\hat{y})$ with respect to $x_1, x_2, \ldots, x_k$ are concentric circles. The reason that the rotatability criterion is so important is that the usual objective of a response surface study is to determine the values of $x_1, x_2, \ldots, x_k$ that optimize the response y and before the experiment is performed the analyst does not know where this point is. Thus all points equidistant from the design center are equally important, regardless of direction. Important references on response surface designs include Box and Behnken [1960], Box and Draper [1959, 1963], and Box and Hunter [1957].

9.9 The relationship between regression and analysis of variance

The analysis of variance is a technique frequently used to analyze data from planned or designed experiments. Although special computing procedures are generally used for the analysis of variance, any analysis of variance problem can also be treated as a linear regression problem. Ordinarily, we do not recommend that regression methods be used for the analysis of variance because the specialized computing techniques are usually quite efficient. However there are some analysis of variance situations, particularly those involving unbalanced designs, where the regression approach is helpful. Furthermore many analysts are unaware of the close connection between the two procedures. In this section we will illustrate this relationship using the one-way classification analysis of variance. For further examples of the relationship between regression and analysis of variance, see Draper and Smith [1981], Neter and Wasserman [1974], Schilling [1974a, b], and Seber [1977].

The model for the one-way classification analysis of variance is

$$y_{ij}=\mu+\tau_i+\varepsilon_{ij}\begin{cases} i=1,2,\ldots,k \\ j=1,2,\ldots,n \end{cases} \tag{9.49}$$

where y_{ij} is the jth observation for the ith treatment or factor level, μ is a parameter common to all k treatments (usually called the *grand mean*), τ_i is a parameter that represents the effect of the ith treatment, and ε_{ij} is a NID$(0, \sigma^2)$ error component. It is customary to define the treatment effects in the balanced case (i.e., an equal number of observations per treatment) as

$$\tau_1+\tau_2+\cdots+\tau_k=0$$

Furthermore the mean of the ith treatment is $\mu_i=\mu+\tau_i$, $i=1,2,\ldots,k$. In the fixed-effects (or model I) case, the analysis of variance is used to test the hypothesis that all k population means are equal, or equivalently,

$$\begin{aligned} &H_0: \tau_1=\tau_2=\cdots=\tau_k=0 \\ &H_1: \tau_i\neq 0, \qquad \text{for at least one } i \end{aligned} \tag{9.50}$$

Table 9.18 One-way analysis of variance

Source of Variation	Sum of Squares	Degrees of Freedom	Mean Square	F_0
Treatments	$n\sum_{i=1}^{k}(\bar{y}_{i.}-\bar{y}_{..})^2$	$k-1$	$\frac{SS_{\text{Treatments}}}{k-1}$	$\frac{MS_{\text{Treatments}}}{MS_E}$
Error	$\sum_{i=1}^{k}\sum_{j=1}^{n}(y_{ij}-\bar{y}_{i.})^2$	$k(n-1)$	$\frac{SS_E}{k(n-1)}$	
Total	$\sum_{i=1}^{k}\sum_{j=1}^{n}(y_{ij}-\bar{y}_{..})^2$	$kn-1$		

Table 9.18 displays the usual analysis of variance calculations. The test statistic F_0 is compared to $F_{\alpha,k-1,k(n-1)}$. If F_0 exceeds this critical value, the null hypothesis H_0 in (9.50) is rejected; that is, we conclude that the k treatment means are not identical. Note that in Table 9.18 we have employed the usual "dot subscript" notation associated with the analysis of variance. That is, the average of the n observations in the ith treatment is

$$\bar{y}_{i.}=\frac{1}{n}\sum_{j=1}^{n}y_{ij}, \qquad i=1,2,\ldots,k$$

and the grand average is

$$\bar{y}_{..}=\frac{1}{kn}\sum_{i=1}^{k}\sum_{j=1}^{n}y_{ij}$$

To illustrate the connection between the one-way classification fixed-effects analysis of variance and regression, suppose that we have $k=3$ treatments, so that (9.49) becomes

$$y_{ij}=\mu+\tau_i+\varepsilon_{ij} \qquad \begin{cases} i=1,2,3 \\ j=1,3,\ldots,n \end{cases}$$

These three treatments may be viewed as three levels of a qualitative factor. In the regression context, qualitative factors are handled using indicator variable. Specifically, a qualitative factor with three levels would require two indicator variables defined as follows:

$$x_1 = \begin{cases} 1 \text{ if the observation is from treatment 1} \\ 0 \text{ otherwise} \end{cases}$$

$$x_2 = \begin{cases} 1 \text{ if the observation is from treatment 2} \\ 0 \text{ otherwise} \end{cases}$$

Therefore the regression model becomes

$$y_{ij} = \beta_0 + \beta_1 x_{1j} + \beta_2 x_{2j} + \varepsilon_{ij} \qquad \begin{cases} i = 1,2,3 \\ j = 1,2,\ldots,n \end{cases} \tag{9.51}$$

where x_{1j} is the value of the indicator variable x_1 for observation j in treatment i and x_{2j} is the value of x_2 for observation j in treatment i.

The relationship between the parameters β_u $(u=0,1,2)$ in the regression model and the parameters μ and τ_i $(i=1,2,\ldots,k)$ in the analysis of variance model is easily determined. Consider the observations from Treatment 1, for which

$$x_{1j} = 1 \qquad \text{and} \quad x_{2j} = 0$$

The regression model (9.51) becomes

$$y_{1j} = \beta_0 + \beta_1(1) + \beta_2(0) + \varepsilon_{1j}$$

$$= \beta_0 + \beta_1 + \varepsilon_{1j}$$

Since in the analysis of variance model an observation from Treatment 1 is represented by $y_{1j} = \mu + \tau_1 + \varepsilon_{1j} = \mu_1 + \varepsilon_{1j}$, this implies that

$$\beta_0 + \beta_1 = \mu_1$$

Similarly, if the observations are from Treatment 2, then $x_{1j} = 0$, $x_{2j} = 1$ and

$$y_{2j} = \beta_0 + \beta_1(0) + \beta_2(1) + \varepsilon_{2j}$$

$$= \beta_0 + \beta_2 + \varepsilon_{2j}$$

Considering the analysis of variance model, $y_{2j} = \mu + \tau_2 + \varepsilon_{2j} = \mu_2 + \varepsilon_{2j}$, so

$$\beta_0 + \beta_2 = \mu_2$$

Finally, consider observations from Treatment 3. Since $x_{1j}=x_{2j}=0$, the regression model becomes

$$y_{3j}=\beta_0+\beta_1(0)+\beta_2(0)+\varepsilon_{3j}$$
$$=\beta_0+\varepsilon_{3j}$$

The corresponding analysis of variance model is $y_{3j}=\mu+\tau_3+\varepsilon_{3j}=\mu_3+\varepsilon_{3j}$ so that

$$\beta_0=\mu_3$$

Thus in the regression model formulation of the one-way classification analysis of variance, the regression coefficients describe comparisons of the first two treatment means μ_1 and μ_2 with the third treatment mean μ_3. That is,

$$\beta_0=\mu_3$$
$$\beta_1=\mu_1-\mu_3$$
$$\beta_2=\mu_2-\mu_3$$

In general if there are k treatments, the regression model for the one-way classification analysis of variance will require $k-1$ indicator variables, for example

$$y_{ij}=\beta_0+\beta_1x_{1j}+\beta_2x_{2j}+\cdots\beta_{k-1}x_{k-1,j}+\varepsilon_{ij} \qquad \begin{cases} i=1,2,\ldots,k \\ j=1,2,\ldots,n \end{cases} \tag{9.52}$$

where

$$x_{ij}=\begin{cases} 1 \text{ if observation } j \text{ is from treatment } i \\ 0 \text{ otherwise} \end{cases}$$

The relationship between the parameters in the regression and analysis of variance models is

$$\beta_0=\mu_k$$
$$\beta_i=\mu_i-\mu_k, \qquad i=1,2,\ldots,k-1$$

Thus β_0 always estimates the mean of the kth treatment and β_i estimates the differences in means between Treatment i and Treatment k.

Now consider fitting the regression model for the one-way analysis of variance. Once again, suppose that we have $k=3$ treatments and now let there be $n=3$ observations per treatment. The **X** matrix and **y** vector are as follows:

$$\mathbf{y}=\begin{bmatrix} y_{11} \\ y_{12} \\ y_{13} \\ y_{21} \\ y_{22} \\ y_{23} \\ y_{31} \\ y_{32} \\ y_{33} \end{bmatrix} \qquad \mathbf{X}=\begin{bmatrix} & x_1 & x_2 \\ 1 & 1 & 0 \\ 1 & 1 & 0 \\ 1 & 1 & 0 \\ 1 & 0 & 1 \\ 1 & 0 & 1 \\ 1 & 0 & 1 \\ 1 & 0 & 0 \\ 1 & 0 & 0 \\ 1 & 0 & 0 \end{bmatrix}$$

Notice that the **X** matrix consists entirely of zeros and ones. This is a characteristic of the regression formulation of any analysis of variance model. The least squares normal equations are

$$(\mathbf{X}'\mathbf{X})\hat{\boldsymbol{\beta}}=\mathbf{X}'\mathbf{y}$$

or

$$\begin{bmatrix} 9 & 3 & 3 \\ 3 & 3 & 0 \\ 3 & 0 & 3 \end{bmatrix}\begin{bmatrix} \hat{\beta}_0 \\ \hat{\beta}_1 \\ \hat{\beta}_2 \end{bmatrix}=\begin{bmatrix} y_{..} \\ y_{1.} \\ y_{2.} \end{bmatrix}$$

where $y_{i.}$ is the total of all observations in treatment i, and $y_{..}$ is the grand total of all 9 observations (i.e., $y_{..}=y_{1.}+y_{2.}+y_{3.}$). The solution to the normal equations is

$$\hat{\beta}_0=\bar{y}_{..}-\bar{y}_{1.}-\bar{y}_{2.}=\bar{y}_{3.}$$

$$\hat{\beta}_1=\bar{y}_{1.}-\bar{y}_{3.}$$

$$\hat{\beta}_2=\bar{y}_{2.}-\bar{y}_{3.}$$

The extra sum of squares method may be used to test for differences in treatment means. For the full model the regression sum of squares is

$$SS_R(\hat{\beta}_0, \hat{\beta}_1, \hat{\beta}_2) = \hat{\boldsymbol{\beta}}'\mathbf{X}'\mathbf{y} = [\bar{y}_{3.}, \bar{y}_{1.} - \bar{y}_{3.}, \bar{y}_{2.} - \bar{y}_{3.}] \begin{bmatrix} y_{..} \\ y_{1.} \\ y_{2.} \end{bmatrix}$$

$$= y_{..}\bar{y}_{3.} + y_{1.}(\bar{y}_{1.} - \bar{y}_{3.}) + y_{2.}(\bar{y}_{2.} - \bar{y}_{3.})$$

$$= (y_{1.} + y_{2.} + y_{3.})\bar{y}_{3.} + y_{1.}(\bar{y}_{1.} - \bar{y}_{3.}) + y_{2.}(\bar{y}_{2.} - \bar{y}_{3.})$$

$$= \bar{y}_{1.}y_{1.} + \bar{y}_{2.}y_{2.} + \bar{y}_{3.}y_{3.}$$

$$= \sum_{i=1}^{3} \frac{y_{i.}^2}{3}$$

with 3 degrees of freedom. The error sum of squares for the full model is

$$SS_E = \sum_{i=1}^{3} \sum_{j=1}^{3} y_{ij}^2 - SS_R(\beta_0, \beta_1, \beta_2)$$

$$= \sum_{i=1}^{3} \sum_{j=1}^{3} y_{ij}^2 - \sum_{i=1}^{3} \frac{y_{i.}^2}{3}$$

$$= \sum_{i=1}^{3} \sum_{j=1}^{3} (y_{ij} - \bar{y}_{i.})^2 \tag{9.53}$$

with $9-3=6$ degrees of freedom. Note that (9.53) is the error sum of squares in the analysis of variance table (Table 9.18) for $k=n=3$.

Testing for differences in treatment means is equivalent to testing

$$H_0: \tau_1 = \tau_2 = \tau_3 = 0$$

$$H_1: \text{at least one } \tau_i \neq 0$$

If H_0 is true, the parameters in the regression model become

$$\beta_0 = \mu$$

$$\beta_1 = 0$$

$$\beta_2 = 0$$

Therefore the reduced model contains only one parameter, that is,

$$y_{ij}=\beta_0+\varepsilon_{ij}$$

The estimate of β_0 in the reduced model is $\hat{\beta}_0=\bar{y}_{..}$, and the single degree of freedom regression sum of squares for this model is

$$SS_R(\beta_0)=\frac{y_{..}^2}{9}$$

The sum of squares for testing for equality of treatment means is the difference in regression sums of squares between the full and reduced models, or

$$\begin{aligned} SS_R(\beta_1,\beta_2|\beta_0) &= SS_R(\beta_0,\beta_1,\beta_2)-SS_R(\beta_0) \\ &= \sum_{i=1}^{3}\frac{y_{i.}^2}{3}-\frac{y_{..}^2}{9} \\ &= 3\sum_{i=1}^{3}(\bar{y}_{i.}-\bar{y}_{..})^2 \qquad (9.54) \end{aligned}$$

This sum of squares has $3-1=2$ degrees of freedom. Note that (9.54) is the treatment sum of squares in Table 9.18 assuming that $k=n=3$. The appropriate test statistic is

$$\begin{aligned} F_0 &= \frac{SS_R(\beta_1,\beta_2|\beta_0)/2}{SS_E/6} \\ &= \frac{3\sum_{i=1}^{3}(\bar{y}_{i.}-\bar{y}_{..})^2/2}{\sum_{i=1}^{3}\sum_{j=1}^{3}(y_{ij}-\bar{y}_{i.})^2/6} \\ &= \frac{MS_{\text{Treatments}}}{MS_E} \end{aligned}$$

If H_0: $\tau_1=\tau_2=\tau_3=0$ is true, then F_0 follows the $F_{2,6}$ distribution. This is the same test statistic given in the analysis of variance table (Table 9.18).

Therefore the regression approach is identical to the one-way analysis of variance procedure outlined in Table 9.18.

Problems

9.1 The data below give the percent share of market of a particular brand of canned peaches for the past 15 months and the relative selling price.

t	x_t	y_t
1	100	15.93
2	98	16.26
3	100	15.94
4	89	16.81
5	95	15.67
6	87	16.47
7	93	15.66
8	82	16.94
9	85	16.60
10	83	17.16
11	81	17.77
12	79	18.05
13	90	16.78
14	77	18.17
15	78	17.25

a. Fit a simple linear regression model to these data. Plot the residuals versus time. Is there any indication of autocorrelation?

b. Use the Durbin–Watson test to determine if there is positive autocorrelation in the errors. What are your conclusions?

c. Use one iteration of the Cochrane–Orcutt procedure to estimate the regression coefficients. Find the standard errors of these regression coefficients.

d. Is there positive autocorrelation remaining after the first iteration? Would you conclude that the iterative parameter estimation technique has been successful?

9.2 The data below give the monthly sales for a cosmetics manufacturer (y_t) and the corresponding monthly sales for the entire industry (x_t). The units of x_t and y_t are millions of dollars.

t	x_t	y_t
1	5.00	.318
2	5.06	.330
3	5.12	.356
4	5.10	.334
5	5.35	.386
6	5.57	.455
7	5.61	.460
8	5.80	.527
9	6.04	.598
10	6.16	.650
11	6.22	.685
12	6.31	.713
13	6.38	.724
14	6.54	.775
15	6.68	.782
16	6.73	.796
17	6.89	.859
18	6.97	.883

a. Build a simple linear regression model relating company sales to industry sales. Plot the residuals against time. Is there any indication of autocorrelation?

b. Use the Durbin–Watson test to determine if there is positive autocorrelation in the errors. What are your conclusions?

c. Use one iteration of the Cochrane–Orcutt procedure to estimate the model parameters. Compare the standard error of these regression coefficients with the standard error of the least squares estimates.

d. Test for positive autocorrelation following the first iteration. Has the iterative procedure been successful?

9.3 Consider the simple linear regression model $y_t = \beta_0 + \beta_1 x_t + \varepsilon_t$, where the errors are generated by the second-order autoregressive process

$$\varepsilon_t = \rho_1 \varepsilon_{t-1} + \rho_2 \varepsilon_{t-2} + a_t$$

Discuss how the Cochrane–Orcutt iterative procedure could be used in this situation. What transformations would be used on the variables y_t and x_t? How would you estimate the parameters ρ_1 and ρ_2?

9.4 Consider the regression model with first-order autoregressive errors defined in (9.2). Derive the mean, variance, and covariance of the errors given in (9.3*a*, *b*, *c*).

9.5 Using the data in Problem 9.1, define a new set of transformed variables as the first difference of the original variables, $x_t' = x_t - x_{t-1}$ and $y_t' = y_t - y_{t-1}$. Regress y_t' on x_t' through the origin. Compare the estimate of the slope from this first difference approach with the estimate obtained from the iterative method in Problem 9.1.

9.6 Suppose that we want to fit the no–intercept model $y = \beta x + \varepsilon$ using weighted least squares. Assume that the observations are uncorrelated but have unequal variances.

a. Find a general formula for the weighted least squares estimator of β.

b. What is the variance of the weighted least squares estimator?

c. Suppose that $V(y_i) = cx_i$; that is, the variance of y_i is proportional to the corresponding x_i. Using the results of Parts a and b, find the weighted least squares estimator of β and the variance of this estimator.

d. Suppose that $V(y_i) = cx_i^2$; that is, the variance of y_i is proportional to the square of the corresponding x_i. Using the results of Parts a and b, find the weighted least squares estimator of β and the variance of this estimator.

9.7 Consider the continuous probability distribution $f(x)$. Suppose that θ is an unknown location parameter, and that the density may be written as $f(x-\theta)$ for $-\infty < \theta < \infty$. Let $x_1, x_2, \ldots, x_n$ be a random sample of size n from the density.

a. Show that the maximum likelihood estimator of θ is the solution to

$$\sum_{i=1}^{n} \psi(x_i - \theta) = 0$$

that maximizes the logarithm of the likelihood function $\ln L(\mu) = \Sigma_{i=1}^{n} \ln f(x_i - \theta)$ where $\psi(x) = \rho'(x)$ and $\rho(x) = -\ln f(x)$.

b. If $f(x)$ is a normal distribution, find $\rho(x)$, $\psi(x)$, and the corresponding maximum likelihood estimator of θ.

c. If $f(x) = (2\sigma)^{-1} e^{-|x|/\sigma}$ (the double exponential distribution), find $\rho(x)$ and $\psi(x)$. Show that the maximum likelihood estimator of θ is the sample median. Compare this estimator with the estimator found in Part b. Does the sample median seem to be a reasonable estimator in this case?

d. If $f(x) = [\pi(1+x^2)]^{-1}$ (the Cauchy distribution), find $\rho(x)$ and $\psi(x)$. How would you solve $\Sigma_{i=1}^{n} \psi(x_i - \theta)$ in this case?

9.8 Suppose that $f(x)$ is a continuous distribution with unknown location parameter θ. Consider a random sample of size n from this distribution, and let the sample values be arranged in ascending order so that $x_{(1)} \leqslant x_{(2)} \leqslant \cdots \leqslant x_{(n)}$. A robust estimator of location is the α-trimmed mean, for example

$$\bar{x}_{\alpha} = \sum_{i} x_{(i)} / (n - 2[n\alpha])$$

where the summation is from $i = [n\alpha + 1]$ to $i = n - [n\alpha], 0 < \alpha < 1$, and [ ] denotes the "greatest integer contained in" operation. How effective do you think trimming is for dealing with data from heavy-tailed distributions? Discuss how you might apply trimming to the regression problem.

9.9 Winsorized regression (Yale and Forsythe [1976]). Winsorization is a robust estimation technique that arrays the sample values in ascending order, that is, $x_{(1)} \leqslant x_{(2)} \leqslant \cdots \leqslant x_{(n)}$, and then redefines the most extreme values (possible outliers) as the next most extreme values. Thus if only the largest and smallest observations are Winsorized, we would set $x'_{(1)} = x_{(2)}$ and $x'_{(n)} = x_{(n-1)}$. Discuss how Winsorization might be applied to the regression problem.

9.10 Consider the stack loss data shown in Table 9.15. Formulate the linear programming problem whose solution is the L_1 norm estimator of the regression coefficients $\boldsymbol{\beta}$. Solve this linear programming problem. Compare the parameter estimates with those given in Example 9.5.

9.11 Find robust estimates of the regression coefficients for the stack loss data in Table 9.15 using the $E_{0.3}$ function defined in Table 9.9. Compare the estimates of the coefficients with those in Example 9.5. How effective was the $E_{0.3}$ function for this problem?

9.12 Find robust estimates of the regression coefficients for the stack loss data in Table 9.15 using Hampel's 17A function defined in Table 9.9. Compare the regression coefficients with those in Example 9.5 obtained using Andrews' wave function. How effective was Hampel's 17A function in this problem?

9.13 Tukey's biweight. A popular ψ-function for robust regression is Tukey's biweight, where

$$\begin{aligned} \psi(z) &= z\left[1-(z/a)^2\right]^2, & |z| \leqslant a \\ &= 0, & |z| > a \end{aligned}$$

with $a = 5$ or 6, usually. Sketch the ψ-function and discuss its behavior. Do you think that Tukey's biweight would give results similar to Andrews' wave function?

9.14 Consider the simple linear regression model for the blood pressure-weight data in Problem 2.12.

a. Find a 95 percent confidence region for β_0 and β_1.

b. Find 95 percent joint confidence intervals for β_0 and β_1 using the Bonferroni method.

c. Plot the Bonferroni confidence intervals and compare them with the confidence ellipse obtained in Part a. Which method is superior?

9.15 Consider the simple linear regression model for the gasoline mileage data developed in Problem 2.4.

a. Find a 95 percent joint confidence region for the intercept and slope.

b. Find 95 percent joint confidence intervals for the intercept and slope using the maximum modulus t method.

c. Plot the maximum modulus t confidence intervals and compare them with the confidence ellipse obtained in Part a. Which method is superior?

9.16 Consider the simple linear regression model for the National Football League data developed in Problem 2.1.

a. Find a 90 percent joint confidence region for the intercept and slope.

b. Find 90 percent joint confidence intervals for the intercept and slope using the Bonferroni method.

c. Interpret the confidence coefficient in these joint confidence statements.

d. Plot both the confidence ellipse and the Bonferroni confidence intervals on the same graph. Which method is most efficient?

9.17 Consider the simple linear regression model for the blood pressure-weight data in Problem 2.12.

a. Find a 95 percent confidence band for the regression line. Plot this band along with the fitted model and discuss the precision of the estimated model.

b. Obtain point estimates of mean systolic blood pressure when the weights are 150, 175, and 200 pounds, respectively. Find a set of 95 percent joint confidence intervals on mean blood pressure at these weights using the Scheffé method.

c. Would the confidence intervals in Part b have been tighter if either the Bonferroni or the maximum modulus t method had been used? If so, construct the tightest confidence intervals.

d. We wish to predict the systolic blood pressure of individuals weighing 150 and 175 pounds, respectively. A pair of predictor intervals having joint confidence at least 0.95 is desired. Will either the Bonferroni or the maximum modulus t method provide shorter intervals than the Scheffé method in this situation? Calculate the desired prediction intervals.

9.18 Consider the simple linear regression model for the gasoline mileage data developed in Problem 2.4.

a. Find a 90 percent confidence band for the regression line. Plot this band along with the fitted model and discuss the precision of the estimated model.

b. Obtain point estimates of mean gasoline mileage if the engine displacement is 200, 250, and 275 $(\text{in.})^3$, respectively. Find a set of 90 percent joint confidence intervals on mean gasoline mileage at these engine displacements using the Scheffé method.

c. Would the confidence intervals in Part b have been tighter if either the Bonferroni or the maximum modulus t method had been used? If so, construct the tightest confidence intervals.

d. Suppose that we wish to construct 90 percent prediction intervals on gasoline mileage for engine displacements of 250 and 275 $(\text{in.})^3$, respectively. Which method, Bonferroni, maximum modulus t, or Scheffé, will produce the tightest intervals? Construct the desired prediction intervals.

9.19 Consider the multiple regression model developed in Problem 4.4 relating gasoline mileage to engine displacement and the number of carburetor barrels.

a. Find a set of 95 percent joint confidence intervals on all three regression coefficients using the Bonferroni method.

b. Find 95 percent joint confidence intervals on the mean gasoline mileage for engines having 2 carburetor barrels and displacements of 200 $(\text{in.})^3$ and engines having 2 carburetor barrels and displacement of 250 $(\text{in.})^3$. Which method, Bonferroni, maximum modulus t, or Scheffé, produces the tightest confidence intervals?

9.20 Consider the regression model in Problem 2.12 relating systolic blood pressure to weight. Suppose that we wish to predict an individual's weight given an observed value of systolic blood pressure. Can this be done using the procedure for predicting x given a value of y described in Section 9.7? In this particular application, how would you respond to the suggestion of building a regression model relating weight to systolic blood pressure?

9.21 Consider the regression model in Problem 2.4 relating gasoline mileage to engine displacement.

a. If a particular car has an observed gasoline mileage of 17 miles per gallon, find a point estimate of the corresponding engine displacement.

b. Find a 95 percent confidence interval on engine displacement.

9.22 Consider the regression model in Problem 2.3 relating total heat flux to radial deflection for the solar energy data in Appendix Table B.2.

a. Suppose that the observed total heat flux is 250 kw. Find a point estimate of the corresponding radial deflection.

b. Construct a 90 percent confidence interval on radial deflection.

9.23 An analyst is fitting a simple linear regression model with the objective of obtaining a minimum variance estimate of the intercept β_0. How should the data collection experiment be designed?

9.24 Suppose that you are fitting a simple linear regression model that will be used to predict the mean response at a particular point such as x_0. How should the data collection experiment be designed so that a minimum variance estimate of the mean of y at x_0 is obtained?

9.25 Consider the linear regression model $y=\beta_0+\beta_1 x_1+\beta_2 x_2+\varepsilon$, where the regressors have been coded so that

$$\sum_{i=1}^{n} x_{i1}=\sum_{i=1}^{n} x_{i2}=0 \qquad \text{and} \qquad \sum_{i=1}^{n} x_{i1}^2=\sum_{i=1}^{n} x_{i2}^2=n$$

a. Show that an orthogonal design ($\mathbf{X}'\mathbf{X}$ diagonal) minimizes the variance of $\hat{\beta}_1$ and $\hat{\beta}_2$.
b. Show that any design for fitting this first-order model that is orthogonal is also rotatable.

9.26 Suppose that a one-way analysis of variance involves four treatments, but that a different number of observations (for example, n_i) has been taken under each treatment. Assuming that $n_1=3$, $n_2=2$, $n_3=4$, and $n_4=3$ write down the $\mathbf{y}$ vector and $\mathbf{X}$ matrix for analyzing this data as a multiple regression model. Are any complications introduced by the unbalanced nature of this data?

9.27 Alternate coding schemes for the regression approach to analysis of variance. Consider (9.51), which represents the regression model corresponding to an analysis of variance with three treatments and n observations per treatment. Suppose that the indicator variables x_1 and x_2 are defined as

$$x_1=\begin{cases} 1 \text{ if observation is from Treatment 1} \\ -1 \text{ if observation is from Treatment 2} \\ 0 \text{ otherwise} \end{cases}$$

$$x_2=\begin{cases} 1 \text{ if observation is from Treatment 2} \\ -1 \text{ if observation is from Treatment 3} \\ 0 \text{ otherwise} \end{cases}$$

a. Show that the relationship between the parameters in the regression and analysis of variance models is

$$\beta_0 = \frac{\mu_1 + \mu_2 + \mu_3}{3} = \bar{\mu}$$

$$\beta_1 = \mu_1 - \bar{\mu}$$

$$\beta_2 = \mu_2 - \bar{\mu}$$

b. Write down the **y** vector and **X** matrix.

c. Develop an appropriate sum of squares for testing the hypothesis H_0: $\tau_1 = \tau_2 = \tau_3 = 0$. Is this the usual treatment sum of squares in the one-way analysis of variance?

9.28 Montgomery [1976] presents data concerning the tensile strength of synthetic fiber used to make cloth for men's shirts. The strength is thought to be affected by the percentage of cotton in the fiber. The data are shown below.

Percentage of Cotton	Tensile Strength				
15	7	7	15	11	9
20	12	17	12	18	18
25	14	18	18	19	19
30	19	25	22	19	23
35	7	10	11	15	11

a. Write down the **y** vector and **X** matrix for the corresponding regression model.

b. Find the least squares estimates of the model parameters.

c. Find a point estimate of the difference in mean strength between 15 percent and 25 percent cotton.

d. Test the hypothesis that the mean tensile strength is the same for all five cotton percentages.

9.29 The two-way analysis of variance Suppose that two different sets of treatments are of interest. Let y_{ijk} be the kth observation in level i of the first treatment type and level j of the second treatment type. The two-way analysis of variance model is

$$y_{ijk} = \mu + \tau_i + \gamma_j + (\tau\gamma)_{ij} + \varepsilon_{ijk} \qquad \begin{cases} i = 1, 2, \ldots, a \\ j = 1, 2, \ldots, b \\ k = 1, 2, \ldots, n \end{cases}$$

where τ_i is the effect of level i of the first treatment type, γ_j is the effect of level j of the second treatment type, $(\tau\gamma)_{ij}$ is an interaction effect between the two treatment types, and ε_{ijk} is an NID$(0, \sigma^2)$ random error component.

a. For the case $a=b=n=2$, write down a regression model that corresponds to the two-way analysis of variance.

b. What are the **y** vector and **X** matrix for this regression model?

c. Discuss how the regression model could be used to test the hypothesis H_0: $\tau_1=\tau_2=0$ (treatment type 1 means are equal), H_0: $\gamma_1=\gamma_2$ (treatment type 2 means are equal), and H_0: $(\tau\gamma)_{11}=(\tau\gamma)_{12}=(\tau\gamma)_{21}=(\tau\gamma)_{22}=0$ (no interaction between treatment types).

10

VALIDATION OF REGRESSION MODELS

10.1 Introduction

Regression models are used extensively for prediction or estimation, data description, parameter estimation, and control. Frequently the user of the regression model is a different individual from the model developer. Before the model is released to the user, some assessment of its validity should be made. We distinguish between model *adequacy checking* and model *validation*. Model adequacy checking includes residual analysis, testing for lack of fit, searching for high-leverage or overly influential observations, and other internal analyses that investigate the fit of the regression model to the available data. Model validation, however, is directed towards determining if the model will function successfully in its intended operating environment.

Since the fit of the model to the available data forms the basis for many of the techniques used in the model development process (such as variable selection), it is tempting to conclude that a model that fits the data well will also be successful in the final application. This is not necessarily so. For example, a model may have been developed primarily for predicting new observations. There is no assurance that the equation that provides the best fit to existing data will be a successful predictor. Influential factors that were unknown during the model building stage may significantly affect the new observations, rendering the predictions almost useless. Furthermore the correlative structure between the regressors may differ in the model building and prediction data. This may result in poor predictive performance for the model. Proper validation of a model developed to predict new observations should involve testing the model in that environment before it is released to the user.

Another critical reason for validation is that the model developer often has little or no control over the model's final use. For example, Snee [1977] observes that although a model has been developed as an interpolation equation, when the user discovers that it is successful in that respect he will also extrapolate with it if the need arises, despite any warnings or cautions from the developer. Furthermore if this extrapolation performs poorly, it is almost always the model *developer* and not the model *user* who is blamed for the failure. Regression model users will also frequently draw conclusions about the process being studied from the signs and magnitudes of the coefficients in their model, even though they have been cautioned about the hazards of interpreting partial regression coefficients. Model validation provides a measure of protection for both model developer and user.

Proper validation of a regression model should include a study of the coefficients to determine if their signs and magnitudes are reasonable. That is, can $\hat{\beta}_j$ be reasonably interpreted as an estimate of the effect of x_j? We should also investigate the *stability* of the regression coefficients. That is, are the $\hat{\beta}_j$ obtained from a new sample likely to be similar to the current coefficients? Finally, validation requires that the model's prediction performance be investigated. Both interpolation and extrapolation modes should be considered.

This chapter will discuss and illustrate several techniques useful in validating regression models. Several references on the general subject of validation are Brown, Durbin and Evans [1975], Geisser [1975], McCarthy [1976], Snee [1977], and Stone [1974]. Snee's paper is particularly recommended.

10.2 Validation techniques

Three types of procedures are useful for validating a regression model. They are

1. Analysis of the model coefficients and predicted values including comparisons with prior experience, physical theory, and other analytical models, or simulation results.

2. Collection of fresh data with which to investigate the model's predictive performance.

3. Data splitting; that is, setting aside some of the original data and using these observations to investigate the model's predictive performance.

The final intended use of the model often indicates the appropriate validation methodology. Thus validation of a model intended for use as a predictive equation should concentrate on determining the model's prediction accuracy. However since the developer often does not control the use of the model, we recommend that whenever possible all the validation techniques above be used. We will now discuss and illustrate these techniques. For some additional examples, see Snee [1977].

10.2.1 Analysis of model coefficients and predicted values

The coefficients in the final regression model should be studied to determine if they are stable and if their signs and magnitudes are reasonable. Previous experience, theoretical considerations, or an analytical model can often provide information concerning the direction and relative size of the effects of the regressors. The coefficients in the estimated model should be compared with this information. Coefficients with unexpected signs or that are too large in absolute value often indicate either an inappropriate model (missing or misspecified regressors) or poor estimates of the effects of the individual regressors. The variance inflation factors (and other multicollinearity diagnostics in Section 8.4) also are an important guide to the validity of the model. If any VIF exceeds 5 or 10, that particular coefficient is poorly estimated or unstable because of near linear dependencies among the regressors. When the data are collected across time, we can examine the stability of the coefficients by fitting the model on shorter time spans. For example, if we had several years of monthly data, we could build a model for each year. Hopefully, the coefficients for each year would be similar.

The predicted values $\hat{y}$ can also provide a measure of model validity. Unrealistic predicted values such as negative predictions of a positive quantity or predictions that fall outside the actual range of the response, indicate poorly estimated coefficients or an incorrect model form. Predicted values inside and on the boundary of the regressor variable hull provide a measure of the model's interpolation performance. Predicted values outside this region are a measure of extrapolation performance.

- **Example 10.1** Consider the acetylene data introduced in Example 8.1. Checking the validity of the nine-term least squares model for this data by studying the coefficients and predicted values reveals several problems. The standardized regression coefficients corresponding to x_1^2, x_3^2, and x_1x_3 are large in absolute value. It is unusual for these coefficients in a quadratic model to be so large. Furthermore the VIFs for those coefficients involving x_1

and x_3 and very large (see Table 8.5), indicating high correlation between x_1 and x_3. The predicted values of percent conversion (Figure 8.3) also indicate potential problems with the nine-term least squares model. The predicted values within the region of the original data are satisfactory, but relatively mild extrapolation produces several negative estimates of percent conversion, a quantity which must be positive. There is evidence that although the nine-term least squares model may be a valid interpolation equation, it extrapolates poorly.

Several alternate models for the acetylene data are developed in Chapter 8, using variable selection and biased estimation techniques to combat the multicollinearity in the data. Both the nine-variable ordinary ridge regression model (Example 8.2) and the five-variable least squares model (Example 8.4) are more reasonable equations, as they reduce the magnitude of the coefficients for x_1^2, x_3^2, and x_1x_3 to more appropriate values and give realistic predicted values for both interpolation and extrapolation. •

10.2.2 Collecting fresh data

The most effective method of validating a regression model with respect to its prediction performance is to collect fresh data and directly compare the model predictions against it. If the model gives accurate predictions of new data the user will have greater confidence in both the model and the model building process. At least 15–20 new observations are required to give a reliable assessment of the model's prediction performance. In situations where two or more alternative regression models have been developed from the data, comparing the prediction performance of these models on fresh data may provide a basis for final model selection.

• **Example 10.2** Consider the delivery time data introduced in Example 4.1. We have previously developed both the least squares fit (Example 4.1) and a robust fit using Ramsay's E_a function (Example 9.5) for these data. The objective of fitting the regression model is to predict new observations. We will investigate the validity of these models as predictors and compare the least squares and robust fits by predicting the delivery time for fresh data.

Recall that the original 25 observations came from four cities; Austin, San Diego, Boston, and Minneapolis. Fifteen new observations from Austin, Boston, San Diego, and a fifth city, Louisville, are shown in Table 10.1, along with the corresponding predicted delivery times and prediction errors from the least squares fit $\hat{y}=2.3412+1.6159x_1+0.0144x_2$ (Columns 5 and 6) and robust fit $\hat{y}=3.8021+1.4894x_1+0.0135x_2$ (Columns 7 and 8). Note that this prediction data set consists of 11 observations from cities used in the original data collection process and four observations from a new city. This mix of old and new cities may provide some information on how well the two models predict at sites where the original data were collected and at new sites.

Table 10.1 Prediction data set for the delivery time example

	(1)	(2)	(3)	(4)	(5)	(6)	(7)	(8)
					Least Squares Fit		Robust Fit	
Observation	City	Cases x_1	Distance x_2	Observed Time, y	$\hat{y}$	$y-\hat{y}$	$\hat{y}$	$y-\hat{y}$
26	San Diego	22	905	51.00	50.9230	.0770	48.8072	2.1928
27	San Diego	7	520	16.80	21.1405	−4.3405	21.2599	−4.4599
28	Boston	15	290	26.16	30.7557	−4.5957	30.0648	−3.9048
29	Boston	5	500	19.90	17.6207	2.2793	18.0106	1.8894
30	Boston	6	1000	24.00	26.4366	−2.4366	26.2615	−2.2615
31	Boston	6	225	18.55	15.2766	3.2734	15.7812	2.7688
32	Boston	10	775	31.93	29.6602	2.2698	29.1764	2.7536
33	Boston	4	212	16.95	11.8576	5.0924	12.6266	4.3234
34	Austin	1	144	7.00	6.0307	0.9693	7.2388	−0.2388
35	Austin	3	126	14.00	9.0033	4.9967	9.9742	4.0258
36	Austin	12	655	37.03	31.1640	5.8660	30.5325	6.4975
37	Louisville	10	420	18.62	24.5482	−5.9282	24.3758	−5.7558
38	Louisville	7	150	16.10	15.8125	.2875	16.2563	−0.1563
39	Louisville	8	360	24.38	20.4524	3.9276	20.5856	3.7944
40	Louisville	32	1530	64.75	76.0820	−11.3320	72.1531	−7.4031

Column 6 of Table 10.1 shows the prediction errors for the least squares model. The average prediction error is 0.4060, which is nearly zero, so that model seems to produce approximately unbiased predictions. There is only one relatively large prediction error, associated with the last observation from Louisville. Checking the original data reveals that this observation is an extrapolation point. Furthermore this point is quite close to Point 9, which we know to be influential. From an overall perspective, these prediction errors increase our confidence in the usefulness of the model. Note that the prediction errors are generally larger than the residuals from the least squares fit. This is easily seen by comparing the residual mean square

$$MS_E = 10.6239$$

from the fitted model and the average squared prediction error

$$\frac{\sum_{i=26}^{40} (y_i - \hat{y}_i)^2}{15} = \frac{332.2809}{15} = 22.1521$$

from the new prediction data. Since MS_E (which may be thought of as the average variance of the residuals from the fit) is smaller than the average squared prediction error, the least squares regression model does not predict new data as well as it fits the existing data. However, the degradation of performance is not severe, and so we conclude that the least squares model is likely to be successful as a predictor. Note also that apart from the one extrapolation point the prediction errors from Louisville are not remarkably different from those experienced in the cities where the original data were collected. While the sample is small, this is an indication that the model may be portable. More extensive data collection at other sites would be helpful in verifying this conclusion.

It is also instructive to compare R^2 from the least squares fit (0.9596, from Example 4.10) to the percent variability in the new data explained by the model, such as

$$R^2_{\text{Prediction}} = 1 - \frac{\sum_{i=26}^{40} (y_i - \hat{y}_i)^2}{\sum_{i=26}^{40} (y_i - \bar{y})^2} = 1 - \frac{332.2809}{3206.2338} = 0.8964$$

Once again, we see that the least squares model does not predict new observations as well as it fits the original data. However, the "loss" in R^2 for prediction is slight.

Columns 7 and 8 of Table 10.1 display the predicted values and prediction errors from the robust fit. The average prediction error is 0.2710, so the robust regression procedure also produces approximately unbiased predictions. However, the prediction errors from the robust fit seem generally smaller than their least squares counterparts. This is confirmed by calculating the average squared prediction error from the robust fit, for example,

$$\frac{\sum_{i=26}^{40} (y_i - \hat{y}_i)^2}{15} = \frac{243.4093}{15} = 16.2273$$

which is smaller than the corresponding quantity for the least squares prediction errors. Similarly, the R^2 for prediction for the robust fit is

$$R^2_{\text{Prediction}} = 1 - \frac{\sum_{i=26}^{40} (y_i - \hat{y}_i)^2}{\sum_{i=26}^{40} (y_i - \bar{y})^2} = 1 - \frac{243.4093}{3206.2338} = 0.9241$$

and this is slightly larger than the corresponding value for the least squares model. Note that the robust fit does not overestimate the delivery time for the extrapolation point as badly as does least squares.

Collecting fresh data has indicated that both the least squares and robust fits for the delivery time data produce reasonably good prediction equations. There is some evidence that the robust fit in this example may be slightly superior to least squares.

10.2.3 Data splitting

In many situations, collecting fresh data for validation purposes is not possible. The data collection budget may already have been spent, the plant may have been converted to the production of other products, or other equipment and resources needed for data collection may be unavailable. When these situations occur, a reasonable procedure is to split the available data into two parts, which Snee [1977] calls the *estimation data* and the *prediction data*. The estimation data is used to build the regression model and the prediction data is then used to study the predictive ability of the model. Sometimes data splitting is called *cross-validation* (see Mosteller and Tukey [1968] and Stone [1974]).

Data splitting may be done in a variety of ways. Allen's [1971b, 1974] prediction error sum of squares or PRESS statistic may be considered as a form of data splitting. To calculate PRESS, select an observation, for example i. Fit the regression model to the remaining $n-1$ observations and use this equation to predict the withheld observation y_i. Denoting this predicted value $\hat{y}_{(i)}$, we may find the prediction error for point i as $e_{(i)}=y_i-\hat{y}_{(i)}$. The prediction error is often called the ith *deleted residual.* This procedure is repeated for each observation $i=1,2,\ldots,n$, producing a set of n deleted residuals $e_{(1)},e_{(2)},\ldots,e_{(n)}$. Then the PRESS statistic is defined as the sum of squares of the n deleted residuals as in

$$\text{PRESS}=\sum_{i=1}^{n} e_{(i)}^2=\sum_{i=1}^{n}\left[y_i-\hat{y}_{(i)}\right]^2 \tag{10.1}$$

Thus PRESS uses each possible subset of $n-1$ observations as the estimation data set, and every observation in turn is used to form the prediction data set.

It would initially seem that calculating PRESS requires fitting n different regressions. However, it is possible to calculate PRESS from the results of a single least squares fit to all n observations. To see how this is accomplished, let $\hat{\boldsymbol{\beta}}_{(i)}$ be the vector of regression coefficients obtained by deleting the ith observation. Then

$$\hat{\boldsymbol{\beta}}_{(i)}=\left[\mathbf{X}'_{(i)}\mathbf{X}_{(i)}\right]^{-1}\mathbf{X}'_{(i)}\mathbf{y}_{(i)} \tag{10.2}$$

where $\mathbf{X}_{(i)}$ and $\mathbf{y}_{(i)}$ are the $\mathbf{X}$ and $\mathbf{y}$ vector with the ith observation deleted.

Thus the ith deleted residual may be written as

$$\begin{aligned} e_{(i)} &= y_i - \hat{y}_{(i)} \\ &= y_i - \mathbf{x}_i'\hat{\boldsymbol{\beta}}_{(i)} \\ &= y_i - \mathbf{x}_i'(\mathbf{X}_{(i)}'\mathbf{X}_{(i)})^{-1}\mathbf{X}_{(i)}'\mathbf{y}_{(i)} \end{aligned}$$

There is a close connection between the $(\mathbf{X}'\mathbf{X})^{-1}$ and $[\mathbf{X}_{(i)}\mathbf{X}_{(i)}]^{-1}$ matrices; namely (see Rao [1965])

$$[\mathbf{X}_{(i)}'\mathbf{X}_{(i)}]^{-1} = (\mathbf{X}'\mathbf{X})^{-1} + \frac{(\mathbf{X}'\mathbf{X})^{-1}\mathbf{x}_i\mathbf{x}_i'(\mathbf{X}'\mathbf{X})^{-1}}{1-h_{ii}} \tag{10.3}$$

where $h_{ii} = \mathbf{x}_i'(\mathbf{X}'\mathbf{X})^{-1}\mathbf{x}_i$. Using (10.3), we may write

$$\begin{aligned} e_{(i)} &= y_i - \mathbf{x}_i'\left[(\mathbf{X}'\mathbf{X})^{-1} + \frac{(\mathbf{X}'\mathbf{X})^{-1}\mathbf{x}_i\mathbf{x}_i'(\mathbf{X}'\mathbf{X})^{-1}}{1-h_{ii}}\right]\mathbf{X}_{(i)}'\mathbf{y}_{(i)} \\ &= y_i - \mathbf{x}_i'(\mathbf{X}'\mathbf{X})^{-1}\mathbf{X}_{(i)}'\mathbf{y}_{(i)} - \frac{\mathbf{x}_i'(\mathbf{X}'\mathbf{X})^{-1}\mathbf{x}_i\mathbf{x}_i'(\mathbf{X}'\mathbf{X})^{-1}\mathbf{X}_{(i)}'\mathbf{y}_{(i)}}{1-h_{ii}} \\ &= \frac{(1-h_{ii})y_i - (1-h_{ii})\mathbf{x}_i'(\mathbf{X}'\mathbf{X})^{-1}\mathbf{X}_{(i)}'\mathbf{y}_{(i)} - h_{ii}\mathbf{x}_i'(\mathbf{X}'\mathbf{X})^{-1}\mathbf{X}_{(i)}'\mathbf{y}_{(i)}}{1-h_{ii}} \\ &= \frac{(1-h_{ii})y_i - \mathbf{x}_i'(\mathbf{X}'\mathbf{X})^{-1}\mathbf{X}_{(i)}'\mathbf{y}_{(i)}}{1-h_{ii}} \end{aligned}$$

Since $\mathbf{X}'\mathbf{y} = \mathbf{X}_{(i)}'\mathbf{y}_{(i)} + \mathbf{x}_i y_i$, this last equation becomes

$$\begin{aligned} e_{(i)} &= \frac{(1-h_{ii})\mathbf{y}_i - \mathbf{x}_i'(\mathbf{X}'\mathbf{X})^{-1}(\mathbf{X}'\mathbf{y} - \mathbf{x}_i y_i)}{1-h_{ii}} \\ &= \frac{(1-h_{ii})y_i - \mathbf{x}_i'(\mathbf{X}'\mathbf{X})^{-1}\mathbf{X}'\mathbf{y} + \mathbf{x}_i'(\mathbf{X}'\mathbf{X})^{-1}\mathbf{x}_i y_i}{1-h_{ii}} \\ &= \frac{(1-h_{ii})y_i - \mathbf{x}_i'\hat{\boldsymbol{\beta}} + h_{ii}y_i}{1-h_{ii}} \\ &= \frac{y_i - \mathbf{x}_i'\hat{\boldsymbol{\beta}}}{1-h_{ii}} \end{aligned} \tag{10.4}$$

Now the numerator of (10.4) is the ordinary residual e_i from a least squares fit to all n observations, so the ith deleted residual is

$$e_{(i)}=\frac{e_i}{1-h_{ii}} \tag{10.5}$$

Thus since PRESS is just the sum of the squares of the deleted residuals, a simple computing formula is

$$\text{PRESS}=\sum_{i=1}^{n}\left(\frac{e_i}{1-h_{ii}}\right)^2 \tag{10.6}$$

In this form, it is easy to see that PRESS is just a weighted sum of squares of the residuals, where the weights are related to the leverage of the observations. PRESS weights the residuals corresponding to high leverage observations more severely than the residuals from less influential points.

- **Example 10.3** Table 10.2 displays the calculation of the PRESS statistic for the delivery time regression model in Example 4.1. The least squares residuals e_i and the diagonal element of the hat matrix h_{ii} were first given in Examples 4.1 and 4.13, respectively. The value of PRESS=457.4000 is almost twice as large as the residual sum of squares $SS_E=233.7260$ for this model. Note that approximately half the PRESS statistic is contributed by Point 9, a relatively remote and influential observation. Thus some degradation in performance of this model as a predictor is anticipated. One way to estimate this decrease in performance is to use PRESS to compute an approximate R^2 for prediction, such as

$$R^2_{\text{Prediction}}=1-\frac{\text{PRESS}}{S_{yy}}=1-\frac{457.4000}{5784.5426}=0.9209$$

Therefore we could expect this model to "explain" about 92.09 percent of the variability in predicting new observations, as compared to the approximately 95.96 percent of the variability in the original data explained by the least squares fit. This estimated loss in R^2 for prediction is roughly comparable to the loss actually observed in Example 10.2. •

The PRESS statistic is also useful in evaluating alternative models when the objective is prediction. A value of PRESS would be calculated for each model under consideration and a model with a small value of PRESS selected.

Table 10.2 Calculation of PRESS for the delivery time data in Example 4.1

Observation, i	e_i	h_{ii}	$[e_i/(1-h_{ii})]^2$
1	−5.0281	0.10180	31.3373
2	1.1464	0.07070	1.5218
3	−0.0498	0.09874	0.0031
4	4.9244	0.05838	27.3499
5	−0.4444	0.07501	0.2308
6	−0.2896	0.04287	0.0915
7	0.8446	0.08180	0.8461
8	1.1566	0.06373	1.5260
9	7.4197	0.49829	218.7093
10	2.3764	0.19630	8.7428
11	2.2375	0.08613	5.9946
12	−0.5930	0.11366	0.4476
13	1.0270	0.06113	1.1965
14	1.0675	0.07824	1.3412
15	0.6712	0.04111	0.4900
16	−0.6629	0.16594	0.6317
17	0.4364	0.05943	0.2153
18	3.4486	0.09626	14.5612
19	1.7932	0.09645	3.9387
20	−5.7880	0.10169	41.5150
21	−2.6142	0.16528	9.8084
22	−3.6865	0.39158	36.7131
23	−4.6076	0.04126	23.0966
24	−4.5728	0.12061	27.0397
25	−0.2126	0.06664	0.0519
			PRESS=457.4000

• **Example 10.4** To illustrate the use of PRESS in choosing between competing models, consider the Hald cement data in Example 7.1. Several candidate models were identified as possibly suitable for these data, including equations involving (x_1, x_2) and (x_1, x_2, x_4). The model involving (x_1, x_2) is the minimum C_p equation and has the second smallest residual mean square, while (x_1, x_2, x_4) is the minimum residual mean square equation and has the smallest C_p statistic. The PRESS statistics for these two models are calculated in Table 10.3. Both models have similar values of PRESS, and it is difficult to establish a clear preference. The model with (x_1, x_2, x_3) has a smaller PRESS statistic, but it requires an additional regressor. Furthermore, x_2 and x_4 are highly correlated and the three-regressor model has stronger multicollinearity. Since the gain in PRESS for the three-regressor model is slight, we recommend the model with (x_1, x_2). •

Table 10.3 Calculation of PRESS for two models for Hald's cement data

Observation	$\hat{y}=52.58+1.468x_1+0.662x_2$			$\hat{y}=71.65+1.452x_1+0.416x_2-0.237x_4$		
i	e_i	h_{ii}	$[e_i/(1-h_{ii})]^2$	e_i	h_{ii}	$[e_i/(1-h_{ii})]^2$
1	−1.5740	0.25119	4.4184	0.0617	0.52058	0.0166
2	−1.0491	0.26189	2.0202	1.4327	0.27670	3.9235
3	−1.5147	0.11890	2.9553	−1.8910	0.13315	4.7588
4	−1.6585	0.24225	4.7905	−1.8016	0.24431	5.6837
5	−1.3925	0.08362	2.3091	0.2562	0.35733	0.1589
6	4.0475	0.11512	20.9221	3.8982	0.11737	19.5061
7	−1.3031	0.36180	4.1627	−1.4287	0.36341	5.0369
8	−2.0754	0.24119	7.4806	−3.0919	0.34522	22.2977
9	1.8245	0.17195	4.9404	1.2818	0.20881	2.6247
10	1.3625	0.55002	9.1683	0.3539	0.65244	1.0368
11	3.2643	0.18402	16.0037	2.0977	0.32105	9.5458
12	0.8628	0.19666	1.1535	1.0556	0.20040	1.7428
13	−2.8934	0.21420	13.5579	−2.2247	0.25923	9.0194
		PRESS x_1, x_2 =	93.8827		PRESSx_1, x_2, x_4 =	85.3516

If the data are collected in a time sequence, then time may be used as the basis of data splitting. That is, a particular time period is identified, and all observations collected before this time period are used to form the estimation data set, while observations collected later than this time period form the prediction data set. Fitting the model to the estimation data and examining its prediction accuracy for the prediction data would be a reasonable validation procedure to determine how the model is likely to perform in the future. This type of validation procedure is relatively common practice in time series analysis for investigating the potential performance of a forecasting model (for some examples, see Montgomery and Johnson [1976]). For examples involving regression models, see Cady and Allen [1972] and Draper and Smith [1981].

In addition to time, other characteristics of the data can often be used for data splitting. For example, consider the delivery time data from Example 4.1, and assume that we had the additional 15 observations in Table 10.1 also available. Since there are five cities represented in the sample, we could use the observations from San Diego, Boston, and Minneapolis (for example) as the estimation data and the observations from Austin and Louisville as the prediction data. This would give 29 observations for estimation and 11 observations for validation. In other problem situations, we may find that operators, batches of raw materials, units of test equipment, laboratories, and so forth can be used to form the estimation and prediction data sets. In cases where no logical basis of data splitting exists, one could randomly assign observations to the estimation and prediction data sets.

A potential disadvantage to these somewhat arbitrary methods of data splitting is that there is often no assurance that the prediction data set "stresses" the model severely enough. For example, a random division of the data would not necessarily ensure that some of the points in the prediction data set are extrapolation points, and the validation effort would provide no information on how well the model is likely to extrapolate. In the absence of an obvious basis for data splitting, it would be helpful to have a formal procedure for choosing the estimation and prediction data sets.

Snee [1977] describes the DUPLEX algorithm for data splitting. He credits the development of the procedure to R. W. Kennard, and notes that it is similar to the CADEX algorithm that Kennard and Stone [1969] proposed for design construction. The algorithm begins with a list of the n observations where the k regressors are standardized to unit length; that is

$$z_{ij}=\frac{x_{ij}-\bar{x}_j}{S_{jj}^{1/2}}\qquad \begin{array}{l} i=1,2,\ldots,n \\ j=1,2,\ldots,k \end{array}$$

where $S_{jj}=\sum_{i=1}^{n}(x_{ij}-\bar{x}_j)^2$ is the corrected sum of squares of the jth regressor. The standardized regressors are then *orthonormalized*. This can be done by factoring the $\mathbf{Z}'\mathbf{Z}$ matrix as

$$\mathbf{Z}'\mathbf{Z}=\mathbf{T}'\mathbf{T} \tag{10.7}$$

where $\mathbf{T}$ is a unique $k\times k$ upper triangular matrix. The elements of $\boldsymbol{T}$ can be found using the square root or Cholesky method (see Graybill [1976, pp. 231–236]). Then make the transformation

$$\mathbf{W}=\mathbf{Z}\mathbf{T}^{-1} \tag{10.8}$$

resulting in a new set of variables (the w's) that are orthogonal and have unit variance. This transformation makes the factor space more spherical.

Using the orthonormalized points, the Euclidean distance between all $\binom{n}{2}$ pairs of points is calculated. The pair of points that are the furthest apart is assigned to the estimation data set. This pair of points is removed from the list of points and the pair of remaining points that are the furthest apart is assigned to the prediction data set. Then this pair of points is removed from the list and the remaining point that is furthest from the pair of points in the estimation data set is included in the estimation data set. At the next step, the remaining unassigned point that is furthest from the two points in the prediction data set is added to the prediction data. The algorithm then continues to alternatively place the remaining points in either the estimation or prediction data sets until all n observations have been assigned.

Snee [1977] suggests measuring the statistical properties of the estimation and prediction data sets by comparing the pth root of the determinants of the $\mathbf{X}'\mathbf{X}$ matrices for these two data sets, where p is the number of parameters in the model. The determinant of $\mathbf{X}'\mathbf{X}$ is related to the volume of the region covered by the points. Thus if $\mathbf{X}_E$ and $\mathbf{X}_P$ denote the $\mathbf{X}$ matrices for points in the estimation and prediction data sets, respectively, then

$$\left(\frac{|\mathbf{X}'_E \mathbf{X}_E|}{|\mathbf{X}'_P \mathbf{X}_P|} \right)^{1/p}$$

is a measure of the relative volumes of the regions spanned by the two data sets. Ideally this ratio should be close to unity. It may also be useful to examine the variance inflation factors for the two data sets and the eigenvalue spectra of $\mathbf{X}'_E \mathbf{X}_E$ and $\mathbf{X}'_P \mathbf{X}_P$ to measure the relative correlation between the regressors.

In using the DUPLEX algorithm, several points should be kept in mind:

1. Some data sets may be too small to effectively use data splitting. Snee [1977] suggests that at least $n \geqslant 2p+25$ observations are required if the estimation and prediction data sets are of equal size, where p is the largest number of parameters likely to be required in the model. This sample size requirement ensures that there are a reasonable number of error degrees of freedom for the model.

2. Although the estimation and prediction data sets are often of equal size, DUPLEX can split the data in any desired ratio. Typically the estimation data set would be larger than the prediction data set. Such splits are found by using the algorithm until the prediction data set contains the required number of points and then placing the remaining unassigned points in the estimation data set. Remember that the prediction data set should contain at least 15 points in order to obtain a reasonable assessment of model performance.

3. Replicates or points that are near-neighbors in the x-space should be eliminated before splitting the data with DUPLEX. Unless these replicates are eliminated, the estimation and prediction data sets may be quite similar and this would not necessarily test the model severely enough. In an extreme case where every point is replicated twice, DUPLEX would form the estimation data set with one replicate and the prediction data set with the other replicate. Snee [1977] suggests using cluster analysis to identify near-neighbors. The algorithm described in Section 4.7.3 may also be helpful. Once a set of near-neighbors is identified, the average of the x-coordinates of these points should be used in the data splitting procedure.

4. A potential disadvantage of data splitting is that it reduces the precision with which regression coefficients are estimated. That is, the standard errors of the regression coefficients obtained from the estimation data set will be larger than they would have been if all the data had been used to estimate the coefficients. In large data sets, the standard errors may be small enough that this loss in precision is unimportant. However, the percentage increase in the standard errors can be large. If the model developed from the estimation data set is a satisfactory predictor, one way to improve the precision of estimation is to re-estimate the coefficients using the entire data set. The estimates of the coefficients in the two analyses should be very similar if the model is an adequate predictor of the prediction data set.

5. *Double* cross-validation may be useful in some problems. This is a procedure in which the data are first split into estimation and prediction data sets, a model developed from the estimation data, and its performance investigated using the prediction data. Then the roles of the two data sets are reversed; a model is developed using the original *prediction* data and it is used to predict the original *estimation* data. The advantage of this procedure is that it provides two evaluations of model performance. The disadvantage is that there are now *three* models to choose from, the two developed via data splitting and the model fitted to all the data. If the model is a good predictor, it will make little difference which one is used, except that the standard errors of the coefficients in the model fitted to the total data set will be smaller. If there are major differences in predictive performance, coefficient estimates, or functional form for these models, then further analysis is necessary to discover the reasons for these differences.

• **Example 10.5** All 40 observations for the delivery time data in Examples 4.1 and 10.2 are shown in Table 10.4. We will assume that these 40 points were collected at one time and use the data set to illustrate data splitting with the DUPLEX algorithm. Since the model will have two regressors, an equal split of the data will give 17 error degrees of freedom for the estimation data. This is adequate, so DUPLEX can be used to generate the estimation and prediction data sets. An x_1-x_2 plot is shown in Figure 10.1. Examination of the data reveals that there are two pairs of points that are near-neighbors in the x-space, Observations 15 and 23 and Observations 16 and 32. These two

Table 10.4 Delivery time data

Observation i	Cases x_1	Distance x_2	Delivery Time y	Estimation (E) or Prediction (P) Data Set
1	7	560	16.68	P
2	3	220	11.50	P
3	3	340	12.03	P
4	4	80	14.88	E
5	6	150	13.75	E
6	7	330	18.11	E
7	2	110	8.00	E
8	7	210	17.83	E
9	30	1460	79.24	E
10	5	605	21.50	E
11	16	688	40.33	P
12	10	215	21.00	P
13	4	255	13.50	E
14	6	462	19.75	P
15	9	448	24.00	E
16	10	776	29.00	P
17	6	200	15.35	P
18	7	132	19.00	E
19	3	36	9.50	P
20	17	770	35.10	E
21	10	140	17.90	E
22	26	810	52.32	E
23	9	450	18.75	E
24	8	635	19.83	E
25	4	150	10.75	E
26	22	905	51.00	P
27	7	520	16.80	E
28	15	290	26.16	P
29	5	500	19.90	E
30	6	1000	24.00	E
31	6	225	18.55	E
32	10	775	31.93	P
33	4	212	16.95	P
34	1	144	7.00	P
35	3	126	14.00	P
36	12	655	37.03	P
37	10	420	18.62	P
38	7	150	16.10	P
39	8	360	24.38	P
40	32	1530	64.75	P

clusters of points are circled in Figure 10.1. The x_1 - x_2 coordinates of these clusters of points are averaged and the list of points for use in the DUPLEX algorithm is shown in columns 1 and 2 of Table 10.5.

The standardized and orthonormalized data is shown in columns 3 and 4 of Table 10.5 and plotted in Figure 10.2. Notice that the region of coverage is more spherical than in Figure 10.1. Figure 10.2 and Tables 10.4 and 10.5 also show how DUPLEX splits the original points into estimation and prediction data. The convex hulls of the two data sets are shown in Figure 10.2. This indicates that the prediction data set contains both interpolation and extrapolation points. For these two data sets we find that $|\mathbf{X}'_E\mathbf{X}_E|=0.44696$ and $|\mathbf{X}'_P\mathbf{X}_P|=0.22441$, thus

$$\left(\frac{|\mathbf{X}'_E\mathbf{X}_E|}{|\mathbf{X}'_P\mathbf{X}_P|}\right)^{1/3}=\left(\frac{0.44696}{0.22441}\right)^{1/3}=1.26$$

indicating that the volumes of the two regions are very similar. The variance inflation factors for the estimation and prediction data are 2.22 and 4.43, respectively, so there is no strong evidence of multicollinearity and both data sets have similar correlative structure.

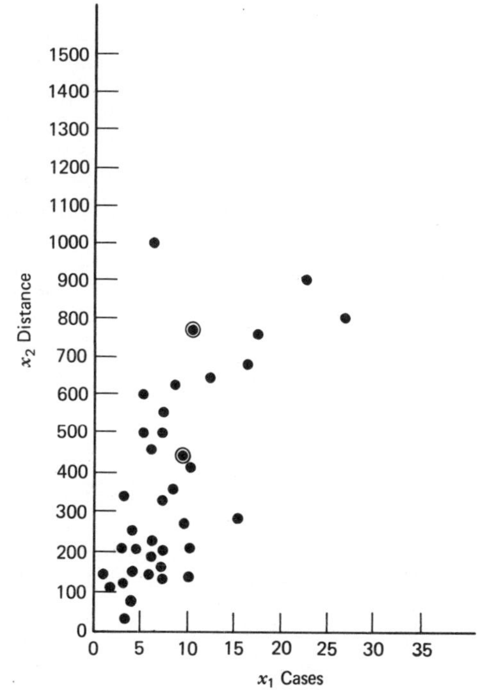

Figure 10.1 Scatter plot of delivery volume x_1 versus distance x_2, Example 10.5.

Table 10.5 Delivery time data with near-neighboring points averaged

Observation	(1) Original Variables	(2)	(3) Standardized, Orthonormalized Data	(4)	Estimation (E) or Prediction (P) Data Set
i	x_1 (cases)	x_2 (distance)	w_1	w_2	
1	7	560	−.047671	.158431	P
2	3	220	−.136037	.013739	P
3	3	340	−.136037	.108082	P
4	4	80	−.113945	−.126981	E
5	6	150	−.069762	−.133254	E
6	7	330	−.047671	−.022393	E
7	2	110	−.158128	−.042089	E
8	7	210	−.047671	−.116736	E
9	30	1460	.460432	.160977	E
10	5	605	−.091854	.255116	E
11	16	688	.151152	−.016816	P
12	10	215	.018603	−.204765	P
13	4	255	−.113945	.010603	E
14	6	462	−.069762	.112038	P
15, 23	9	449	−.003488	.009857	E
16, 32	10	775.5	.018603	.235895	P
17	6	200	−.069762	−.093945	P
18	7	132	−.047671	−.178059	E
19	3	36	−.136037	−.130920	P
20	17	770	.173243	.016998	E
21	10	140	.018603	−.263729	E
22	26	810	.372066	−.227434	E
24	8	635	−.025580	.186742	E
25	4	150	−.113945	−.071948	E
26	22	905	.283700	−.030133	P
27	7	520	−.047671	.126983	E
28	15	290	.129060	−.299067	P
29	5	500	−.091854	.172566	E
30	6	1000	−.069762	.535009	E
31	6	225	−.069762	−.074290	E
33	4	212	−.113945	−.023204	P
34	1	144	−.180219	.015295	P
35	3	126	−.136037	−.060163	P
36	12	655	.062786	.079853	P
37	10	420	.018603	−.043596	P
38	7	150	−.047671	−.163907	P
39	8	360	−.025580	−.029461	P
40	32	1530	.504614	.154704	P

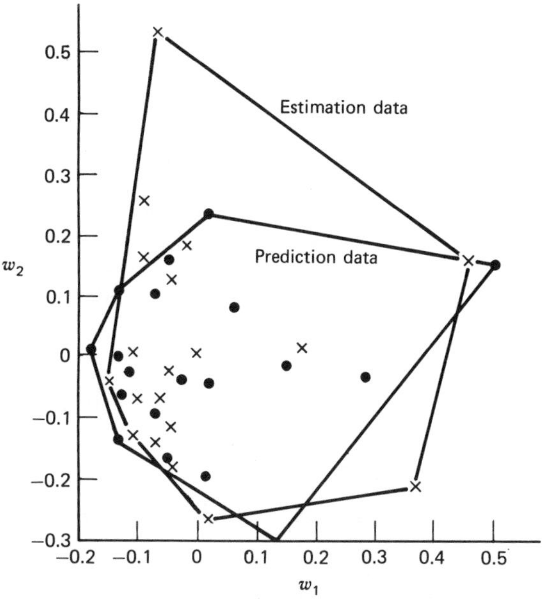

Figure 10.2 Estimation data (×) and prediction data (•) using orthonormalized regressors.

Panel A of Table 10.6 summarizes a least squares fit to the estimation data. The parameter estimates in this model exhibit reasonable signs and magnitudes, and the VIFs are acceptably small. Analysis of the residuals (not shown) reveals no severe model inadequacies, except that the normal probability plot indicates that the error distribution has heavier tails than the normal. Checking Table 10.4, we see that Point 9, which has previously been shown to be influential, is in the estimation data set. Apart from our concern

Table 10.6 Summary of least squares fit to the delivery time data

A. Analysis Using Estimation Data				B. Analysis Using All Data			
Variable	Coefficient Estimate	Standard Error	t_0	Variable	Coefficient Estimate	Standard Error	t_0
Intercept	2.4123	1.4165	1.70	Intercept	3.9840	0.9861	4.04
x_1	1.6392	0.1769	9.27	x_1	1.4877	0.1376	10.81
x_2	0.0136	0.0036	3.78	x_2	0.0134	0.0028	4.72
$MS_E = 13.9154$, $R^2 = .952$				$MS_E = 13.6841$, $R^2 = .944$			

about the normality assumption and the influence of Point 9, we conclude that the least squares fit to the estimation data is not unreasonable.

Columns 2 and 3 of Table 10.7 show the results of predicting the observations in the predicting data set using the least squares model developed from the estimation data. We see that the predicted values generally correspond closely to the observed values. The only unusually large prediction error is for point 40, which has the largest observed time in the prediction data. This point also has the largest values of x_1 (32 cases) and x_2 (1530 ft) in the entire data set. It is very similar to Point 9 in the estimation data ($x_1=30$, $x_2=1460$), but represents an extrapolation for the model fit to the estimation data. The sum of squares of the prediction errors is $\Sigma e_i^2=322.4452$, and the approximate R^2 for prediction is

$$R^2_{\text{Prediction}}=1-\frac{\Sigma e_i^2}{S_{yy}}=1-\frac{322.4452}{4113.5442}=0.922$$

Table 10.7 Prediction performance for model developed from the estimation data

	(1)	(2) Least Squares Fit	(3) Least Squares Fit	(4) Robust Fit (Ramsay's $E_{0.3}$)	(5) Robust Fit (Ramsay's $E_{0.3}$)
Observation i	Observed y_i	Predicted $\hat{y}_i$	Prediction Error $e_i=y_i-\hat{y}_i$	Predicted $\hat{y}_i$	Prediction Error $e_i=y_i-\hat{y}_i$
1	16.68	21.4976	−4.8176	21.5240	−4.8440
2	11.50	10.3199	1.1801	10.7482	0.7518
3	12.03	11.9508	0.0792	12.3055	−0.2755
11	40.33	37.9901	2.3399	37.5032	2.8268
12	21.00	21.7264	−0.7264	21.8197	−0.8197
14	19.75	18.5265	1.2235	18.6614	1.0886
16	29.00	29.3509	−0.3509	29.0998	−0.0098
17	15.35	14.9657	0.3843	15.2614	0.0886
19	9.50	7.8192	1.6808	8.3605	1.1395
26	51.00	50.7746	0.2254	49.8646	1.1354
28	26.16	30.9417	−4.7817	30.7474	−4.5874
32	31.93	29.3373	2.5927	29.0868	2.8432
33	16.95	11.8504	5.0996	12.2353	4.7147
34	7.00	6.0086	0.9914	6.5802	0.4198
35	14.00	9.0424	4.9576	9.5284	4.4716
36	37.03	30.9848	6.0452	30.7113	6.3187
37	18.62	24.5125	−5.8925	24.4799	−5.8599
38	16.10	15.9254	0.1746	16.2034	−0.1034
39	24.38	20.4187	3.9613	20.5195	3.8605
40	64.75	75.6609	−10.9109	73.8842	−9.1342

where $S_{yy}=4113.5442$ is the corrected sum of squares of the responses in the prediction data set. Thus we might expect this model to "explain" about 92.2 percent of the variability in fresh data, as compared to the 95.2 percent of the variability explained by the least squares fit to the estimation data. This loss in R^2 is small, so there is reasonably strong evidence that the least squares model will be a satisfactory predictor.

Panel B of Table 10.6 shows the results of a least squares fit to all 40 observations. Notice that the estimates of the intercept and β_1 are somewhat different than were obtained from the fit to the estimation data. Furthermore as expected, the standard errors of the regression coefficients for the fit using all the data are smaller than their counterparts in the fit using only the estimation data. The estimation data set contains Point 9, which we know is influential. Deleting Point 9 will make the intercept larger and β_1 smaller, as occurs in the least squares fit to the entire data set. A robust fit to the estimation data would produce a similar effect. The full data set also includes Point 40 (which is similar to 9). Possibly the larger data set dilutes the effect of these influential points.

Because there is some evidence that the least squares fit to the estimation data may be influenced by Point 9, it would be of interest to examine the predictive performance of a robust fit to these data. A robust fit to the estimation data using Ramsay's $E_{0.3}$ function is

$$\hat{y}=3.1206+1.5909x_1+0.0130$$

Columns 4 and 5 of Table 10.7 show the results of applying this model to the prediction data. The sum of squares of the prediction errors is $\Sigma e_i^2=280.6992$, which is smaller than the prediction error sum of squares for the least squares fit. The approximate R^2 for prediction for the robust fit is

$$R^2_{\text{Prediction}}=1-\frac{\Sigma e_i^2}{S_{yy}}=1-\frac{280.6992}{4113.5442}=0.932$$

which is only slightly larger than that obtained from the least squares model. The robust fit is also likely to be a satisfactory predictor.

Which model should be recommended for use in practice? In terms of their predictive performance, there is little difference between the least squares and robust fits to the estimation data. Because the robust fit is not affected by remote observations as severely as least squares, it may be viewed by some analysts as slightly preferable. These are essentially the same conclusions reached by other validation studies for this data set in Section 10.2.2. •

10.3 Data from planned experiments

Most of the validation techniques discussed in this chapter assume that the model has been developed from unplanned data. While the techniques could also be applied in situations where a designed experiment has been used to collect the data, usually validation of a model developed from such data is somewhat easier. Many experimental designs result in regression coefficients that are nearly uncorrelated, so multicollinearity is not usually a problem. An important aspect of experimental design is selection of the factors to be studied and identification of the ranges they are to be varied over. If done properly, this helps ensure that all the important regressors are included in the data, and that an appropriate range of values has been obtained for each regressor. Furthermore, in designed experiments considerable effort is usually devoted to the data collection process itself. This helps to minimize problems with "wild" or dubious observations and yields data with relatively small measurement errors.

When planned experiments are used to collect data, it is usually desirable to include a set of extra points for use in testing the predictive performance of the model. A widely used approach is to include the points that would allow fitting a model one degree higher than presently employed. Thus if we are contemplating fitting a first-order model, the design should include enough points to fit a second-order model.

Problems

10.1 Consider the regression model developed for the National Football League data in Problem 4.1.

a. Calculate the PRESS statistic for this model. What comments can you make about the likely predictive performance of this model?

b. Delete half the observations (chosen at random), and refit the regression model. Have the regression coefficients changed dramatically? How well does this model predict the number of games won for the deleted observations?

c. Delete the observation for Dallas, Los Angeles, Houston, San Francisco, Chicago, and Atlanta and refit the model. How well does this model predict the number of games won by these teams?

10.2 Using the DUPLEX procedure, split the National Football League data used in Problem 4.1 into estimation and prediction data sets. Evaluate the statistical properties of these two data sets. Develop a model from the estimation data and evaluate its performance on the prediction data. Discuss the predictive performance of this model.

10.3 Calculate the PRESS statistic for the model developed from the estimation data in Problem 10.2. How well is the model likely to predict? Compare this indication of prediction performance with the actual performance observed in Problem 10.2.

10.4 Consider the delivery time data discussed in Example 10.5. Find the PRESS statistic for the model developed from the estimation data. How well is the model likely to perform as a predictor? Compare this with the observed performance in prediction.

10.5 Consider the delivery time data discussed in Example 10.5.

a. Develop a regression model using the prediction data set.

b. How do the estimates of the parameters in this model compare with those from the model developed from the estimation data? What does this imply about model validity?

c. Use the model developed in Part a to predict the delivery times for the observations in the original estimation data. Are your results consistent with those obtained in Example 10.5?

10.6 In Problem 4.4 a regression model was developed for the gasoline mileage data using the regressor engine displacement x_1 and number of carburetor barrels x_6. Calculate the PRESS statistic for this model. What conclusions can you draw about the model's likely predictive performance?

10.7 In Problem 4.6 a regression model was developed for the gasoline mileage data using the regressors vehicle length x_8 and vehicle weight x_{10}. Calculate the PRESS statistic for this model. What conclusions can you draw about the potential performance of this model as a predictor?

10.8 PRESS statistics for two different models for the gasoline mileage data were calculated in Problems 10.6 and 10.7. On the basis of the PRESS statistics, which model do you think is the best predictor?

10.9 Consider the gasoline mileage data in Appendix Table B.3. Delete 8 observations (chosen at random) from the data, and develop an appropriate regression model. Use this model to predict the 8 withheld observations. What assessment would you make of this model's predictive performance?

10.10 Consider the gasoline mileage data in Appendix Table B.3. Use the DUPLEX algorithm to split the data into estimation and prediction sets.

a. Evaluate the statistical properties of these data sets.

b. Fit a model involving x_1 and x_6 to the estimation data. Do the coefficients and fitted values from this model seem reasonable?

c. Use this model to predict the observations in the prediction data set. What is your evaluation of this model's predictive performance?

10.11 Refer to Problem 10.2. What are the standard errors of the regression coefficients for the model developed from the estimation data? How do they compare with the standard errors for the model in Problem 4.1 developed using all the data?

10.12 Refer to Problem 10.2. Develop a model for the National Football League data using the prediction data set.

a. How do the coefficients and estimated values compare with those quantities for the models developed from the estimation data?

b. How well does this model predict the observations in the original estimation data set?

10.13 What difficulties do you think would be encountered in developing a computer program to implement the DUPLEX algorithm? For example, how efficient is the procedure likely to be for large sample sizes? What modifications in the procedure would you suggest to overcome those difficulties?

10.14 If $\mathbf{Z}$ is the $n \times k$ matrix of standardized regressors and $\mathbf{T}$ is the $k \times k$ upper triangular matrix in (10.3), show that the transformed regressors $\mathbf{W} = \mathbf{Z}\mathbf{T}^{-1}$ are orthogonal and have unit variance.

10.15 Show that the least squares estimate of $\boldsymbol{\beta}$ (for example $\hat{\boldsymbol{\beta}}_{(i)}$) with the ith observation deleted can be written in terms of the estimate based on all n points as

$$\hat{\boldsymbol{\beta}}_{(i)} = \hat{\boldsymbol{\beta}} - \frac{e_i}{1 - h_{ii}} (\mathbf{X}'\mathbf{X})^{-1} \mathbf{x}_i$$

10.16 Show that an estimate of σ^2 with the ith observation deleted can be written in terms of the estimate based on all n points as

$$\hat{\sigma}^2_{(i)} = \frac{n - p - f_i^2}{n - p - 1} \hat{\sigma}^2$$

where f_i is the ith studentized residual.

REFERENCES

Adichie, J. N. [1967], "Estimates of regression parameters based on rank tests," *Ann. Math. Statist.*, **38**, 894–904.

Aitkin, M. A. [1974], "Simultaneous inference and the choice of variable subsets," *Technometrics*, **16**, 221–227.

Allen, D. M. [1971a], "Mean square error of prediction as a criterion for selecting variables," *Technometrics*, **13**, 469–475.

Allen, D. M. [1971b], *The prediction sum of squares as a criterion for selecting prediction variables*, Technical Report No. 23, Dept. of Statistics, University of Kentucky.

Allen, D. M. [1974], "The relationship between variable selection and data augmentation and a method for prediction," *Technometrics*, **16**, 125–127.

Andrews, D. F. [1971], "Significance tests based on residuals," *Biometrika*, **58**, 139–148.

Andrews, D. F. [1974], "A robust method for multiple linear regression," *Technometrics*, **16**, 523–531.

Andrews, D. F. [1979], "The robustness of residual displays," in R. L. Launer and G. N. Wilkinson (Eds.), *Robustness in Statistics*, Academic Press, New York, pp. 19–32.

Andrews, D. F., P. J. Bickel, F. R. Hampel, P. J. Huber, W. H. Rogers, and J. W. Tukey [1972], *Robust Estimates of Location*, Princeton University Press, Princeton, N.J.

Andrews, D. F. and D. Pregibon [1978], "Finding the outliers that matter," *J. R. Statist. Soc. Ser. B*, **40**, 85–93.

Anscombe, F. J. [1960], "Rejection of outliers," *Technometrics*, **2**, 123–167.

Anscombe, F. J. [1961], "Examination of residuals," in *Proceedings of the Fourth Berkeley Symposium on Mathematical Statistics and Probability*, Vol. 1. Berkeley: University of California, 1–36.

Anscombe, F. J. [1967], "Topics in the investigation of linear relations fitted by the method of least squares," *J. R. Statist. Soc. Ser. B*, **29**, 1–52.

Anscombe, F. J. [1973], "Graphs in statistical analysis," *Amer. Statist.*, **27**, 1, 17–21.

Anscombe, F. J. and J. W. Tukey [1963], "The examination and analysis of residuals," *Technometrics*, **5**, 141–160.

Askin, R. G. and D. C. Montgomery [1980], "Augmented robust estimators," *Technometrics*, **22**, 333–341.

Barnett, V. and T. Lewis [1978], *Outliers in Statistical Data*, Wiley, New York.

Barrodale, I. [1968], "L_1 approximations and the analysis of data," *Appl. Statist.* **17**, 51–57.

Barrodale, I. and F. D. K. Roberts [1973], "An improved algorithm for discrete L_1 linear approximation," *SIAM J. Numer. Anal.*, **10**, 839–848.

Bartlett, M. S. [1947], "The use of transformations," *Biometrics*, **3**, 39–52.

Beale, E. M. L., M. G. Kendall, and D. G. Mann [1967], "The discarding of variables in multivariate analysis," *Biometrika*, **54**, 357–366.

Beaton, A. E. [1964], *The Use of Special Matrix Operators in Statistical Calculus*, Research Bulletin RB-64-51, Educational Testing Service, Princeton, N.J.

Beaton, A. E. and J. W. Tukey [1974], "The fitting of power series, meaning polynomials, illustrated on band spectroscopic data," *Technometrics*, **16**, 147–185.

Behnken, D. W. and N. R. Draper [1972], "Residuals and their variance patterns," *Technometrics*, **14**, 1, 101–111.

Belsley, D. A., E. Kuh, and R. E. Welsch [1980], *Regression Diagnostics: Identifying Influential Data and Sources of Collinearity*, Wiley, New York.

Bendel, R. B. and A. A. Afifi [1974], "Comparison of stopping rules in forward stepwise regression," presented at the Joint Statistical Meeting, St. Louis, Missouri.

Berk, K. N. [1978], "Comparing subset regression procedures," *Technometrics*, **20**, 1–6.

Berkson, J. [1950], "Are there two regressions?," *J. Amer. Statist. Assoc.*, **45**, 164–180.

Berkson, J. [1969], "Estimation of a linear function for a calibration line; consideration of a recent proposal," *Technometrics*, **11**, 649–660.

Book, D., J. Booker, H. O. Hartley, and R. L. Sielken, Jr. [1980], *Unbiased L_1 Estimators and Their Covariances*, ONR THEMIS Technical Report No. 64, Institute of Statistics, Texas A&M University.

Box, G. E. P. [1966], "Use and abuse of regression," *Technometrics*, **8**, 625–629.

Box, G. E. P. and D. W. Behnken [1960], "Some new three level designs for the study of quantitative variables," *Technometrics*, **2**, 455–475.

Box, G. E. P. and D. R. Cox [1964], "An analysis of transformations," *J. R. Statist. Soc. Ser. B*, **26**, 211–243.

Box, G. E. P. and N. R. Draper [1959], "A basis for the selection of a response surface design," *J. Amer. Statist. Assoc.*, **54**, 622–654.

Box, G. E. P. and N. R. Draper [1963], "The choice of a second-order rotatable design," *Biometrika*, **50**, 335–352.

Box, G. E. P. and J. S. Hunter [1957], "Multifactor experimental designs for exploring response surfaces," *Ann. Math. Statist.*, **28**, 195–242.

Box, G. E. P, W. G. Hunter and J. S. Hunter [1978], *Statistics for Experimenters*, Wiley, New York.

Box, G. E. P. and G. M. Jenkins [1976], *Time Series Analysis, Forecasting, and Control*, rev. ed., Holden-Day, San Francisco.

Box, G. E. P. and P. W. Tidwell [1962], "Transformation of the independent variables," *Technometrics*, **4**, 531–550.

Box, G. E. P. and J. M. Wetz [1973], *Criterion for Judging the Adequacy of Estimation by an Approximating Response Polynomial*, Technical Report No. 9, Dept. of Statistics, University of Wisconsin, Madison.

Bradley, R. A. and S. S. Srivastava [1979], "Correlation and polynomial regression," *Amer. Statist.*, **33**, 11–14.

Brown, P. J. [1977], "Centering and scaling in ridge regression," *Technometrics*, **19**, 35–36.

Brown, R. L., J. Durbin, and J. M. Evans [1975], "Techniques for testing the constancy of regression relationships over time (with discussion)," *J. R. Statist. Soc. Ser. B*, **37**, 149–192.

Brownlee, K. A. [1965], *Statistical Theory and Methodology in Science and Engineering*, 2nd ed. Wiley, New York.

Buse, A. and L. Lim [1977], "Cubic splines as a special case of restricted least squares," *J. Amer. Statist. Assoc.*, **72**, 64–68.

Cady, F. B. and D. M. Allen [1972], "Combining experiments to predict future yield data," *Agronomy J.*, **64**, 211–214.

Chatterjee, S. and B. Price [1977], *Regression Analysis by Example*, Wiley, New York.

Cochrane, D. and G. H. Orcutt [1949], "Application of least squares regression to relationships containing autocorrelated error terms," *J. Amer. Statist. Assoc.*, **44**, 32–61.

Conniffe, D. and J. Stone [1973], "A critical view of ridge regression," *Statist.*, **22**, 181–187.

Conniffe, D. and J. Stone [1975], "A reply to Smith and Goldstein," *Statist.*, **24**, 67–68.

Cook, R. D. [1977], "Detection of influential observation in linear regression," *Technometrics*, **19**, 15–18.

Cook, R. D. [1979], "Influential observations in linear regression." *J. Amer. Statist. Assoc.*, **74**, 169–174.

Cook, R. D. and P. Prescott [1981], "On the accuracy of Bonferroni significance levels for detecting outliers in linear models," *Technometrics*, **22**, 59–63.

Cook, R. D. and S. Weisberg [1980], "Characterizations of an empirical influence function for detecting influential cases in regression," *Technometrics*, **22**, 495–508.

Cox, D. R. [1968], "Notes on some aspects of regression analysis," *J. R. Statist. Soc. Ser. A*, **131**, 265–279.

Cox, D. R. [1970], *The Analysis of Binary Data*, Methuen, London.

Cox, D. R. and E. J. Snell [1974], "The choice of variables in observational studies," *Appl. Statist.*, **23**, 51–59.

Curry, H. B. and I. J. Schoenberg [1966], "On Polya frequency functions IV: The fundamental spline functions and their limits," *J. Anal. Math.*, **17**, 71–107.

Daniel, C. [1976], *Applications of Statistics to Industrial Experimentation*, Wiley, New York.

Daniel, C. and F. S. Wood [1980], *Fitting Equations to Data*, 2nd ed. Wiley, New York.

David, H. A. [1970], *Order Statistics*, Wiley, New York.

Davies, R. B. and B. Hutton [1975], "The effects of errors in the independent variables in linear regression," *Biometrika*, **62**, 383–391.

DeLury, D. B. [1960], *Values and Integrals of the Orthogonal Polynomials up to $N=26$*, University of Toronto Press, Toronto.

Dempster, A. P., M. Schatzoff, and N. Wermuth [1977]. "A simulation study of alternatives to ordinary least squares," *J. Amer. Statist. Assoc.*, **72**, 77–90.

Denby, L. and W. A. Larson [1977], "Robust regression estimators compared via Monte Carlo," *Commun. Statist.*, **A6**, 335–362.

Dixon, W. J. (Ed.) [1977], *BMDP-77: Biomedical Computer Programs P-Series*, University of California Press, Berkeley.

Dolby, G. R. [1976], "The ultrastructural relation: A synthesis of the functional and structural relations," *Biometrika*, **63**, 39–50.

Dolby, J. L. [1963], "A quick method for choosing a transformation," *Technometrics*, **5**, 317–325.

Draper, N. R., J. Guttman, and H. Kanemasa [1971], "The distribution of certain regression statistics," *Biometrika*, **58**, 295–298.

Draper, N. R. and J. A. John [1981], "Influential observations and outliers in regression," *Technometrics*, **23**, 21–26.

Draper, N. R. and H. Smith [1981], *Applied Regression Analysis*, 2nd ed., Wiley, New York.

Draper, N. R. and R. C. Van Nostrand [1977a], *Shrinkage Estimators: Review and Comments*, Technical Report No. 500, Dept. of Statistics, University of Wisconsin, Madison.

Draper, N. R. and R. C. Van Nostrand [1977b], *Ridge Regression: Is It Worthwhile?*, Technical Report No. 501, Dept. of Statistics, University of Wisconsin, Madison.

Draper, N. R. and R. C. Van Nostrand [1979], "Ridge regression and James–Stein estimators: Review and comments," *Technometrics*, **21**, 451–466.

Dunn, O. J. [1961], "Multiple comparisons among means," *J. Amer. Statist. Assoc.*, **56**, 52–64.

Durbin, J. [1970], "Testing for serial correlation in least squares regression when some of the regressors are lagged dependent variables," *Econometrica*, **38**, 410–421.

Durbin, J. and G. S. Watson [1950], "Testing for serial correlation in least squares regression I," *Biometrika*, **37**, 409–438.

Durbin, J. and G. S. Watson [1951], "Testing for serial correlation in least squares regression II," *Biometrika*, **38**, 159–178.

Durbin, J. and G. S. Watson [1971], "Testing for serial correlation in least squares regression III," *Biometrika*, **58**, 1–19.

Dutter, R. [1977], "Numerical solution of robust regression problems: Computational aspects, a comparison," *J. Statist. Comput. Simul.*, **5**, 207–238.

Dykstra, O., Jr. [1971], "The augmentation of experimental data to maximize $|\mathbf{X}'\mathbf{X}|$," *Technometrics*, **13**, 682–688.

Edwards, J. B. [1969], "The relation between the F–test and R^2," *The Amer. Statist.*, **23**, 28.

Efroymson, M. A. [1960], "Multiple regression analysis," in A. Ralston and H. S. Wilf (Eds.), *Mathematical Methods for Digital Computers*, Wiley, New York.

Ellenberg, J. H. [1976], "Testing for a single outlier from a general regression," *Biometrics*, **32**, 637–645.

Ellerton, R. R. W. [1978], "Is the regression equation adequate—A generalization," *Technometrics*, **20**, 313–316.

Ezekiel, M. [1930], *Methods of Correlation Analysis*, Wiley, New York.

Ezekiel, M. and K. A. Fox [1959], *Methods of Correlation and Regression Analysis*, Wiley, New York.

Farrar, D. E. and R. R. Glauber [1967], "Multicollinearity in regression analysis: The problem revisited," *Rev. Econ. Statist.*, **49**, 92–107.

Feder, P. I. [1974]. "Graphical techniques in statistical data analysis—Tools for extracting information from data," *Technometrics*, **16**, 287–299.

Finney, D. J. [1952], *Probit Analysis*, Cambridge University Press, Cambridge.

Forsythe, A. B. [1972], "Robust estimation of straight-line regression coefficients by minimizing pth power deviations," *Technometrics*, **14**, 159–166.

Forsythe, G. E. [1957], "Generation and use of orthogonal polynomials for data-fitting with a digital computer," *J. Soc. Ind. Appl. Math.*, **5**, 74–87.

Fuller, W. A. [1976], *Introduction to Statistical Time Series*, Wiley, New York.

Furnival, G. M. [1971]. "All possible regression with less computation," *Technometrics*, **13**, 403–408.

Furnival, G. M. and R. W. M. Wilson, Jr. [1974], "Regression by leaps and bounds," *Technometrics*, **16**, 499–511.

Gallant, A. R. and W. A. Fuller [1973], "Fitting segmented polynomial regression models whose join points have to be estimated," *J. Amer. Statist. Assoc.*, **63**, 144–147.

Garside, M. J. [1965], "The best subset in multiple regression analysis," *Appl. Statist.*, **14**, 196–200.

Garside, M. J. [1971], "Some computational procedures for the best subset problem," *Appl. Statist.*, **20**, 8–15.

Gartside, P. S. [1972], "A study of methods for comparing several variances," *J. Amer. Statist. Assoc.*, **67**, 342–346.

Gaylor, D. W. and J. A. Merrill [1968], "Augmenting existing data in multiple regression," *Technometrics*, **10**, 73–81.

Geisser, S. [1975], "The predictive sample reuse method with applications," *J. Amer. Statist. Assoc.*, **70**, 320–328.

Gentle, J. M., W. J. Kennedy, and V. A. Sposito [1977], "On least absolute deviations estimators," *Commun. Statist.*, **A6**, 839–845.

Gibbons, D. G. [1979], *A Simulation Study of Some Ridge Estimators*, General Motors Research Laboratories, Mathematics Dept., GMR-2659 (rev. ed.), Warren, Michigan.

Gnanadesikan, R. [1977], *Methods for Statistical Analysis of Multivariate Data*, Wiley, New York.

Goldberger, A. S. [1964], *Econometric Theory*, Wiley, New York.

Goldstein, M. and A. F. M. Smith [1974], "Ridge-type estimators for regression analysis," *J. R. Statist. Soc. Ser. B*, **36**, 284–291.

Golub, G. H. [1969], "Matrix decompositions and statistical calculations," in R. C. Milton and J. A. Welder, (Eds.), *Statistical Computation*, Academic, New York.

Gorman, J. W. and R. J. Toman [1966], "Selection of variables for fitting equations to data," *Technometrics*, **8**, 27–51.

Graybill, F. A. [1961], *An Introduction to Linear Statistical Models*, Vol. 1, McGraw-Hill, New York.

Graybill, F. A. [1976], *Theory and Application of the Linear Model*, Duxbury, North Scituate, Mass.

Guilkey, D. K. and J. L. Murphy [1975], "Directed ridge regression techniques in cases of multicollinearity," *J. Amer. Statist. Assoc.*, **70**, 769–775.

Gunst, R. F. [1979], "Similarities among least squares, principal component, and latent root regression estimators," presented at the Washington, D. C. Joint Statistical Meetings.

Gunst, R. F. and R. L. Mason [1977], "Biased estimation in regression: An evaluation using mean squared error," *J. Amer. Statist. Assoc.*, **72**, 616–628.

Gunst, R. F. and R. L. Mason [1979], "Some considerations in the evaluation of alternative prediction equations," *Technometrics*, **21**, 55–63.

Gunst, R. F., J. T. Webster, and R. L. Mason [1976], "A comparison of least squares and latent root regression estimators," *Technometrics*, **18**, 75–83.

Hahn, G. J. [1972], "Simultaneous prediction intervals for a regression model," *Technometrics*, **14**, 203–214.

Hahn, G. J. [1973], "The coefficient of determination exposed!," *Chem. Technol.*, **3**, 609–614.

Hahn, G. J. [1979], "Fitting regression models with no intercept term," *J. Qual. Technol.*, **9**, 2, 56–61.

Hahn, G. J. and R. W. Hendrickson [1971], "A table of percentage points of the largest absolute value of k student t variates and its applications," *Biometrika*, **58**, 323–332.

Haitovski, Y. [1969], "A note on the maximization of $\bar{R}^2$," *Amer. Statist.*, **23**, 1, 20–21.

Hald, A [1952], *Statistical Theory With Engineering Applications*, Wiley, New York.

Halperin, M. [1961], "Fitting of straight lines and prediction when both variables are subject to error," *J. Amer. Statist. Assoc.*, **56**, 657–669.

Halperin, M. [1970], "On inverse estimation in linear regression," *Technometrics*, **12**, 727–736.

Hawkins, D. M. [1973], "On the investigation of alternative regressions by principal components analysis," *Appl. Statist.*, **22**, 275–286.

Hawkins, D. M. [1975], "Relations between ridge regression and eigenanalysis of the augmented correlation matrix," *Technometrics*, **17**, 477–480.

Hayes, J. G. (Ed.) [1970], *Numerical Approximations to Functions and Data*, Athlone Press, London.

Hayes, J. G. [1974], "Numerical methods for curve and surface fitting," *J. Inst. Math. Appl.*, **10**, 144–152.

Hemmerle, W. J. [1975], "An explicit solution for generalized ridge regression," *Technometrics*, **17**, 309–314.

Hemmerle, W. J. and T. F. Brantle [1978], "Explicit and constrained generalized ridge regression," *Technometrics*, **20**, 109–120.

Hill, R. C., G. G. Judge, and T. B. Fomby [1978], "On testing the adequacy of a regression model," *Technometrics*, **20**, 491–494.

Hill, R. W. [1979], "On estimating the covariance matrix of robust regression M-estimates," *Commun. Statist.*, **A8**, 1183–1196.

Hill, R. W. and P. W. Holland [1977], "Two robust alternatives to least squares regression," *J. Amer. Statist. Assoc.*, **72**, 828–833.

Himmelblau, D. M. [1970], *Process Analysis by Statistical Methods*, Wiley, New York.

Hoadley, B. [1970], "A Bayesian look at inverse linear regression," *J. Amer. Statist. Assoc.*, **65**, 356–369.

Hoaglin, D. C. and R. E. Welsch [1978], "The hat matrix in regression and ANOVA," *Amer. Statist.*, **32**, 1, 17–22.

Hocking, R. R. [1972], "Criteria for selection of a subset regression: Which one should be used," *Technometrics*, **14**, 967–970.

Hocking, R. R. [1974], "Misspecification in regression," *Amer. Statist.*, **28**, 39–40.

Hocking, R. R. [1976], "The analysis and selection of variables in linear regression," *Biometrics*, **32**, 1–49.

Hocking, R. R. [1978], "The regression dilemma: Variable elimination, coefficient shrinkage, or robust estimation," presented at the ASQC 1978 Fall Technical Conference, Rochester, N.Y.

Hocking, R. R. and L. R. LaMotte [1973], "Using the SELECT program for choosing subset regressions," in *Proceedings of the University of Kentucky Conference on Regression with a Large Number of Predictor Variables*, Thompson, W. O. and F. B. Cady (Eds.), Dept. of Statistics, University of Kentucky, Lexington.

Hocking, R. R. and R. N. Leslie [1967], "Selection of the best subset in regression analysis," *Technometrics*, **9**, 531–540.

Hocking, R. R., F. M. Speed, and M. J. Lynn [1976], "A class of biased estimators in linear regression," *Technometrics*, **18**, 425–437.

Hodges, S. D. and P. G. Moore [1972], "Data uncertainties and least squares regression," *Appl. Statist.*, **21**, 185–195.

Hoerl, A. E. [1959], "Optimum solution of many variable equations," *Chem. Eng. Prog.*, **55**, 69.

Hoerl, A. E. and R. W. Kennard [1970a], "Ridge regression: Biased estimation for nonorthogonal problems," *Technometrics*, **12**, 55–67.

Hoerl, A. E. and R. W. Kennard [1970b], "Ridge regression: Applications to nonorthogonal problems," *Technometrics*, **12**, 69–82.

Hoerl, A. E. and R. W. Kennard [1976], "Ridge regression: Iterative estimation of the biasing parameter," *Commun. Statist.*, **A5**, 77–88.

Hoerl, A. E., R. W. Kennard, and K. F. Baldwin [1975], "Ridge regression: Some simulations," *Commun. Statist.*, **4**, 105–123.

Hogg, R. V. [1974], "Adaptive robust procedures: A partial review and some suggestions for future applications and theory," *J. Amer. Statist. Assoc.*, **69**, 909–925.

Hogg, R. V. [1979a], "Statistical robustness: One view of its use in applications today," *Amer. Statist.*, **33**, 3, 108–115.

Hogg, R. V. [1979b], "An introduction to robust estimation," in R. L. Launer and G. N. Wilkinson (Eds.), *Robustness in Statistics*, Academic, New York, 1–18.

Hogg, R. V. and R. H. Randles [1975], "Adaptive distribution-free regression methods and their applications," *Technometrics*, **17**, 399–407.

Holland, P. W. and R. E. Welsch [1977], "Robust regression using iteratively reweighted least squares," *Commun. Statist.*, **A6**, 813–828.

Huber, P. J. [1964], "Robust estimation of a location parameter," *Ann. Math. Statist.*, **35**, 73–101.

Huber, P. J. [1972], "Robust statistics: A review," *Ann. Math. Statist.*, **43**, 1041–1067.

Huber, P. J. [1973], "Robust regression: Asymptotics, conjectures, and Monte Carlo," *Ann. Statist.*, **1**, 799–821.

Huber, P. J. [1975], "Robustness and designs," *A Survey of Statistical Design and Linear Models*, North-Holland, Amsterdam.

Huber, P. J. [1981], *Robust Statistics*, Wiley, New York.

Jaeckel, L. A. [1972], "Estimating regression coefficients by minimizing the dispersion of the residuals," *Ann. Math. Statist.*, **43**, 1449–1458.

Johnston, J. [1972], *Econometric Methods*, McGraw-Hill, New York.

Jurečková, J. [1977], "Asymptotic relations of M-estimates and *R*-estimates in linear regression models," *Ann. Statist.*, **5**, 464–472.

Kalotay, A. J. [1971], "Structural solution to the linear calibration problem," *Technometrics*, **13**, 761–769.

Kendall, M. G. and A. Stuart [1968], *The Advanced Theory of Statistics*, Vol. 3, Charles Griffin, London.

Kendall, M. G. and G. U. Yule [1950], *An Introduction to the Theory of Statistics*, Charles Griffin, London.

Kennard, R. L. and L. Stone [1969], "Computer aided design of experiments," *Technometrics*, **11**, 137–148.

Kennedy, W. J. and T. A. Bancroft [1971], "Model-building for prediction in regression using repeated significance tests," *Ann. Math. Statist.*, **42**, 1273–1284.

Krutchkoff, R. G. [1967], "Classical and inverse regression methods of calibration," *Technometrics*, **9**, 425–439.

Krutchkoff, R. G. [1969], "Classical and inverse regression methods of calibration in extrapolation," *Technometrics*, **11**, 605–608.

Kunugi, T., T. Tamura, and T. Naito [1961], "New acetylene process uses hydrogen dilution," *Chem. Eng. Prog.*, **57**, 43–49.

LaMotte, L. R. [1972], "The SELECT routines: A program for identifying best subset regression," *Appl. Statist.*, **21**, 92–93.

LaMotte, L. R. and R. R. Hocking [1970], "Computational efficiency in the selection of regression variables," *Technometrics*, **13**, 83–93.

Land, C. E. [1974], "Confidence interval estimation for means after data transformation to normality," *J. Amer. Statist. Assoc.*, **69**, 795–802 (Correction, ibid, **71**, 255).

Larsen, W. A. and S. J. McCleary [1972], "The use of partial residual plots in regression analysis," *Technometrics*, **14**, 781–790.

Lawless, J. F. [1978], "Ridge and related estimation procedures: Theory and practice," *Commun. Statist.*, **A7**, 139–164.

Lawless, J. F. and P. Wang [1976], "A simulation of ridge and other regression estimators," *Commun. Statist.*, **A5**, 307–323.

Lawson, C. R. and R. J. Hanson [1974], *Solving Least Squares Problems*, Prentice-Hall, Englewood Cliffs, N.J.

Leamer, E. E. [1973], "Multicollinearity: A Bayesian interpretation," *Rev. Econ. Statist.*, **55**, 371–380.

Leamer, E. E. [1978], *Specification Searches: Ad Hoc Inference With Nonexperimental Data*, Wiley, New York.

Lieberman, G. J., R. G. Miller, Jr., and M. A. Hamilton [1967], "Unlimited simultaneous discrimination intervals in regression," *Biometrika*, **54**, 133–145.

Lindley, D. V. [1947], "Regression lines and the linear functional relationship," *J. R. Statist. Soc. Suppl.*, **9**, 218–244.

Lindley, D. V. and A. F. M. Smith [1972], "Bayes estimates for the linear model (with discussion)," *J. R. Statist. Soc. Ser. B*, **34**, 1–41.

LINPACK [1979], *Linpack Users Guide*, by J. J. Dangarra, J. R. Bunch, C. B. Moler, and G. W. Stewart, *Soc. Ind. Appl. Math.*, Philadelphia.

Longley, J. W. [1967], "An appraisal of least-squares programs for the electronic computer from the point of view of the user," *J. Amer. Statist. Assoc.*, **62**, 819–841.

Lowerre, J. M. [1974], "On the mean square error of parameter estimates for some biased estimators," *Technometrics*, **16**, 461–464.

Mallows, C. L. [1964], "Choosing variables in a linear regression: A graphical aid," Presented at the Central Regional Meeting of the Institute of Mathematical Statistics, Manhattan, Kansas.

Mallows, C. L. [1966], "Choosing a subset regression," Presented at the Joint Statistical Meetings, Los Angeles.

Mallows, C. L. [1973], "Some comments on C_p," Technometrics, **15**, 661–675.

Mandansky, A. [1959], "The fitting of straight lines when both variables are subject to error," *J. Amer. Statist. Assoc.*, **54**, 173–205.

Mansfield, E. R., J. T. Webster, and R. F. Gunst [1977], "An analytic variable selection procedure for principal component regression," *Appl. Statist.*, **26**, 34–40.

Mantel, N. [1970], "Why stepdown procedures in variable selection," *Technometrics*, **12**, 621–625.

Marquardt, D. W. [1970], "Generalized inverses, ridge regression, biased linear estimation, and nonlinear estimation," *Technometrics*, **12**, 591–612.

Marquardt, D. W. and R. D. Snee [1975], "Ridge regression in practice," *Amer. Statist.*, **29**, 1, 3–20.

Mason, R. L., R. F. Gunst, and J. T. Webster [1975], "Regression analysis and problems of multicollinearity," *Commun. Statist.*, **4**, 3, 277–292.

Mayer, L. S. and T. A. Willke [1973], "On biased estimation in linear models," *Technometrics*, **16**, 494–508.

McCarthy, P. J. [1976], "The use of balanced half-sample replication in cross-validation studies," *J. Amer. Statist. Assoc.*, **71**, 596–604.

McDonald, G. C. and D. I. Galarneau [1975], "A Monte Carlo evaluation of some ridge-type estimators," *J. Amer. Statist. Assoc.*, **70**, 407–416.

Miller, R. G., Jr. [1966], *Simultaneous Statistical Inference*, McGraw-Hill, New York.

Montgomery, D. C. [1976], *Design and Analysis of Experiments*, Wiley, New York.

Montgomery, D. C. and L. A. Johnson [1976], *Forecasting and Time Series Analysis*, McGraw-Hill, New York.

Montgomery, D. C., E. W. Martin, and E. A. Peck [1980], "Interior analysis of the observations in multiple linear regression," *J. Qual. Technol.*, **12**, 3, 165–173.

Morgan, J. A. and J. F. Tatar [1972], "Calculation of the residual sum of squares for all possible regressions," *Technometrics*, **14**, 317–325.

Mosteller, F. and J. W. Tukey [1968], "Data analysis including statistics," in G. Lindzey and E. Aronson, (Eds.), *Handbook of Social Psychology*, 2, Addison-Wesley, Reading, Mass.

Mosteller, F. and J. W. Tukey [1977], *Data Analysis and Regression: A Second Course in Statistics*, Addison-Wesley, Reading, Mass.

Moussa-Hamouda, E. and F. C. Leone [1974], "The 0-blue estimators for complete and censored samples in linear regression," *Technometrics*, **16**, 441–446.

Moussa-Hamouda, E. and F. C. Leone [1977a], "The robustness of efficiency of adjusted trimmed estimators in linear regression," *Technometrics*, **19**, 19–34.

Moussa-Hamouda, E. and F. C. Leone [1977b], "Efficiency of ordinary least squares from trimmed and winsorized samples in linear regression," *Technometrics*, **19**, 265–273.

Mullet, G. M. [1976], "Why regression coefficients have the wrong sign," *J. Qual. Technol.*, **8**, 121–126.

Mullet, G. M. and T. W. Murray [1971], "A new method for examining rounding error in least squares regression computer programs," *J. Amer. Statist. Assoc.*, **66**, 496–498.

Myers, R. H. [1971], *Response Surface Methodology*, Allyn and Bacon, Boston.

Narula, S. and J. S. Ramberg [1972], Letter to the Editor, *Amer. Statist.*, **26**, 42.

Naszodi, L. J. [1978], "Elimination of the bias in the course of calibration," *Technometrics*, **20**, 201–205.

Neter, J. and W. Wasserman [1974], *Applied Linear Statistical Models*, Richard D. Irwin, Homewood, Ill.

Neyman, J. and E. L. Scott [1960], "Correction for bias introduced by a transformation of variables," *Ann. Math. Statist.*, **31**, 643–655.

Obenchain, R. L. [1975], "Ridge analysis following a preliminary test of the shrunken hypothesis," *Technometrics*, **17**, 431–441.

Obenchain, R. L. [1977], "Classical *F*-tests and confidence intervals for ridge regression," *Technometrics*, **19**, 429–439.

Ott, R. L. and R. H. Myers [1968], "Optimal experimental designs for estimating the independent variable in regression," *Technometrics*, **10**, 811–823.

Pearson, E. S. and H. O. Hartley [1966], *Biometrika Tables for Statisticians*, Vol. 1, 3rd Ed., Cambridge University Press, Cambridge.

Perng, S. K. and Y. L. Tong [1974], "A sequential solution to the inverse linear regression problem," *Ann. Statist.*, **2**, 535–539.

Pesaran, M. H. and L. J. Slater [1980], *Dynamic Regression: Theory and Algorithms*, Halsted Press, New York.

Poirier, D. J. [1973], "Piecewise regression using cubic splines," *J. Amer. Statist. Assoc.*, **68**, 515–524.

Poirier, D. J. [1975], "On the use of bilinear splines in economics," *J. Econ.*, **3**, 23–24.

Pope, P. T. and J. T. Webster [1972], "The use of an F-statistic in stepwise regression procedures," *Technometrics*, **14**, 327–340.

Ramsay, J. O. [1977], "A comparative study of several robust estimates of slope, intercept, and scale in linear regression," *J. Amer. Statist. Assoc.*, **72**, 608–615.

Rao, C. R. [1965], *Linear Statistical Inference and Its Applications*, Wiley, New York.

Rao, P. [1971], "Some notes on misspecification in regression," *Amer. Statist.*, **25**, 37–39.

Rosenberg, S. H. and P. S. Levy [1972], "A characterization on misspecification in the general linear regression model," *Biometrics*, **28**, 1129–1132.

ROSEPACK [1980], "A system of subroutines for iteratively reweighted least squares computations," to appear in *ACM Trans. Math. Softw.*

Rosner, B. [1975], "On the detection of many outliers," *Technometrics*, **17**, 221–229.

St. John, R. C. and N. R. Draper [1975], "D-optimality for regression designs: A review." *Technometrics*, **17**, 15–23.

Schatzoff, M., R. Tsao, and S. Fienberg [1968], "Efficient calculation of all possible regressions," *Technometrics*, **10**, 769–779.

Scheffé, H. [1953], "A method for judging all contrasts in the analysis of variance," *Ann. Math. Statist.*, **40**, 87–104.

Scheffé, H. [1959], *The Analysis of Variance*, Wiley, New York.

Scheffé, H. [1973], "A statistical theory of calibration," *Ann. Statist.*, **1**, 1–37.

Schilling, E. G. [1974a], "The relationship of analysis of variance to regression. Part I. Balanced designs," *J. Qual. Technol.*, **6**, 74–83.

Schilling, E. G. [1974b], "The relationship of analysis of variance to regression. Part II. Unbalanced designs," *J. Qual. Technol.*, **6**, 146–153.

Sclove, S. L. [1968], "Improved estimators for coefficients in linear regression," *J. Amer. Statist. Assoc.*, **63**, 596–606.

Searle, S. R. [1971], *Linear Models*, Wiley, New York.

Searle, S. R. and J. G. Udell [1970], "The use of regression on dummy variables in market research," *Management Science, B*, **16**, 397–409.

Seber, G. A. F. [1977], *Linear Regression Analysis*, Wiley, New York.

Shukla, G. K. [1972], "On the problem of calibration," *Technometrics*, **14**, 547–553.

Sielken, R. L., Jr. and H. O. Hartley [1973], "Two linear programming algorithms for unbiased estimation of linear models," *J. Amer. Statist. Assoc.*, **68**, 639–641.

Silvey, S. D. [1969], "Multicollinearity and imprecise estimation," *J. R. Statist. Soc. Ser. B*, **31**, 539–552.

Smith, A. F. M. and M. Goldstein [1975], "Ridge regression: Some comments on a paper of Conniffe and Stone," *Statist.*, **24**, 61–66.

Smith, B. T., J. M. Boyle, B. S. Garbow, Y. Ikebe, V. C. Klema, and C. B. Moler [1974], *Matrix Eigensystem Routines*, Springer-Verlag, Berlin.

Smith, G. and F. Campbell [1980], "A critique of some ridge regression methods (with discussion)," *J. Amer. Statist.*, Assoc. **75**, 74–103.

Smith, J. H. [1972}, "Families of transformations for use in regression analysis," *Amer. Statist.*, **26**, 3, 59–61.

Smith, P. L. [1979], "Splines as a useful and convenient statistical tool," *Amer. Statist.*, **33**, 2, 57–62.

Snee, R. D. [1973], "Some aspects of nonorthogonal data analysis, Part I. Developing prediction equations." *J. Qual. Technol.*, **5**, 67–79.

Snee, R. D. [1977], "Validation of regression models: Methods and examples," *Technometrics*, **19**, 415–428.

Sprent, P. [1969], *Models in Regression and Related Topics*, Methuen, London.

Stefansky, W. [1971], "Rejecting outliers by maximum normed residual," *Ann. Math. Statist.*, **42**, 35–45.

Stefansky, W. [1972], "Rejecting outliers in factorial designs," *Technometrics*, **14**. 469–479.

Stein, C. [1960]. "Multiple regression," Ingrim Olkin, (Ed.), *Contributions to Probability and Statistics: Essays in Honor of Harold Hotelling*, Sanford University Press, Stanford, California.

Stewart, G. W. [1973], *Introduction to Matrix Computations*, Academic, New York.

Stone, M. [1974], "Cross-validating choice and assessment of statistical predictions (with discussion)," *J. R. Statist. Soc. Ser. B*, **36**, 111–147.

Suich, R. and G. C. Derringer [1977], "Is the regression equation adequate—one criterion," *Technometrics*, **19**, 213–216.

Theil, H. [1963], "On the use of incomplete prior information in regression analysis," *J. Amer. Statist. Assoc.*, **58**, 401–414.

Theil, H. and A. S. Goldberger [1961], "On pure and mixed statistical estimation in economics," *Int. Econ. Rev.*, **2**, 65–78.

Theobald, C. M. [1974], "Generalizations of mean square error applied to ridge regression," *J. R. Statist. Soc. Ser. B*, **36**, 103–106.

Thompson, M. L. [1978a], "Selection of variables in multiple regression: Part I. A review and evaluation," *Int. Statist. Rev.*, **46**, 1–19.

Thompson, M. L. [1978b], "Selection of variables in multiple regression: Part II. Chosen procedures, computations and examples," *Int. Statist. Rev.*, **46**, 129–146.

Tucker, W. T. [1980], "The linear calibration problem revisited," presented at the ASQC Fall Technical Conference, Cincinnati, Ohio.

Tufte, E. R. [1974], *Data Analysis for Politics and Policy*, Prentice-Hall, Englewood Cliffs, N. J.

Tukey, J. W. [1957], "On the comparative anatomy of transformations," *Ann. Math. Statist.*, **28**, 602–632.

Wagner, H. M. [1959], "Linear programming techniques for regression analysis," *J. Amer. Statist. Assoc.*, **54**, 206–212.

Walls, R. E. and D. L. Weeks [1969], "A note on the variance of a predicted response in regression," *Amer. Statist.*, **23**, 24–26.

Webster, J. T., R. F. Gunst, and R. L. Mason [1974], "Latent root regression analysis," *Technometrics*, **16**, 513–522.

Weisberg, S. [1980], *Applied Linear Regression*, Wiley, New York.

Welsch, R. E. [1975], "Confidence regions for robust regression," *Statistical Computing Section Proceedings of the American Statistical Association*, Washington, D. C.

Welsch, R. E. and E. Kuh [1977], "Linear regression diagnostics," Working Paper No. 173, National Bureau of Economic Research, Cambridge, Mass.

Welsch, R. E. and S. C. Peters [1978], "Finding influential subsets of data in regression models," *Proceedings of the Eleventh Interface Symposium on Computer Science and Statistics*, A. R. Gallant and T. M. Gerig, (Eds.), Institute of Statistics, North Carolina State University, 240–244.

White, J. W. and R. F. Gunst [1979], "Latent root regression: Large sample analysis," *Technometrics*, **21**, 481–488.

Wichern, D. W. and G. A. Churchill [1978], "A comparison of ridge estimators," *Technometrics*, **20**, 301–311.

Wilde, D. J. and C. S. Beightler [1967], *Foundations of Optimization*, Prentice–Hall, Englewood Cliffs, N.J.

Wilkinson, J. W. [1965], *The Algebraic Eigenvalue Problem*, Oxford University Press.

Willan, A. R. and D. G. Watts [1978], "Meaningful multicollinearity measures," *Technometrics*, **20**, 407–412.

Williams, D. A. [1973], Letter to the Editor, *Appl. Statist.*, **22**, 407–408.

Williams, E. J. [1969], "A note on regression methods in calibration," *Technometrics*, **11**, 189–192.

Wold, S. [1974], "Spline functions in data analysis," *Technometrics*, **16**, 1–11.

Wonnacott, R. J. and T. H. Wonnacott [1970], *Econometrics*, Wiley, New York.

Working, H. and H. Hotelling [1929], "Application of the theory of error to the interpretation of trends," *J. Amer. Statist. Assoc., Suppl.* (*Proceedings*), **24**, 73–85.

Wood, F. S. [1973], "The use of individual effects and residuals in fittings equations to data," *Technometrics*, **15**, 677–695.

Yale, C. and A. B. Forsythe [1976], "Winsorized regression," *Technometrics*, **18**, 291–300.

Younger, M. S. [1979], *A Handbook for Linear Regression*, Duxbury Press, North Scituate, Mass.

Zellner, A. [1971], *An Introduction to Bayesian Inference in econometrics*, Wiley, New York.

APPENDIX A

STATISTICAL TABLES

Table A.1 Cumulative normal distribution
Table A.2 Percentage points of the χ^2 distribution
Table A.3 Percentage points of the t distribution
Table A.4 Percentage points of the F distribution
Table A.5 Orthogonal polynomials
Table A.6 Critical values of the Durbin–Watson statistic
Table A.7 Percentage points of the Bonferroni t statistic
Table A.8 Percentage points of the maximum modulus t statistic

Table A.1 Cumulative standard normal distribution

$$\Phi(z)=\int_{-\infty}^{z}\frac{1}{\sqrt{2\pi}}e^{-u^2/2}\,du$$

z	.00	.01	.02	.03	.04	z
.0	.50000	.50399	.50798	.51197	.51595	.0
.1	.53983	.54379	.54776	.55172	.55567	.1
.2	.57926	.58317	.58706	.59095	.59483	.2
.3	.61791	.62172	.62551	.62930	.63307	.3
.4	.65542	.65910	.66276	.66640	.67003	.4
.5	.69146	.69497	.69847	.70194	.70540	.5
.6	.72575	.72907	.73237	.73565	.73891	.6
.7	.75803	.76115	.76424	.76730	.77035	.7
.8	.78814	.79103	.79389	.79673	.79954	.8
.9	.81594	.81859	.82121	.82381	.82639	.9
1.0	.84134	.84375	.84613	.84849	.85083	1.0
1.1	.86433	.86650	.86864	.87076	.87285	1.1
1.2	.88493	.88686	.88877	.89065	.89251	1.2
1.3	.90320	.90490	.90658	.90824	.90988	1.3
1.4	.91924	.92073	.92219	.92364	.92506	1.4
1.5	.93319	.93448	.93574	.93699	.93822	1.5
1.6	.94520	.94630	.94738	.94845	.94950	1.6
1.7	.95543	.95637	.95728	.95818	.95907	1.7
1.8	.96407	.96485	.96562	.96637	.96711	1.8
1.9	.97128	.97193	.97257	.97320	.97381	1.9
2.0	.97725	.97778	.97831	.97882	.97932	2.0
2.1	.98214	.98257	.98300	.98341	.98382	2.1
2.2	.98610	.98645	.98679	.98713	.98745	2.2
2.3	.98928	.98956	.98983	.99010	.99036	2.3
2.4	.99180	.99202	.99224	.99245	.99266	2.4
2.5	.99379	.99396	.99413	.99430	.99446	2.5
2.6	.99534	.99547	.99560	.99573	.99585	2.6
2.7	.99653	.99664	.99674	.99683	.99693	2.7
2.8	.99744	.99752	.99760	.99767	.99774	2.8
2.9	.99813	.99819	.99825	.99831	.998.36	2.9
3.0	.99865	.99869	.99874	.99878	.99882	3.0
3.1	.99903	.99906	.99910	.99913	.99916	3.1
3.2	.99931	.99934	.99936	.99938	.99940	3.2
3.3	.99952	.99953	.99955	.99957	.99958	3.3
3.4	.99966	.99968	.99969	.99970	.99971	3.4
3.5	.99977	.99978	.99978	.99979	.99980	3.5
3.6	.99984	.99985	.99985	.99986	.99986	3.6
3.7	.99989	.99990	.99990	.99990	.99991	3.7
3.8	.99993	.99993	.99993	.99994	.99994	3.8
3.9	.99995	.99995	.99996	.99996	.99996	3.9

Table A.1 Cumulative standard normal distribution (*Continued*)

$$\Phi(z)=\int_{-\infty}^{z} \frac{1}{\sqrt{2\pi}} e^{-u^2/2}\, du$$

z	.05	.06	.07	.08	.09	z
.0	.51994	.52392	.52790	.53188	.53586	.0
.1	.55962	.56356	.56749	.57142	.57534	.1
.2	.59871	.60257	.60642	.61026	.61409	.2
.3	.63683	.64058	.64431	.64803	.65173	.3
.4	.67364	.67724	.68082	.68438	.68793	.4
.5	.70884	.71226	.71566	.71904	.72240	.5
.6	.74215	.74537	.74857	.75175	.75490	.6
.7	.77337	.77637	.77935	.78230	.78523	.7
.8	.80234	.80510	.80785	.81057	.81327	.8
.9	.82894	.83147	.83397	.83646	.83891	.9
1.0	.85314	.85543	.85769	.85993	.86214	1.0
1.1	.87493	.87697	.87900	.88100	.88297	1.1
1.2	.89435	.89616	.89796	.89973	.90147	1.2
1.3	.91149	.91308	.91465	.91621	.91773	1.3
1.4	.92647	.92785	.92922	.93056	.93189	1.4
1.5	.93943	.94062	.94179	.94295	.94408	1.5
1.6	.95053	.95154	.95254	.95352	.95448	1.6
1.7	.95994	.96080	.96164	.96246	.96327	1.7
1.8	.96784	.96856	.96926	.96995	.97062	1.8
1.9	.97441	.97500	.97558	.97615	.97670	1.9
2.0	.97982	.98030	.98077	.98124	.98169	2.0
2.1	.98422	.98461	.98500	.98537	.98574	2.1
2.2	.98778	.98809	.98840	.98870	.98899	2.2
2.3	.99061	.99086	.99111	.99134	.99158	2.3
2.4	.99286	.99305	.99324	.99343	.99361	2.4
2.5	.99461	.99477	.99492	.99506	.99520	2.5
2.6	.99598	.99609	.99621	.99632	.99643	2.6
2.7	.99702	.99711	.99720	.99728	.99736	2.7
2.8	.99781	.99788	.99795	.99801	.99807	2.8
2.9	.99841	.99846	.99851	.99856	.99861	2.9
3.0	.99886	.99889	.99893	.99897	.99900	3.0
3.1	.99918	.99921	.99924	.99926	.99929	3.1
3.2	.99942	.99944	.99946	.99948	.99950	3.2
3.3	.99960	.99961	.99962	.99964	.99965	3.3
3.4	.99972	.99973	.99974	.99975	.99976	3.4
3.5	.99981	.99981	.99982	.99983	.99983	3.5
3.6	.99987	.99987	.99988	.99988	.99989	3.6
3.7	.99991	.99992	.99992	.99992	.99992	3.7
3.8	.99994	.99994	.99995	.99995	.99995	3.8
3.9	.99996	.99996	.99996	.99997	.99997	3.9

Source: Reproduced with permission from *Probability and Statistics in Engineering and Management Science*, 2nd ed., by W. W. Hines and D. C. Montgomery, Wiley, New York, 1980.

Table A.2 Percentage points of the χ^2 distribution

ν \ α	.995	.990	.975	.950	.900	.500	.100	.050	.025	.010	.005
1	.00+	.00+	.00+	.00+	.02	.45	2.71	3.84	5.02	6.63	7.88
2	.01	.02	.05	.10	.21	1.39	4.61	5.99	7.38	9.21	10.60
3	.07	.11	.22	.35	.58	2.37	6.25	7.81	9.35	11.34	12.84
4	.21	.30	.48	.71	1.06	3.36	7.78	9.49	11.14	13.28	14.86
5	.41	.55	.83	1.15	1.61	4.35	9.24	11.07	12.83	15.09	16.75
6	.68	.87	1.24	1.64	2.20	5.35	10.65	12.59	14.45	16.81	18.55
7	.99	1.24	1.69	2.17	2.83	6.35	12.02	14.07	16.01	18.48	20.28
8	1.34	1.65	2.18	2.73	3.49	7.34	13.36	15.51	17.53	20.09	21.96
9	1.73	2.09	2.70	3.33	4.17	8.34	14.68	16.92	19.02	21.67	23.59
10	2.16	2.56	3.25	2.94	4.87	9.34	15.99	18.31	20.48	23.21	25.19
11	2.60	3.05	3.82	4.57	5.58	10.34	17.28	19.68	21.92	24.72	26.76
12	3.07	3.57	4.40	5.23	6.30	11.34	18.55	21.03	23.34	26.22	28.30
13	3.57	4.11	5.01	5.89	7.04	12.34	19.81	22.36	24.74	27.69	29.82
14	4.07	4.66	5.63	6.57	7.79	13.34	21.06	23.68	26.12	29.14	31.32
15	4.60	5.23	6.27	7.26	8.55	14.34	22.31	25.00	27.49	30.58	32.80
16	5.14	5.81	6.91	7.96	9.31	15.34	23.54	26.30	28.85	32.00	34.27
17	5.70	6.41	7.56	8.67	10.09	16.34	24.77	27.59	30.19	33.41	35.72
18	6.26	7.01	8.23	9.39	10.87	17.34	25.99	28.87	31.53	34.81	37.16
19	6.84	7.63	8.91	10.12	11.65	18.34	27.20	30.14	32.85	36.19	38.58
20	7.43	8.26	9.59	10.85	12.44	19.34	28.41	31.41	34.17	37.57	40.00
21	8.03	8.90	10.28	11.59	13.24	20.34	29.62	32.67	35.48	38.93	41.40
22	8.64	9.54	10.98	12.34	14.04	21.34	30.81	33.92	36.78	40.29	42.80
23	9.26	10.20	11.69	13.09	14.85	22.34	32.01	35.17	38.08	41.64	44.18
24	9.89	10.86	12.40	13.85	15.66	23.34	33.20	36.42	39.36	42.98	45.56
25	10.52	11.52	13.12	14.61	16.47	24.34	34.28	37.65	40.65	44.31	46.93
26	11.16	12.20	13.84	15.38	17.29	25.34	35.56	38.89	41.92	45.64	48.29
27	11.81	12.88	14.57	16.15	18.11	26.34	36.74	40.11	43.19	46.96	49.65
28	12.46	13.57	15.31	16.93	18.94	27.34	37.92	41.34	44.46	48.28	50.99
29	13.12	14.26	16.05	17.71	19.77	28.34	39.09	42.56	45.72	49.59	52.34
30	13.79	14.95	16.79	18.49	20.60	29.34	40.26	43.77	46.98	50.89	53.67
40	20.71	22.16	24.43	26.51	29.05	39.34	51.81	55.76	59.34	63.69	66.77
50	27.99	29.71	32.36	34.76	37.69	49.33	63.17	67.50	71.42	76.15	79.49
60	35.53	37.48	40.48	43.19	46.46	59.33	74.40	79.08	83.30	88.38	91.95
70	43.28	45.44	48.76	51.74	55.33	69.33	85.53	90.53	95.02	100.42	104.22
80	51.17	53.54	57.15	60.39	64.28	79.33	96.58	101.88	106.63	112.33	116.32
90	59.20	61.75	65.65	69.13	73.29	89.33	107.57	113.14	118.14	124.12	128.30
100	67.33	70.06	74.22	77.93	82.36	99.33	118.50	124.34	129.56	135.81	140.17

ν = degrees of freedom.

Source: Reproduced with permission from *Probability and Statistics in Engineering and Management Science*, 2nd ed., by W. W. Hines and D. C. Montgomery, Wiley, New York, 1980.

Table A.3 Percentage points of the *t* distribution

ν \ α	.40	.25	.10	.05	.025	.01	.005	.0025	.001	.0005
1	.325	1.000	3.078	6.314	12.706	31.821	63.657	127.32	318.31	636.62
2	.289	.816	1.886	2.920	4.303	6.965	9.925	14.089	23.326	31.598
3	.277	.765	1.638	2.353	3.182	4.541	5.841	7.453	10.213	12.924
4	.271	.741	1.533	2.132	2.776	3.747	4.604	5.598	7.173	8.610
5	.267	.727	1.476	2.015	2.571	3.365	4.032	4.773	5.893	6.869
6	.265	.718	1.440	1.943	2.447	3.143	3.707	4.317	5.208	5.959
7	.263	.711	1.415	1.895	2.365	2.998	3.499	4.029	4.785	5.408
8	.262	.706	1.397	1.860	2.306	2.896	3.355	2.833	4.501	5.041
9	.261	.703	1.383	1.833	2.262	2.821	3.250	3.690	4.297	4.781
10	.260	.700	1.372	1.812	2.228	2.764	3.169	3.581	4.144	4.587
11	.260	.697	1.363	1.796	2.201	2.718	3.106	3.497	4.025	4.437
12	.259	.695	1.356	1.782	2.179	2.681	3.055	3.428	3.930	4.318
13	.259	.694	1.350	1.771	2.160	2.650	3.012	3.372	3.852	4.221
14	.258	.692	1.345	1.761	2.145	2.624	2.977	3.326	3.787	4.140
15	.258	.691	1.341	1.753	2.131	2.602	2.947	3.286	3.733	4.073
16	.258	.690	1.337	1.746	2.120	2.583	2.921	3.252	3.686	4.015
17	.257	.689	1.333	1.740	2.110	2.567	2.898	3.222	3.646	3.965
18	.257	.688	1.330	1.734	2.101	2.552	2.878	3.197	3.610	3.922
19	.257	.688	1.328	1.729	2.093	2.539	2.861	3.174	3.579	3.883
20	.257	.687	1.325	1.725	2.086	2.528	2.845	3.153	3.552	3.850
21	.257	.686	1.323	1.721	2.080	2.518	2.831	3.135	3.527	3.819
22	.256	.686	1.321	1.717	2.074	2.508	2.819	3.119	3.505	3.792
23	.256	.685	1.319	1.714	2.069	2.500	2.807	3.104	3.485	2.767
24	.256	.685	1.318	1.711	2.064	2.492	2.797	3.091	3.467	3.745
25	.256	.684	1.316	1.708	2.060	2.485	2.787	3.078	3.450	3.725
26	.256	.684	1.315	1.706	2.056	2.479	2.779	3.067	3.435	3.707
27	.256	.684	1.314	1.703	2.052	2.473	2.771	3.057	3.421	3.690
28	.256	.683	1.313	1.701	2.048	2.467	2.763	3.047	3.408	2.674
29	.256	.683	1.311	1.699	2.045	2.462	2.756	3.308	3.396	3.659
30	.256	.683	1.310	1.697	2.042	2.457	2.750	3.030	3.385	3.646
40	.255	.681	1.303	1.648	2.021	2.423	2.704	2.971	3.307	3.551
60	.254	.679	1.296	1.671	2.000	2.390	2.660	2.915	3.232	3.460
120	.254	.677	1.289	1.658	1.980	2.358	2.617	2.860	3.160	3.373
∞	.253	.674	1.282	1.645	1.960	2.326	2.576	2.807	3.090	3.291

Source: This table is adapted from *Biometrika Tables for Statisticians*, Vol. 1, 3rd ed., 1966, by E. S. Pearson and H. O. Hartley, with permission of the Biometrika Trustees.

Table A.4 Percentage points of the *F* distribution

$F_{.25, \nu_1, \nu_2}$

ν_2 \ ν_1 — Degrees of freedom for the denominator (ν_2)	Degrees of freedom for the numerator (ν_1) 1	2	3	4	5	6	7	8	9
1	5.83	7.50	8.20	8.58	8.82	8.98	9.10	9.19	9.26
2	2.57	3.00	3.15	3.23	3.28	3.31	3.34	3.35	3.37
3	2.02	2.28	2.36	2.39	2.41	2.42	2.43	2.44	2.44
4	1.81	2.00	2.05	2.06	2.07	2.08	2.08	2.08	2.08
5	1.69	1.85	1.88	1.89	1.89	1.89	1.89	1.89	1.89
6	1.62	1.76	1.78	1.79	1.79	1.78	1.78	1.78	1.77
7	1.57	1.70	1.72	1.72	1.71	1.71	1.70	1.70	1.70
8	1.54	1.66	1.67	1.66	1.66	1.65	1.64	1.64	1.63
9	1.51	1.62	1.63	1.63	1.62	1.61	1.60	1.60	1.59
10	1.49	1.60	1.60	1.59	1.59	1.58	1.57	1.56	1.56
11	1.47	1.58	1.58	1.57	1.56	1.55	1.54	1.53	1.53
12	1.46	1.56	1.56	1.55	1.54	1.53	1.52	1.51	1.51
13	1.45	1.55	1.55	1.53	1.52	1.51	1.50	1.49	1.49
14	1.44	1.53	1.53	1.52	1.51	1.50	1.49	1.48	1.47
15	1.43	1.52	1.52	1.51	1.49	1.48	1.47	1.46	1.46
16	1.42	1.51	1.51	1.50	1.48	1.47	1.46	1.45	1.44
17	1.42	1.51	1.50	1.49	1.47	1.46	1.45	1.44	1.43
18	1.41	1.50	1.49	1.48	1.46	1.45	1.44	1.43	1.42
19	1.41	1.49	1.49	1.47	1.46	1.44	1.43	1.42	1.41
20	1.40	1.49	1.48	1.47	1.45	1.44	1.43	1.42	1.41
21	1.40	1.48	1.48	1.46	1.44	1.43	1.42	1.41	1.40
22	1.40	1.48	1.47	1.45	1.44	1.42	1.41	1.40	1.39
23	1.39	1.47	1.47	1.45	1.43	1.42	1.41	1.40	1.39
24	1.39	1.47	1.46	1.44	1.43	1.41	1.40	1.39	1.38
25	1.39	1.47	1.46	1.44	1.42	1.41	1.40	1.39	1.38
26	1.38	1.46	1.45	1.44	1.42	1.41	1.39	1.38	1.37
27	1.38	1.46	1.45	1.43	1.42	1.40	1.39	1.38	1.37
28	1.38	1.46	1.45	1.43	1.41	1.40	1.39	1.38	1.37
29	1.38	1.45	1.45	1.43	1.41	1.40	1.38	1.37	1.36
30	1.38	1.45	1.44	1.42	1.41	1.39	1.38	1.37	1.36
40	1.36	1.44	1.42	1.40	1.39	1.37	1.36	1.35	1.34
60	1.35	1.42	1.41	1.38	1.37	1.35	1.33	1.32	1.31
120	1.34	1.40	1.39	1.37	1.35	1.33	1.31	1.30	1.29
∞	1.32	1.39	1.37	1.35	1.33	1.31	1.29	1.28	1.27

10	12	15	20	24	30	40	60	120	∞
9.32	9.41	9.49	9.58	9.63	9.67	9.71	9.76	9.80	9.85
3.38	3.39	3.41	3.43	3.43	3.44	3.45	3.46	3.47	3.48
2.44	2.45	2.46	2.46	2.46	2.47	2.47	2.47	2.47	2.47
2.08	2.08	2.08	2.08	2.08	2.08	2.08	2.08	2.08	2.08
1.89	1.89	1.89	1.88	1.88	1.88	1.88	1.87	1.87	1.87
1.77	1.77	1.76	1.76	1.75	1.75	1.75	1.74	1.74	1.74
1.69	1.68	1.68	1.67	1.67	1.66	1.66	1.65	1.65	1.65
1.63	1.62	1.62	1.61	1.60	1.60	1.59	1.59	1.58	1.58
1.59	1.58	1.57	1.56	1.56	1.55	1.54	1.54	1.53	1.53
1.55	1.54	1.53	1.52	1.52	1.51	1.51	1.50	1.49	1.48
1.52	1.51	1.50	1.49	1.49	1.48	1.47	1.47	1.46	1.45
1.50	1.49	1.48	1.47	1.46	1.45	1.45	1.44	1.43	1.42
1.48	1.47	1.46	1.45	1.44	1.43	1.42	1.42	1.41	1.40
1.46	1.45	1.44	1.43	1.42	1.41	1.41	1.40	1.39	1.38
1.45	1.44	1.43	1.41	1.41	1.40	1.39	1.38	1.37	1.36
1.44	1.43	1.41	1.40	1.39	1.38	1.37	1.36	1.35	1.34
1.43	1.41	1.40	1.39	1.38	1.37	1.36	1.35	1.34	1.33
1.42	1.40	1.39	1.38	1.37	1.36	1.35	1.34	1.33	1.32
1.41	1.40	1.38	1.37	1.36	1.35	1.34	1.33	1.32	1.30
1.40	1.39	1.37	1.36	1.35	1.34	1.33	1.32	1.31	1.29
1.39	1.38	1.37	1.35	1.34	1.33	1.32	1.31	1.30	1.28
1.39	1.37	1.36	1.34	1.33	1.32	1.31	1.30	1.29	1.28
1.38	1.37	1.35	1.34	1.33	1.32	1.31	1.30	1.28	1.27
1.38	1.36	1.35	1.33	1.32	1.31	1.30	1.29	1.28	1.26
1.37	1.36	1.34	1.33	1.32	1.31	1.29	1.28	1.27	1.25
1.37	1.35	1.34	1.32	1.31	1.30	1.29	1.28	1.26	1.25
1.36	1.35	1.33	1.32	1.31	1.30	1.28	1.27	1.26	1.24
1.36	1.34	1.33	1.31	1.30	1.29	1.28	1.27	1.25	1.24
1.35	1.34	1.32	1.31	1.30	1.29	1.27	1.26	1.25	1.23
1.35	1.34	1.32	1.30	1.29	1.28	1.27	1.26	1.24	1.23
1.33	1.31	1.30	1.28	1.26	1.25	1.24	1.22	1.21	1.19
1.30	1.29	1.27	1.25	1.24	1.22	1.21	1.19	1.17	1.15
1.28	1.26	1.24	1.22	1.21	1.19	1.18	1.16	1.13	1.10
1.25	1.24	1.22	1.19	1.18	1.16	1.14	1.12	1.08	1.00

Table A.4 Percentage points of the *F* distribution (*Continued*)

$F_{.10, \nu_1, \nu_2}$

Degrees of freedom for the denominator (ν_2) \ Degrees of freedom for the numerator (ν_1)	1	2	3	4	5	6	7	8	9
1	39.86	49.50	53.59	55.83	57.24	58.20	58.91	59.44	59.86
2	8.53	9.00	9.16	9.24	9.29	9.33	9.35	9.37	9.38
3	5.54	5.46	5.39	5.34	5.31	5.28	5.27	5.25	5.24
4	4.54	4.32	4.19	4.11	4.05	4.01	3.98	3.95	3.94
5	4.06	3.78	3.62	3.52	3.45	3.40	3.37	3.34	3.32
6	3.78	3.46	3.29	3.18	3.11	3.05	3.01	2.98	2.96
7	3.59	3.26	3.07	2.96	2.88	2.83	2.78	2.75	2.72
8	3.46	3.11	2.92	2.81	2.73	2.67	2.62	2.59	2.56
9	3.36	3.01	2.81	2.69	2.61	2.55	2.51	2.47	2.44
10	3.29	2.92	2.73	2.61	2.52	2.46	2.41	2.38	2.35
11	3.23	2.86	2.66	2.54	2.45	2.39	2.34	2.30	2.27
12	3.18	2.81	2.61	2.48	2.39	2.33	2.28	2.24	2.21
13	3.14	2.76	2.56	2.43	2.35	2.28	2.23	2.20	2.16
14	3.10	2.73	2.52	2.39	2.31	2.24	2.19	2.15	2.12
15	3.07	2.70	2.49	2.36	2.27	2.21	2.16	2.12	2.09
16	3.05	2.67	2.46	2.33	2.24	2.18	2.13	2.09	2.06
17	3.03	2.64	2.44	2.31	2.22	2.15	2.10	2.06	2.03
18	3.01	2.62	2.42	2.29	2.20	2.13	2.08	2.04	2.00
19	2.99	2.61	2.40	2.27	2.18	2.11	2.06	2.02	1.98
20	2.97	2.59	2.38	2.25	2.16	2.09	2.04	2.00	1.96
21	2.96	2.57	2.36	2.23	2.14	2.08	2.02	1.98	1.95
22	2.95	2.56	2.35	2.22	2.13	2.06	2.01	1.97	1.93
23	2.94	2.55	2.34	2.21	2.11	2.05	1.99	1.95	1.92
24	2.93	2.54	2.33	2.19	2.10	2.04	1.98	1.94	1.91
25	2.92	2.53	2.32	2.18	2.09	2.02	1.97	1.93	1.89
26	2.91	2.52	2.31	2.17	2.08	2.01	1.96	1.92	1.88
27	2.90	2.51	2.30	2.17	2.07	2.00	1.95	1.91	1.87
28	2.89	2.50	2.29	2.16	2.06	2.00	1.94	1.90	1.87
29	2.89	2.50	2.28	2.15	2.06	1.99	1.93	1.89	1.86
30	2.88	2.49	2.28	2.14	2.03	1.98	1.93	1.88	1.85
40	2.84	2.44	2.23	2.09	2.00	1.93	1.87	1.83	1.79
60	2.79	2.39	2.18	2.04	1.95	1.87	1.82	1.77	1.74
120	2.75	2.35	2.13	1.99	1.90	1.82	1.77	1.72	1.68
∞	2.17	2.30	2.08	1.94	1.85	1.77	1.72	1.67	1.63

10	12	15	20	24	30	40	60	120	∞
60.19	60.71	61.22	61.74	62.00	62.26	62.53	62.79	63.06	63.33
9.39	9.41	9.42	9.44	9.45	9.46	9.47	9.47	9.48	9.49
5.23	5.22	5.20	5.18	5.18	5.17	5.16	5.15	5.14	5.13
3.92	3.90	3.87	3.84	3.83	3.82	3.80	3.79	3.78	2.76
3.30	3.27	3.24	3.21	3.19	3.17	3.16	3.14	3.12	3.10
2.94	2.90	2.87	2.84	2.82	2.80	2.78	2.76	2.74	2.72
2.70	2.67	2.63	2.59	2.58	2.56	2.54	2.51	2.49	2.47
2.54	2.50	2.46	2.42	2.40	2.38	2.36	2.34	2.32	2.29
2.42	2.38	2.34	2.30	2.28	2.25	2.23	2.21	2.18	2.16
2.32	2.28	2.24	2.20	2.18	2.16	2.13	2.11	2.08	2.06
2.25	2.21	2.17	2.12	2.10	2.08	2.05	2.03	2.00	1.97
2.19	2.15	2.10	2.06	2.04	2.01	1.99	1.96	1.93	1.90
2.14	2.10	2.05	2.01	1.98	1.96	1.93	1.90	1.88	1.85
2.10	2.05	2.01	1.96	1.94	1.91	1.89	1.86	1.83	1.80
2.06	2.02	1.97	1.92	1.90	1.87	1.85	1.82	1.79	1.76
2.03	1.99	1.94	1.89	1.87	1.84	1.81	1.78	1.75	1.72
2.00	1.96	1.91	1.86	1.84	1.81	1.78	1.75	1.72	1.69
1.98	1.93	1.89	1.84	1.81	1.78	1.75	1.72	1.69	1.66
1.96	1.91	1.86	1.81	1.79	1.76	1.73	1.70	1.67	1.63
1.94	1.89	1.84	1.79	1.77	1.74	1.71	1.68	1.64	1.61
1.92	1.87	1.83	1.78	1.75	1.72	1.69	1.66	1.62	1.59
1.90	1.86	1.81	1.76	1.73	1.70	1.67	1.64	1.60	1.57
1.89	1.84	1.80	1.74	1.72	1.69	1.66	1.62	1.59	1.55
1.88	1.83	1.78	1.73	1.70	1.67	1.64	1.61	1.57	1.53
1.87	1.82	1.77	1.72	1.69	1.66	1.63	1.59	1.56	1.52
1.86	1.81	1.76	1.71	1.68	1.65	1.61	1.58	1.54	1.50
1.85	1.80	1.75	1.70	1.67	1.64	1.60	1.57	1.53	1.49
1.84	1.79	1.74	1.69	1.66	1.63	1.59	1.56	1.52	1.48
1.83	1.78	1.73	1.68	1.65	1.62	1.58	1.55	1.51	1.47
1.82	1.77	1.72	1.67	1.64	1.61	1.57	1.54	1.50	1.46
1.76	1.71	1.66	1.61	1.57	1.54	1.51	1.47	1.42	1.38
1.71	1.66	1.60	1.54	1.51	1.48	1.44	1.40	1.35	1.29
1.65	1.60	1.55	1.48	1.45	1.41	1.37	1.32	1.26	1.19
1.60	1.55	1.49	1.42	1.38	1.34	1.30	1.24	1.17	1.00

Table A.4 Percentage points of the *F* distribution (*Continued*)

$F_{.05, \nu_1, \nu_2}$

ν_2 \ ν_1 — Degrees of freedom for the denominator (ν_2)	Degrees of freedom for the numerator (ν_1) 1	2	3	4	5	6	7	8	9
1	161.4	199.5	215.7	224.6	230.2	234.0	236.8	238.9	240.5
2	18.51	19.00	19.16	19.25	19.30	19.33	19.35	19.37	19.38
3	10.13	9.55	9.28	9.12	9.01	8.94	8.89	8.85	8.81
4	7.71	6.94	6.59	6.39	6.26	6.16	6.09	6.04	6.00
5	6.61	5.79	5.41	5.19	5.05	4.95	4.88	4.82	4.77
6	5.99	5.14	4.76	4.53	4.39	4.28	4.21	4.15	4.10
7	5.59	4.74	4.35	4.12	3.97	3.87	3.79	3.73	3.68
8	5.32	4.46	4.07	3.84	3.69	3.58	3.50	3.44	3.39
9	5.12	4.26	3.86	3.63	3.48	3.37	3.29	3.23	3.18
10	4.96	4.10	3.71	3.48	3.33	3.22	3.14	3.07	3.02
11	4.48	3.98	3.59	3.36	3.20	3.09	3.01	2.95	2.90
12	4.75	3.89	3.49	3.26	3.11	3.00	2.91	2.85	2.80
13	4.67	3.81	3.41	3.18	3.03	2.92	2.83	2.77	2.71
14	4.60	3.74	3.34	3.11	2.96	2.85	2.76	2.70	2.65
15	4.54	3.68	3.29	3.06	2.90	2.79	2.71	2.64	2.59
16	4.49	3.63	3.24	3.01	2.85	2.74	2.66	2.59	2.54
17	4.45	3.59	3.20	2.96	2.81	2.70	2.61	2.55	2.49
18	4.41	3.55	3.16	2.93	2.77	2.66	2.58	2.51	2.46
19	4.38	3.52	3.13	2.90	2.74	2.63	2.54	2.48	2.42
20	4.35	3.49	3.10	2.87	2.71	2.60	2.51	2.45	2.39
21	4.32	3.47	3.07	2.84	2.68	2.57	2.49	2.42	2.37
22	4.30	3.44	3.05	2.82	2.66	2.55	2.46	2.40	2.34
23	4.28	3.42	3.03	2.80	2.64	2.53	2.44	2.37	2.32
24	4.26	3.40	3.01	2.78	2.62	2.51	2.42	2.36	2.30
25	4.24	3.39	2.99	2.76	2.60	2.49	2.40	2.34	2.28
26	4.23	3.37	2.98	2.74	2.59	2.47	2.39	2.32	2.27
27	4.21	3.35	2.96	2.73	2.57	2.46	2.37	2.31	2.25
28	4.20	3.34	2.95	2.71	2.56	2.45	2.36	2.29	2.24
29	4.18	3.33	2.93	2.70	2.55	2.43	2.35	2.28	2.22
30	4.17	3.32	2.92	2.69	2.53	2.42	2.33	2.27	2.21
40	4.08	3.23	2.84	2.61	2.45	2.34	2.25	2.18	2.12
60	4.00	3.15	2.76	2.53	2.37	2.25	2.17	2.10	2.04
120	3.92	3.07	2.68	2.45	2.29	2.17	2.09	2.02	1.96
∞	3.84	3.00	2.60	2.37	2.21	2.10	2.01	1.94	1.88

10	12	15	20	24	30	40	60	120	∞
241.9	243.9	245.9	248.0	249.1	250.1	251.1	252.2	253.3	254.3
19.40	19.41	19.43	19.45	19.45	19.46	19.47	19.48	19.49	19.50
8.79	8.74	8.70	8.66	8.64	8.62	8.59	8.57	8.55	8.53
5.96	5.91	5.86	5.80	5.77	5.75	5.72	5.69	5.66	5.63
4.74	4.68	4.62	4.56	4.53	4.50	4.46	4.43	4.40	4.36
4.06	4.00	3.94	3.87	3.84	3.81	3.77	3.74	3.70	3.67
3.64	3.57	3.51	3.44	3.41	3.38	3.34	3.30	3.27	3.23
3.35	3.28	3.22	3.15	3.12	3.08	3.04	3.01	2.97	2.93
3.14	3.07	3.01	2.94	2.90	2.86	2.83	2.79	2.75	2.71
2.98	2.91	2.85	2.77	2.74	2.70	2.66	2.62	2.58	2.54
2.85	2.79	2.72	2.65	2.61	2.57	2.53	2.49	2.45	2.40
2.75	2.69	2.62	2.54	2.51	2.47	2.43	2.38	2.34	2.30
2.67	2.60	2.53	2.46	2.42	2.38	2.34	2.30	2.25	2.21
2.60	2.53	2.46	2.39	2.35	2.31	2.27	2.22	2.18	2.13
2.54	2.48	2.40	2.33	2.29	2.25	2.20	2.16	2.11	2.07
2.49	2.42	2.35	2.28	2.24	2.19	2.15	2.11	2.06	2.01
2.45	2.38	2.31	2.23	2.19	2.15	2.10	2.06	2.01	1.96
2.41	2.34	2.27	2.19	2.15	2.11	2.06	2.02	1.97	1.92
2.38	2.31	2.23	2.16	2.11	2.07	2.03	1.98	1.93	1.88
2.35	2.28	2.20	2.12	2.08	2.04	1.99	1.95	1.90	1.84
2.32	2.25	2.18	2.10	2.05	2.01	1.96	1.92	1.87	1.81
2.30	2.23	2.15	2.07	2.03	1.98	1.94	1.89	1.84	1.78
2.27	2.20	2.13	2.05	2.01	1.96	1.91	1.86	1.81	1.76
2.25	2.18	2.11	2.03	1.98	1.94	1.89	1.84	1.79	1.73
2.24	2.16	2.09	2.01	1.96	1.92	1.87	1.82	1.77	1.71
2.22	2.15	2.07	1.99	1.95	1.90	1.85	1.80	1.75	1.69
2.20	2.13	2.06	1.97	1.93	1.88	1.84	1.79	1.73	1.67
2.19	2.12	2.04	1.96	1.91	1.87	1.82	1.77	1.71	1.65
2.18	2.10	2.03	1.94	1.90	1.85	1.81	1.75	1.70	1.64
2.16	2.09	2.01	1.93	1.89	1.84	1.79	1.74	1.68	1.62
2.08	2.00	1.92	1.84	1.79	1.74	1.69	1.64	1.58	1.51
1.99	1.92	1.84	1.75	1.70	1.65	1.59	1.53	1.47	1.39
1.91	1.83	1.75	1.66	1.61	1.55	1.55	1.43	1.35	1.25
1.83	1.75	1.67	1.57	1.52	1.46	1.39	1.32	1.22	1.00

Table A.4 Percentage points of the *F* distribution (*Continued*)

$F_{.025, \nu_1, \nu_2}$

ν_2 \ ν_1	Degrees of freedom for the numerator (ν_1) 1	2	3	4	5	6	7	8	9
1	647.8	799.5	864.2	899.6	921.8	937.1	948.2	956.7	963.3
2	38.51	39.00	39.17	39.25	39.30	39.33	39.36	39.37	39.39
3	17.44	16.04	15.44	15.10	14.88	14.73	14.62	14.54	14.47
4	12.22	10.65	9.98	9.60	9.36	9.20	9.07	8.98	8.90
5	10.01	8.43	7.76	7.39	7.15	6.98	6.85	6.76	6.68
6	8.81	7.26	6.60	6.23	5.99	5.82	5.70	5.60	5.52
7	8.07	6.54	5.89	5.52	5.29	5.12	4.99	4.90	4.82
8	7.57	6.06	5.42	5.05	4.82	4.65	4.53	4.43	4.36
9	7.21	5.71	5.08	4.72	4.48	4.32	4.20	4.10	4.03
10	6.94	5.46	4.83	4.47	4.24	4.07	3.95	3.85	3.78
11	6.72	5.26	4.63	4.28	4.04	3.88	3.76	3.66	3.59
12	6.55	5.10	4.47	4.12	3.89	3.73	3.61	3.51	3.44
13	6.41	4.97	4.35	4.00	3.77	3.60	3.48	3.39	3.31
14	6.30	4.86	4.24	3.89	3.66	3.50	3.38	3.29	3.21
15	6.20	4.77	4.15	3.80	3.58	3.41	3.29	3.20	3.12
16	6.12	4.69	4.08	3.73	3.50	3.34	3.22	3.12	3.05
17	6.04	4.62	4.01	3.66	3.44	3.28	3.16	3.06	2.98
18	5.98	4.56	3.95	3.61	3.38	3.22	3.10	3.01	2.93
19	5.92	4.51	3.90	3.56	3.33	3.17	3.05	2.96	2.88
20	5.87	4.46	3.86	3.51	3.29	3.13	3.01	2.91	2.84
21	5.83	4.42	3.82	3.48	3.25	3.09	2.97	2.87	2.80
22	5.79	4.38	3.78	3.44	3.22	3.05	2.93	2.84	2.76
23	5.75	4.35	3.75	3.41	3.18	3.02	2.90	2.81	2.73
24	5.72	4.32	3.72	3.38	3.15	2.99	2.87	2.78	2.70
25	5.69	4.29	3.69	3.35	3.13	2.97	2.85	2.75	2.68
26	5.66	4.27	3.67	3.33	3.10	2.94	2.82	2.73	2.65
27	5.63	4.24	3.65	3.31	3.08	2.92	2.80	2.71	2.63
28	5.61	4.22	3.63	3.29	3.06	2.90	2.78	2.69	2.61
29	5.59	4.20	3.61	3.27	3.04	2.88	2.76	2.67	2.59
30	5.57	4.18	3.59	3.25	3.03	2.87	2.75	2.65	2.57
40	5.42	4.05	3.46	3.13	2.90	2.74	2.62	2.53	2.45
60	5.29	3.93	3.34	3.01	2.79	2.63	2.51	2.41	2.33
120	5.15	3.80	3.23	2.89	2.67	2.52	2.39	2.30	2.22
∞	5.02	3.69	3.12	2.79	2.57	2.41	2.29	2.19	2.11

Degrees of freedom for the denominator (ν_2)

10	12	15	20	24	30	40	60	120	∞
968.6	976.7	984.9	993.1	997.2	1001	1006	1010	1014	1018
39.40	39.41	39.43	39.45	39.46	39.46	39.47	29.48	39.49	39.50
14.42	14.34	14.25	14.17	14.12	14.08	14.04	13.99	13.95	13.90
8.84	8.75	8.66	8.56	8.51	8.46	8.41	8.36	8.31	8.26
6.62	6.52	6.43	6.33	6.28	6.23	6.18	6.12	6.07	6.02
5.46	5.37	5.27	5.17	5.12	5.07	5.01	4.96	4.90	4.85
4.76	4.67	4.57	4.47	4.42	4.36	4.31	4.25	4.20	4.14
4.30	4.20	4.10	4.00	3.95	3.89	3.84	3.78	3.73	3.67
3.96	3.87	3.77	3.67	3.61	3.56	3.51	3.45	3.39	3.33
3.72	3.62	3.52	3.42	3.37	3.31	3.26	3.20	3.14	3.08
3.53	3.43	3.33	3.23	3.17	3.12	3.06	3.00	2.94	2.88
3.37	3.28	3.18	3.07	3.02	2.96	2.91	2.85	2.79	2.72
3.25	3.15	3.05	2.95	2.89	2.84	2.78	2.72	2.66	2.60
3.15	3.05	2.95	2.84	2.79	2.73	2.67	2.61	2.55	2.49
3.06	2.96	2.86	2.76	2.70	2.64	2.59	2.52	2.46	2.40
2.99	2.89	2.79	2.68	2.63	2.57	2.51	2.45	2.38	2.32
2.92	2.82	2.72	2.62	2.56	2.50	2.44	2.38	2.32	2.25
2.87	2.77	2.67	2.56	2.50	2.44	2.38	2.32	2.26	2.19
2.82	2.72	2.62	2.51	2.45	2.39	2.33	2.27	2.20	2.13
2.77	2.68	2.57	2.46	2.41	2.35	2.29	2.22	2.16	2.09
2.73	2.64	2.53	2.42	2.37	2.31	2.25	2.18	2.11	2.04
2.70	2.60	2.50	2.39	2.33	2.27	2.21	2.14	2.08	2.00
2.67	2.57	2.47	2.36	2.30	2.24	2.18	2.11	2.04	1.97
2.64	2.54	2.44	2.33	2.27	2.21	2.15	2.08	2.01	1.94
2.61	2.51	2.41	2.30	2.24	2.18	2.12	2.05	1.98	1.91
2.59	2.49	2.39	2.28	2.22	2.16	2.09	2.03	1.95	1.88
2.57	2.47	2.36	2.25	2.19	2.13	2.07	2.00	1.93	1.85
2.55	2.45	2.34	2.23	2.17	2.11	2.05	1.98	1.91	1.83
2.53	2.43	2.32	2.21	2.15	2.09	2.03	1.96	1.89	1.81
2.51	2.41	2.31	2.20	2.14	2.07	2.01	1.94	1.87	1.79
2.39	2.29	2.18	2.07	2.01	1.94	1.88	1.80	1.72	1.64
2.27	2.17	2.06	1.94	1.88	1.82	1.74	1.67	1.58	1.48
2.16	2.05	1.94	1.82	1.76	1.69	1.61	1.53	1.43	1.31
2.05	1.94	1.83	1.71	1.64	1.57	1.48	1.39	1.27	1.00

Table A.4 Percentage points of the *F* distribution (*Continued*)

$F_{.01,\nu_1,\nu_2}$

ν_2 \ ν_1	Degrees of freedom for the numerator (ν_1) 1	2	3	4	5	6	7	8	9
1	4052	4999.5	5403	5625	5764	5859	5928	5982	6022
2	98.50	99.00	99.17	99.25	99.30	99.33	99.36	99.37	99.39
3	34.12	30.82	29.46	28.71	28.24	27.91	27.67	27.49	27.35
4	21.20	18.00	16.69	15.98	15.52	15.21	14.98	14.80	14.66
5	16.26	13.27	12.06	11.39	10.97	10.67	10.46	10.29	10.16
6	13.75	10.92	9.78	9.15	8.75	8.47	8.26	8.10	7.98
7	12.25	9.55	8.45	7.85	7.46	7.19	6.99	6.84	6.72
8	11.26	8.65	7.59	7.01	6.63	6.37	6.18	6.03	5.91
9	10.56	8.02	6.99	6.42	6.06	5.80	5.61	5.47	5.35
10	10.04	7.56	6.55	5.99	5.64	5.39	5.20	5.06	4.94
11	9.65	7.21	6.22	5.67	5.32	5.07	4.89	4.74	4.63
12	9.33	6.93	5.95	5.41	5.06	4.82	4.64	4.50	4.39
13	9.07	6.70	5.74	5.21	4.86	4.62	4.44	4.30	4.19
14	8.86	6.51	5.56	5.04	4.69	4.46	4.28	4.14	4.03
15	8.68	6.36	5.42	4.89	4.36	4.32	4.14	4.00	3.89
16	8.53	6.23	5.29	4.77	4.44	4.20	4.03	3.89	3.78
17	8.40	6.11	5.18	4.67	4.34	4.10	3.93	3.79	3.68
18	8.29	6.01	5.09	4.58	4.25	4.01	3.84	3.71	3.60
19	8.18	5.93	5.01	4.50	4.17	3.94	3.77	3.63	3.52
20	8.10	5.85	4.94	4.43	4.10	3.87	3.70	3.56	3.46
21	8.02	5.78	4.87	4.37	4.04	3.81	3.64	3.51	3.40
22	7.95	5.72	4.82	4.31	3.99	3.76	3.59	3.45	3.35
23	7.88	5.66	4.76	4.26	3.94	3.71	3.54	3.41	3.30
24	7.82	5.61	4.72	4.22	3.90	3.67	3.50	3.36	3.26
25	7.77	5.57	4.68	4.18	3.85	3.63	3.46	3.32	3.22
26	7.72	5.53	4.64	4.14	3.82	3.59	3.42	3.29	3.18
27	7.68	5.49	4.60	4.11	3.78	3.56	3.39	3.26	3.15
28	7.64	5.45	4.57	4.07	3.75	3.53	3.36	3.23	3.12
29	7.60	5.42	4.54	4.04	3.73	3.50	3.33	3.20	3.09
30	7.56	5.39	4.51	4.02	3.70	3.47	3.30	3.17	3.07
40	7.31	5.18	4.31	3.83	3.51	3.29	3.12	2.99	2.89
60	7.08	4.98	4.13	3.65	3.34	3.12	2.95	2.82	2.72
120	6.85	4.79	3.95	3.48	3.17	2.96	2.79	2.66	2.56
∞	6.63	4.61	3.78	3.32	3.02	2.80	2.64	2.51	2.41

Degrees of freedom for the denominator (ν_2)

10	12	15	20	24	30	40	60	120	∞
6056	6106	6157	6209	6235	6261	6287	6313	6339	6366
99.40	99.42	99.43	99.45	99.46	99.47	99.47	99.48	99.49	99.50
27.23	27.05	26.87	26.69	26.00	26.50	26.41	26.32	26.22	26.13
14.55	14.37	14.20	14.02	13.93	13.84	13.75	13.65	13.56	13.46
10.05	9.89	9.72	9.55	9.47	9.38	9.29	9.20	9.11	9.02
7.87	7.72	7.56	7.40	7.31	7.23	7.14	7.06	6.97	6.88
6.62	6.47	6.31	6.16	6.07	5.99	5.91	5.82	5.74	5.65
5.81	5.67	5.52	5.36	5.28	5.20	5.12	5.03	4.95	4.86
5.26	5.11	4.96	4.81	4.73	4.65	4.57	4.48	4.40	4.31
4.85	4.71	4.56	4.41	4.33	4.25	4.17	4.08	4.00	3.91
4.54	4.40	4.25	4.10	4.02	3.94	3.86	3.78	3.69	3.60
4.30	4.16	4.01	3.86	3.78	3.70	3.62	3.54	3.45	3.36
4.10	3.96	3.82	3.66	3.59	3.51	3.43	3.34	3.25	3.17
3.94	3.80	3.66	3.51	3.43	3.35	3.27	3.18	3.09	3.00
3.80	3.67	3.52	3.37	3.29	3.21	3.13	3.05	2.96	2.87
3.69	3.55	3.41	3.26	3.18	3.10	3.02	2.93	2.84	2.75
3.59	3.46	3.31	3.16	3.08	3.00	2.92	2.83	2.75	2.65
3.51	3.37	3.23	3.08	3.00	2.92	2.84	2.75	2.66	2.57
3.43	3.30	3.15	3.00	2.92	2.84	2.76	2.67	2.58	2.59
3.37	3.23	3.09	2.94	2.86	2.78	2.69	2.61	2.52	2.42
3.31	3.17	3.03	2.88	2.80	2.72	2.64	2.55	2.46	2.36
3.26	3.12	2.98	2.83	2.75	2.67	2.58	2.50	2.40	2.31
3.21	3.07	2.93	2.78	2.70	2.62	2.54	2.45	2.35	2.26
3.17	3.03	2.89	2.74	2.66	2.58	2.49	2.40	2.31	2.21
3.13	2.99	2.85	2.70	2.62	2.54	2.45	2.36	2.27	2.17
3.09	2.96	2.81	2.66	2.58	2.50	2.42	2.33	2.23	2.13
3.06	2.93	2.78	2.63	2.55	2.47	2.38	2.29	2.20	2.10
3.03	2.90	2.75	2.60	2.52	2.44	2.35	2.26	2.17	2.06
3.00	2.87	2.73	2.57	2.49	2.41	2.33	2.23	2.14	2.03
2.98	2.84	2.70	2.55	2.47	2.39	2.30	2.21	2.11	2.01
2.80	2.66	2.52	2.37	2.29	2.20	2.11	2.02	1.92	1.80
2.63	2.50	2.35	2.20	2.12	2.03	1.94	1.84	1.73	1.60
2.47	2.34	2.19	2.03	1.95	1.86	1.76	1.66	1.53	1.38
2.32	2.18	2.04	1.88	1.79	1.70	1.59	1.47	1.32	1.00

Source: Adapted with permission from *Biometrika Tables for Statisticians*, Vol. 1, 3rd ed., by E. S. Pearson and H. O. Hartley, Cambridge University Press, Cambridge, 1966.

Table A.5 Orthogonal polynomials

	$n=3$		$n=4$			$n=5$			
X_j	P_1	P_2	P_1	P_2	P_3	P_1	P_2	P_3	P_4
1	−1	1	−3	1	−1	−2	2	−1	1
2	0	−2	−1	−1	3	−1	−1	2	−4
3	1	1	1	−1	−3	0	−2	0	6
4			3	1	1	1	−1	−2	−4
5						2	2	1	1
6									
7									
$\sum_{j=1}^{n} \{P_i(X_j)\}^2$	2	6	20	4	20	10	14	10	70
λ	1	3	2	1	$\frac{10}{3}$	1	1	$\frac{5}{6}$	$\frac{35}{12}$

	$n=8$							
X_j	P_1	P_2	P_3	P_4	P_5	P_6	P_1	P_2
1	−7	7	−7	7	−7	1	−4	28
2	−5	1	5	−13	23	−5	−3	7
3	−3	−3	7	−3	−17	9	−2	−8
4	−1	−5	3	9	−15	−5	−1	−17
5	1	−5	−3	9	15	−5	0	−20
6	3	−3	−7	−3	17	9	1	−17
7	5	1	−5	−13	−23	−5	2	−8
8	7	7	7	7	7	1	3	7
9							4	28
10								
$\sum_{j=1}^{n} \{P_i(X_j)\}^2$	168	168	264	616	2,184	264	60	2,772
λ	2	1	$\frac{2}{3}$	$\frac{7}{12}$	$\frac{7}{10}$	$\frac{11}{60}$	1	3

$n=6$					$n=7$					
P_1	P_2	P_3	P_4	P_5	P_1	P_2	P_3	P_4	P_5	P_6
−5	5	−5	1	−1	−3	5	−1	3	−1	1
−3	−1	7	−3	5	−2	0	1	−7	4	−6
−1	−4	4	2	−10	−1	−3	1	1	−5	15
1	−4	−4	2	10	0	−4	0	6	0	−20
3	−1	−7	−3	−5	1	−3	−1	1	5	15
5	5	5	1	1	2	0	−1	−7	−4	−6
					3	5	1	3	1	1
70	84	180	28	252	28	84	6	154	84	924
2	$\frac{3}{2}$	$\frac{5}{3}$	$\frac{7}{12}$	$\frac{21}{10}$	1	1	$\frac{1}{6}$	$\frac{7}{12}$	$\frac{7}{20}$	$\frac{77}{60}$

$n=9$				$n=10$					
P_3	P_4	P_5	P_6	P_1	P_2	P_3	P_4	P_5	P_6
−14	14	−4	4	−9	6	−42	18	−6	3
7	−21	11	−17	−7	2	14	−22	14	−11
13	−11	−4	22	−5	−1	35	−17	−1	10
9	9	−9	1	−3	−3	31	3	−11	6
0	18	0	−20	−1	−4	12	18	−6	−8
−9	9	9	1	1	−4	−12	18	6	−8
−13	−11	4	22	3	−3	−31	3	11	6
−7	−21	−11	−17	5	−1	−35	−17	1	10
14	14	4	4	7	2	−14	−22	−14	−11
				9	6	42	18	6	3
990	2,002	468	1,980	330	132	8,580	2,860	780	660
$\frac{5}{6}$	$\frac{7}{12}$	$\frac{3}{20}$	$\frac{11}{60}$	2	$\frac{1}{2}$	$\frac{5}{3}$	$\frac{5}{12}$	$\frac{1}{10}$	$\frac{11}{240}$

Source: Adapted with permission from *Biometrika Tables for Statisticians*, Vol. 1, 3rd ed., by E. S. Pearson and H. O. Hartley, Cambridge University Press, Cambridge, 1966.

Table A.6 Critical values of the Durbin-Watson statistic

		k=Number of Regressors (Excluding the Intercept)									
	Probability in Lower Tail	1		2		3		4		5	
Sample Size	(Significance Level=α)	d_L	d_U	d_L	d_U	d_L	d_U	d_L	d_U	d_L	d_U
	.01	.81	1.07	.70	1.25	.59	1.46	.49	1.70	.39	1.96
15	.025	.95	1.23	.83	1.40	.71	1.61	.59	1.84	.48	2.09
	.05	1.08	1.36	.95	1.54	.82	1.75	.69	1.97	.56	2.21
	.01	.95	1.15	.86	1.27	.77	1.41	.63	1.57	.60	1.74
20	.025	1.08	1.28	.99	1.41	.89	1.55	.79	1.70	.70	1.87
	.05	1.20	1.41	1.10	1.54	1.00	1.68	.90	1.83	.79	1.99
	.01	1.05	1.21	.98	1.30	.90	1.41	.83	1.52	.75	1.65
25	.025	1.13	1.34	1.10	1.43	1.02	1.54	.94	1.65	.86	1.77
	.05	1.20	1.45	1.21	1.55	1.12	1.66	1.04	1.77	.95	1.89
	.01	1.13	1.26	1.07	1.34	1.01	1.42	.94	1.51	.88	1.61
30	.025	1.25	1.38	1.18	1.46	1.12	1.54	1.05	1.63	.98	1.73
	.05	1.35	1.49	1.28	1.57	1.21	1.65	1.14	1.74	1.07	1.83
	.01	1.25	1.34	1.20	1.40	1.15	1.46	1.10	1.52	1.05	1.58
40	.025	1.35	1.45	1.30	1.51	1.25	1.57	1.20	1.63	1.15	1.69
	.05	1.44	1.54	1.39	1.60	1.34	1.66	1.29	1.72	1.23	1.79
	.01	1.32	1.40	1.28	1.45	1.24	1.49	1.20	1.54	1.16	1.59
50	.025	1.42	1.50	1.38	1.54	1.34	1.59	1.30	1.64	1.26	1.69
	.05	1.50	1.59	1.46	1.63	1.42	1.67	1.38	1.72	1.34	1.77
	.01	1.38	1.45	1.35	1.48	1.32	1.52	1.28	1.56	1.25	1.60
60	.025	1.47	1.54	1.44	1.57	1.40	1.61	1.37	1.65	1.33	1.69
	.05	1.55	1.62	1.51	1.65	1.48	1.69	1.44	1.73	1.41	1.77
	.01	1.47	1.52	1.44	1.54	1.42	1.57	1.39	1.60	1.36	1.62
80	.025	1.54	1.59	1.52	1.62	1.49	1.65	1.47	1.67	1.44	1.70
	.05	1.61	1.66	1.59	1.69	1.56	1.72	1.53	1.74	1.51	1.77
	.01	1.52	1.56	1.50	1.58	1.48	1.60	1.45	1.63	1.44	1.65
100	.025	1.59	1.63	1.57	1.65	1.55	1.67	1.53	1.70	1.51	1.72
	.05	1.65	1.69	1.63	1.72	1.61	1.74	1.59	1.76	1.57	1.78

Source: Adapted from "Testing for Serial Correlation in Least Squares Regression II," by J. Durbin and G. S. Watson, *Biometrika*, Vol. 38, 1951, with permission of the publisher.

Table A.7 Percentage points of the Bonferroni t-statistic

Entries in the table are $t_{\nu,\alpha/(2p)}$ where $P[T \geqslant t_{\nu,\alpha/(2p)}] = \alpha/(2p)$ and T has the t-distribution with ν degrees of freedom. This table was computed on a CDC Cyber 74 Computer at the Georgia Institute of Technology using IMSL subroutine MDSTI. See Section 9.6 in the text.

	$\alpha=0.10$								
ν/p	2	3	4	5	6	7	8	9	10
5	2.5706	2.9117	3.1634	3.3649	3.5341	3.6805	3.8100	3.9264	4.0322
6	2.4469	2.7491	2.9687	3.1427	3.2875	3.4119	3.5212	3.6190	3.7074
7	2.3646	2.6419	2.8412	2.9980	3.1276	3.2383	3.3353	3.4216	3.4995
8	2.3060	2.5660	2.7515	2.8965	3.0158	3.1174	3.2060	3.2846	3.3554
9	2.2622	2.5096	2.6850	2.8214	2.9333	3.0283	3.1109	3.1841	3.2498
10	2.2281	2.4660	2.6338	2.7638	2.8701	2.9601	3.0382	3.1073	3.1693
11	2.2010	2.4313	2.5931	2.7181	2.8200	2.9062	2.9809	3.0468	3.1058
12	2.1788	2.4030	2.5600	2.6810	2.7795	2.8626	2.9345	2.9978	3.0545
13	2.1604	2.3796	2.5326	2.6503	2.7459	2.8265	2.8961	2.9575	3.0123
14	2.1448	2.3598	2.5096	2.6245	2.7178	2.7862	2.8640	2.9236	2.9768
15	2.1314	2.3429	2.4899	2.6025	2.6937	2.7705	2.8366	2.8948	2.9467
16	2.1199	2.3283	2.4729	2.5835	2.6730	2.7482	2.8131	2.8700	2.9208
17	2.1098	2.3156	2.4581	2.5669	2.6550	2.7289	2.7925	2.8484	2.8882
18	2.1009	2.3043	2.4450	2.5524	2.6391	2.7119	2.7745	2.8295	2.8784
19	2.0930	2.2944	2.4334	2.5395	2.6251	2.6969	2.7586	2.8127	2.8609
20	2.0860	2.2855	2.4231	2.5280	2.6126	2.6834	2.7444	2.7978	2.8453
21	2.0796	2.2775	2.4138	2.5176	2.6013	2.6714	2.7316	2.7844	2.8314
22	2.0739	2.2703	2.4055	2.5083	2.5912	2.6606	2.7201	2.7723	2.8188
23	2.0687	2.2637	2.3979	2.4999	2.5820	2.6507	2.7097	2.7614	2.8073
24	2.0639	2.2577	2.3909	2.4922	2.5736	2.6418	2.7002	2.7514	2.7969
25	2.0595	2.2523	2.3846	2.4851	2.5660	2.6336	2.6916	2.7423	2.7874
30	2.0423	2.2306	2.3596	2.4573	2.5357	2.6012	2.6574	2.7064	2.7500
35	2.0301	2.2154	2.3420	2.4377	2.5145	2.5786	2.6334	2.6813	2.7238
40	2.0211	2.2041	2.3289	2.4233	2.4989	2.5618	2.6157	2.6627	2.7045
45	2.0141	2.1954	2.3189	2.4121	2.4868	2.5489	2.6021	2.6485	2.6696
50	2.0086	2.1885	2.3109	2.4033	2.4772	2.5387	2.5913	2.6372	2.6778
60	2.0003	2.1782	2.2990	2.3901	2.4630	2.5235	2.5752	2.6203	2.6603
70	1.9944	2.1709	2.2906	2.3808	2.4529	2.5128	2.5639	2.6085	2.6479
80	1.9901	2.1654	2.2844	2.3739	2.4454	2.5047	2.5554	2.5996	2.6387
90	1.9867	2.1612	2.2795	2.3685	2.4395	2.4985	2.5489	2.5928	2.6316
100	1.9840	2.1579	2.2757	2.3642	2.4349	2.4936	2.5437	2.5873	2.6209

Table A.7 Percentage points of the Bonferroni t-statistic (*Continued*)

	$\alpha=0.05$								
ν/p	2	3	4	5	6	7	8	9	10
5	3.1634	3.5341	3.8100	4.0322	4.2193	4.3818	4.5257	4.6553	4.7733
6	2.9687	3.2875	3.5212	3.7074	3.8630	3.9971	4.1152	4.2209	4.3168
7	2.8412	3.1276	3.3353	3.4995	3.6358	3.7527	3.8552	3.9467	4.0293
8	2.7515	3.0158	3.2060	3.3554	3.4789	3.5844	3.6766	3.7586	3.8325
9	2.6850	2.9333	3.1109	3.2498	3.3642	2.4616	3.5465	3.6219	3.6897
10	2.6338	2.8701	3.0382	3.1693	3.2768	3.3682	3.4477	3.5182	3.5814
11	2.5931	2.8200	2.9809	3.1058	3.2081	3.2949	3.3702	3.4368	3.4966
12	2.5600	2.7795	2.9345	3.0545	3.1527	3.2357	3.3078	3.3714	3.4284
13	2.5326	2.7459	2.8961	3.0123	3.1070	3.1871	3.2565	3.3177	3.3725
14	2.5096	2.7178	2.8640	2.9768	3.0688	3.1464	3.2135	3.2727	3.3257
15	2.4899	2.6937	2.8366	2.9467	3.0363	3.1118	3.1771	3.2346	3.2860
16	2.4729	2.6730	2.8131	2.9208	3.0083	3.0821	3.1458	3.2019	3.2520
17	2.4581	2.6550	2.7925	2.8982	2.9840	3.0563	3.1186	3.1735	3.2224
18	2.4450	2.6391	2.7745	1.8784	2.9627	3.0336	3.0948	3.1486	3.1966
19	2.4334	2.6251	2.7586	2.8609	2.9439	3.0136	3.0738	3.1266	3.1737
20	2.4231	2.6126	2.7444	2.8453	2.9271	2.9958	3.0550	3.1070	3.1534
21	2.4138	2.6013	2.7316	2.8314	2.9121	2.9799	3.0382	3.0895	3.1352
22	2.4055	2.5912	2.7201	2.8188	2.8985	2.9655	3.0231	3.0737	3.1188
23	2.3979	2.5820	2.7097	2.8073	2.8863	2.9525	3.0095	3.0595	3.1040
24	2.3909	2.5736	2.7002	2.7969	2.8751	2.9406	2.9970	3.0465	3.0905
25	2.3846	2.5660	2.6916	2.7874	2.8649	2.9298	2.9856	3.0346	3.0782
30	2.3586	2.5357	2.6574	2.7500	2.8247	2.8872	2.9409	2.9880	3.0298
35	2.3420	2.5145	2.6334	2.7238	2.7966	2.8575	2.9097	2.9554	2.9860
40	2.3289	2.4989	2.6157	2.7045	2.7759	2.8355	2.8867	2.9314	2.9712
45	2.3189	2.4868	2.6021	2.6896	2.7599	2.8187	2.8690	2.9130	2.9521
50	2.3109	2.4772	2.5913	2.6776	2.7473	2.8053	2.8550	2.8984	2.9370
60	2.2990	2.4630	2.5752	2.6603	2.7286	2.7855	2.8342	2.8768	2.9146
70	2.2906	2.4529	2.5639	2.6479	2.7153	2.7715	2.8195	2.8615	2.8987
80	2.2844	2.4454	2.5554	2.6387	2.7054	2.7610	2.8086	2.8502	2.8870
90	2.2795	2.4395	2.5489	2.6316	2.6978	2.7530	2.8002	2.8414	2.8779
100	2.2757	2.4349	2.5437	2.6259	2.6918	2.7466	2.7935	2.8344	2.8707

Table A.7 Percentage points of the Bonferroni t-statistic (*Continued*)

	$\alpha=0.01$								
ν/p	2	3	4	5	6	7	8	9	10
5	4.7733	5.2474	5.6042	5.8934	6.1384	6.3518	6.5414	6.7126	6.8688
6	4.3168	4.6979	4.9807	5.2076	5.3982	5.5632	5.7090	5.8399	5.9588
7	4.0293	4.3553	4.5946	4.7853	4.9445	5.0815	5.2022	5.3101	5.4079
8	3.8325	4.1224	4.3335	4.5008	4.6398	4.7590	4.8636	4.9570	5.0413
9	3.6897	3.9542	4.1458	4.2968	4.4219	4.5288	4.6224	4.7058	4.7809
10	3.5814	3.8273	4.0045	4.1437	4.2586	4.3567	4.4423	4.5184	4.5869
11	3.4966	3.7283	3.8945	4.0247	4.1319	4.2232	4.3028	4.3735	4.4370
12	3.4284	3.6489	3.8065	3.9296	4.0308	4.1169	4.1918	4.2582	4.3178
13	3.3725	3.5838	3.7345	3.8520	3.9484	4.0302	4.1013	4.1643	4.2208
14	3.3257	3.5296	3.6746	3.7874	3.8798	3.9582	4.0263	4.0865	4.1405
15	3.2860	3.4837	3.6239	3.7328	3.8220	3.8975	3.9630	4.0209	4.0728
16	3.2520	3.4443	3.5805	3.6862	3.7725	3.8456	3.9089	3.9649	4.0150
17	3.2224	3.4102	3.5429	3.6458	3.7297	3.8007	3.8623	3.9165	3.9651
18	3.1966	3.3804	3.5101	3.6105	3.6924	2.7616	3.8215	3.8744	3.9216
19	3.1737	3.3540	3.4812	3.5794	3.6595	3.7271	3.7857	3.8373	3.8834
20	3.1534	3.3306	3.4554	3.5518	3.6303	3.6966	3.7539	3.8044	3.8495
21	3.1352	3.3097	3.4325	3.5272	3.6043	3.6693	3.7255	3.7750	3.8193
22	3.1188	3.2909	3.4118	3.5050	3.5808	3.6448	3.7000	3.7487	3.7921
23	3.1040	3.2739	3.3931	3.4850	3.5597	3.6226	3.6770	3.7249	3.7676
24	3.0905	3.2584	3.3761	3.4668	3.5405	3.6025	3.6561	3.7033	3.7454
25	3.0782	3.2443	3.3606	3.4502	3.5230	3.5842	3.6371	3.6836	3.7251
30	3.0298	3.1888	3.2999	3.3852	3.4544	3.5125	3.5626	3.6067	3.6460
35	2.9960	3.1502	3.2577	3.3400	3.4068	3.4628	3.5110	3.5534	3.5911
40	2.9712	3.1218	3.2266	3.3069	3.3718	3.4263	3.4732	3.5143	3.5510
45	2.9521	3.1000	3.2028	3.2815	3.3451	3.3984	3.4442	3.4845	3.5203
50	2.9370	3.0828	3.1840	3.2614	3.3238	3.3763	3.4214	3.4609	3.4960
60	2.9146	3.0573	3.1562	3.2317	3.2927	3.3437	3.3876	3.4260	3.4602
70	2.8987	3.0393	3.1366	3.2108	3.2707	3.3208	3.3638	2.4015	3.4350
80	2.8870	3.0259	3.1220	3.1953	3.2543	3.3037	3.3462	3.3833	3.4163
90	2.8779	3.0156	3.1108	3.1833	3.2417	3.2906	3.3326	3.3693	3.4019
100	2.8707	3.0073	3.1018	3.1737	3.2317	3.2802	3.3218	3.3582	3.3905

Table A.8 Percentage points of the maximum modulus t–statistic

Entries in the table are $u_{\alpha,k,\nu}$ where $P[U \geqslant u_{\alpha,k,\nu}]=\alpha$ and U is the maximum absolute value of k uncorrelated student–t random variables, each having ν degrees of freedom.

	$\alpha=0.10$										
$\nu \backslash k$	1	2	3	4	5	6	8	10	12	15	20
3	2.353	2.989	3.369	3.637	3.844	4.011	4.272	4.471	4.631	4.823	5.066
4	2.132	2.662	2.976	3.197	3.368	3.506	3.722	3.887	4.020	4.180	4.383
5	2.015	2.491	2.769	2.965	3.116	3.239	3.430	3.576	3.694	3.837	4.018
6	1.943	2.385	2.642	2.822	2.961	3.074	3.249	3.384	3.493	3.624	3.790
7	1.895	2.314	2.556	2.725	2.856	2.962	3.127	3.253	3.355	3.478	3.635
8	1.860	2.262	2.494	2.656	2.780	2.881	3.038	3.158	3.255	3.373	3.522
9	1.833	2.224	2.447	2.603	2.723	2.819	2.970	3.086	3.179	3.292	3.436
10	1.813	2.193	2.410	2.562	2.678	2.771	2.918	3.029	3.120	3.229	3.368
11	1.796	2.169	2.381	2.529	2.642	2.733	2.875	2.984	3.072	3.178	3.313
12	1.782	2.149	2.357	2.501	2.612	2.701	2.840	2.946	3.032	3.136	3.268
15	1.753	2.107	2.305	2.443	2.548	2.633	2.765	2.865	2.947	3.045	3.170
20	1.725	2.065	2.255	2.386	2.486	2.567	2.691	2.786	2.863	2.956	3.073
25	1.708	2.041	2.226	2.353	2.450	2.528	2.648	2.740	2.814	2.903	3.016
30	1.697	2.025	2.207	2.331	2.426	2.502	2.620	2.709	2.781	2.868	2.978
40	1.684	2.006	2.183	2.305	2.397	2.470	2.585	2.671	2.741	2.825	2.931
60	1.671	1.986	2.160	2.278	2.368	2.439	2.550	2.634	2.701	2.782	2.884

Table A.8 Percentage points of the maximum modulus t-statistic (*Continued*)

$\alpha=0.05$

$\nu \backslash k$	1	2	3	4	5	6	8	10	12	15	20
3	3.183	3.960	4.430	4.764	5.023	5.233	5.562	5.812	6.015	6.259	6.567
4	2.777	3.382	3.745	4.003	4.203	4.366	4.621	4.817	4.975	5.166	5.409
5	2.571	3.091	3.399	3.619	3.789	3.928	4.145	4.312	4.447	4.611	4.819
6	2.447	2.916	3.193	3.389	3.541	3.664	3.858	4.008	4.129	4.275	4.462
7	2.365	2.800	3.056	3.236	3.376	3.489	3.668	3.805	3.916	4.051	4.223
8	2.306	2.718	2.958	3.128	3.258	3.365	3.532	3.660	3.764	3.891	4.052
9	2.262	2.657	2.885	3.046	3.171	3.272	3.430	3.552	3.651	3.770	3.923
10	2.228	2.609	2.829	2.984	3.103	3.199	3.351	3.468	3.562	3.677	3.823
11	2.201	2.571	2.784	2.933	3.048	3.142	3.288	3.400	3.491	3.602	3.743
12	2.179	2.540	2.747	2.892	3.004	3.095	3.236	3.345	3.433	3.541	3.677
15	2.132	2.474	2.669	2.805	2.910	2.994	3.126	3.227	3.309	3.409	3.536
20	2.086	2.411	2.594	2.722	2.819	2.898	3.020	3.114	3.190	3.282	3.399
25	2.060	2.374	2.551	2.673	2.766	2.842	2.959	3.048	3.121	3.208	3.320
30	2.042	2.350	2.522	2.641	2.732	2.805	2.918	3.005	3.075	3.160	3.267
40	2.021	2.321	2.488	2.603	2.690	2.760	2.869	2.952	3.019	3.100	3.203
60	2.000	2.292	2.454	2.564	2.649	2.716	2.821	2.900	2.964	3.041	3.139

$\alpha=0.01$

$\nu \backslash k$	1	2	3	4	5	6	8	10	12	15	20
3	5.841	7.127	7.914	8.479	8.919	9.277	9.838	10.269	10.616	11.034	11.559
4	4.604	5.462	5.985	6.362	6.656	6.897	7.274	7.565	7.801	8.087	8.451
5	4.032	4.700	5.106	5.398	5.625	5.812	6.106	6.333	6.519	6.744	7.050
6	3.707	4.271	4.611	4.855	5.046	5.202	5.449	5.640	5.796	5.985	6.250
7	3.500	3.998	4.296	4.510	4.677	4.814	5.031	5.198	5.335	5.502	5.716
8	3.355	3.809	4.080	4.273	4.424	4.547	4.742	4.894	5.017	5.168	5.361
9	3.250	3.672	3.922	4.100	4.239	4.353	4.532	4.672	4.785	4.924	5.103
10	3.169	3.567	3.801	3.969	4.098	4.205	4.373	4.503	4.609	4.739	4.905
11	3.106	3.485	3.707	3.865	3.988	4.087	4.247	4.370	4.470	4.593	4.750
12	3.055	3.418	3.631	3.782	3.899	3.995	4.146	4.263	4.359	4.475	4.625
15	2.947	3.279	3.472	3.608	3.714	3.800	3.935	4.040	4.125	4.229	4.363
20	2.845	3.149	3.323	3.446	3.541	3.617	3.738	3.831	3.907	3.999	4.117
25	2.788	3.075	3.239	3.354	3.442	3.514	3.626	3.713	3.783	3.869	3.978
30	2.750	3.027	3.185	3.295	3.379	3.448	3.555	3.637	3.704	3.785	3.889
40	2.705	2.969	3.119	3.223	3.303	3.367	3.468	3.545	3.607	3.683	3.780
60	2.660	2.913	3.055	3.154	3.229	3.290	3.384	3.456	3.515	3.586	3.676

Source: Adapted from "A Table of Percentage Points of the Distribution of the Largest Absolute Value of k Student-t Variates and its Applications", by G. J. Hahn and R. W. Hendrickson, *Biometrika*, Vol. 58, 1971, with permission of the publisher.

APPENDIX B

DATA SETS FOR EXERCISES

Table B.1 National Football League 1976 team performance
Table B.2 Solar thermal energy test data
Table B.3 Gasoline mileage performance for 32 automobiles
Table B.4 Property valuation data
Table B.5 Belle Ayr liquefaction runs
Table B.6 Tube-flow reactor data
Table B.7 Factors affecting CPU time
Table B.8 Effectiveness of insecticides on cricket eggs

Table B.1 National Football League 1976 team performance

Team	y	x_1	x_2	x_3	x_4	x_5	x_6	x_7	x_8	x_9
Washington	10	2113	1985	38.9	64.7	+4	868	59.7	2205	1917
Minnesota	11	2003	2855	38.8	61.3	+3	615	55.0	2096	1575
New England	11	2957	1737	40.1	60.0	+14	914	65.6	1847	2175
Oakland	13	2285	2905	41.6	45.3	−4	957	61.4	1903	2476
Pittsburgh	10	2971	1666	39.2	53.8	+15	836	66.1	1457	1866
Baltimore	11	2309	2927	39.7	74.1	+8	786	61.0	1848	2339
Los Angeles	10	2528	2341	38.1	65.4	+12	754	66.1	1564	2092
Dallas	11	2147	2737	37.0	78.3	−1	761	58.0	1821	1909
Atlanta	4	1689	1414	42.1	47.6	−3	714	57.0	2577	2001
Buffalo	2	2566	1838	42.3	54.2	−1	797	58.9	2476	2254
Chicago	7	2363	1480	37.3	48.0	+19	984	67.5	1984	2217
Cincinnati	10	2109	2191	39.5	51.9	+6	700	57.2	1917	1758
Cleveland	9	2295	2229	37.4	53.6	−5	1037	58.8	1761	2032
Denver	9	1932	2204	35.1	71.4	+3	986	58.6	1709	2025
Detroit	6	2213	2140	38.8	58.3	+6	819	59.2	1901	1686
Green Bay	5	1722	1730	36.6	52.6	−19	791	54.4	2288	1835
Houston	5	1498	2072	35.3	59.3	−5	776	49.6	2072	1914
Kansas City	5	1873	2929	41.1	55.3	+10	789	54.3	2861	2496
Miami	6	2118	2268	38.2	69.6	+6	582	58.7	2411	2670
New Orleans	4	1775	1983	39.3	78.3	+7	901	51.7	2289	2202
New York Giants	3	1904	1792	39.7	38.1	−9	734	61.9	2203	1988
New York Jets	3	1929	1606	39.7	68.8	−21	627	52.7	2592	2324
Philadelphia	4	2080	1492	35.5	68.8	−8	722	57.8	2053	2550
St. Louis	10	2301	2835	35.3	74.1	+2	683	59.7	1979	2110
San Diego	6	2040	2416	38.7	50.0	0	576	54.9	2048	2628
San Francisco	8	2447	1638	39.9	57.1	−8	848	65.3	1786	1776
Seattle	2	1416	2649	37.4	56.3	−22	684	43.8	2876	2524
Tampa Bay	0	1503	1503	39.3	47.0	−9	875	53.5	2560	2241

y: Games won (per 14 game season)
x_1: Rushing yards (season)
x_2: Passing yards (season)
x_3: Punting average (Yds/Punt)
x_4: Field goal percentage (Fgs made/Fgs attempted–season)
x_5: Turnover differential (turnovers acquired–turnovers lost)
x_6: Penalty yards (season)
x_7: Percent rushing (rushing plays/total plays)
x_8: Opponents' rushing yards (season)
x_9: Opponents' passing yards (season)

Table B.2 Solar thermal energy test data

y	x_1	x_2	x_3	x_4	x_5
271.8	783.35	33.53	40.55	16.66	13.20
264.0	748.45	36.50	36.19	16.46	14.11
238.8	684.45	34.66	37.31	17.66	15.68
230.7	827.80	33.13	32.52	17.50	10.53
251.6	860.45	35.75	33.71	16.40	11.00
257.9	875.15	34.46	34.14	16.28	11.31
263.9	909.45	34.60	34.85	16.06	11.96
266.5	905.55	35.38	35.89	15.93	12.58
229.1	756.00	35.85	33.53	16.60	10.66
239.3	769.35	35.68	33.79	16.41	10.85
258.0	793.50	35.35	34.72	16.17	11.41
257.6	801.65	35.04	35.22	15.92	11.91
267.3	819.65	34.07	36.50	16.04	12.85
267.0	808.55	32.20	37.60	16.19	13.58
259.6	774.95	34.32	37.89	16.62	14.21
240.4	711.85	31.08	37.71	17.37	15.56
227.2	694.85	35.73	37.00	18.12	15.83
196.0	638.10	34.11	36.76	18.53	16.41
278.7	774.55	34.79	34.62	15.54	13.10
272.3	757.90	35.77	35.40	15.70	13.63
267.4	753.35	36.44	35.96	16.45	14.51
254.5	704.70	37.82	36.26	17.62	15.38
224.7	666.80	35.07	36.34	18.12	16.10
181.5	568.55	35.26	35.90	19.05	16.73
227.5	653.10	35.56	31.84	16.51	10.58
253.6	704.05	35.73	33.16	16.02	11.28
263.0	709.60	36.46	33.83	15.89	11.91
265.8	726.90	36.26	34.89	15.83	12.65
263.8	697.15	37.20	36.27	16.71	14.06

y: Total heat flux (kwatts)
x_1: Insolation (watts/m^2)
x_2: Position of focal point in east direction (inches)
x_4: Position of focal point in north direction (inches)
x_5: Time of day

Table B.3 Gasoline mileage performance for 32 automobiles

Automobile	y	x_1	x_2	x_3	x_4	x_5	x_6	x_7	x_8	x_9	x_{10}	x_{11}
Apollo	18.90	350	165	260	8.0:1	2.56:1	4	3	200.3	69.9	3910	A
Omega	17.00	350	170	275	8.5:1	2.56:1	4	3	199.6	72.9	3860	A
Nova	20.00	250	105	185	8.25:1	2.73:1	1	3	196.7	72.2	3510	A
Monarch	18.25	351	143	255	8.0:1	3.00:1	2	3	199.9	74.0	3890	A
Duster	20.07	225	95	170	8.4:1	2.76:1	1	3	194.1	71.8	3365	M
Jenson Conv.	11.2	440	215	330	8.2:1	2.88:1	4	3	184.5	69	4215	A
Skyhawk	22.12	231	110	175	8.0:1	2.56:1	2	3	179.3	65.4	3020	A
Monza	21.47	262	110	200	8.5:1	2.56:1	2	3	179.3	65.4	3180	A
Scirocco	34.70	89.7	70	81	8.2:1	3.90:1	2	4	155.7	64	1905	M
Corolla SR-5	30.40	96.9	75	83	9.0:1	4.30:1	2	5	165.2	65	2320	M
Camaro	16.50	350	155	250	8.5:1	3.08:1	4	3	195.4	74.4	3885	A
Datsun B210	36.50	85.3	80	83	8.5:1	3.89:1	2	4	160.6	62.2	2009	M
Capri II	21.50	171	109	146	8.2:1	3.22:1	2	4	170.4	66.9	2655	M
Pacer	19.70	258	110	195	8.0:1	3.08:1	1	3	171.5	77	3375	A
Bobcat	20.30	140	83	109	8.4:1	3.40:1	2	4	168.8	69.4	2700	M
Granada	17.80	302	129	220	8.0:1	3.0:1	2	3	199.9	74	3890	A
Eldorado	14.39	500	190	360	8.5:1	2.73:1	4	3	224.1	79.8	5290	A
Imperial	14.89	440	215	330	8.2:1	2.71:1	4	3	231.0	79.7	5185	A
Nova LN	17.80	350	155	250	8.5:1	3.08:1	4	3	196.7	72.2	3910	A
Valiant	16.41	318	145	255	8.5:1	2.45:1	2	3	197.6	71	3660	A
Starfire	23.54	231	110	175	8.0:1	2.56:1	2	3	179.3	65.4	3050	A
Cordoba	21.47	360	180	290	8.4:1	2.45:1	2	3	214.2	76.3	4250	A
Trans Am	16.59	400	185	NA	7.6:1	3.08:1	4	3	196	73	3850	A
Corolla E-5	31.90	96.9	75	83	9.0:1	4.30:1	2	5	165.2	61.8	2275	M
Astre	29.40	140	86	NA	8.0:1	2.92:1	2	4	176.4	65.4	2150	M
Mark IV	13.27	460	223	366	8.0:1	3.00:1	4	3	228	79.8	5430	A
Celica GT	23.90	133.6	96	120	8.4:1	3.91:1	2	5	171.5	63.4	2535	M
Charger SE	19.73	318	140	255	8.5:1	2.71:1	2	3	215.3	76.3	4370	A
Cougar	13.90	351	148	243	8.0:1	3.25:1	2	3	215.5	78.5	4540	A
Elite	13.27	351	148	243	8.0:1	3.26:1	2	3	216.1	78.5	4715	A
Matador	13.77	360	195	295	8.25:1	3.15:1	4	3	209.3	77.4	4215	A
Corvette	16.50	350	165	255	8.5:1	2.73:1	4	3	185.2	69	3660	A

y: Miles/gallon
x_1: Displacement (cubic in.)
x_2: Horsepower (ft-lb)
x_3: Torque (ft-lb)
x_4: Compression ratio
x_5: Rear axle ratio
x_6: Carburator (barrels)
x_7: No. of transmission speeds
x_8: Overall length (in.)
x_9: Width (in)
x_{10}: Weight (lbs)
x_{11}: Type of transmission (A–automatic, M–manual)

Source: *Motor Trend*, 1975.

Table B.4 Property valuation data

y	x_1	x_2	x_3	x_4	x_5	x_6	x_7	x_8	x_9
25.9	4.9176	1.0	3.4720	0.9980	1.0	7	4	42	0
29.5	5.0208	1.0	3.5310	1.5000	2.0	7	4	62	0
27.9	4.5429	1.0	2.2750	1.1750	1.0	6	3	40	0
25.9	4.5573	1.0	4.0500	1.2320	1.0	6	3	54	0
29.9	5.0597	1.0	4.4550	1.1210	1.0	6	3	42	0
29.9	3.8910	1.0	4.4550	0.9880	1.0	6	3	56	0
30.9	5.8980	1.0	5.8500	1.2400	1.0	7	3	51	1
28.9	5.6039	1.0	9.5200	1.5010	0.0	6	3	32	0
35.9	5.8282	1.0	6.4350	1.2250	2.0	6	3	32	0
31.5	5.3003	1.0	4.9883	1.5520	1.0	6	3	30	0
31.0	6.2712	1.0	5.5200	0.9750	1.0	5	2	30	0
30.9	5.9592	1.0	6.6660	1.1210	2.0	6	3	32	0
30.0	5.0500	1.0	5.0000	1.0200	0.0	5	2	46	1
36.9	8.2464	1.5	5.1500	1.6640	2.0	8	4	50	0
41.9	6.6969	1.5	6.9020	1.4880	1.5	7	3	22	1
40.5	7.7841	1.5	7.1020	1.3760	1.0	6	3	17	0
43.9	9.0384	1.0	7.8000	1.5000	1.5	7	3	23	0
37.5	5.9894	1.0	5.5200	1.2560	2.0	6	3	40	1
37.9	7.5422	1.5	5.0000	1.6900	1.0	6	3	22	0
44.5	8.7951	1.5	9.8900	1.8200	2.0	8	4	50	1
37.9	6.0831	1.5	6.7265	1.6520	1.0	6	3	44	0
38.9	8.3607	1.5	9.1500	1.7770	2.0	8	4	48	1
36.9	8.1400	1.0	8.0000	1.5040	2.0	7	3	3	0
45.8	9.1416	1.5	7.3262	1.8310	1.5	8	4	31	0

y: Sale price of the house/1000

x_1: Taxes (local, school, county)/1000

x_2: Number of baths

x_3: Lot size (sq ft × 1000)

x_4: Living space (sq ft × 1000)

x_5: Number of garage stalls

x_6: Number of rooms

x_7: Number of bedrooms

x_8: Age of the home (years)

x_9: Number of fireplaces

Source: "Prediction, Linear Regression and Minimum Sum of Relative Errors," by S. C. Narula and J. F. Wellington, *Technometrics*, **19**, 1977. Also see "Letter to the Editor," *Technometrics*, **22**, 1980.

Table B.5 Belle Ayr liquefaction runs

Run No.	y	x_1	x_2	x_3	x_4	x_5	x_6	x_7
1	36.98	5.1	400	51.37	4.24	1484.83	2227.25	2.06
2	13.74	26.4	400	72.33	30.87	289.94	434.90	1.33
3	10.08	23.8	400	71.44	33.01	320.79	481.19	0.97
4	8.53	46.4	400	79.15	44.61	164.76	247.14	0.62
5	36.42	7.0	450	80.47	33.84	1097.26	1645.89	0.22
6	26.59	12.6	450	89.90	41.26	605.06	907.59	0.76
7	19.07	18.9	450	91.48	41.88	405.37	608.05	1.71
8	5.96	30.2	450	98.6	70.79	253.70	380.55	3.93
9	15.52	53.8	450	98.05	66.82	142.27	213.40	1.97
10	56.61	5.6	400	55.69	8.92	1362.24	2043.36	5.08
11	26.72	15.1	400	66.29	17.98	507.65	761.48	0.60
12	20.80	20.3	400	58.94	17.79	377.60	566.40	0.90
13	6.99	48.4	400	74.74	33.94	158.05	237.08	0.63
14	45.93	5.8	425	63.71	11.95	130.66	1961.49	2.04
15	43.09	11.2	425	67.14	14.73	682.59	1023.89	1.57
16	15.79	27.9	425	77.65	34.49	274.20	411.30	2.38
17	21.60	5.1	450	67.22	14.48	1496.51	2244.77	0.32
18	35.19	11.7	450	81.48	29.69	652.43	978.64	0.44
19	26.14	16.7	450	83.88	26.33	458.42	687.62	8.82
20	8.60	24.8	450	89.38	37.98	312.25	468.38	0.02
21	11.63	24.9	450	79.77	25.66	307.08	460.62	1.72
22	9.59	39.5	450	87.93	22.36	193.61	290.42	1.88
23	4.42	29.0	450	79.50	31.52	155.96	233.95	1.43
24	38.89	5.5	460	72.73	17.86	1392.08	2088.12	1.35
25	11.19	11.5	450	77.88	25.20	663.09	994.63	1.61
26	75.62	5.2	470	75.50	8.66	1464.11	2196.17	4.78
27	36.03	10.6	470	83.15	22.39	720.07	1080.11	5.88

y: CO_2
x_1: Space time, min.
x_2: Temperature, °C
x_3: Percent solvation
x_4: Oil yield (g/100g of MAF)
x_5: Coal total
x_6: Solvent total
x_7: Hydrogen consumption

Source: *Industrial Chemical Process Design Development*, **17**, No. 3, 1978, "Belle Ayr Liquefaction Runs with Solvent."

Table B.6 Tube-flow reactor data

Run No.	y	x_1	x_2	x_3	x_4
1	0.000450	0.0105	90.9	0.0164	0.0177
2	0.000450	0.0110	84.6	0.0165	0.0172
3	0.000473	0.0106	88.9	0.0164	0.0157
4	0.000507	0.0116	488.7	0.0187	0.0082
5	0.000457	0.0121	454.4	0.0187	0.0070
6	0.000452	0.0123	439.2	0.0187	0.0065
7	0.000453	0.0122	447.1	0.0186	0.0071
8	0.000426	0.0122	451.6	0.0187	0.0062
9	0.001215	0.0123	487.8	0.0192	0.0153
10	0.001256	0.0122	467.6	0.0192	0.0129
11	0.001145	0.0094	95.4	0.0163	0.0354
12	0.001085	0.0100	87.1	0.0162	0.0342
13	0.001066	0.0101	82.7	0.0162	0.0323
14	0.001111	0.0099	87.0	0.0163	0.0337
15	0.001364	0.0110	516.4	0.0190	0.0161
16	0.001254	0.0117	488.0	0.0189	0.0149
17	0.001396	0.0110	534.5	0.0189	0.0163
18	0.001575	0.0104	542.3	0.0189	0.0164
19	0.001615	0.0067	98.8	0.0163	0.0379
20	0.001733	0.0066	84.8	0.0162	0.0360
21	0.002753	0.0044	69.6	0.0163	0.0327
22	0.003186	0.0073	436.9	0.0189	0.0263
23	0.003227	0.0078	406.3	0.0192	0.0200
24	0.003469	0.0067	447.9	0.0192	0.0197
25	0.001911	0.0091	58.5	0.0164	0.0331
26	0.002588	0.0079	394.3	0.0177	0.0674
27	0.002635	0.0068	461.0	0.0174	0.0770
28	0.002725	0.0065	469.2	0.0173	0.0780

y: $NbOCl_3$ concentration (g-mol/l)
x_1: $COCl_2$ concentration (g-mol/l)
x_2: Space time (sec)
x_3: Molar density (g-mol/l)
x_4: Mole fraction CO_2
Source: *Industrial and Engineering Chemistry, Process Design Development*, **11**, No. 2, 1972, "Kinetics of Chlorination of Niobium Oxychloride by Phosgene in a Tube-Flow Reactor."

Table B.7 Factors affecting CPU time

y	x_1	x_2	x_3	x_4
.4190	297	11714	15	0
.4678	387	13901	15	0
.0543	349	791	3	3
.0590	468	965	3	3
.0360	12	254	3	3
.0676	713	1411	3	3
.0893	18	266	2	1
.0386	11	120	1	1
.0988	18	245	2	1
.0260	11	102	1	1
.0410	18	307	2	1
.0196	12	143	1	1
.1653	17	4299	5	2
.1421	17	925	2	1
.0620	17	3076	5	2
.0560	17	914	2	1
.1405	17	3846	5	2
.0316	14	429	1	2
.2260	91	4781	13	3
.0417	14	1329	1	2
.4217	238	14872	13	3
.0278	14	249	1	2
.1873	100	4006	13	3
.2035	65	1234	6	8
.0410	17	152	1	1
.2042	84	1366	8	8
.2172	65	1228	6	8
.2005	65	1231	6	8
.2622	51	4098	13	3
.0537	14	1869	1	2
.3073	200	6691	13	3
.0430	14	1102	1	2
.2483	88	5028	13	3
.0718	14	2716	1	2
.4587	273	12016	13	3
.2477	50	3934	13	3
.2990	245	7214	13	3
.1927	65	1230	6	8

y: CPU time in seconds
x_1: Cards In—total no. of punch cards read into the computer system (includes control and data cards)
x_2: Lines Out—total lines generated by program execution
x_3: Steps—number of computer programs executed to perform the entire job function
x_4: Mounted Devices—the number of devices which are mounted to perform the job function.

Table B.8 Effectiveness of insecticides on cricket eggs

y	x_1	x_2	x_3	x_4
0	3	4	0	.6
0	3	4	0	.8
0	3	4	0	1.0
99	3	4	1	.6
0	3	4	1	.8
0	3	4	1	1.0
70	4	4	0	.6
0	4	4	0	.8
0	4	4	0	1.0
74	4	4	1	.6
30	4	4	1	.8
20	4	4	1	1.0
30	3	5	0	.6
43	3	5	0	.8
43	3	5	0	1.0
80	3	5	1	.6
49	3	5	1	.8
55	3	5	1	1.0
72	4	5	0	.6
74	4	5	0	.8
36	4	5	0	1.0
55	4	5	1	.6
44	4	5	1	.8
45	4	5	1	1.0
35	5	5	0	.6
48	5	5	0	.8
26	5	5	0	1.0
55	5	5	1	.6
27	5	5	1	.8
25	5	5	1	1.0
98	3	6	0	.6
78	3	6	0	.8
110	3	6	0	1.0
75	3	6	1	.6
99	3	6	1	.8
105	3	6	1	1.0
102	4	6	0	.6
102	4	6	0	.8
76	4	6	0	1.0
80	4	6	1	.6
69	4	6	1	.8

Table B.8 **Effectiveness of insecticides on cricket eggs (*Continued*)**

y	x_1	x_2	x_3	x_4
69	4	6	1	1.0
60	5	6	0	.6
62	5	6	0	.8
60	5	6	0	1.0
78	5	6	1	.6
60	5	6	1	.8
85	5	6	1	1.0
52	6	6	0	.6
25	6	6	0	.8
23	6	6	0	1.0
44	6	6	1	.6
45	6	6	1	.8
20	6	6	1	1.0
80	3	7	0	.6
98	3	7	0	.8
115	3	7	0	1.0
90	3	7	1	.6
100	3	7	1	.8
76	3	7	1	1.0
95	4	7	0	.6
85	4	7	0	.8
105	4	7	0	1.0
95	4	7	1	.6
98	4	7	1	.8
60	4	7	1	1.0
73	5	7	0	.6
82	5	7	0	.8
95	5	7	0	1.0
81	5	7	1	.6
83	5	7	1	.8
80	5	7	1	1.0
95	6	7	0	.6
49	6	7	0	.8
38	6	7	0	1.0
73	6	7	1	.6
75	6	7	1	.8
58	6	7	1	1.0
43	7	7	0	.6
49	7	7	0	.8
51	7	7	0	1.0
74	7	7	1	.6

Table B.8 **Effectiveness of insecticides on cricket eggs (*Continued*)**

y	x_1	x_2	x_3	x_4
55	7	7	1	.8
56	7	7	1	1.0
101	3	8	0	.6
95	3	8	0	.8
130	3	8	0	1.0
130	3	8	1	.6
120	3	8	1	.8
125	3	8	1	1.0
107	4	8	0	.6
95	4	8	0	.8
107	4	8	0	1.0
105	4	8	1	.6
101	4	8	1	.8
105	4	8	1	1.0
100	5	8	0	.6
98	5	8	0	.8
108	5	8	0	1.0
110	5	8	1	.6
117	5	8	1	.8
105	5	8	1	1.0
80	6	8	0	.6
110	6	8	0	.8
110	6	8	0	1.0
102	6	8	1	.6
95	6	8	1	.8
80	6	8	1	1.0
83	7	8	0	.6
98	7	8	0	.8
83	7	8	0	1.0
70	7	8	1	.6
102	7	8	1	.8
105	7	8	1	1.0
57	8	8	0	.6
70	8	8	0	.8
70	8	8	0	1.0
55	8	8	1	.6
46	8	8	1	.8
47	8	8	1	1.0
120	3	9	0	.6
120	3	9	0	.8
118	3	9	0	1.0

Table B.8 Effectiveness of insecticides on cricket eggs (*Continued*)

y	x_1	x_2	x_3	x_4
105	3	9	1	.6
108	3	9	1	.8
105	3	9	1	1.0
103	4	9	0	.6
103	4	9	0	.8
101	4	9	0	1.0
100	4	9	1	.6
100	4	9	1	.8
100	4	9	1	1.0
101	5	9	0	.6
101	5	9	0	.8
101	5	9	0	1.0
105	5	9	1	.6
101	5	9	1	.8
108	5	9	1	1.0
105	6	9	0	.6
100	6	9	0	.8
105	6	9	0	1.0
105	6	9	1	.6
99	6	9	1	.8
105	6	9	1	1.0
98	7	9	0	.6
100	7	9	0	.8
98	7	9	0	1.0
105	7	9	1	.6
97	7	9	1	.8
99	7	9	1	1.0
102	8	9	0	.6
100	8	9	0	.8
100	8	9	0	1.0
98	8	9	1	.6
100	8	9	1	.8
96	8	9	1	1.0

y: Activity of cholinestrase in the eggs, as a percent of normal
x_1: Age in days of the eggs at the time of the treatment
x_2: Age of the eggs at the time of observation
x_3: Indicator variable for the type of chemical; 0=carbaryl, 1=propoxur
x_4: Dosage level of the treatment, in micrograms per egg
Source: "Insecticide Action of Carbamates of the Eggs of the House Cricket", by Margaret I. Hartman, *J. Econ. Entomol.*, **65**, No. 6, June, 1972.

AUTHOR INDEX

Adichie, J. L., 382, 447
Affifi, A. A., 279, 447, 448
Aitkin, M. A., 250, 447
Allen, D. M., 255, 430, 434, 447, 449
Andrews, D. F., 61, 70, 165, 365, 371, 378, 379, 380, 381, 382, 447
Anscombe, F. J., 70, 71, 149, 447, 448
Askin, R. G., 383, 448

Baldwin, K. F., 321, 340, 454
Bancroft, T. A., 279, 448, 454
Barnett, V., 70, 448
Barrodale, I., 366, 448
Bartlett, M. S., 448
Beale, E. M. L., 266, 448
Beaton, A. E., 262, 368, 448
Behnken, D. W., 150, 408, 448
Beightler, C. S., 356, 459
Belsley, D. A., 163, 165, 176, 303, 304, 306, 320, 341, 448
Bendel, R. B., 279, 448
Berk, K. N., 278, 448
Berkson, J., 388, 402, 448
Bickel, P. J., 447
Book, D., 366, 448
Booker, J., 366, 448
Box, G. E. P., 38, 78, 82, 94, 96, 98, 199, 249, 348, 408, 448, 449, 451
Boyle, J. M., 458
Bradley, R. A., 184, 449
Brantle, T. F., 330, 340, 453
Brown, P. J., 341, 449
Brown, R. L., 425, 449
Brownlee, K. A., 449
Bunch, J. R., 455
Buse, A., 193, 449

Cady, F. B., 434, 449
Campbell, F., 341, 458
Chatterjee, S., 356, 449
Churchill, G. A., 340, 459
Cochrane, D., 355, 449
Conniffe, D., 340, 449
Cook, R. D., 71, 143, 163, 164, 165, 449
Cox, D. R., 94, 240, 245, 448, 449, 450
Curry, H. B., 193, 450

Daniel, C., 60, 71, 80, 123, 148, 156, 160, 256, 266, 268, 270, 378, 379, 450
David, H. A., 59, 450
Davies, R. B., 150, 388, 450
DeLury, D. B., 209, 450
Dempster, A. P., 322, 340, 450
Denby, L., 382, 450
Derringer, G. C., 78, 458
Dixon, W. J., 262, 282, 450
Dolby, G. R., 82, 388, 450
Dongarra, J. R., 455
Draper, N. R., 70, 75, 150, 165, 256, 258, 278, 341, 407, 408, 434, 448, 450, 457
Dunn, O. J., 395, 450
Durbin, J., 349, 350, 353, 425, 449, 450, 451
Dutter, R., 351, 371
Dykstra, O., Jr., 407, 451

Edwards, J. B., 251, 451
Efroymson, M. A., 275, 451
Ellenberg, J. H., 71, 451
Ellerton, R. R. W., 78, 451
Evans, J. M., 425, 449
Ezekiel, M., 147, 251, 451

Farrar, D. E., 306, 451
Feder, P. I., 70, 451
Fienberg, S., 457
Finney, D. J., 238, 451
Fomby, T. B., 78, 453
Forsythe, A. B., 184, 366, 418, 451, 459
Fox, K. A., 147, 451
Fuller, W. A., 193, 348, 451
Furnival, G. M., 262, 266, 451

Galarneau, D. I., 321, 340, 456
Gallant, A. R., 193, 451
Garbow, B. S., 458
Garside, M. J., 262, 451
Gaylor, D. W., 407, 451
Geisser, S., 425, 451
Gentle, J. M., 366, 451
Gibbons, D. G., 340, 452
Glauber, R. R., 306, 451
Gnanadesikan, R., 61, 452
Goldberger, A. S., 320, 353, 452, 458
Goldstein, M., 322, 341, 452, 458
Golub, G. H., 175, 452
Gorman, J. W., 267, 270, 452
Graybill, F. A., 45, 139, 231, 388, 401, 403, 435, 452
Guilkey, D. K., 345
Gunst, R. F., 78, 289, 290, 297, 299, 336, 339, 340, 452, 455, 456, 459
Guttman, I., 450

Hahn, G. J., 34, 395, 452
Haitovski, Y., 251, 452
Hald, A., 256, 452
Halperin, M., 388, 402, 452
Hamilton, M. A., 402, 455
Hampel, F. R., 447
Hanson, R. J., 175, 455
Hartley, H. O., 209, 366, 448, 457, 458
Hawkins, D. M., 339, 452, 453
Hayes, J. G., 193, 453
Hemmerle, W. J., 329, 330, 340, 453
Hendrickson, R. W., 395, 452
Hill, R. C., 78, 453
Hill, R. W., 365, 371, 453
Himmelblau, D. M., 293, 453
Hoadley, B., 402, 403, 453
Hoaglin, D. C., 163, 453
Hocking, R. R., 245, 247, 266, 318, 329, 330, 339, 371, 453, 455

Hodges, S. D., 388, 453
Hoerl, A. E., 312, 321, 322, 323, 327, 328, 340, 341, 453, 454
Hogg, R. V., 365, 371, 382, 454
Holland, P. W., 365, 368, 454
Hotelling, H., 397, 459
Huber, P. J., 150, 365, 370, 371, 447, 454
Hunter, J. S., 38, 82, 199, 249, 408, 449
Hunter, W. G., 38, 82, 199, 249, 449
Hutton, B., 150, 388, 450

Ikebe, Y., 458

Jaeckel, L. A., 382, 454
Jenkins, G. M., 348, 449
John, J. A., 165, 450
Johnson, L. A., 434, 456
Johnston, J., 348, 354, 388, 454
Judge, G. G., 78, 453
Jurečková, J., 382, 454

Kalotay, A. J., 403, 454
Kanemasa, H., 450
Kendall, M. G., 37, 448, 454
Kennard, R. W., 312, 321, 322, 323, 327, 328, 340, 341, 431, 453, 454
Kennedy, W. J., 279, 366, 451, 454
Klema, V. C., 458
Krutchkoff, R. G., 402, 454
Kuh, E., 163, 165, 176, 303, 304, 306, 320, 341, 448, 459
Kunugi, T., 293, 454

LaMotte, L. R., 266, 453, 455
Land, C. E., 90, 455
Larsen, W. A., 147, 148, 382, 450, 455
Lawless, J. F., 322, 340, 455
Lawson, C. R., 175, 455
Leamer, E. E., 319, 455
Leone, F. C., 382, 456
Leslie, R. N., 266, 453
Levy, P. S., 247, 457
Lewis, T., 70, 448
Lieberman, G. J., 402, 455
Lim, L., 193, 449
Lindley, D. V., 320, 322, 388, 455
Longley, J. W., 173, 386, 455
Lowerre, J. M., 318, 455
Lynn, M. J., 453

McCarthy, P. J., 456
McCleary, S. J., 147, 148, 455
McDonald, G. C., 321, 340, 456
Mallows, C. L., 252, 321, 455
Mandansky, A., 388, 455
Mann, D. G., 448
Mansfield, E. R., 336, 455
Mantel, N., 278, 455
Marquardt, D. W., 293, 294, 296, 300, 313, 316, 322, 326, 338, 341, 342, 456
Martin, E. W., 158, 456
Mason, R. L., 78, 289, 290, 297, 299, 336, 339, 340, 452, 456, 459
Mayer, L. S., 318, 456
Merrill, J. A., 407, 451
Miller, R. G., Jr., 402, 455, 456
Moler, C. B., 458
Montgomery, D. C., 158, 199, 231, 383, 407, 434, 448, 456
Moore, P. G., 388, 453
Morgan, J. A., 262, 456
Mosteller, F., 82, 430, 456
Moussa-Hamouda, E., 382, 456
Mullet, G. M., 173, 383, 386, 456
Murphy, J. L., 345, 452
Murray, T. W., 173, 386, 456
Myers, R. H., 199, 403, 407, 456, 457

Naito, T., 454
Narula, S., 247, 456
Naszodi, L. J., 403, 456
Neter, J., 231, 357, 408, 456
Neyman, J., 89, 456

Obenchain, R. L., 319, 322, 457
Orcutt, G. H., 355, 449
Ott, R. L., 403, 457

Pearson, E. S., 209, 457
Peck, E. A., 158, 457
Perng, S. K., 403, 457
Pesaran, M. H., 354, 457
Peters, S. C., 165, 459
Poirier, D. J., 193, 457
Pope, P. T., 278, 457
Pregibon, D., 165, 447
Prescott, P., 71, 449
Price, B., 356, 449

Ramberg, J. S., 247, 456
Ramsay, J. O., 371, 457
Randles, R. H., 382, 454
Rao, C. R., 431, 457
Rao, P., 247, 457
Roberts, F. D. K., 366, 448
Rogers, W. H., 447
Rosenberg, S. H., 247, 457
Rosner, B., 71, 457

St. John, R. C., 407, 457
Schatzoff, M., 262, 450, 457
Scheffé, H., 395, 403, 457
Schilling, E. G., 408, 457
Schoenberg, I. J., 193, 450
Sclove, S. L., 345, 457
Scott, E. L., 89, 456
Searle, S. R., 45, 139, 232, 457
Seber, G. A. F., 45, 139, 148, 151, 175, 209, 231, 251, 256, 262, 388, 397, 401, 408, 457
Shukla, G. K., 403, 458
Sielken, R. L., Jr., 366, 448, 458
Silvey, S. D., 306, 458
Slater, L. J., 355, 457
Smith, A. F. M., 321, 322, 341, 452, 455, 459
Smith, B. T., 301, 458
Smith, G., 341, 458
Smith, H., 70, 75, 256, 258, 408, 434, 450
Smith, J. H., 82, 320, 458
Smith, P. L., 191, 458
Snee, R. D., 293, 294, 296, 313, 316, 326, 341, 342, 425, 426, 430, 431, 436, 437, 456, 458
Snell, E. J., 245, 450
Speed, F. M., 453
Sposito, V. A., 366, 451
Sprent, P., 388, 458
Srivastava, S. S., 184, 449
Stefansky, W., 71, 458
Stein, C., 345, 458
Stewart, G. W., 301, 454, 458
Stone, J., 340, 449
Stone, L., 435, 454
Stone, M., 425, 430, 458
Stuart, A., 454
Suich, R., 78, 458

Tamura, T., 454
Tatar, J. F., 262, 456

Theil, H., 320, 458
Theobald, C. M., 318, 458
Thompson, M. L., 245, 458
Tidwell, P. W., 96, 98, 449
Toman, R. J., 267, 270, 452
Tong, Y. L., 403, 457
Tsao, R., 457
Tucker, W. T., 403, 458
Tufte, E. R., 37, 458
Tukey, J. W., 70, 71, 82, 368, 430, 447, 448, 456, 459

Udell, J. G., 232, 457

Van Nostrand, R. C., 341, 450

Wagner, H. M., 366, 459
Walls, R. E., 247, 459
Wang, P., 322, 340, 455
Wasserman, W., 231, 357, 408, 456
Watson, G. S., 349, 350, 450, 451
Watts, D. G., 305, 306, 459
Webster, J. T., 278, 289, 290, 297, 299, 339, 452, 455, 456, 457, 459
Weeks, D. L., 247, 459
Weisberg, S., 143, 144, 165, 363, 449, 459
Welsch, R. E., 163, 165, 176, 303, 304, 306, 320, 341, 368, 371, 448, 453, 454, 459
Wermuth, N., 450
Wetz, J. M., 78, 449
White, J. W., 339, 459
Wichern, D. W., 340, 459
Wilde, D. J., 356, 459
Wilkinson, J. W., 301, 459
Willan, A. R., 305, 306, 459
Williams, D. A., 71, 459
Williams, E. J., 402, 459
Willke, T. A., 318, 456
Wilson, R. W. M., Jr., 262, 266, 451
Wold, S., 190, 193, 459
Wonnacott, R. J., 348, 388, 459
Wonnacott, T. H., 348, 388, 459
Wood, F. S., 60, 80, 123, 148, 156, 160, 256, 266, 270, 378, 379, 450, 459
Working, H., 397, 459

Yale, C., 418, 459
Younger, M. S., 282, 459
Yule, G. U., 37, 454

Zellner, A., 319, 459

SUBJECT INDEX

Absolute errors, minimizing sum of, 366
Adjusted R^2, 251
Alias matrix, 247
All possible regressions, 256
Analysis of variance:
  for multiple regression model, 129
  one-way model, 409
  for polynomial models, 186-188, 194, 211
  regression methods for, 408
  for straight line, 24
Arc-sine transformation, 88
Assumptions for regression model, 8, 20, 111, 124
Autocorrelation, 67, 347

Backward elimination, 273
Bayes' estimation, ridge estimator obtained by, 319
Biased estimation, 310-342
  fractional rank, 338
  generalized ridge, 327
  latent root regression, 339
  principal components, 334
  ridge regression, 310
Bias in regression estimates, 247
BMDP programs, all possible regressions, 263
Bonferroni inequality, 394
Bonferroni *t*-intervals, 393

Calibration problem, 400-405
Centering data, 168, 341
Coefficient of multiple determination, *see* R^2
Condition index, 303
Condition number, 301
Confidence interval:
  β_j, 125
  in correlation, 49
  for E(y), 30, 127
  intercept, β_0, 27
  slope, β_1, 27
  transformation parameter, 94
  variance σ^2, 28
Confidence region, for several parameters, 126, 389-400
Corrected sum:
  of products, 11
  of squares, 11
Correlation, 45
  confidence interval, 49
  hypothesis tests, 48
  matrix, 169
  partial, 271
  and regression, 47
  between residuals, 150, 349
  serial, 67, 347
  between x and y, 48
Correlation form of regression, 169
Covariance of $\bar{y}$, $\hat{\beta}_1$, 16
  matrix, 120
C_p statistic, 252

Dependent variable, 3
Designed experiments, 405, 443
Diagram, scatter, 1, 122
Directed search on *t*, 266
Discrimination, 400
Double exponential distribution, 365
Dummy variables, 216
DUPLEX algorithm, 435

Durbin-Watson test, 349

Errors in regressors, 386
Error sum of squares, 18
Estimation:
  linear least squares, 9, 111
  maximum likelihood, 43, 46, 121
  mixed, 320
Estimator:
  bias, 347
  minimum variance unbiased, 16
Extra sum of squares method, 132

F_{IN}, 271
F_{OUT}, 273
F and $\bar{R}_p^2$, 251
$F = t^2$, 26
F-test:
  assumptions, 20
  for general linear hypothesis, 139
  for lack of fit, 77
  partial, 134, 271-282
  for significance of regression, 24, 129
F-to-enter, 271
F-to-remove, 273

Gauss-Markoff theorem, 16
Generalized least squares, 360
Generalized ridge regression, 327
Generating subset regressions, 255-282
Geometry of least squares, 118
Goodness of fit for regression, 75, 154

Hat matrix, 115
Hidden extrapolation, 142

Ill-conditioning, 184, 288
Independent variable, 3
Indicator variables, 216
Influence function, 370
Influential observations, 160
Interactions, 111, 223, 229
Inverse regression, 400
Iteratively reweighted least squares, 368

Joint confidence region, 126, 389

Knot, 190

Lack of fit, 75, 154
Latent root regression, 339
Least squares, 9
  assumptions, 8
  generalized, 360
  iteratively reweighted, 368
  maximum likelihood link, 44
  properties, 15-17
  restricted, 180, 318
  weighted, 99, 100, 363
Linear hypothesis, 139
Linear model, 111
Linear programming, 366
Logarithmic transformation, 89
Logistic regression, 238

Minimum variance unbiased estimator, 16
Mixed estimation, 320
Multicollinearity, 149, 287
Multiple correlation coefficient, *see* R^2
Multiple regression, 109
  correlation form, 167
  geometry of, 118
  normal equations, 112, 114
  for one-way anova, 408

Near-neighbor analysis, 154
Normal equations:
  multiple regression, 112, 114
  straight line regression, 10
Normal probability plot, 59

One-way classification, 408
Orthogonal columns in $\underset{\sim}{X}$, 137, 287
Orthogonal polynomials, 206
Outliers, 36, 70

Partial correlations, 271
Partial F-test:
  definition, 134
  in selection procedures, 271-282
  true probability levels, 278
Partial regression coefficients, 110
Piecewise regression, 243
Polynomial models, 181
  examples, 184, 193, 199, 209
  response surface example, 199
Predicted value, 10
Prediction interval, 30, 141
Predictor variables:
  errors in, 386

space, 160
PRESS, 255
Probability paper, 59
Probit analysis, 238
Proportions as responses, 64
arc-sine transformation, 88
Pure error, 76
approximate repeats, 154

QR decomposition, 175

R^2, 33, 146
adjusted, 251
use in variable selection, 249
Reciprocal transformations, 89
Regression analysis:
assumptions, 8, 20, 111, 124
inverse, 400
model-building, 244
multiple, 109
polynomials, 181
Regression equation, examination of:
extra sum of squares, 132
general, 57
R^2, 33, 146
residual analyses, 58, 147
standard error of $\hat{\beta}_j$, 125
Repeat runs, 75
approximate, 156
Residual MS (mean square), 18
Residuals, 11
Cook's statistic, 163
correlation between, 150
deleted, 430
Durbin-Watson test, 349
examination of, 75
maximum normed, 71
other types, 149
outliers, 36, 70
PRESS, 255, 430
serial correlation, 67, 347
standardized, 58
statistics for checking, 70, 349
studentized, 150
vs. time, 67, 348
vs. x's, 66
vs. $\hat{y}$, 63
Residual SS (sum of squares), 18, 36
Response surfaces, 198
Response variable, 3
transformation of, 79, 88, 94
Restricted least squares, 180, 318
Ridge regression, 310
Bayesian view, 319
choice of k, 313, 320-322
generalized, 327
restricted least squares view, 318
ridge trace, 313
variable selection, 323
Robust regression, 364
L estimators, 381
M estimators, 367
R estimators, 381
Rotatability, 407
Rounding errors, 173

SAS Programs:
GLM, 50
stepwise, 271, 273, 275, 280
Scaling and centering, 167, 341
unit length scaling, 168
unit normal scaling, 167
Scatter diagram, 1
multiple regression, 122
Second order model, 181, 198
Segmented lines, 189, 243
Selection procedures:
all possible regressions, 256
backward elimination, 273
criteria, 249
directed search on t, 266
forward selection, 271
stepwise, 275
variations, 279
Serial correlation, 67, 347
Significant regression, 22, 128
Splines, 189
Squared multiple correlation, *see* R^2
Square root transformations, 88
Stabilizing variance, 88, 94
Standard error:
of $\hat{\beta}_0$, 27
of $\hat{\beta}_1$, 27
of $\hat{\beta}_j$, 125
Standardized regression coefficients, 170
Stepwise regression, 275
computations, 275
printout (SAS), 276
Straight line regression, 8, 9
anova, 24

assumptions, 8
computer printout (SAS), 51
experimental design, 405
inverse regression, 400
normal equations, 10
residuals, 11
Sum of squares:
for pure error, 76
for regression, 24
for residual, 18
S_{xx}, S_{xy}, 11
S_{yy}, 18

$t^2 = F$, 26
Transformations:
analytical selection, 93
arc-sine, 88
logarithmic, 89
maximum likelihood selection, 94
on proportions, 88
reciprocal, 89
on regressors, 79, 88, 96
on response variables, 79, 88, 94
square root, 88
to stabilize variance, 88, 94
t-test:
for intercept, 22
for regression coefficient, 132
for slope, 21

Useful regression, 77, 78

Validation of model:
coefficients and, 426
data splitting, 430
DUPLEX algorithm, 435
fresh data, 427
PRESS, 430
Variables, 3
dummy, 216
indicator, 216
predictor, 3, 160, 386
regressor, 3
selection, 244
response, 3
transformation of, 79, 88, 94
selection procedures, variations in, 249, 279
two regressor case, 8
x-, *see* x-variables
Variance:
of intercept, 16
of predicted value, 29, 127
of slope, 16
Variance inflation factor (VIF), 299
Variance stabilizing transformation, 88, 94
Variations in variable selection procedures, 279

Weighted least squares, 99, 363
example, 100
Working-Hotelling confidence band, 397

x-variables:
centered and scaled, 168, 341
data space, 149, 160
orthogonal columns, 137, 287
transformations, 79, 96